I0072230

Advances in Microbiology

Volume I

Advances in Microbiology
Volume I

Edited by **Lucy Phillip**

New York

Published by Callisto Reference,
106 Park Avenue, Suite 200,
New York, NY 10016, USA
www.callistoreference.com

Advances in Microbiology: Volume I
Edited by Lucy Phillip

International Standard Book Number: 978-1-63239-047-9 (Hardback)

Printed in the United States of America.

Contents

Preface

It is said that even before human beings had appeared on the earth, microorganisms were already occupying it. Their association with the earth and its changing atmosphere goes back to millions of years. Hence this species offers a distinct field of study that has far reaching implications on the rest of the inhabitants.

The study pertaining to these microorganisms that are unseen through naked eye forms the sphere of microbiology. These comprise bacteria, viruses, archaea, fungi and protozoa. This subject deals with fundamental research on the biochemistry, physiology, cell biology, ecology, evolution and clinical aspects of microorganisms, including the host response to these agents. From single cell organisms to multi-cellular ones and also a-cellular which means no cell organisms.

The branches of microbiology can be classified into pure and applied sciences. Microbiology can be also classified based on taxonomy, in the cases of bacteriology, mycology, protozoology, and phycology. There is considerable overlap between the specific branches of microbiology with each other and with other disciplines, and certain aspects of these branches can extend beyond the traditional scope of microbiology. Hence, microbiology in its purest form like parasitology, bacteriology; integrative form like cellular microbiology, molecular microbiology as well as applied form like food microbiology, agricultural microbiology, environmental microbiology lead up to various intensive studies with microbiology at their core.

Given its immense utility, it's obvious that microbiology yields huge benefits as well. Microbes or microorganisms are used for industrial fermentation, vehicles for cloning, production of few biopolymers, and recent research studies are indicating that they could be important for cancer treatment as well. With such advantages, it's certainly right to conclude that microbiology is an extremely important field of study which has opened up new avenues for research and learning.

I would like to thank all the authors who have contributed their time and knowledge for this book. I would also like to thank my publisher for supporting me at every step and my family for their faith in me.

Editor

1

Coleopteran Antimicrobial Peptides: Prospects for Clinical Applications

Monde Ntwasa,[1] Akira Goto,[2] and Shoichiro Kurata[2]

[1] *School of Molecular and Cell Biology, University of the Witwatersrand, Wits 2050, South Africa*
[2] *Graduate School of Pharmaceutical Sciences, Tohoku University, Aoba 6-3, Aramaki, Aoba-ku, Sendai 980-8578, Japan*

Correspondence should be addressed to Shoichiro Kurata, kurata@mail.pharm.tohoku.ac.jp

Academic Editor: Paul Cotter

Antimicrobial peptides (AMPs) are activated in response to septic injury and have important roles in vertebrate and invertebrate immune systems. AMPs act directly against pathogens and have both wound healing and antitumor activities. Although coleopterans comprise the largest and most diverse order of eukaryotes and occupy an earlier branch than *Drosophila* in the holometabolous lineage of insects, their immune system has not been studied extensively. Initial research reports, however, indicate that coleopterans possess unique immune response mechanisms, and studies of these novel mechanisms may help to further elucidate innate immunity. Recently, the complete genome sequence of *Tribolium* was published, boosting research on coleopteran immunity and leading to the identification of *Tribolium* AMPs that are shared by *Drosophila* and mammals, as well as other AMPs that are unique. AMPs have potential applicability in the development of vaccines. Here, we review coleopteran AMPs, their potential impact on clinical medicine, and the molecular basis of immune defense.

1. Overview

Research on innate immunity has led to an accumulation of information that offers prospects for the development of antimicrobial therapeutic drugs and vaccines. The low rate of discovery of new antibiotics, the emergence of multiple-drug resistance, and the alarming death rate due to infection indicate a clear need for the development of alternative means to combat infections. A highlight of the 20th century was the discovery of vaccines that led to the eradication of diseases such as polio, small pox, and others. Even after more than two decades, however, a vaccine against the highly mutable human immune-deficiency virus remains to be developed, illustrating the need for new strategies to produce vaccines. A better understanding of innate immunity has revealed important links between innate and adaptive immune systems that could lead to effective approaches in vaccine development.

Coleopterans comprise 40% of the 360,000 currently known insect species and are therefore the largest and most diverse order of eukaryotic organisms [1]. *Tribolium*, the coleopteran model, is proposed to be a better model than *Drosophila*, especially for evolutionary studies, as it is acknowledged to be the most evolutionarily successful metazoan and to be more representative of insects in general than *Drosophila* [1, 2]. Coleopterans, with no adaptive immunity, thrive on this planet. Studies of the molecular basis of coleopteran immunity could therefore lead to a better understanding of the evolution of the innate immune system. Much of the work on innate immunity and studies of the functional aspects of antimicrobial peptides (AMPs) has been performed using *Drosophila*, which represents dipterans, while studies on coleopterans lag behind. Insects and humans share innate immunity, but humans also have adaptive immunity. Some of the conserved molecular signaling pathways that are used by insects and humans for immune defense are also used for early embryonic development in insects, but there are notable differences, probably due to the fact that the innate immune systems of invertebrates and vertebrates diverged some 800 million years ago, and adaptive immunity appeared in the vertebrate branch only about 500 million years ago [3, 4]. The divergence of dipterans and coleoptera occurred some 284 million years ago, and *Drosophila*, in the dipteran branch, exhibits a remarkably accelerated protein evolution [5]. Furthermore, despite these separate evolutionary paths, molecular coevolution could have occurred between coleoptera and mammals due to interdependence, that is, sharing common habitats and resources.

While the majority of the work on immunity has been conducted using *Drosophila* as a model, there is evidence that coleoptera has retained many ancestral vertebrate genes, suggesting that studies of coleoptera could provide more insight into the properties and evolution of innate immunity. For example, *Tribolium* has many ancestral genes that are present in vertebrates and absent in *Drosophila* [6]. Similarly, the sequenced *Tribolium* genome revealed that ancestral genes involved in cell-cell communication and development are retained in *Tribolium,* but not in *Drosophila* [2]. Furthermore, in homology searches, human genes compare significantly better with *Tribolium* than *Drosophila* [5].

AMPs are small peptides characterized by an overall positive charge (cationic), hydrophobicity, and amphipathicity. Structurally, they fall into two broad groups: linear α-helical and cysteine-containing forms with one or more disulfide bridges and β-hairpin-like, β-sheet, or mixed α-helical/β-sheet structures. The peptides assume these conformations upon contact with the target membranes [7–9]. Their characteristic physicochemical properties facilitate interactions with the phospholipid bilayer in the cell membranes of pathogens [10–12]. AMPs have been shown to kill pathogens directly by disrupting their membranes using mechanisms that are not fully understood. Several models, however, have been proposed. First, there is the "barrel-stave" model whereby a transmembrane pore is created by amphipathic α-helical peptides, disrupting the cell membrane of a pathogen. Second, the "carpet" model proposes that the peptides solubilize the membrane by interacting with the lipid head groups on the pathogen cell surface. This model was also proposed for viral killing [13]. Another is the aggregation model that is exhibited by sapecin from *Sarcophaga peregrina*, based on the existence of hydrophobic and hydrophilic domains on the AMPs. These structural features allow the peptides to form pores with hydrophilic walls and hydrophobic regions facing the acyl side chains of pathogen membrane phospholipids, thus facilitating movement of hydrophilic molecules through the pore [14]. Finally, the toroidal model, a subtle variation of the aggregation model, involves the formation of a dynamic lipid-layer core by hydrophilic regions of the peptide and lipid head groups and is induced by magainins, melittin, and protegrins [15–17]. While the indispensability of the structural features of cationic peptides in pathogen killing is under debate, charge differences between cationic peptides and lipids on the membrane are considered crucial. This may be the basis for their selective activity as nonhemolytic peptides have a high net positive charge distributed along the peptide length, whereas hemolytic peptides have a low negative charge [10, 11]. Evidence suggests that AMPs have intracellular targets. This is exemplified by elafin, a cationic and α-helical human innate defense AMP that does not lyse the bacterial membrane and is translocated into the cytoplasm. *In vitro* analysis using a mobility shift assay revealed that elafin binds DNA [18]. The histone-derived peptide buforin II binds nucleic acids in gel retardation assays and rapidly kills *Escherichia coli* by translocating into the cytoplasm of the pathogen and probably interfering with the functions of DNA or RNA. The structurally similar magainin 2 also kills *E. coli* but does not enter the cytoplasm [19]. Similarly, cationic antibacterial peptides enter the cytoplasm of *Aspergillus nidulans* and kill the fungus by targeting intracellular molecules whose identity has not been verified [20]. An excellent review of the intracellular targets of AMPs was recently published [21]. More studies are required, however, to confirm the existence and actual mode of action of AMPs with intracellular targets.

Insects produce AMPs constitutively at local sites or the AMPs are released systemically upon pathogenic infection to initiate pathogen-killing activities. In addition to the well-characterized *Drosophila* and mouse innate immune signaling pathways, the sequencing of the *Tribolium* genome has boosted research progress because bioinformatics analyses revealed putative immune-related genes based on comparisons with the genomes of other species [22].

AMPs are multifunctional molecules that, in addition to their well-known role as effectors of the innate immune system, are involved in several biologic processes and pathologic conditions, such as immune modulation, angiogenesis, and cytokine and histamine release [23–27]. Probably due to the negative charge in the plasma membrane of many cancer cells, some cationic peptides also have anticancer activity [28, 29]. These properties can be potentially exploited for clinical purposes [12, 30]. Cecropins are selectively cytotoxic to cancer cells, preventing their proliferation in bladder cancer, and are therefore likely candidates in strategies for the development of anticancer drugs [31]. In addition to antimicrobial activity, defensins facilitate the induction of adaptive immunity and promote cell proliferation and wound healing. Defensins show chemotactic activity whereby dendritic cells, monocytes, and T cells are recruited to the site of infection. Moreover, human β-defensins and the cathelicidin LL-37 stimulate the production of pruritogenic cytokines, such as interleukin-31, leukotrienes, prostaglandin E2, and others, suggesting an important role in allergic reactions [32–34]. AMPs also form the basis of the potentially lucrative commercial area of "cosmeceuticals"-products with beneficial topical activities that are delivered by rubbing, sprinkling, spraying, and so forth [35].

Here, we review the progress made in discovery of coleopteran AMPs, the molecular basis of *Tribolium* innate immunity, and prospects for the application of antimicrobial peptides in medicine.

2. The Discovery Process

2.1. Antimicrobial Peptides in Tribolium. The first wide-scale study of *Tribolium* immunity was conducted by Zou et al. in 2007 [22]. Taking advantage of the fully sequenced *Tribolium* genome to predict putative immune genes using bioinformatics techniques and real-time polymerase chain reaction (PCR), Zou et al. [22] predicted 12 AMPs in *Tribolium* compared to 20 in *Drosophila*, the most studied invertebrate. Another study using suppression subtractive hybridization led to the addition of a few more AMPs to this list [36] (see Table 1). Both studies identified four defensins in *Tribolium*, and phylogenetic analysis indicated that three of these are found in the evolutionary branch comprising only coleopterans. The fourth defensin (Def4) is found in a mixed branch that includes hymenopterans. A search of the Defensins Knowledgebase [37] revealed that the sequence information of this defensin is not available in the public domain, although its existence has been reported [22]. Attacins, which were identified in lepidopterans, were found in a cluster of three genes. Attacins are rich in glycine and proline, are structurally similar to coleoptericins, and are inducible by bacteria. Furthermore, *Drosophila* studies demonstrated that the induction of attacin is reduced in both *imd* and Tl^{-} mutants [38]. Coleoptericins were first isolated from the larvae of *Allomyrina dichotoma* beetles immunized with *E. coli*. Coleoptericins also show activity against *Staphylococcus aureus*, methicillin-resistant *S. aureus*, and *Bacillus subtilis*. Like attacins, but unlike cecropins, coleoptericins do not form pores on the bacterial membrane, but do cause defects in cell division, as liposomes containing *E. coli* or *S. aureus* membrane constituents do not leak upon treatment with the recombinant form of coleoptericin, but instead form chains [39].

Tribolium cecropins are predicted to be pseudogenes because of a shift in the open reading frame; some cecropin-related proteins with an unusual structure, however, have been reported [22]. A cecropin has been reported in at least one coleopteran, *Acalolepta luxuriosa* [44].

Four thaumatin-like genes were found in *Tribolium* using suppression subtractive hybridization and genome search. Experimentally, septic injury induces thaumatin-1 and defensins in *Tribolium* [36]. Sterile wounding also induces thaumatin-1 and defensin-2. Furthermore, recombinant thaumatin-1 heterologously overexpressed in *E. coli* is active against fungi [36]. Coleopteran cationic peptides might be remarkably different from other known peptides and are therefore not readily identified by homology searches. A clear homolog of the *Drosophila* antifungal drosomycin could not be found in the *Tribolium* genome, but a weakly homologous protein with a cysteine-rich sequence was detected [22]. An overview of *Tribolium* AMPs indicates similarities with other coleopterans, but some differences with *Drosophila*. The work reported by these groups provides a good basis for advancing research on coleopteran AMPs.

2.2. Other Antimicrobial Peptides Identified in Coleopterans. A number of AMPs present in certain coleopterans have not yet been identified in *Tribolium* (see Table 2). One of these is an interesting class of insect peptides that adopts the knottin fold and was first identified in 2003 from the harlequin beetle, *Acrocinus longimanus*. Members of this class include Alo-1, Alo-2, and Alo3 [42]. Psacotheasin from the yellow star longhorn beetle *Psacothea hilaris* has also been identified as a member of this class [43, 45]. Alo-3 is active against fungi, while psacotheasin is active against bacteria and fungi. The knottin fold is characterized by a disulfide topology of the "*abcabc*" type, in which disulfide bridges are formed between the first cysteine and the fourth, second, and fifth cysteines, and the third and sixth cysteines [46]. Disulfide bridge formation may confer important properties to the peptides, such as stability and resistance to protease cleavage. Members of the knottin family in general have low sequence similarity, reducing their chances of identification by homology searches [46]. In contrast, however, the coleopteran knottin fold AMPs share sequence similarities with several plant antifungal peptides [42]. Although the mechanism by which these peptides function is not fully understood, psacotheasin kills *Candida albicans* by inducing apoptosis [47]. This has clinical significance as *C. albicans* can cause mild superficial to severe infections in immunocompromised patients. A better understanding of the molecular events that are critical to the induction of apoptosis by cationic peptides could lead to new targets for antifungal drug development. Alarmingly, candidemia, a systemic *Candida* infection, is on the increase and is accompanied by the reemergence of resistance against common drugs, pointing to the urgency of finding alternative means of treating fungal infections [48, 49].

2.3. Databases. The Antimicrobial Peptides database, a comprehensive and searchable database for AMPs was established based on information from literature surveys [50, 51]. Currently, an updated version on the website indicates that there are 1773 cationic peptides in the database, including antiviral (5.8%), antibacterial (78.56%), antifungal (31.19%), and antitumor (6.14%) peptides. Some of these peptides function against more than one type of pathogens. The structures of 231 of these peptides have been determined by nuclear magnetic resonance and X-ray diffraction studies. Another useful database is the Defensins Knowledgebase, which allows text-based searches for information on this large family of AMPs [37]. It is a manually curated and specialized database similar to the shrimp penaeidin database, PenBase [52]. We have also started molecular studies of another coleopteran, the dung beetle *Euoniticellus intermedius*, and sequenced the adult transcriptome with a view to study its immune system [53]. These databases serve as useful tools for the discovery and design of new peptides. Indeed, key features upon which antimicrobial activity is based have been studied using the Antimicrobial Peptides database [54, 55]. Such analyses generate an important information pool for drug design.

3. Regulation of AMP Expression by Coleopterans

The signaling pathways that mediate the immune response in *Tribolium castaneum* were initially predicted based on a combination of *in silico* studies and experimental work by Zou et al. [22] and more recently another study involving the burying beetle *Nicrophorus vespilloides* [56]. In addition, studies using adult beetles exposed to *E. coli*, *M. luteus*, *C. albicans*, and *S. cerevisiae* have provided information on the signaling pathways. Accordingly, large-scale studies using real-time PCR revealed the presence of innate immune genes, such as PGRP-LA, PGRP-LE, PGRP-SA, PGRP-SB,

TABLE 1: Antimicrobial peptides currently predicted or identified in *Tribolium*.

Antimicrobial peptide	Accession number	Reference	Target	Method of identification
Attacin1	GLEAN_07737	[22]		Homology searches
Attacin2	GLEAN_07738	[22]		Homology searches
Attacin3	GLEAN_07739	[22]		Homology searches
Cecropin1	GLEAN_00499	[22, 31]	Antibacterial, antitumor	Homology searches
Cecropin2	Cec2	[22, 31]	Antibacterial, antitumor	Homology searches
Cecropin3	GLEAN_00500	[22, 31]	Antibacterial, antitumor	Homology searches
Defensin1	GLEAN_06250; XM_962101	[22, 36]	Antibacterial	Homology searches and suppression subtractive hybridization
Defensin2	GLEAN_10517; XM_963144	[22, 36]	Antibacterial	Homology searches and suppression subtractive hybridization
Defensin3	GLEAN_12469; XM_968482	[22, 36]	Antibacterial	Homology searches and suppression subtractive hybridization
Defensin4	Def4	[22]		Homology searches
Coleoptericin1	GLEAN_05093	[22]	Antibacterial	Homology searches
Coleoptericin2	GLEAN_05096	[22]	Antibacterial	Homology searches
Similar to thaumatin family	XM_963631	[36]	Antifungal	Suppression subtractive hybridization
Probable antimicrobial peptide	Tc11324	[22]		Homology searches
Putative antimicrobial peptide	AM712902	[36]		Suppression subtractive hybridization

TABLE 2: Antimicrobial peptides expressed in other coleopterans not yet identified in *Tribolium*.

Antimicrobial peptide	Organism	Accession no.	Reference
Diptericin A	*S. zeamis, (G. morsitans)*	Q8WTD5	[40]
Acaloleptin A	*S. zeamis (A. luxuriosa)*	Q76K70	[40]
Sarcotoxin II-1	*S. zeamis, (S. peregrina)*	P24491	[40]
Tenecin-1	*S. zeamis, (T. molitor)*	Q27023	[40, 41]
Tenecin-2	*T. molitor*		[41]
Luxuriosin	*S. zeamis, (A. luxuriosa)*	Q60FC9	[40]
Alo-3 (knottin type)	*A. longimanus*	P83653	[42]
Psacotheasin (Knottin type)	*P. hilaris*		[43]

several Toll proteins, and the immune deficiency (IMD) protein. Notably, some of the PGRPs had no orthologs in *Drosophila,* indicating a diversity of specificity. Recent biochemical studies using the large beetles *Tenebrio molitor* and *Holotrichia diomphalia* further elucidated the extracellular signaling network involved in responses to fungal and bacterial infections [41, 57]. Overall, coleopteran signaling appears to occur via the Toll and IMD pathways (Figure 1).

The Toll pathway is activated by PAMPS such as β-1,3-glucans, found in fungi, and by Lys-type peptidoglycans (PGN), found primarily in Gram-positive bacteria. A complex of the PAMPS and pathogen recognition receptors (PRRs) activates an apical protease, leading to a three-step serine protease cascade that culminates in the generation of active spaëtzle, the ligand of the transmembrane receptor Toll. Subsequent intracellular signaling leads to the transcriptional activation of genes that encode antimicrobial peptides.

Activation of the immune response by DAP-type PGN found primarily in Gram-negative bacteria and Gram-positive bacilli is still poorly understood in flies and beetles. Generally, it is understood that Gram-negative bacteria require the IMD pathway because imd^{-} mutants cannot express antimicrobial peptides against Gram-negative bacteria. In *Drosophila,* candidates for the signal transduction-activated Gram-negative bacteria are the transmembrane receptor PGRP-LC and PGRPP-LE. Both molecules can activate the IMD pathway [3, 58]. Because these molecules are present in beetles and PGRP-LE is orthologous to the *Drosophila* protein, it is likely that the corresponding pathways are conserved. In *Tribolium,* PGRP-LA and PGRP-LE are activated by bacterial infection, but poorly activated by *C. albicans* and *M. luteus* [22]. Other *Tribolium* studies show that the IMD pathway is activated by two Gram-negative bacteria, *Xenorhabdus nematophila* and *E. coli,* inducing 12 AMPs of which 5 are significantly dependent on the IMD pathway as demonstrated by RNA interference studies [59]. The same study, however, demonstrated that two Gram-positive bacteria with different peptidoglycans expressed the same AMPs with only defensin-1 being dependent on Toll. Taken together, these studies show that while the pathways

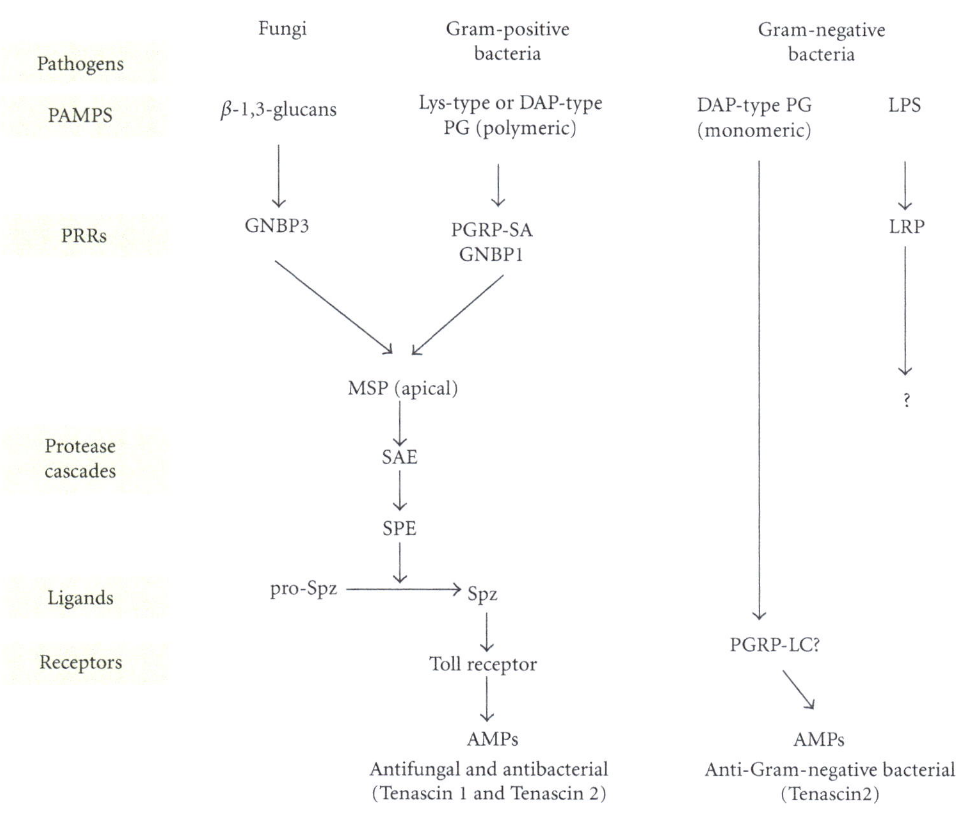

FIGURE 1: Activation mechanisms in the coleopteran immune system. Immune response pathways activated by bacteria and fungi showing a pathogen-associated recognition pattern (PAMP), pattern recognition receptors (PRRs), and downstream signaling molecules. The protease cascade in the Toll pathway involves the modular apical modular serine protease (MSP), the Spz-processing enzyme-activating enzyme (SAE), and the spaëtzle processing enzyme (SPE). GNBP3: glucan binding protein 3; PGRP: peptidoglycan recognition protein.

may be conserved, differences in PAMPS recognition and signal transduction exist between *Tribolium* and *Drosophila*.

The discovery of another PRR known as the LPS recognition protein (LRP) based on its *E. coli* agglutinating properties suggests the existence of an LPS pathway. LRP circulates in the hemolymph and does not agglutinate *S. aureus* or *C. albicans*. Interestingly, LRP comprises six repeats of an epidermal-growth-factor- (EGF)-like domain, an unusual structural feature for PRRs [60]. The downstream events in this pathway remain unclear.

4. Antimicrobial Peptides in Clinical Medicine

Cationic peptides have emerged as important targets for the development of therapeutics against bacteria, fungi, viruses, and parasites. They are key effector molecules in host defense through direct and indirect antimicrobial activity. Furthermore, in vertebrates, these peptides mediate a variety of cellular processes such as immunomodulation, wound healing, and tumorigenesis. These roles provide opportunities for the development of therapeutic products and vaccines. AMPs are attractive molecules for the development of clinical and veterinary therapeutics because they are fast acting and effective against susceptible pathogens, are less likely to cause the emergence of resistance compared to traditional antibiotics, have low toxicity to mammalian cells, and their mode of action tends to be more physical rather than targeted at metabolic pathways. A search of the FreePatentsOnline database using the word "antimicrobial peptide" produced more than 66.000 hits, and a number of AMPs have undergone clinical development [30]. A recent review of cationic peptides lists the peptides that are in various stages of clinical trials [29].

As mentioned above, the predicted *Tribolium* AMPs include defensins, attacin, coleoptericin, thaumatin, and cecropin. Defensins exhibit a broad spectrum of antimicrobial activity directed at bacteria, fungi, and viruses and are probably the most studied class of AMPs. Many therapeutic products have been modeled on them. The different types of defensins are either expressed constitutively or induced by infections to control the composition of microorganisms on surfaces such as the small and large intestines [61].

Many challenges remain that hamper the development of commercially viable peptides. The pressing issues concern pharmacokinetics (how the body deals with peptide drugs). When peptides are administered orally, the gastrointestinal tract may prevent their reabsorption into the systemic circulation. Furthermore, peptides may elicit an antigenic response when injected directly in the blood. This leaves topical medication the most feasible formulation while more research is being pursued to address the remaining obstacles. Despite these obstacles, the prospects for AMPs are not bleak because some have proceeded to clinical application. There is some optimism that these obstacles may soon be overcome by new strategies that combine natural cationic peptides and stable synthetic immunomodulatory peptides [29]. In this regard, peptide drugs such as Polymyxin B and gramicidin that are used for the treatment of Gram-negative bacterial infections are reported to be safe and effective, and peptides such as the indolicidin-derived CLS001 (previously known as MX594AN) have reached phase III clinical trials with promising prospects [62–64]. Because of their evolutionary distance, during which their survival against microbes has been solely dependent on innate immunity, insects provide interesting models for novel AMP drug design [65, 66].

5. Conclusions

The emergence of multidrug-resistant pathogens threatens human health globally and presents an urgent need to find antimicrobials with a reduced chance of inducing resistance. Cationic peptides for which the mechanism of action involves targeting the plasma membrane in a nonspecific manner, but does not involve specific proteins, offer good prospects. Admittedly, more work is needed to elucidate the mechanism of action of these peptides, as there is some evidence for intracellular targets. The importance of cationic peptides is further highlighted by their emerging prospects in other aspects of medicine, such as cancer treatment and vaccine development. Coleopterans are the most evolutionarily successful group of insects and are more representative of insects than *Drosophila*. In addition, human genes are more comparable to those of *Tribolium* than those of *Drosophila*. Thus, coleopterans are emerging as an important species for study as, like vertebrates, they have retained ancestral genes that are not present in *Drosophila*. Indeed, there is overwhelming evidence that coleopterans are more suitable for comparative studies between phyla than the commonly used dipterans. Here, we suggest that perhaps the outstanding evolutionary success of coleopterans is consistent with a robust immune system that warrants more attention than it has received to date.

Acknowledgments

All the authors are supported by the South Africa/Japan Joint Science and Technology Research Collaboration through the National Research Foundation and the Japan Society for the Promotion of Science. This work was supported by the Strategic International Cooperative program from the Japan Science and Technology Agency; Grants-in-Aid for Scientific Research from the Ministry of Education, Culture, Sports, Science, and Technology of Japan; the Japan Society for the Promotion of Science; the Program for the Promotion of Basic Research Activities for Innovative Biosciences (PRO-BRAIN); the National Institutes of Health (AI07495); the Takeda Science Foundation; the Mitsubishi Foundation; the Naito Foundation; Astellas Foundation for Research on Metabolic Disorders; a Global COE Research Grant (Tohoku University Ecosystem Adaptability); the National Research Foundation (NRF), South Africa.

References

[1] D. Tautz, "Insects on the rise," *Trends in Genetics*, vol. 18, no. 4, pp. 179–180, 2002.

[2] S. Richards, R. A. Gibbs, G. M. Weinstock et al., "The genome of the model beetle and pest *Tribolium castaneum*," *Nature*, vol. 452, no. 7190, pp. 949–955, 2008.

[3] J. A. Hoffmann, "The immune response of *Drosophila*," *Nature*, vol. 426, no. 6962, pp. 33–38, 2003.

[4] J. A. Hoffmann, "Primitive immune systems," *Immunological Reviews*, vol. 198, pp. 5–9, 2004.

[5] J. Savard, D. Tautz, and M. J. Lercher, "Genome-wide acceleration of protein evolution in flies (Diptera)," *BMC Evolutionary Biology*, vol. 6, p. 7, 2006.

[6] M. van der Zee, R. N. da Fonseca, and S. Roth, "TGFβ signaling in *Tribolium*: vertebrate-like components in a beetle," *Development Genes and Evolution*, vol. 218, no. 3-4, pp. 203–213, 2008.

[7] P. Bulet and R. Stöcklin, "Insect antimicrobial peptides: structures, properties and gene regulation," *Protein and Peptide Letters*, vol. 12, no. 1, pp. 3–11, 2005.

[8] P. Bulet, R. Stöcklin, and L. Menin, "Anti-microbial peptides: from invertebrates to vertebrates," *Immunological Reviews*, vol. 198, no. 1, pp. 169–184, 2004.

[9] A. Tossi, L. Sandri, and A. Giangaspero, "Amphipathic, α-helical antimicrobial peptides," *Peptide Science*, vol. 55, no. 1, pp. 4–30, 2000.

[10] Y. Shai, "Mechanism of the binding, insertion and destabilization of phospholipid bilayer membranes by α-helical antimicrobial and cell non-selective membrane-lytic peptides," *Biochimica et Biophysica Acta*, vol. 1462, no. 1-2, pp. 55–70, 1999.

[11] Z. Oren and Y. Shai, "Mode of action of linear amphipathic α-helical antimicrobial peptides," *Peptide Science*, vol. 47, no. 6, pp. 451–463, 1998.

[12] K. V. R. Reddy, R. D. Yedery, and C. Aranha, "Antimicrobial peptides: premises and promises," *International Journal of Antimicrobial Agents*, vol. 24, no. 6, pp. 536–547, 2004.

[13] R. E. Dean, L. M. O'Brien, J. E. Thwaite, M. A. Fox, H. Atkins, and D. O. Ulaeto, "A carpet-based mechanism for direct antimicrobial peptide activity against vaccinia virus membranes," *Peptides*, vol. 31, no. 11, pp. 1966–1972, 2010.

[14] K. Takeuchi, H. Takahashi, M. Sugai et al., "Channel-forming membrane permeabilization by an antibacterial protein, sapecin," *Journal of Biological Chemistry*, vol. 279, no. 6, pp. 4981–4987, 2004.

[15] K. Matsuzaki, "Magainins as paradigm for the mode of action of pore forming polypeptides," *Biochimica et Biophysica Acta*, vol. 1376, no. 3, pp. 391–400, 1998.

[16] K. A. Brogden, "Antimicrobial peptides: pore formers or metabolic inhibitors in bacteria?" *Nature Reviews Microbiology*, vol. 3, no. 3, pp. 238–250, 2005.

[17] R. Mani, S. D. Cady, M. Tang, A. J. Waring, R. I. Lehrer, and M. Hong, "Membrane-dependent oligomeric structure and pore formation of a β-hairpin antimicrobial peptide in lipid bilayers from solid-state NMR," *Proceedings of the National Academy of Sciences of the United States of America*, vol. 103, no. 44, pp. 16242–16247, 2006.

[18] A. Bellemare, N. Vernoux, S. Morin, S. M. Gagné, and Y. Bourbonnais, "Structural and antimicrobial properties of human pre-elafin/trappin-2 and derived peptides against *Pseudomonas aeruginosa*," *BMC Microbiology*, vol. 10, no. 1, p. 253, 2010.

[19] C. B. Park, H. S. Kim, and S. C. Kim, "Mechanism of action of the antimicrobial peptide buforin II: buforin II kills microorganisms by penetrating the cell membrane and inhibiting cellular functions," *Biochemical and Biophysical Research Communications*, vol. 244, no. 1, pp. 253–257, 1998.

[20] D. Mania, K. Hilpert, S. Ruden, R. Fischer, and N. Takeshita, "Screening for antifungal peptides and their modes of action in *Aspergillus nidulans*," *Applied and Environmental Microbiology*, vol. 76, no. 21, pp. 7102–7108, 2010.

[21] J. D. F. Hale and R. E. W. Hancock, "Alternative mechanisms of action of cationic antimicrobial peptides on bacteria," *Expert Review of Anti-Infective Therapy*, vol. 5, no. 6, pp. 951–959, 2007.

[22] Z. Zou, J. D. Evans, Z. Lu et al., "Comparative genomic analysis of the *Tribolium* immune system," *Genome Biology*, vol. 8, no. 8, p. R177, 2007.

[23] R. Bals and J. M. Wilson, "Cathelicidins—a family of multifunctional antimicrobial peptides," *Cellular and Molecular Life Sciences*, vol. 60, no. 4, pp. 711–720, 2003.

[24] M. G. Scott and R. E. W. Hancock, "Cationic antimicrobial peptides and their multifunctional role in the immune system," *Critical Reviews in Immunology*, vol. 20, no. 5, pp. 407–431, 2000.

[25] J. E. Meyer and J. Harder, "Antimicrobial peptides in oral cancer," *Current Pharmaceutical Design*, vol. 13, no. 30, pp. 3119–3130, 2007.

[26] A. F. Gombart, N. Borregaard, and H. P. Koeffler, "Human cathelicidin antimicrobial peptide (CAMP) gene is a direct target of the vitamin D receptor and is strongly up-regulated in myeloid cells by 1,25-dihydroxyvitamin D3," *FASEB Journal*, vol. 19, no. 9, pp. 1067–1077, 2005.

[27] B. de Yang, Q. Chen, A. P. Schmidt et al., "LL-37, the neutrophil granule-and epithelial cell-derived cathelicidin, utilizes formyl peptide receptor-like 1 (FPRL1) as a receptor to chemoattract human peripheral blood neutrophils, monocytes, and T cells," *Journal of Experimental Medicine*, vol. 192, no. 7, pp. 1069–1074, 2000.

[28] F. Schweizer, "Cationic amphiphilic peptides with cancer-selective toxicity," *European Journal of Pharmacology*, vol. 625, no. 1–3, pp. 190–194, 2009.

[29] A. T. Y. Yeung, S. L. Gellatly, and R. E. W. Hancock, "Multifunctional cationic host defence peptides and their clinical applications," *Cellular and Molecular Life Sciences*, vol. 68, no. 13, pp. 2161–2176, 2011.

[30] E. Andrès and J. L. Dimarcq, "Clinical development of antimicrobial peptides," *International Journal of Antimicrobial Agents*, vol. 25, no. 5, pp. 448–449, 2005.

[31] H. Suttmann, M. Retz, F. Paulsen et al., "Antimicrobial peptides of the Cecropin-family show potent antitumor activity against bladder cancer cells," *BMC Urology*, vol. 8, no. 1, p. 5, 2008.

[32] D. Yang, O. Chertov, S. N. Bykovskaia et al., "β-Defensins: linking innate and adaptive immunity through dendritic and T cell CCR6," *Science*, vol. 286, no. 5439, pp. 525–528, 1999.

[33] F. Niyonsaba, H. Ushio, M. Hara et al., "Antimicrobial peptides human β-defensins and cathelicidin LL-37 induce the secretion of a pruritogenic cytokine IL-31 by human mast cells," *Journal of Immunology*, vol. 184, no. 7, pp. 3526–3534, 2010.

[34] S. Dressel, J. Harder, J. Cordes et al., "Differential expression of antimicrobial peptides in margins of chronic wounds," *Experimental Dermatology*, vol. 19, no. 7, pp. 628–632, 2010.

[35] L. Zhang and T. J. Falla, "Cosmeceuticals and peptides," *Clinics in Dermatology*, vol. 27, no. 5, pp. 485–494, 2009.

[36] B. Altincicek, E. Knorr, and A. Vilcinskas, "Beetle immunity: identification of immune-inducible genes from the model insect *Tribolium castaneum*," *Developmental and Comparative Immunology*, vol. 32, no. 5, pp. 585–595, 2008.

[37] S. Seebah, A. Suresh, S. Zhuo et al., "Defensins knowledgebase: a manually curated database and information source focused on the defensins family of antimicrobial peptides," *Nucleic Acids Research*, vol. 35, supplement 1, pp. D265–D268, 2007.

[38] M. Hedengren, K. Borge, and D. Hultmark, "Expression and evolution of the *Drosophila* Attacin/Diptericin gene family," *Biochemical and Biophysical Research Communications*, vol. 279, no. 2, pp. 574–581, 2000.

[39] A. Sagisaka, A. Miyanoshita, J. Ishibashi, and M. Yamakawa, "Purification, characterization and gene expression of a glycine and proline-rich antibacterial protein family from larvae of a beetle, Allomyrina dichotoma," *Insect Molecular Biology*, vol. 10, no. 4, pp. 293–302, 2001.

[40] C. Anselme, V. Pérez-Brocal, A. Vallier et al., "Identification of the weevil immune genes and their expression in the bacteriome tissue," *BMC Biology*, vol. 6, no. 1, p. 43, 2008.

[41] Y. Yu, J.-W. Park, H.-M. Kwon et al., "Diversity of innate immune recognition mechanism for bacterial polymeric meso-diaminopimelic acid-type peptidoglycan in insects," *Journal of Biological Chemistry*, vol. 285, no. 43, pp. 32937–32945, 2010.

[42] F. Barbault, C. Landon, M. Guenneugues et al., "Solution structure of Alo-3: a new knottin-type antifungal peptide from the insect *Acrocinus longimanus*," *Biochemistry*, vol. 42, no. 49, pp. 14434–14442, 2003.

[43] J. S. Hwang, J. Lee, B. Hwang et al., "Isolation and characterization of psacotheasin, a novel knottin-type antimicrobial peptide, from *Psacothea hilaris*," *Journal of Microbiology and Biotechnology*, vol. 20, no. 4, pp. 708–711, 2010.

[44] A. Saito, K. Ueda, M. Imamura et al., "Purification and cDNA cloning of a cecropin from the longicorn beetle, *Acalolepta luxuriosa*," *Comparative Biochemistry and Physiology B*, vol. 142, no. 3, pp. 317–323, 2005.

[45] B. Hwang, J. S. Hwang, J. Lee, and D. G. Lee, "Antifungal properties and mode of action of psacotheasin, a novel knottin-type peptide derived from *Psacothea hilaris*," *Biochemical and Biophysical Research Communications*, vol. 400, no. 3, pp. 352–357, 2010.

[46] S. Lu, P. Deng, X. Liu et al., "Solution structure of the major α-amylase inhibitor of the crop plant amaranth," *Journal of Biological Chemistry*, vol. 274, no. 29, pp. 20473–20478, 1999.

[47] B. Hwang, J. S. Hwang, J. Lee, and D. G. Lee, "The antimicrobial peptide, psacotheasin induces reactive oxygen species and triggers apoptosis in *Candida albicans*," *Biochemical and Biophysical Research Communications*, vol. 405, no. 2, pp. 267–271, 2011.

[48] R. E. Lewis and M. E. Klepser, "The changing face of nosocomial candidemia: epidemiology, resistance, and drug therapy," *American Journal of Health-System Pharmacy*, vol. 56, no. 6, pp. 525–533, 1999.

[49] M. H. Nguyen, J. E. Peacock, A. J. Morris et al., "The changing face of candidemia: emergence of non-*Candida albicans* species and antifungal resistance," *American Journal of Medicine*, vol. 100, no. 6, pp. 617–623, 1996.

[50] Z. Wang and G. Wang, "APD: the antimicrobial peptide database," *Nucleic Acids Research*, vol. 32, supplement 1, pp. D590–D592, 2004.

[51] G. Wang, X. Li, and Z. Wang, "APD2: the updated antimicrobial peptide database and its application in peptide design," *Nucleic Acids Research*, vol. 37, supplement 1, pp. D933–D937, 2009.

[52] Y. Gueguen, J. Garnier, L. Robert et al., "PenBase, the shrimp antimicrobial peptide penaeidin database: sequence-based classification and recommended nomenclature," *Developmental and Comparative Immunology*, vol. 30, no. 3, pp. 283–288, 2006.

[53] L. M. Khanyile, R. Hull, and M. Ntwasa, "Dung beetle database: comparison with other invertebrate transcriptomes," *Bioinformation*, vol. 3, no. 4, pp. 159–161, 2008.

[54] S. Lata, N. K. Mishra, and G. P. S. Raghava, "AntiBP2: improved version of antibacterial peptide prediction," *BMC Bioinformatics*, vol. 11, no. 1, p. S19, 2010.

[55] S. Lata, B. K. Sharma, and G. P. S. Raghava, "Analysis and prediction of antibacterial peptides," *BMC Bioinformatics*, vol. 8, p. 263, 2007.

[56] H. Vogel, C. Badapanda, and A. Vilcinskas, "Identification of immunity-related genes in the burying beetle Nicrophorus vespilloides by suppression subtractive hybridization," *Insect Molecular Biology*, vol. 20, no. 6, pp. 787–800, 2011.

[57] J.-W. Park, C.-H. Kim, J. Rui et al., *Beetle Immunity Invertebrate Immunity*, Springer, New York, NY, USA, 2010.

[58] A. Takehana, T. Katsuyama, T. Yano et al., "Overexpression of a pattern-recognition receptor, peptidoglycan-recognition pro-tein-LE, activates imd/relish-mediated antibacterial defense and the prophenoloxidase cascade in *Drosophila* larvae," *Proceedings of the National Academy of Sciences of the United States of America*, vol. 99, no. 21, pp. 13705–13710, 2002.

[59] S. Shrestha and Y. Kim, "Activation of immune-associated phospholipase A2 is functionally linked to Toll/Imd signal pathways in the red flour beetle, *Tribolium castaneum*," *Developmental and Comparative Immunology*, vol. 34, no. 5, pp. 530–537, 2010.

[60] J. S. Ju, M. H. Cho, L. Brade et al., "A novel 40-kDa protein containing six repeats of an epidermal growth factor-like domain functions as a pattern recognition protein for lipopolysaccharide," *Journal of Immunology*, vol. 177, no. 3, pp. 1838–1845, 2006.

[61] J. Wehkamp, E. F. Stange, and K. Fellermann, "Defensin-immunology in inflammatory bowel disease," *Gastroentérologie Clinique et Biologique*, vol. 33, supplement 3, pp. S137–S144, 2009.

[62] M. E. Evans, D. J. Feola, and R. P. Rapp, "Polymyxin B sulfate and colistin: old antibiotics for emerging multiresistant gram-negative bacteria," *Annals of Pharmacotherapy*, vol. 33, no. 9, pp. 960–967, 1999.

[63] A. K. Marr, W. J. Gooderham, and R. E. Hancock, "Antibacterial peptides for therapeutic use: obstacles and realistic outlook," *Current Opinion in Pharmacology*, vol. 6, no. 5, pp. 468–472, 2006.

[64] S. A. Okorochenkov, G. A. Zheltukhina, and V. E. Nebol'sin, "Antimicrobial peptides: the mode of action and perspectives of practical application," *Biochemistry (Moscow) Supplement Series B*, vol. 5, no. 2, pp. 95–102, 2011.

[65] M. Zasloff, "Antimicrobial peptides of multicellular organisms," *Nature*, vol. 415, no. 6870, pp. 389–395, 2002.

[66] R. E. W. Hancock, K. L. Brown, and N. Mookherjee, "Host defence peptides from invertebrates—emerging antimicrobial strategies," *Immunobiology*, vol. 211, no. 4, pp. 315–322, 2006.

2

Polymorphic Amplified Typing Sequences and Pulsed-Field Gel Electrophoresis Yield Comparable Results in the Strain Typing of a Diverse Set of Bovine *Escherichia coli* O157:H7 Isolates

Indira T. Kudva,[1] Margaret A. Davis,[2,3] Robert W. Griffin,[4] Jeonifer Garren,[4] Megan Murray,[4,5] Manohar John,[4,6,7] Carolyn J. Hovde,[8] and Stephen B. Calderwood[4,6,9]

[1] *Food Safety and Enteric Pathogens Research Unit, National Animal Disease Center, Agricultural Research Service, U.S. Department of Agriculture, Ames, IA 50010, USA*
[2] *Department of Microbiology, Molecular Biology and Biochemistry, University of Idaho, Moscow, ID 83843, USA*
[3] *Department of Veterinary Microbiology and Pathology, College of Veterinary Medicine, Washington State University, Pullman, WA 99164, USA*
[4] *Division of Infectious Diseases, Massachusetts General Hospital, Boston, MA 02114, USA*
[5] *Department of Epidemiology, Harvard School of Public Health, Boston, MA 02115, USA*
[6] *Department of Medicine, Harvard Medical School, Boston, MA 02115, USA*
[7] *Pathovacs Inc., Ames, IA 50010, USA*
[8] *School of Food Sciences, University of Idaho, Moscow, ID 83843, USA*
[9] *Department of Microbiology and Immunobiology, Harvard Medical School, Boston, MA 02115, USA*

Correspondence should be addressed to Indira T. Kudva, ikudva2002@yahoo.com and Stephen B. Calderwood, scalderwood@partners.org

Academic Editor: Giuseppe Comi

Polymorphic amplified typing sequences (PATS), a PCR-based *Escherichia coli* O157:H7 (O157) strain typing system, targets insertions-deletions and single nucleotide polymorphisms at *Xba*I and *Avr*II restriction enzyme sites, respectively, and the virulence genes (*stx1*, *stx2*, *eae*, *hlyA*) in the O157 genome. In this study, the ability of PATS to discriminate O157 isolates associated with cattle was evaluated. An in-depth comparison of 25 bovine O157 isolates, from different geographic locations across Northwest United States, showed that about 85% of these isolates shared the same dendogram clade by PATS and pulsed-field gel electrophoresis (PFGE), irrespective of the restriction enzyme sites targeted. The Pearson's correlation coefficient, r, calculated at about 0.4, 0.3, and 0.4 for *Xba*I-based, *Avr*II-based and combined-enzymes PATS and PFGE similarities, respectively, indicating that these profiles shared a good but not high correlation, an expected inference given that the two techniques discriminate differently. Isolates that grouped differently were better matched to their locations using PATS. Overall, PATS discriminated the bovine O157 isolates without interpretive biases or sophisticated analytical software, and effectively complemented while not duplicating PFGE. With its quick turnaround time, PATS has excellent potential as a convenient tool for early epidemiological or food safety investigations, enabling rapid notification/implementation of quarantine measures.

1. Introduction

Escherichia coli O157:H7 (O157) causes an estimated 63,153 domestically acquired foodborne illnesses, 2,138 hospitalizations and 20 deaths annually in the United States [1–7]. Although a 44% decline in O157 cases was reported for the year 2010, over the past six years at least 13 different multistate O157 outbreaks have occurred, many of which have had a direct link to beef or produce possibly contaminated with manure [6, 7]. In fact, with cattle being the primary reservoir for this human pathogen [2–4], most human infections occur through food sources that are

Table 1: Summary of bovine *E. coli* O157:H7 isolates used in this study.

Isolates	Source	Location	Collection dates
5	Dairy cattle	Yakima Valley, Washington	7/22/1991
168	Dairy cattle	Northwest Idaho	6/7/1993
214	Dairy cattle	Northwest Washington	6/30/1993
268	Dairy cattle	West Oregon	8/3/1993
425	Dairy cattle	Northwest Washington	10/26/1993
528	Dairy calf	Central Washington	4/21/1994
757	Dairy cattle	Northwest Washington	6/21/1994
806	Dairy cattle	South Idaho	7/5/1994
817	Dairy cattle	West Oregon	7/6/1994
807	Dairy cattle	Central Idaho	7/6/1994
827	Dairy cattle	West Oregon	7/11/1994
908	Dairy cattle	South Idaho	7/12/1994
928	Dairy cattle	Northwest Oregon	7/13/1994
935	Dairy cattle	Northwest Oregon	7/13/1994
961	Dairy cattle	Northwest Oregon	7/18/1994
977	Dairy cattle	Central Idaho	7/25/1994
1015	Dairy cattle	Northwest Oregon	7/25/1994
1016	Dairy cattle	South Central Washington	7/27/1994
1041	Dairy cattle	Northwest Oregon	8/1/1994
1273	Dairy cattle	Southwest Oregon	8/30/1994
1286	Dairy cattle	Western Washington	9/2/1994
1328	Dairy cattle	Washington Basin	9/13/1994
1492	Feedlot	Unknown	10/3/1994
5671	Mill feed	WSU[a] Feedmill	1/5/1999
5974	Dairy cattle	WSU Dairy	9/1/1999

[a]WSU: Washington State University, Pullman, WA.

cattle derived (undercooked hamburger) or contaminated by cattle feces, such as salad vegetables, water, apple cider, and unpasteurized milk. With the current mechanization and globalization trends in food production and distribution, the need to monitor produce for foodborne pathogens such as O157 continues to remain critical to the prevention of extensive outbreaks, as is rapid epidemiological surveillance to identify and eliminate potential sources from the food chain.

Pulsed-field gel electrophoresis (PFGE) is the bacterial strain typing method of choice, regularly used by diagnostic and epidemiological laboratories to type O157 strains. To overcome the drawbacks of standard PFGE methodology, several modifications have been implemented that seek to address issues of, restriction enzyme inhibition, DNA degradation, variations in electrophoretic patterns between gels, improper resolution of digested DNA, subjective interpretation of these patterns even with sophisticated pattern-recognition computer software, and most importantly to decrease the turnaround time from 3 to 4 days to within 24 h, so data can be made available in a timely manner [8–12]. Even with all the modifications, it has been noted that single-restriction enzyme PFGE gives a poor measure of genetic relatedness as it does not resolve the entire repertoire of DNA fragments generated following restriction digestion [13].

Consequently, this has led to the incorporation of other genome-sequence-based techniques, such as multilocus sequence typing (MLST) and/or multilocus variable-number tandem repeat analysis (MLVA), either in conjunction with PFGE or by themselves, to type O157 isolates. However, even these methodologies cannot speed up the process as they rely primarily on generation of sequencing quality DNA, analysis of multiple genes or distances between tandem repeat sequences, which require complex instrumentation, and interpreting software [14, 15]. Hence, all these techniques would be useful in detailed, comprehensive analysis for followup cross-referencing, and banking purposes, rather than being the "first response" tools to rapidly sort out linked and unlinked cases/sources in an outbreak situation.

In previous studies, a touchdown PCR-based O157 strain typing system that incorporated polymorphisms at the *Xba*I- and *Avr*II-restriction sites, and amplified four virulence genes in the O157 genome was standardized against 46 O157 isolates from different sources and outbreaks [16–18]. This system termed the polymorphic amplified typing sequences (PATS) was not only able to provide a DNA fingerprint but also provide virulence profiles of the examined O157 isolates. PATS was less discriminatory when only one of the restriction enzyme sites was targeted but in the combination indicated above, PATS matched related isolates better than PFGE while differentiating between the unrelated isolates [16–18].

Table 2: *XbaI* -based PATS profiles of the 25 bovine O157 isolates. *AvrII*-based PATS profiles of the 25 bovine O157 isolates.

(a)

PATS type[a]	PCR amplification and restriction digestion patterns of amplicons obtained using 8 PATS primer pairs[b]								Isolates[c]
	IK8	IK19	IK25	IK114	IK118	IK123	IKB3	IKB5	
1	2	2	0	2	2	2	2	2	168, 908, 928,
2	0	2	0	2	2	2	2	2	977
3	2	2	0	2	2	2	0	2	5, 757, 806, 817, 1015, 1492
4	0	2	0	2	2	2	0	2	807
5	2	2	0	2	2	2	0	0	268, 827, 935
6	2	2	0	2	2	0	0	0	1041, 5671, 5974
7	2	2	0	0	2	2	2	0	1273
8	2	2	0	0	2	2	0	0	214, 528, 1016, 1286, 1328
9	2	2	0	2	0	2	0	0	425
10	0	2	0	0	2	0	0	0	961

(b)

PATS type[a]	PCR amplification and restriction digestion patterns of amplicons obtained using 7 PATS primer pairs[d]							Isolates[c]
	IKNR3	IKNR7	IKNR10	IKNR12	IKNR16	IKNR27	IKNR33	
1	2	2	2	2	2	2	2	5, 168, 817, 908, 928, 977, 1015, 1492
2	2	2	2	0	2	2	2	268, 757, 806, 807
3	2	2	2	2	1	2	2	1041
4	0	2	2	2	1	2	2	827, 5671, 5974
5	1	2	2	2	1	2	0	214, 528, 961, 1016, 1273, 1286, 1328
6	2	1	1	2	1	2	2	425, 935

[a]PATS types are designated arbitrarily with different numbers.
[b]Prefixes of each PATS primer pair A/B are indicated. 0: no amplicon; 2: amplicon with one *XbaI* site.
[c]Bovine *E. coli* O157:H7 isolates from different locations that fell within a given PATS type.
[d]Prefixes of each PATS primer pair A/B are indicated. 0: no amplicon; 1: amplicon without *Avr*II site; 2: amplicon with one *Avr*II site.

In this study, we decided to evaluate PATS against a diverse set of bovine O157 isolates and compare the profiles generated, with the PFGE patterns for the same, at length. For this, we targeted the same combinations of restriction enzyme sites. Although PATS directly sorts the polymorphisms at the restriction enzyme sites, and PFGE analyzes the DNA fragments generated as a result of these polymorphisms, we wanted to identify the degree of similarity between the two techniques and also ascertain if PATS would continue to maintain its ability to relate/discriminate bovine isolates as it did for human isolates in earlier studies [18].

2. Materials and Methods

2.1. Bacteria. Twenty-five O157 bovine isolates from various farms along the northwest region of United States (Idaho, Washington, and Oregon states) were obtained from collections maintained at the Field Disease Investigation Unit, College of Veterinary Medicine, Washington State University, Pullman, WA. The identification code used for each of these isolates is as indicated in Table 1.

2.2. PATS. PCR was done using conditions and primers as described previously [16–18]. Briefly, colony lysate of each O157 strain was tested against individual primer pairs, using the hot start PCR technique [19] in combination with a touchdown PCR profile [20]. To create this profile, an amplification segment of 20 cycles was set where the annealing temperature started at 73°C to touchdown at 53°C at the end of those cycles. Then, another amplification segment of 10 cycles was set, using the last annealing temperature of 53°C. Each reaction was done in triplicate to confirm profiles generated. Primer pairs targeting the 8 polymorphic *Xba*I- and 7 polymorphic *Avr*II-restriction enzyme sites, and the four virulence genes encoding the Shiga toxin 1 (stx_1), shiga toxin 2 (stx_2), Intimin-γ (*eae*), and hemolysin-A (*hlyA*), were used [16–18]. PCR reactions were purified using the QIAquick PCR purification kit (Qiagen, Valencia, Ca.), and all reactions, except for those amplifying virulence genes, were digested with the appropriate restriction enzyme (New England Biolabs, Beverly, Ma.) to confirm the presence of the restriction site within amplicons prior to resolution on a 4% agarose gel.

Presence or absence of amplicons was recorded as before [17, 18]. Briefly, for the virulence genes, the presence of an

Table 3: Combined-PATS profiles for the 25 bovine O157 isolates.

PATS type[a]	PCR amplification and restriction digestion patterns of amplicons obtained using 15 PATS—4 virulence gene primer pairs[b]																				Isolates[c]
	Polymorphic *Xba*I sites								Polymorphic *Avr*II sites							Virulence genes					
	IK8	IK19	IK25	IK114	IK118	IK123	IKB3	IKB5	IKNR3	IKNR7	IKNR10	IKNR12	IKNR16	IKNR27	IKNR33	stx_1	stx_2	*eae*	*hlyA*	
1 (previous 19)[d]	2	2	0	2	2	2	2	2	2	2	2	2	2	2	2	1	1	1	1	168, 908, 928
2 (previous 2)	0	2	0	2	2	2	2	2	2	2	2	2	2	2	2	1	1	1	1	977
3 (previous 18)	2	2	0	2	2	2	0	2	2	2	2	2	2	2	2	1	1	1	1	5, 817, 1015, 1492
4 (previous 16)	2	2	0	2	2	2	0	2	2	2	2	0	2	2	2	1	1	1	1	757, 806
5	0	2	0	2	2	2	0	2	2	2	2	0	2	2	2	1	1	1	1	807
6	2	2	0	2	2	2	0	0	2	2	2	0	2	2	2	1	1	1	1	268
7	2	2	0	2	2	0	0	0	2	2	2	2	1	2	2	1	1	1	1	1041
8	2	2	0	2	2	0	0	0	0	2	2	2	1	2	2	1	1	1	1	5671, 5974
9	2	2	0	0	2	2	2	0	1	2	2	2	1	2	0	1	0	1	1	1273
10	2	2	0	2	2	2	0	0	0	2	2	2	1	2	2	0	1	1	1	827
11 (previous 8)	2	2	0	2	2	2	0	0	2	1	1	2	1	2	2	0	1	1	1	935
12	2	2	0	0	2	2	0	0	1	2	2	2	1	2	0	0	1	1	1	214, 528, 1016, 1328
13	2	2	0	2	0	2	0	0	2	1	1	2	1	2	2	0	1	1	1	425
14	2	2	0	0	2	2	0	0	1	2	2	2	1	2	0	0	1	1	0	1286
15	0	2	0	0	2	0	0	0	1	2	2	2	1	2	0	0	1	1	1	961

[a]PATS types are designated arbitrarily with different numbers.
[b]Prefixes of each PATS primer pair A/B and virulence gene primer pair F/R are indicated. 0: no amplicon; 1: amplicon without *Avr*II site; 2: amplicon with one *Xba*I or *Avr*II site.
[c]Bovine *E. coli* O157:H7 isolates from different locations that fell within a given PATS type.
[d]Identical profiles observed in previous study [18].

amplicon was recorded as "1" (as these lacked either of the restriction enzyme sites being tested) and "0" for the absence of an amplicon. For PCR targeting the polymorphic *Xba*I restriction sites, presence of an amplicon was recorded as a "2" (as all amplicons could be digested into 2 fragments following enzymatic cleavage by the *Xba*I restriction enzyme) and as "0" in the absence of an amplicon. Likewise, for PCR targeting the *Avr*II restriction site, presence of an amplicon was recorded as "1" if the amplicon had no *Avr*II site, "2" if the amplicon had an *Avr*II site that resulted in it being digested into 2 fragments following enzymatic cleavage by the *Avr*II restriction enzyme and "0" in the absence of an amplicon [18].

2.3. PFGE. Standard PFGE methods were used to analyze the 25 bovine isolates as previously described [13, 18]. Briefly, the genomic DNA of each isolate was embedded in separate agarose plugs and digested at 37°C for 2 h with 30 U of *Xba*I or *Bln*I (*Avr*II) (Gibco BRL, Grand Island, N.Y.) per plug. The plugs were loaded onto a 1% agarose-tris buffer gel (SeaKem Gold Agarose; BioWhittaker Molecular Applications, Rockland, Maine), and PFGE was performed with a CHEF Mapper XA apparatus (Bio-Rad Laboratories, Hercules, Calif.). DNA was electrophoresed for 18 h at a constant voltage of 200 V (6 V/cm), with a pulse time of 2.2 to 54.2 s, an electric field angle of 120°, and a temperature of 14°C, before being stained with ethidium bromide. Gel images were analyzed using Bionumerics (Applied Maths, Saint-Martens-Latem, Belgium).

2.4. Data Analysis. (i) PATS dendograms: dendograms were constructed by coding molecular data as described previously [16–18]. Briefly, the presence or absence of each of the 15 amplicons (representing 8 polymorphic *Xba*I and 7 polymorphic *Avr*II sites) was coded as a dichotomous variable. Characters representing gain or loss of a restriction site recognized by *Avr*II were weighted to reflect the increased probability of losing a site than gaining one; however, such weighting had no impact on the resulting dendogram. Trees were constructed using the unweighted pair-group method with arithmetic means (UPGMA) option in the phylogenetic analysis using parsimony (PAUP; Sinauer Associates, Inc., Publishers, Sunderland, Ma.). (ii) PFGE dendograms: Dendograms were constructed using UPGMA cluster analysis based on Dice coefficients performed in Bionumerics, for *Xba*I- and *Avr*II-based PFGE patterns [21]. Combined Dice coefficients were then used to generate the *Xba*I and *Avr*II-combined-PFGE dendogram. (iii) Similarities: Bionumerics was used to determine Dice similarity coefficients for PFGE banding patterns as described previously, using the formula, $2n/a + b$, where n = number of matching bands and a + b = total number of bands

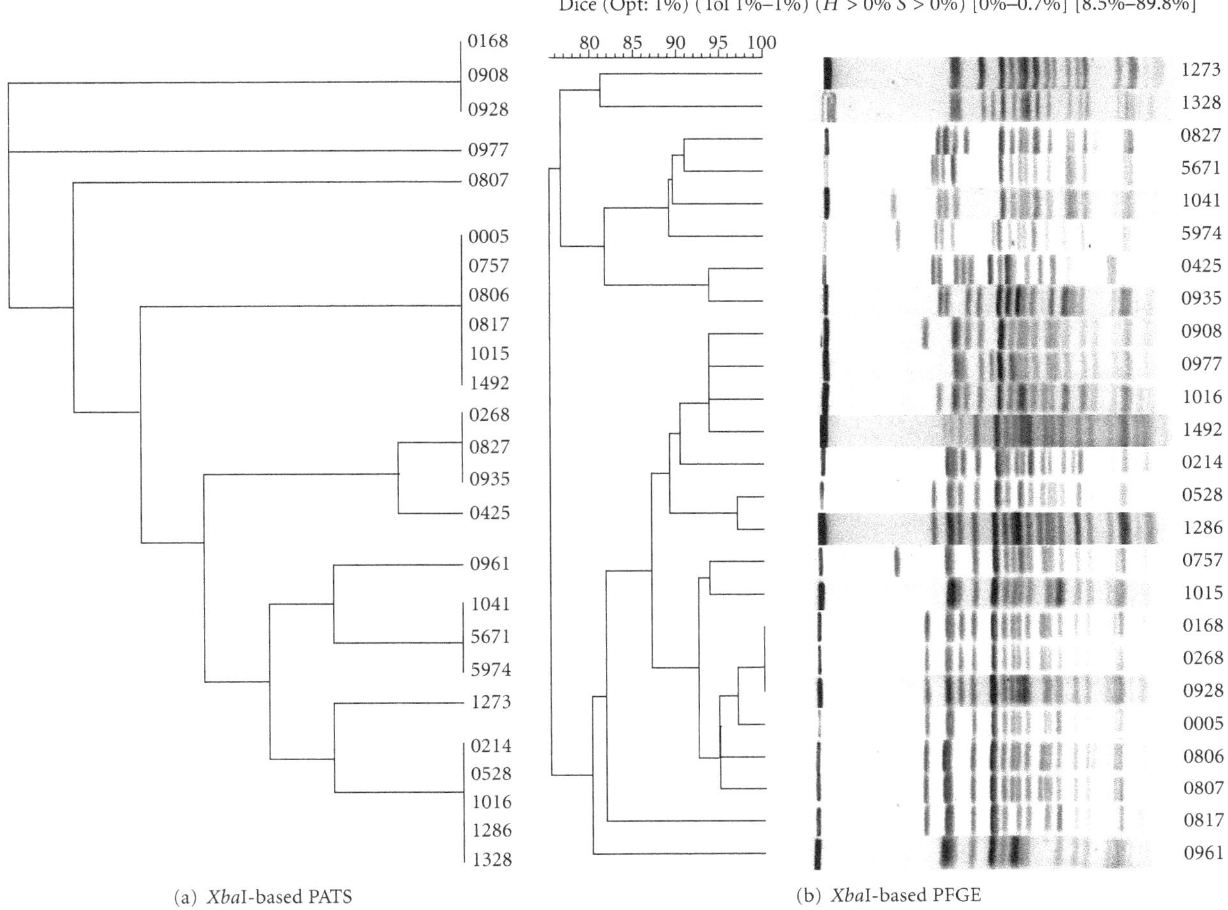

(a) *Xba*I-based PATS

(b) *Xba*I-based PFGE

FIGURE 1: Analysis of relatedness between 25 bovine O157 isolates. (a) Dendogram for *Xba*I-based PATS profiles was constructed using the UPGMA option in the phylogenetic analysis using parsimony (PAUP; Sinauer Associates, Inc., Publishers, Sunderland, Ma.). (b) Dendogram for *Xba*I-based PFGE profiles was constructed using the UPGMA cluster analysis based on Dice coefficients performed in Bionumerics (Applied Maths, Saint-Martens-Laatem, Belgium). Percent tolerance used is shown above the dendogram.

(matching and nonmatching) being compared between a pair of O157 isolates [13, 21]. For PATS, this coefficient was manually derived for each isolate pair using the modified formula, {2 × the number of concordant markers} ÷ {total number of markers being compared}. The total number of markers being compared between a pair of O157 isolates was 12 for *Xba*I-based PATS, and 11 for *Avr*II-based PATS, including the virulence genes. The Dice similarity coefficients for PFGE and PATS were then used to calculate Pearson's correlation coefficients as shown in the scatter plots.

3. Results

3.1. PATS Screening of the 25 Bovine O157 Isolates. All O157 isolates were analyzed for polymorphic *Xba*I- and *Avr*II-restriction sites, along with virulence genes. Independent of each other, *Xba*I-based PATS generated 10 different profiles, while *Avr*II-based PATS generated 6 different profiles (Tables 2(a) and 2(b)). However, in combination along with the four virulence genes, PATS analysis resulted in 15 distinct profiles demonstrating an increase in discrimination as observed previously (Table 3) [18]. These distinct profiles were clustered into smaller, related clades in the dendograms generated in PAUP. As shown in Figures 1(a), 2(a), and 3(a), the *Xba*I-based, *Avr*II-based, and combined PATS generated 5, 3, and 7 clades, respectively. Interestingly, five PATS profiles, 1, 2, 3, 4, and 11 (Table 3), were identical to the PATS profiles 19, 2, 18, 16, and 8, respectively, observed in a previous analysis of 46 unrelated O157 isolates associated with human disease [18], which may be reflective of the clonality of O157 isolates despite its divergence into multiple strain types [22].

3.2. PFGE Analysis of the 25 Bovine O157 Isolates. As shown in Figures 1(b), 2(b), and 3(b), PFGE analysis of the 25 bovine O157 isolates yielded complex genomic DNA electrophoresis patterns. Whether the isolates were grouped traditionally based on band differences (identical, closely related with 1–3 band differences, more distantly related with

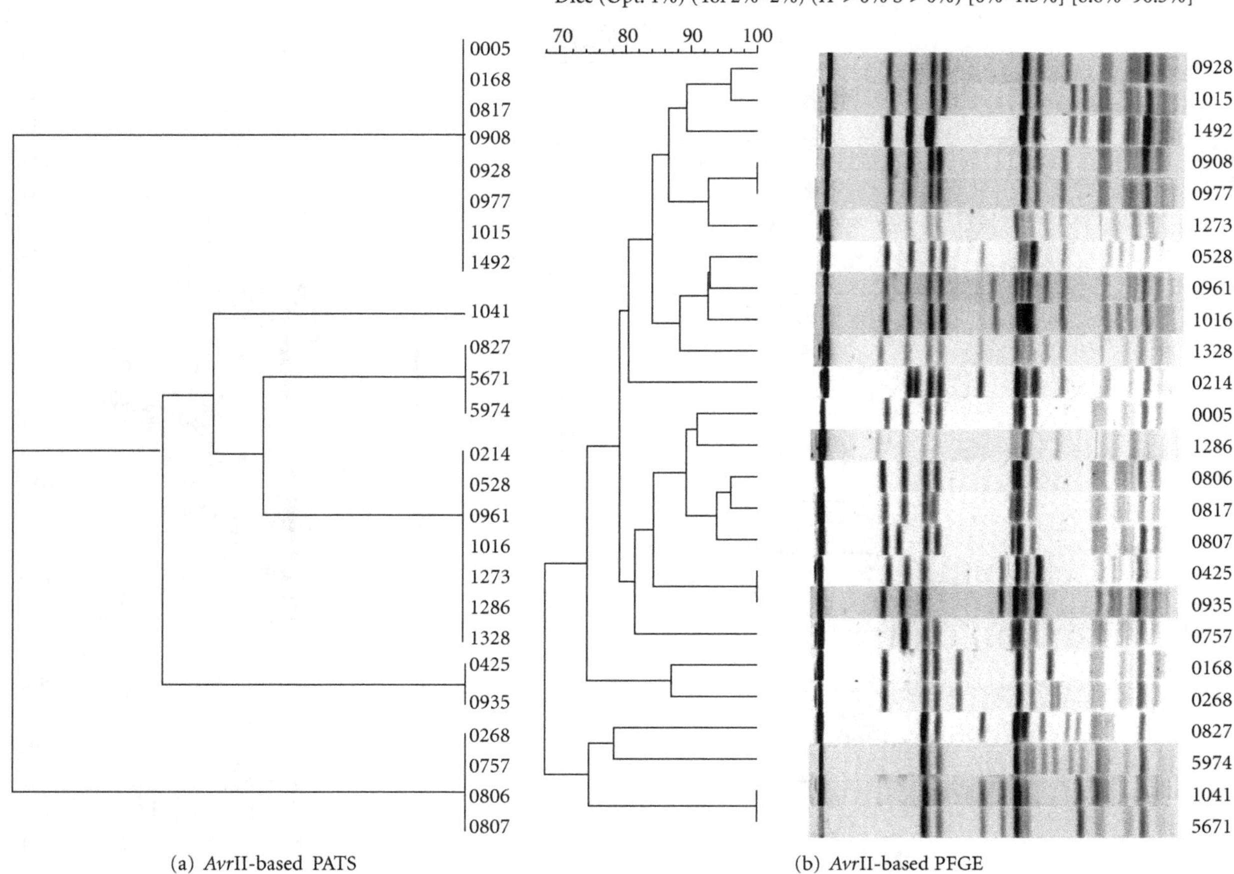

(a) *Avr*II-based PATS (b) *Avr*II-based PFGE

FIGURE 2: Analysis of relatedness between the 25 bovine O157 isolates. (a) Dendogram for *Avr*II-based PATS profiles was constructed using the UPGMA option in the phylogenetic analysis using parsimony (PAUP; Sinauer Associates, Inc., Publishers, Sunderland, Ma.). (b) Dendogram for *Avr*II-based PFGE profiles was constructed using the UPGMA cluster analysis based on Dice coefficients performed in Bionumerics (Applied Maths, Saint-Martens-Laatem, Belgium). Percent tolerance used is shown above the dendogram.

4–6 band differences) (data not shown), or grouped based on Dice similarity coefficients, the dendograms generated in Bionumerics indicated a high similarity among the isolates. Using the latter configurations, *Xba*I-based PFGE generated 6 clades from 23 different DNA banding patterns, *Avr*II-based PFGE generated 7 clades from 22 different DNA banding patterns, and combined PFGE generated 9 clades.

3.3. PATS Has Good Correlation with PFGE While Maintaining Its Distinctive Discriminating Features. Comparison of the dendograms generated showed that about 85% of the bovine O157 isolates formed similar groups by PATS and PFGE, irrespective of whether it was *Xba*I-based (84% similar groups), *Avr*II-based (84% similar groups), or based on a combination of enzymes (88% similar groups). However, the inherent differences between the two techniques was reflected when the Dice similarity coefficients were subjected to Pearson's correlation coefficient analysis. The Pearson's correlation coefficient, r, was calculated at about 0.4, 0.3, and 0.4 for *Xba*I-based, *Avr*II-based and combined PATS and PFGE similarities, respectively (Figure 4). This clearly indicated that these profiles shared a good if not high correlation. This may be reflective of the two techniques assessing the same restriction site polymorphisms in a different manner; PATS directly ascertains the presence/absence/other variations at these sites, while PFGE evaluates the resulting variations in the number and sizes of the genomic DNA fragments generated post-digestion with the same restriction enzymes.

A more positive correlation was observed for the *Xba*I-based, and *Xba*I and *Avr*II-combined-PFGE and PATS profiles than for the *Avr*II-based profiles, suggesting that perhaps the latter was more discriminatory. In fact, polymorphisms at the *Avr*II restriction sites and the presence/absence of virulence genes increased the discriminatory ability of PATS. Although we cannot rule out the inability to resolve ambiguous patterns due to comigrating bands or incompletely digested spurious bands by PFGE, this observation with PATS lends support to some of the discrimination seen with PFGE as well [17, 18]. Thus, the two typing techniques seem to complement each other while maintaining their own discriminative features.

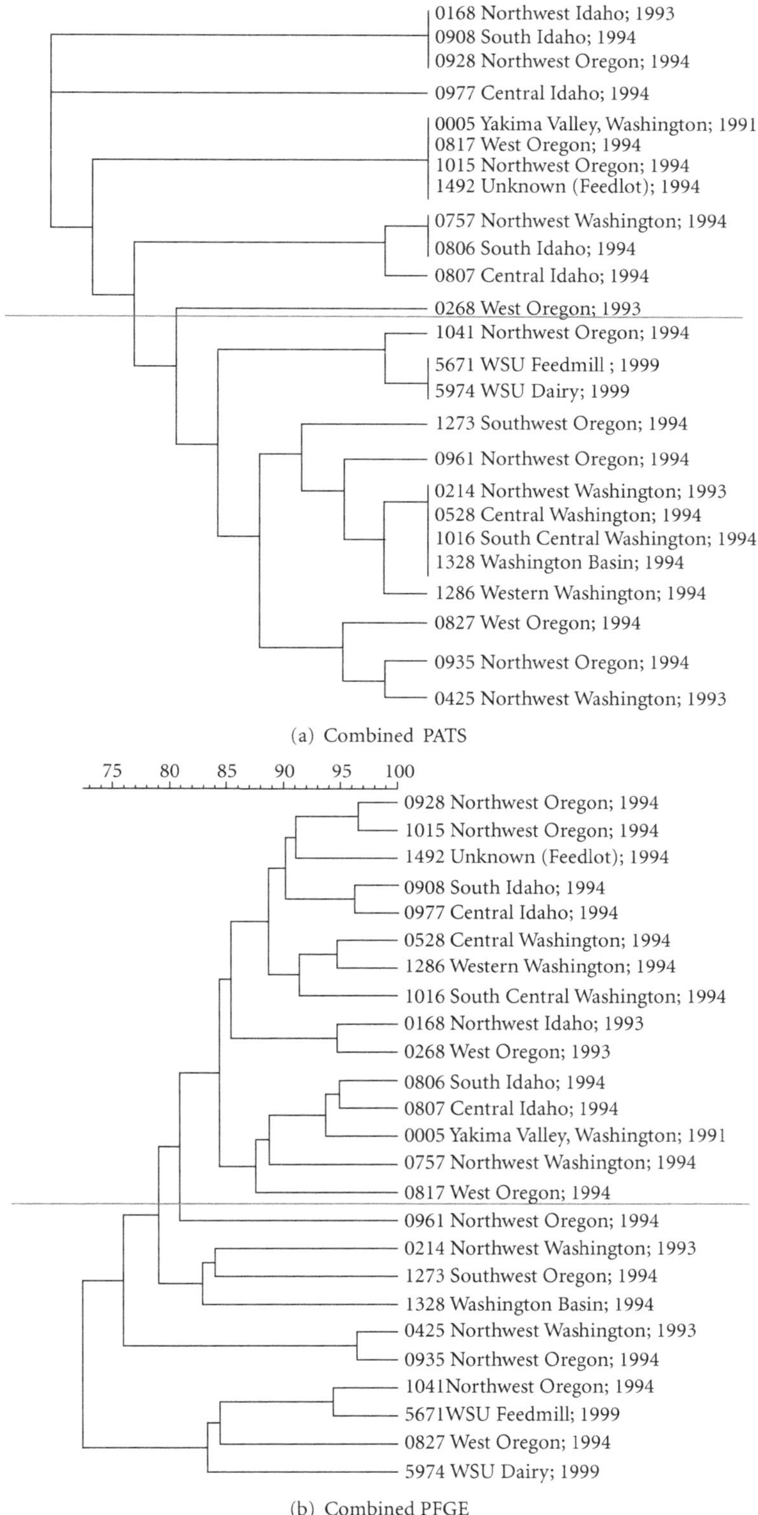

Figure 3: Analysis of relatedness between the 25 bovine O157 isolates. (a) Dendogram for combined-PATS (*Xba*I-, *Avr*II- and virulence genes-based) profiles was constructed using the UPGMA option in the phylogenetic analysis using parsimony (PAUP; Sinauer Associates, Inc., Publishers, Sunderland, Ma.). (b) Dendogram for combined-PFGE resulting from cluster analysis of combined Dice coefficients from PFGE following digestion with *Xba*I and PFGE following digestion with *Avr*II. Red line on each dendogram delineates isolates similarly grouped by combined-PATS and combined-PFGE.

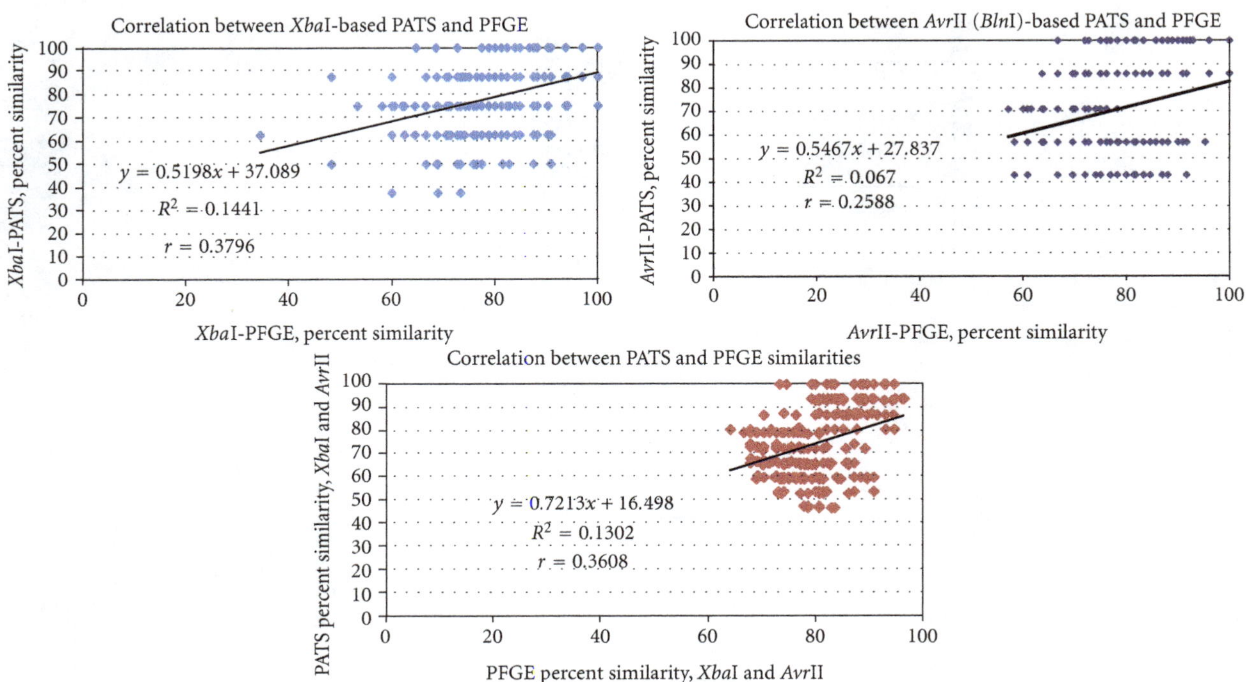

Figure 4: Scatter plots comparing PATS and PFGE profiles generated for 25 bovine O157 isolates. The derivation of the Pearson correlation coefficient is shown within each graph.

4. Discussion

The goal of this study was to compare the results of *Xba*I-based-, *Avr*II-based-, and combined-PATS with similar analyses done using PFGE, on 25 bovine O157 collected from different geographic locations to determine if these two strain typing techniques would relate and discriminate between bovine isolates in a similar manner as previously reported with human isolates. *Xba*I-based, *Avr*II-based, and combined PATS generated 5, 3, and 7 clades, respectively, reflecting the clonality of O157 (Figures 1(a) and 2(a)). A similar tendency was observed with *Xba*I-based (6 clades), *Avr*II-based (7 clades), and combined (9 clades) PFGE profiles based on the Dice coefficient similarities (Figures 1(b) and 2(b)). Clades generated with combined-PATS grouped the majority of isolates from Washington State and Idaho State in separate clusters interspersed with isolates from Oregon State (Figure 3(a)). While combined-PFGE-generated clades did not distribute the isolates in the exact manner as combined-PATS, about 85% of the isolates maintained a similar distribution (Figure 3(b)). As this was a random study of isolates, we were unable to determine if there was any transfer of animals, feed or other farm related goods between Oregon and the other 2 states that may have caused the O157 isolates to be closely related [21]. However, in an epidemiological situation this would be a reasonable cause to verify such an exchange.

PFGE is currently the standard strain typing technique used by various epidemiological and diagnostic laboratories to estimate the relatedness of outbreak or nosocomial O157 and other bacterial isolates [12, 23]. Compared to other strain typing methodologies being used, PFGE does provide relatively distinctive profiles for strains in several serotypes making it a popular strain typing tool. However, challenges in using PFGE are well known especially, the improper digestion and resolution of DNA bands, comigration of similar sized DNA fragments and nonhomologous DNA, or changes in electrophoretic conditions or analytical software, resulting in "untypeable" or incorrect profiles that cannot be interpreted or compared [8–11, 13, 24]. These drawbacks have led to several measures to make the PFGE protocol more uniform across laboratories, along with suggestions to use additional restriction enzymes, speed up turnaround time [13, 23, 25–29], and use other DNA sequence-based typing systems in parallel to confirm the validity of observations made with PFGE. Yet these variations continue to impede streamlining this process. Assessing genetic relatedness in a timely manner is crucial to any epidemiological survey. PFGE and DNA-sequence-based typing systems rely on expensive instrumentation and software to interpret data, which make them more useful in detailed analysis and banking of pathogens. In the field, however, a rapid and straightforward technique with sufficient power to discriminate between isolates without technical and subjective biases would help track down sources and speed up the process of sorting out linked and unlinked cases/sources in an outbreak situation. Such a technique would need to complement PFGE and not duplicate it or render the process more cumbersome.

In this study, both combined-PATS and PFGE had comparable discriminatory abilities, and the use of two restriction enzymes may have added to the observed similarities. O157 isolates that shared the same PATS group also fell into the same clade by PFGE. Some O157 isolates that fell into different groups/clades by the two techniques were better linked to their location using PATS which supports possible over-discrimination by PFGE. As seen previously [18], PATS was reliable, simple, user-friendly, and easy to perform and interpret in this instance as well. Because PATS is a PCR-based technique that directly addresses polymorphisms at restriction enzyme sites (Indels or SNPs), it eliminates the need for extensive electrophoresis, sequence analysis, and software to interpret results, thereby making it cost effective as well. PATS continued to maintain its high typeability and reproducibility as observed previously [18]. Based on all these observations, it appears that PATS would make an ideal "first response" epidemiological tool. We are in the process of evaluating the reliability of PATS in a "blind study" where the details of the O157 being typed will be withheld until the end of the study to ensure a typical field situation. We are also expanding application of PATS to other human pathogens of bovine origin such as, Shiga-toxin producing *Escherichia coli* (STECs), while seeking options to automate the process to further reduce the turnaround time to less than 6–8 hrs.

Acknowledgments

This study was supported in part by Grant Cooperative Agreement Number U60-CCU303019-15 from the Centers for Disease Control-Association of Public Health Laboratories to S. B. Calderwood and I. T. Kudva This work was also supported in part by the Center for Integration of Medicine and Innovative Technology (CIMIT), through U.S. Army Medical Research and Materiel Command Cooperative Agreement Number DAMD17-02-2-0006, award to S. B. Calderwood, I. T. Kudva, and M. John U.S. Utility Patent Pending for PATS-a bacterial strain typing technology (Filed Nov. 1, 2001). The work of M. A. Davis and C. J. Hovde was supported, in part, by the Idaho Agriculture Experiment Station, USDA Grant 04–04562, and Public Health Service NIH Grants P20-RR16454 (NCRR) and P20-GM103408 (NIGMS). Mention of trade names or commercial products in this paper is solely for the purpose of providing specific information and does not imply recommendation or endorsement by the U.S. Department of Agriculture. USDA is an equal opportunity provider and employer.

References

[1] P. M. Griffin, S. M. Ostroff, R. V. Tauxe et al., "Illness associated with *Escherichia coli* 0157:H7 infections. A broad clinical spectrum," *Annals of Internal Medicine*, vol. 109, no. 9, pp. 705–712, 1988.

[2] P. M. Griffin, *Infections of the Gastrointestinal Tract*, Raven Press, New York, NY, USA, 1995.

[3] J. B. Kaper, "The locus of enterocyte effacement pathogenicity island of Shiga toxin-producing *Escherichia coli* O157:H7 and other attaching and effacing *E. coli*," *Japanese Journal of Medical Science and Biology*, vol. 51, no. 1, pp. S101–S107, 1998.

[4] J. C. Paton and A. W. Paton, "Pathogenesis and diagnosis of Shiga toxin-producing *Escherichia coli* infections," *Clinical Microbiology Reviews*, vol. 11, no. 3, pp. 450–479, 1998.

[5] P. M. Griffin, P. S. Mead, T. vanGilder, S. B. Hunter, N. A. Strockbine, and R. V. Tauxe, "Shiga-toxin producing *Escherichia coli* infections in the United States: current status and challenges," in *Proceedings of the 4th International Symposium and Workshop on Shiga Toxin (Verocytotoxin)-Producing Escherichia coli Infections (VTEC '00)*, 2000.

[6] E. Scallan, R. M. Hoekstra, F. J. Angulo et al., "Foodborne illness acquired in the United States-Major pathogens," *Emerging Infectious Diseases*, vol. 17, no. 1, pp. 7–15, 2011.

[7] D. Gilliss, A. Cronquist, M. Cartter et al., "Vital signs: incidence and trends of infection with pathogens transmitted commonly through food—foodborne diseases active surveillance network, 10 U.S. sites, 1996–2010," *Morbidity and Mortality Weekly Report*, vol. 60, no. 22, pp. 749–755, 2011.

[8] B. Birren and E. Lai, *Pulsed-Field Gel Electrophoresis—A Practical Guide*, Academic Press, San Diego, Calif, USA, 1993.

[9] K. D. Harsono, C. W. Kaspar, and J. B. Luchansky, "Comparison and genomic sizing of *Escherichia coli* O157:H7 isolates by pulsed-field gel electrophoresis," *Applied and Environmental Microbiology*, vol. 59, no. 9, pp. 3141–3144, 1993.

[10] J. M. Johnson, S. D. Weagant, K. C. Jinneman, and J. L. Bryant, "Use of pulsed-field gel electrophoresis for epidemiological study of *Escherichia coli* O157:H7 during a food-borne outbreak," *Applied and Environmental Microbiology*, vol. 61, no. 7, pp. 2806–2808, 1995.

[11] T. Murase, S. Yamai, and H. Watanabe, "Changes in pulsed-field gel electrophoresis patterns in clinical isolates of enterohemorrhagic *Escherichia coli* O157:H7 associated with loss of Shiga toxin genes," *Current Microbiology*, vol. 38, no. 1, pp. 48–50, 1999.

[12] D. M. Olive and P. Bean, "Principles and applications of methods for DNA-based typing of microbial organisms," *Journal of Clinical Microbiology*, vol. 37, no. 6, pp. 1661–1669, 1999.

[13] M. A. Davis, D. D. Hancock, T. E. Besser, and D. R. Call, "Evaluation of pulsed-field gel electrophoresis as a tool for determining the degree of genetic relatedness between strains of *Escherichia coli* O157:H7," *Journal of Clinical Microbiology*, vol. 41, no. 5, pp. 1843–1849, 2003.

[14] A. C. Noller, M. C. McEllistrem, O. C. Stine et al., "Multilocus sequence typing reveals a lack of diversity among *Escherichia coli* O157:H7 isolates that are distinct by pulsed-field gel electrophoresis," *Journal of Clinical Microbiology*, vol. 41, no. 2, pp. 675–679, 2003.

[15] A. C. Noller, M. C. McEllistrem, A. G. F. Pacheco, D. J. Boxrud, and L. H. Harrison, "Multilocus variable-number tandem repeat analysis distinguishes outbreak and sporadic *Escherichia coli* O157:H7 isolates," *Journal of Clinical Microbiology*, vol. 41, no. 12, pp. 5389–5397, 2003.

[16] I. T. Kudva, P. S. Evans, N. T. Perna et al., "Strains of *Escherichia coli* O157:H7 differ primarily by insertions or deletions, not single-nucleotide polymorphisms," *Journal of Bacteriology*, vol. 184, no. 7, pp. 1873–1879, 2002.

[17] I. T. Kudva, P. S. Evans, N. T. Perna et al., "Polymorphic amplified typing sequences provide a novel approach to *Escherichia coli* O157:H7 strain typing," *Journal of Clinical Microbiology*, vol. 40, no. 4, pp. 1152–1159, 2002.

[18] I. T. Kudva, R. W. Griffin, M. Murray et al., "Insertions, deletions, and single-nucleotide polymorphisms at rare restriction enzyme sites enhance discriminatory power of polymorphic amplified typing sequences, a novel strain typing system for

Escherichia coli O157:H7," *Journal of Clinical Microbiology*, vol. 42, no. 6, pp. 2388–2397, 2004.

[19] C. W. Dieffenbach and G. S. Dveksler, *PCR Primer-A Laboratory Manual*, Cold Spring Harbor Press, New York, NY, USA, 1995.

[20] R. H. Don, P. T. Cox, B. J. Wainwright, K. Baker, and J. S. Mattick, "'Touchdown' PCR to circumvent spurious priming during gene amplification," *Nucleic Acids Research*, vol. 19, no. 14, p. 4008, 1991.

[21] M. A. Davis, D. D. Hancock, T. E. Besser et al., "Correlation between geographic distance and genetic similarity in an international collection of bovine faecal *Escherichia coli* O157:H7 isolates," *Epidemiology and Infection*, vol. 131, no. 2, pp. 923–930, 2003.

[22] A. N. Wetzel and J. T. LeJeune, "Clonal dissemination of *Escherichia coli* O157:H7 subtypes among dairy farms in Northeast Ohio," *Applied and Environmental Microbiology*, vol. 72, no. 4, pp. 2621–2626, 2006.

[23] E. M. Ribot, M. A. Fair, R. Gautom et al., "Standardization of pulsed-field gel electrophoresis protocols for the subtyping of *Escherichia coli* O157:H7, *Salmonella*, and *Shigella* for PulseNet," *Foodborne Pathogens and Disease*, vol. 3, no. 1, pp. 59–67, 2006.

[24] H. Bohm and H. Karch, "DNA fingerprinting of *Escherichia coli* O157:H7 strains by pulsed-field gel electrophoresis," *Journal of Clinical Microbiology*, vol. 30, no. 8, pp. 2169–2172, 1992.

[25] F. C. Tenover, R. D. Arbeit, R. V. Goering et al., "Interpreting chromosomal DNA restriction patterns produced by pulsed-field gel electrophoresis: criteria for bacterial strain typing," *Journal of Clinical Microbiology*, vol. 33, no. 9, pp. 2233–2239, 1995.

[26] R. K. Gautom, "Rapid pulsed-field gel electrophoresis protocol for typing of *Escherichia coli* O157:H7 and other gram-negative organisms in 1 day," *Journal of Clinical Microbiology*, vol. 35, no. 11, pp. 2977–2980, 1997.

[27] J. M. K. Koort, S. Lukinmaa, M. Rantala, E. Unkila, and A. Siitonen, "Technical improvement to prevent DNA degradation of enteric pathogens in pulsed-field gel electrophoresis," *Journal of Clinical Microbiology*, vol. 40, no. 9, pp. 3497–3498, 2002.

[28] S. B. Hunter, P. Vauterin, M. A. Lambert-Fair et al., "Establishment of a universal size standard strain for use with the pulsenet standardized pulsed-field gel electrophoresis protocols: converting the national databases to the new size standard," *Journal of Clinical Microbiology*, vol. 43, no. 3, pp. 1045–1050, 2005.

[29] A. Rementeria, L. Gallego, G. Quindós, and J. Garaizar, "Comparative evaluation of three commercial software packages for analysis of DNA polymorphism patterns," *Clinical Microbiology and Infection*, vol. 7, no. 6, pp. 331–336, 2001.

3

Prevalence and Antimicrobial Resistance of Thermophilic *Campylobacter* Isolated from Chicken in Côte d'Ivoire

Goualié Gblossi Bernadette,[1,2] Akpa Eric Essoh,[1] Kakou-N'Gazoa Elise Solange,[2] Guessennd Natalie,[2] Bakayoko Souleymane,[2] Niamké Lamine Sébastien,[1] and Dosso Mireille[2]

[1] *Laboratoire de Biotechnologies, Filière Biochimie-Microbiologie, Unité de Formation et de Recherche en Biosciences, Université de Cocody-Abidjan, 01 BP 582, Abidjan, Cote d'Ivoire*
[2] *Institut Pasteur de Côte d'ivoire, 01 BP 490, Abidjan, Cote d'Ivoire*

Correspondence should be addressed to Goualié Gblossi Bernadette, bettygoualie@yahoo.fr

Academic Editor: Hugh W. Morgan

Thermophilic *Campylobacters* are major causes of gastroenteritis in human. The main risk factor of infection is consumption of contaminated or by cross-contaminated poultry meat. In Côte d'Ivoire, gastroenteritis is usually observed but no case of human campylobacteriosis has been formally reported to date. The aims of this study were to determine prevalence and antimicrobial resistance of *Campylobacter jejuni* and *Campylobacter coli* isolated from chickens ceaca in commercial slaughter in Abidjan. Between May and November 2009, one hundred and nineteen (119) chicken caeca samples were collected and analyzed by passive filtration method followed by molecular identification (PCR). From these 119 samples, 76 (63.8%) were positive to *Campylobacter* tests. Among the positive colonies, 51.3% were *C. jejuni* and 48.7% were *C. coli*. Of the 39 *C. jejuni* isolates, 79.5%, 38.5%, 17.9%, 10.3%, and 7.7% were, respectively, resistant, to nalidixic acid, ciprofloxacin, amoxicillin, erythromycin, and gentamicin. Among the 37 isolates of *C. coli*, 78.4%, 43.2%, 13.5%, 8.1%, and 0% were resistant, respectively, to the same antibiotics. In conclusion, we reported in this study the presence of high *Campylobacter* contamination of the studied chickens. Molecular identification of the bacteria was performed and determination of high resistance to antimicrobials of the fluoroquinolone family was revealed.

1. Introduction

Campylobacter bacteria are Gram negative, curved, highly mobile, and microaerophilic. Thermophilic *Campylobacter* species, particularly *Campylobacter jejuni* and *Campylobacter coli*, have been recognized as a major cause of acute bacterial gastroenteritis in humans since 1970 and it is estimated that *Campylobacter* spp. are responsible for 400–500 millions cases of diarrhea each year worldwide [1–4]. In developing countries, incidence of children under 5 years old is estimated to 40 000 cases for 100 000 persons per year [5, 6] and according to World Health Organization (WHO) this incidence is underestimated.

Unpasteurized milk, water, and foods of animal origin are potential sources of contaminations [7–9], but the major risk factor for campylobacteriosis for humans is the consumption of undercooked poultry and the handling of raw poultry [9–16].

Most *Campylobacter* infections do not need to be treated with antimicrobial agents. However, in a subset of patients *Campylobacter* may cause severe complications and increased risk for death and therefore requires treatment [17–19].

When clinical treatment is necessary, fluoroquinolones (Ciprofloxacin, e.g.) or macrolides (Erythromycin) are currently used because of their large spectra activity on enteric pathogens [20, 21].

However, antimicrobial drug resistance in *Campylobacter* infections has increased dramatically in many countries during the 1990s. This emergence of *strains* resistance to antimicrobials provoked controversy over the use of antimicrobials in animal food production [22, 23].

Because of the increasing concerns of the public regarding the risk of exposure to antibiotic-resistant bacteria through food, monitoring programs for antimicrobial resistances in indicator bacteria isolated from food animals have been developed in a number of countries [24–27].

In Côte d'Ivoire, although poultry meat and particularly broiler chicken is a major proteins source for the population, no case of human campylobacteriosis has been published to date. However, our study in 2005 [28] showed a high prevalence (67%) of *Campylobacter* spp. in chickens. This situation can be explained by the fact that *Vibrio cholerae*, *Salmonella*, and many other microorganisms, out of *Campylobacter*, are the priorities of health authorities of Côte d'Ivoire. Therefore, survey data on the prevalence of *Campylobacter* spp. mainly *C. jejuni* and *C. coli* (as they are responsible of more than 98% cases of campylobacteriosis) [29] are needed. In addition, determination of antimicrobial resistance of *Campylobacter* isolates from chickens is important.

The aims of this study were, on one hand, to estimate prevalence of thermophilic *Campylobacters* in slaughtered chickens in Abidjan by biochemical and molecular technical and, on the other hand, to evaluate their antimicrobial resistance.

2. Material and Methods

2.1. Study Area and Sample Preparation. This study was conducted in Adjamé, a municipality in the north part of Abidjan. Adjamé is the principal market of Abidjan with more than 3 million visitors per day [30].

From May to November 2009, 119 samples of chicken caeca were collected in a commercial poultry processing plant. This plant is one of the primary poultry processing of this municipality with visitors of diverse origin particularly restaurateurs.

Each randomly selected sample was collected during evisceration and put into a stomacher bag, then rapidly transported to the laboratory in a cooler.

2.2. Isolation and Biochemical Identification. Isolation of *Campylobacter* was performed with passive filtration method preceded by enrichment as proposed by Federighi [31].

Briefly, 1 g of caeca contents was transferred to 9 mL of Preston enrichment broth base (CM 0067 Oxoid, OXOID LTD., Basingstoke, Hampshire,UK) containing *Campylobacter* growth factor (SR 0232E Oxoid, OXOID LTD., Basingstoke, Hampshire, UK) and 7% (v/v) defibrinated sheep blood. Incubation was performed in an anaerobic jar containing a packet generator microaerophilic atmosphere (5% oxygen, 10% carbon dioxide, 85% nitrogen) type CAMPYGen (CN0025A Oxoid, Basingstoke, Hampshire, UK) during 24 hours at 37°C. After enrichment, 300 μL of broth was filtered through acetate cellulose filter (0.45 μm) on Columbia agar (Sharlau; Barcelona, Spain) containing 5% (v/v) fresh sheep blood at aerobic conditions during one hour.

The filter was removed from agar and *Campylobacter* was isolated at 42°C during 48 hours under microaerophilic atmosphere.

One presumptive *Campylobacter* colony from each agar plate was subcultured and identified by Gram staining reaction, and biochemical pattern for oxidase, catalase, indoxyl acetate hydrolysis, and hippurate hydrolysis [32]. All isolates were stored in 25% (v/v) glycerol-peptone broth at −70°C.

2.3. DNA Extraction and PCR Conditions. Genomic DNAs from all isolated colonies were obtained by treatment with dodecyl sulfate sodium and proteinase K followed by extraction with phenol-chloroform and precipitation with ethanol [33]. PCR procedures used in this study have been described previously by Linton et al. [34]. Two genes for the identification of *Campylobacter jejuni* and *Campylobacter coli* were used: *hipo* gene (encoding *C. jejuni* hippurase) and *asp* gene (encoding *C. coli* aspartokinase). Sequences of the two sets of primers used for gene amplification are *asp* (F 5′-GGT ATG ATT TCT ACA AAG CGA G-3′ and R 5′-ATA TAT CGT CGC GTG AAAGAC-3′) and *hipo* (F 5′-GAA GAG GGT TTG GGT GGT G-3′ and R 5′-AGC CGC ATA ATA ACT TAGCTTTG-3′).

PCR was performed in final volume of 50 μL mix containing 2.5 μL of each deoxynucleoside triphosphate (10 mM), 2.5 μL of $MgCl_2$ (25 mM), 10 μL of Buffer 5X DNA Taq polymerase, 0.4 μL of Taq polymerase, 1.5 μL of each primer *asp* (10 μM), and 5 μL of each primer *hipo* (10 μM). Amplification reactions were carried out using thermal cycler (Gene Amp PCR system type 9700, Applied Biosystems, Villebon-sur-yvette, France) with the following program: an initial denaturation at 94°C for 15 min followed by 30 cycles of denaturation at 94°C for 1 min, annealing at 49°C for 1 min and polymerization at 72°C for 1 min. A final extension was performed at 72°C for 7 min.

The amplification generated 735 bp and 500 bp DNA fragments corresponding, respectively, to *Campylobacter jejuni* and *Campylobacter coli*. The PCR products were stained with a 0.6% solution of ethidium bromide and were visualized under UV light after gel electrophoresis on 1.5% agarose.

2.4. Antimicrobial Susceptibility Testing. Antimicrobial susceptibility testing was performed by disc diffusion method using Mueller-Hinton agar (Oxoid) supplemented with 5% defibrinated sheep blood, according to Clinical Laboratory Standards Institute (CLSI) guidelines [35]. Disks impregnated with antibiotics (Biomerieux, Marcy-l'Etoile, France) and their corresponding concentration are the followings: ciprofloxacin (CIP: 5 μg); nalidixic acid (NA: 30 μg); erythromycin (E: 15 μg); gentamicin (GM: 10 UI); amoxicillin (AMX: 25 μg).

Briefly, well-isolated colonies of same morphological type were selected from an agar plate culture and transferred into 10 mL of sterile saline buffer (NaCl 0.9%). After homogenization, 2 mL of the mixture were flooded onto the surface of a Mueller-Hinton agar (Oxoid) containing 5% defibrinated sheep blood. The inoculum was allowed to dry for 5 min and antibiotic discs were placed on the plate. After 48 h of microaerobic incubation at 37°C, diameters of

TABLE 1: Antimicrobial susceptibility of *Campylobacter jejuni* and *Campylobacter coli* isolated from chickens in 2009 in Abidjan (Cote d'Ivoire).

Campylobacter strains	Antibiotic disks				
	NA	CIP	AMX	E	GM
Campylobacter spp. ($n = 76$)					
R	60 (78.9%)	38 (50%)	9 (11.8)	10 (13.5%)	3 (3.9%)
I	7 (9.2%)	10 (13.2%)	14 (18.4%)	12 (15.8%)	0 (0%)
S	9 (11.9%)	28 (36.8%)	53 (69.8%)	54 (70.7%)	73 (96.1%)
C. jejuni ($n = 39$)					
R	31 (79.5%)	15 (38.5%)	4 (10.2%)	7 (17.9%)	3 (7.7%)
I	2 (12.8%)	6 (15.4%)	9 (23.1%)	8 (20.5%)	0 (0%)
S	6 (7.7%)	18 (46.1%)	26 (66.7%)	24 (61.6%)	36 (92.3%)
C. coli ($n = 37$)					
R	29 (78.4%)	16 (43.2%)	5 (13.5%)	3 (8.1%)	0 (0%)
I	2 (5.4%)	4 (10.8%)	5 (13.5%)	4 (10.8%)	0 (0%)
S	6 (16.2%)	17 (46%)	27 (73%)	30 (81.1%)	37 (100%)

S: sensitive, I: intermediate, R: resistant, NA: Nalidixic acid, CIP: Ciprofloxacin, AMX: Amoxicillin, E: Erythromycin, GM: Gentamicin.

the inhibition zones were measured with calipers. *Escherichia coli* ATCC 25922 and *Staphylococcus aureus* ATCC 25923 were used as reference strains. The isolates were classified as sensitive, intermediate, and resistant according to the guidelines prepared by CLSI [35].

3. Results

3.1. Prevalence. Overall, 119 chickens ceaca were purchased from one commercial slaughter and cultured for thermophilic *Campylobacters*. From these 119 samples, 76 (63,8%) were positives and 43 (36.2%) were negatives to *Campylobacter* tests. Among the *Campylobacter* positives, 51.3% were *Campylobacter jejuni* and 48.7% were *Campylobacter coli*.

3.2. Antimicrobial Susceptibility. The results of antimicrobial susceptibility testing for *Campylobacter* spp. isolated from chickens to nalidixic acid, ciprofloxacin, amoxicillin, erythromycin, and gentamicin are presented in Table 1.

For all isolates, antimicrobial resistance to nalidixic acid and ciprofloxacin was observed in 78.9% and 50%, respectively.

Among the 39 isolates of *C. jejuni* 79.5% ; 38.5% ; 17.9%; 7.7% were resistant to nalidixic acid, ciprofloxacin, amoxicillin, erythromycin, and gentamicin, respectively.

Among the 37 isolates, 78.4%; 43.2%; 13.5%; 8.1% were resistant to nalidixic acid, ciprofloxacin, amoxicillin, and erythromycin, respectively. In this study all isolates of *C. coli* were sensitive to gentamicin.

4. Discussion

Case-control studies of foodborne infection rates have estimated that 50 to 70% of *Campylobacter* illness is due to consumption of contaminated poultry and their products [9, 17, 36]. Several studies examined thermophilic *Campylobacter* in poultry, and the findings indicated prevalence ranges of the bacteria from 3% to 98% with *C. jejuni* or *C. coli* as the main isolates [36, 37].

Poultry and their products are commonly consumed in modern Ivoirians diets, but campylobacteriosis is not reported in Côte d'Ivoire up to now. Therefore, the primarily objective of this study was to determine prevalence of thermophilic species including *C. jejuni* and *C. coli,* in poultry in Abidjan.

This work demonstrated high prevalence of thermophilic *Campylobacter* in the studied chickens (63.8%). This proportion is similar to the results obtained in previous study (66.6%) [28]. Among the *Campylobacter* isolates 51.3% were *C. jejuni* and 48.7% were *C. coli.*

Such a high isolation rate of thermophilic *Campylobacter* in chickens has also been reported [37–42]. However, prevalence of *C. coli* in chicken is higher in this study than others findings [39, 42].

Poor hygiene and sanitation in poultry farms in Côte d'Ivoire could explain this high level of prevalence of *C. coli.* Indeed, most farms do not have security fence to prevent penetration of other animals including pigs, which are good carriers of *C. coli.* In some cases flocks of sheep or cattle and poultry take place at the same sites contributing like that to the contamination of chicken flocks with *C. coli.*

Antibiogram test indicated higher resistance of the microorganisms to ciprofloxacin and nalidixic acid. Recent studies reported that fluoroquinolone-resistant *Campylobacter* spp. rapidly emerged among poultry flocks [19, 24, 43, 44].

Alfredson and Korolik [45] advanced an hypothesis to explain this observation. They suggested that the use of enrofloxacin (derivates close to the fluoroquinolones used in human medicine) in animals flocks has probably exerted a selection pressure in animal reservoirs [22]. The absence of fluoroquinolone resistance of *Campylobacter* in Australia, country that has never used enrofloxacin, is a strong argument in favor of this hypothesis.

Resistance to erythromycin and amoxicillin, two antibiotics widely used to treat illness in Côte d'Ivoire, is not negligible. However, rates of that resistance are lower compared to those obtained with the fluoroquinolones.

These results are similar to previous observations [19, 20, 45–48]. The relatively high percentages of resistance to most antimicrobial agents tested in our study may be due to high usage of these agents as growth promoters or in animal treatment. In fact, in Cote d'Ivoire, as in most of developing countries [49], the use of antibiotics for humans and animals is relatively unrestricted.

Furthermore, no measures of hygiene are observed in both farms and in the process of slaughter which could cause contamination of poultry carcasses. Since campylobacteriosis is transmitted by consumption of contaminated food, preferably poultry, the presence of strains of antibiotic-resistant *Campylobacter* on chicken meat can be a real public health problem in Côte d'Ivoire. Indeed, the situation could deteriorate more rapidly in the study area, where there is an expansion of poultry farms with widespread and uncontrolled use of antibiotics. Therefore, surveillance of resistance pattern is necessary to guide rational use of antimicrobial agents in poultry farms.

In conclusion, we reported the presence of high contamination in poultry in Côte d'Ivoire by thermophilic *Campylobacter* and high resistance to antimicrobials of the fluoroquinolone family. These results show the need to strengthen the implementation of specific control procedures to decrease the contamination of poultry meat by *Campylobacter* and the necessity to reduce using of antibiotics in poultry sector. This study also shows the need to establish an efficient system for the control of *Campylobacter* in chickens.

Acknowledgments

The authors acknowledge Mr. Fofana Kouakou and Mr. Tiekoura Kouakou of the "Monitoring of resistance of microorganisms to antibiotics and natural substances" Unit of Pasteur Institute of Côte d'Ivoire for their help on this study, particularly concerning the antibiotics susceptibility testing.

References

[1] C. R. Friedman, J. Neimann, H. C. Wegener, and R. V. Tauxe, "Epidemiology of *Campylobacter jejuni* infections in the United States and other industrialized nations," in *Campylobacter*, Nachamkin and M. J. Blaser, Eds., pp. 121–138, ASM press, Washington, DC, USA, 2nd edition, 2000.

[2] World Health Organization WHO Department of Communicable Disease Surveillance and Response, "The increasing incidence of human campylobacteriosis report and proceeding of a WHO consultation of experts," 2001, http://whqlibdoc.who.int/hq/.

[3] C. C. Tam, S. J. O'Brien, G. K. Adak, S. M. Meakins, and J. A. Frost, "*Campylobacter coli*—an important foodborne pathogen," *Journal of Infection*, vol. 47, no. 1, pp. 28–32, 2003.

[4] G. M. Ruiz-Palacios, "The health burden of Campylobacter infection and the impact of antimicrobial resistance: playing chicken," *Clinical Infectious Diseases*, vol. 44, no. 5, pp. 701–703, 2007.

[5] R. A. Oberhelman and D. N. Taylor, "Campylobacter infections in developing countries," in *Campylobacter*, I. Nachamkin and M. J. Blaser, Eds., pp. 139–153, American Society for Microbiology, Washington, DC, USA, 2nd edition, 2000.

[6] M. R. Rao, A. B. Naficy, S. J. Savarino et al., "Pathogenicity and convalescent excretion of Campylobacter in rural Egyptian children," *American Journal of Epidemiology*, vol. 154, no. 2, pp. 166–173, 2001.

[7] M. L. Hänninen, H. Haajanen, T. Pummi et al., "Detection and typing of *Campylobacter jejuni* and *Campylobacter coli* and analysis of indicator organisms in three waterborne outbreaks in Finland," *Applied and Environmental Microbiology*, vol. 69, no. 3, pp. 1391–1396, 2003.

[8] E. Litrup, M. Torpdahl, and E. M. Nielsen, "Multilocus sequence typing performed on *Campylobacter coli* isolates from humans, broilers, pigs and cattle originating in Denmark," *Journal of Applied Microbiology*, vol. 103, no. 1, pp. 210–218, 2007.

[9] R. V. Tauxe, M. S. Deming, and P. A. Blake, "*Campylobacter jejuni* infections on college campuses: a national survey," *American Journal of Public Health*, vol. 75, no. 6, pp. 659–660, 1985.

[10] N. V. Harris, N. S. Weiss, and C. M. Nolan, "The role of poultry and meats in the etiology of *Campylobacter jejuni*/coli enteritis," *American Journal of Public Health*, vol. 76, no. 4, pp. 407–411, 1986.

[11] G. Kapperud, E. Skjerve, N. H. Bean, S. M. Ostroff, and J. Lassen, "Risk factors for sporadic Campylobacter infections: results of a case- control study in southeastern Norway," *Journal of Clinical Microbiology*, vol. 30, no. 12, pp. 3117–3121, 1992.

[12] M. J. Blaser, "Epidemiologic and clinical features of *Campylobacter jejuni* infections," *Journal of Infectious Diseases*, vol. 176, no. 6, pp. S103–S105, 1997.

[13] J. M. Cappelier, C. Magras, J. L. Jouve, and M. Federighi, "Recovery of viable but non-culturable *Campylobacter jejuni* cells in two animal models," *Food Microbiology*, vol. 16, no. 4, pp. 375–383, 1999.

[14] C. Zhao, B. Ge, J. De Villena et al., "Prevalence of *Campylobacter* spp., *Escherichia coli*, and *Salmonella* serovars in retail chicken, Turkey, pork, and beef from the greater Washington, D.C., area," *Applied and Environmental Microbiology*, vol. 67, no. 12, pp. 5431–5436, 2001.

[15] B. Ge, D. G. White, P. F. McDermott et al., "Antimicrobial-resistant *Campylobacter* species from retail raw meats," *Applied and Environmental Microbiology*, vol. 69, no. 5, pp. 3005–3007, 2003.

[16] R. J. Meldrum and I. G. Wilson, "*Salmonella* and *Campylobacter* in United Kingdom retail raw chicken in 2005," *Journal of Food Protection*, vol. 70, no. 8, pp. 1937–1939, 2007.

[17] B. M. Allos, "*Campylobacter jejuni* infections: update on emerging issues and trends," *Clinical Infectious Diseases*, vol. 32, no. 8, pp. 1201–1206, 2001.

[18] M. J. Blaser and J. Engberg, "Clinical aspects of *Campylobacter jejuni* and *Campylobacter coli* infections," in *Campylobacter*, I. Nachamkin, C. M. Szymanski, and M. J. Blaser, Eds., vol. 3, pp. 99–121, ASM Press, Washington, DC, USA, 2008.

[19] I. Nachamkin, H. Ung, and M. Li, "Increasing fluoroquinolone resistance in *Campylobacter jejuni*, Pennsylvania, USA, 1982–2001," *Emerging Infectious Diseases*, vol. 8, no. 12, pp. 1501–1503, 2002.

[20] F. Jorgensen, R. Bailey, S. Williams et al., "Prevalence and numbers of *Salmonella* and *Campylobacter* spp. on raw, whole chickens in relation to sampling methods," *International Journal of Food Microbiology*, vol. 76, no. 1-2, pp. 151–164, 2002.

[21] J. Ruiz, P. Goñi, F. Marco et al., "Increased resistance to quinolones in *Campylobacter jejuni*: a genetic analysis of gyrA gene mutations in quinolone-resistant clinical isolates," *Microbiology and Immunology*, vol. 42, no. 3, pp. 223–226, 1998.

[22] M. Teuber, "Veterinary use and antibiotic resistance," *Current Opinion in Microbiology*, vol. 4, no. 5, pp. 493–499, 2001.

[23] A. Caprioli, L. Busani, J. L. Martel, and R. Helmuth, "Monitoring of antibiotic resistance in bacteria of animal origin: epidemiological and microbiological methodologies," *International Journal of Antimicrobial Agents*, vol. 14, no. 4, pp. 295–301, 2000.

[24] T. Chuma, T. Ikeda, T. Maeda, H. Niwa, and K. Okamoto, "Antimicrobial susceptibilities of *Campylobacter* strains isolated from broilers in the Southern Part of Japan from 1995 to 1999," *Journal of Veterinary Medical Science*, vol. 63, no. 9, pp. 1027–1029, 2001.

[25] T. Asai, K. Harada, K. Ishihara et al., "Association of antimicrobial resistance in *Campylobacter* isolated from food-producing animals with antimicrobial use on farms," *Japanese Journal of Infectious Diseases*, vol. 60, no. 5, pp. 290–294, 2007.

[26] K. Han, S. S. Jang, E. Choo, S. Heu, and S. Ryu, "Prevalence, genetic diversity, and antibiotic resistance patterns of *Campylobacter jejuni* from retail raw chickens in Korea," *International Journal of Food Microbiology*, vol. 114, no. 1, pp. 50–59, 2007.

[27] X. Chen, G. W. Naren, C. M. Wu et al., "Prevalence and antimicrobial resistance of *Campylobacter* isolates in broilers from China," *Veterinary Microbiology*, vol. 144, no. 1-2, pp. 133–139, 2010.

[28] G. B. Goualié, G. T. Karou, S. Bakayoko et al., "Prévalence de *Campylobacter* chez les poulets vendus dans les marchés d'Abidjan. Étude pilote réalisée dans la commune d'Adjamé en 2005," RASPA, 2010, Vol. 8 N0S.

[29] I. D. Ogden, J. F. Dallas, M. MacRae et al., "*Campylobacter* excreted into the environment by animal sources: prevalence, concentration shed, and host association," *Foodborne Pathogens and Disease*, vol. 6, no. 10, pp. 1161–1170, 2009.

[30] "Historique de la commune d'Adjamé," 2011, http://www.rezoivoire.net/cotedivoire/ville/64/historique-de-la-commune-d-adjame.html.

[31] M. Federighi, *Campylobacter et Hygiène des Aliments*, Polytechnica, 1999.

[32] F. J. Bolton, D. R. Wareing, M. B. Skirrow, and D. N. Hutchinson, "Identification and biotyping of *Campylobacter*," in *Identification Methods in Applied and Environmental Microbiology*, G. R. Board, D. Jones, and F. A. Skinner, Eds., pp. 151–161, Blackwell Scientific Publications, Oxford, UK, 1992.

[33] J. Sambrook and D. W. Russel, "the basic polymerase chain reaction," in *Molecular Cloning-a Laboratory Manual*, vol. 3, pp. 19–824, Cold Spring Harbor Laboratory Press, New York, NY, USA, 3rd edition, 2001.

[34] D. Linton, A. J. Lawson, R. J. Owen, and J. Stanley, "PCR detection, identification to species level, and fingerprinting of *Campylobacter jejuni* and *Campylobacter coli* direct from diarrheic samples," *Journal of Clinical Microbiology*, vol. 35, no. 10, pp. 2568–2572, 1997.

[35] CLSI, *Performance Standards for Antimicrobial Disk Susceptibility Tests. Approved Standard*, Clinical Laboratory Standards Institute, Wayne, Pa, USA, 8th edition, 2003, CLSI Document M2-A8.

[36] G. Kapperud, G. Espeland, E. Wahl et al., "Factors associated with increased and decreased risk of *Campylobacter* infection: a prospective case-control study in Norway," *American Journal of Epidemiology*, vol. 158, no. 3, pp. 234–242, 2003.

[37] E. Cardinale, J. D. Perrier-Gros-Claude, F. Tall, M. Cissé, E. F. Guèye, and G. Salvat, "Prevalence of *Salmonella* and *Campylobacter* in retail chickens carcasses in Sénégal," *Revue d'Elevage et de medecine veterinaire des Pays Tropicaux*, vol. 56, no. 1-2, pp. 13–16, 2003.

[38] D. G. Newell and J. A. Wagenaar, "Poultry infections and their control at the farm level," in *Campylobacter*, I. Na-chamkin and M. J. Blaser, Eds., pp. 497–509, American Society for Microbiology Press, Washington, DC, USA, 2nd edition, 2000.

[39] D. G. Newell and C. Fearnley, "Sources of *Campylobacter* colonization in broiler chickens," *Applied and Environmental Microbiology*, vol. 69, no. 8, pp. 4343–4351, 2003.

[40] P. A. Magistrado, M. M. Garcia, and A. K. Raymundo, "Isolation and polymerase chain reaction-based detection of *Campylobacter jejuni* and *Campylobacter coli* from poultry in the Philippines," *International Journal of Food Microbiology*, vol. 70, no. 1-2, pp. 197–206, 2001.

[41] D. D. Ringoir and V. Korolik, "Colonisation phenotype and colonisation potential differences in *Campylobacter jejuni* strains in chickens before and after passage in vivo," *Veterinary Microbiology*, vol. 92, no. 3, pp. 225–235, 2003.

[42] A. A. Saleha, "Epidemiological study on the colonization of chickens with *Campylobacter* in broiler farms in Malaysia: possible risk and management factors," *International Journal of Poultry Sciences*, vol. 3, pp. 129–134, 2004.

[43] T. Kassa, S. Gebre-Selassie, and D. Asrat, "Antimicrobial susceptibility patterns of thermotolerant *Campylobacter* strains isolated from food animals in Ethiopia," *Veterinary Microbiology*, vol. 119, no. 1, pp. 82–87, 2007.

[44] E. Rahimi, H. Momtaz, M. Ameri, H. Ghasemian-Safaei, and M. Ali-Kasemi, "Prevalence and antimicrobial resistance of *Campylobacter* species isolated from chicken carcasses during processing in Iran," *Poultry Science*, vol. 89, no. 5, pp. 1015–1020, 2010.

[45] D. A. Alfredson and V. Korolik, "Antibiotic resistance and resistance mechanisms in *Campylobacter jejuni* and *Campylobacter coli*," *FEMS Microbiology Letters*, vol. 277, no. 2, pp. 123–132, 2007.

[46] D. Ewnetu and A. Mihret, "Prevalence and antimicrobial resistance of *Campylobacter* isolates from humans and chickens in Bahir Dar, Ethiopia," *Foodborne Pathogens and Disease*, vol. 7, no. 6, pp. 667–670, 2010.

[47] I. Hein, C. Schneck, M. Knögler et al., "*Campylobacter jejuni* isolated from poultry and humans in Styria, Austria: epidemiology and ciprofloxacin resistance," *Epidemiology and Infection*, vol. 130, no. 3, pp. 377–386, 2003.

[48] C. Gaudreau and H. Gilbert, "Antimicrobial resistance of *Campylobacter jejuni* subsp. jejuni strains isolated from humans in 1998 to 2001 in Montréal, Canada," *Antimicrobial Agents and Chemotherapy*, vol. 47, no. 6, pp. 2027–2029, 2003.

[49] H. J. Kim, J. H. Kim, Y. I. Kim et al., "Prevalence and characterization of *Campylobacter* spp. isolated from domestic and imported poultry meat in Korea, 2004–2008," *Foodborne Pathogens and Disease*, vol. 7, no. 10, pp. 1203–1209, 2010.

4

Gut Microbial Translocation in Critically Ill Children and Effects of Supplementation with Pre and Pro Biotics

Paola Papoff,[1] Giancarlo Ceccarelli,[2] Gabriella d'Ettorre,[2] Carla Cerasaro,[1] Elena Caresta,[1] Fabio Midulla,[1] and Corrado Moretti[1]

[1] *Pediatric Emergency and Intensive Care Division, Department of Pediatrics, Sapienza University of Rome, Viale Regina Elena 324, 00161 Rome, Italy*
[2] *Department of Public Health and Infectious Diseases, Sapienza University of Rome, 100161 Rome, Italy*

Correspondence should be addressed to Paola Papoff, p.papoff@libero.it

Academic Editor: Vincenzo Vullo

Bacterial translocation as a direct cause of sepsis is an attractive hypothesis that presupposes that in specific situations bacteria cross the intestinal barrier, enter the systemic circulation, and cause a systemic inflammatory response syndrome. Critically ill children are at increased risk for bacterial translocation, particularly in the early postnatal age. Predisposing factors include intestinal obstruction, obstructive jaundice, intra-abdominal hypertension, intestinal ischemia/reperfusion injury and secondary ileus, and immaturity of the intestinal barrier per se. Despite good evidence from experimental studies to support the theory of bacterial translocation as a cause of sepsis, there is little evidence in human studies to confirm that translocation is directly correlated to bloodstream infections in critically ill children. This paper provides an overview of the gut microflora and its significance, a focus on the mechanisms employed by bacteria to gain access to the systemic circulation, and how critical illness creates a hostile environment in the gut and alters the microflora favoring the growth of pathogens that promote bacterial translocation. It also covers treatment with pre- and pro biotics during critical illness to restore the balance of microbial communities in a beneficial way with positive effects on intestinal permeability and bacterial translocation.

1. Introduction

Despite advances in diagnosis and treatment, bacterial sepsis remains a significant cause of pediatric morbidity and mortality, particularly among critically ill children. Sepsis is the consequence of microbial invasion, or microbial products release, into the bloodstream, which result in systemic inflammatory response syndrome (SIRS). Bloodstream infection may arise through multiple routes, including bacterial translocation across the epithelial-mucosa as in the airways, gastrointestinal tract, kidney or genital tract, and skin breaks as in wounds and during insertion of central venous catheters or other medical devices [1]. Among these different possibilities, bacterial translocation across the gastrointestinal tract has been suggested as one of the principal pathogenetic mechanisms of sepsis and organ dysfunction among critically ill children [2]. There are a number of reasons why bacterial translocation may be relevant to the development of sepsis in children requiring intensive care: the majority of infections diagnosed in children in intensive care unit are due to microorganisms already present in the patients' admission flora of the throat and gut [3]; critical illness, coupled with intensive care treatment, results in a high reduction in microbiota biodiversity with a massive increase of enterococci [4]; gut permeability alterations to large molecules have been documented in critically ill patients [5]; selective gut decontamination seems to reduce infections in a subset of patients admitted to intensive care unit [6]; early enteral feeding is associated with reduced incidence of infections in critically ill patients [7]. Although clinical evidence suggests the importance of the gastrointestinal tract in the development of sepsis syndrome, bacterial translocation itself may not be the primary cause [8].

The purpose of this paper is, therefore, to discuss bacterial translocation, including its definition and role in causing sepsis syndrome and organ failure in critically ill children. The present paper also includes an analysis of the potential benefits of prebiotic and probiotics supplementation for the prevention of sepsis.

2. Gut Microflora and Its Significance

Human gut contains ~500 different species of microbes as commensals, including obligate anaerobes (about 95%) and facultative anaerobes (1–10%). Obligate anaerobes include *Bifidobacterium*, *Clostridium*, *Eubacterium*, *Fusobacterium*, *Peptococcus*, *Peptostreptococcus*, and *Bacteroides*, and facultative anaerobes *Lactobacillus*, *Bacillus*, *Streptococcus*, *Staphylococcus*, *E. coli*, *Klebsiella*, and *P. aeruginosa*. Bifidobacteria are the predominant microbes representing up to 80% of the cultivable fecal bacteria in infants and 25% in adults [9]. Although commensal bacteria are present in extremely high numbers, they rarely cause local or systemic disease, while have several important physiologic effects in the distal intestinal tract [10]. They directly activate the development and differentiation of the intestinal epithelium and its immune system and contribute to maintain an immunologically balanced inflammatory response. Microflora has nutritive functions, as well. It produces several enzymes for fermentation of nondigestible dietary residue and for secretion of endogenous mucus and helps in recovering lost energy in the form of short-chain fatty acids. It also plays a part in synthesis of vitamins, and in absorption of calcium, magnesium, and iron [11]. Finally, the gut microflora provides a physical barrier against invading pathogens, the so-called "colonization resistance" [12], through a competition for epithelial cell adhesion sites and available nutrients, and by releasing antibacterial substances (e.g., bacteriocins and lactic acid).

Prolonged critical care therapy may disrupt the balance between the host and the gut commensal flora in several ways [13]. Virtually all critically ill patients receive antibiotics which may profoundly affect the intestinal commensal microflora. Using molecular biology techniques, Iapichino et al. evaluated the intestinal microbiota composition of previously healthy patients on admission to intensive care unit [4]. While the first faecal samples showed a banding pattern that was similar to that of healthy subjects, after one week of critical illness and intensive care treatment, including antibiotics, a well definite alteration in the overall microbiota composition was evident, with the presence of a dominant band related to *Enterococcus*. An alteration of intestinal microbial flora has also been observed in premature neonates admitted to intensive care unit. Due to widespread use of broad spectrum antibiotics and late feeding, newborn infants present a delayed colonization with Lactobacillus and Bifidobacteria and a rapid appearance of enterococci, including *Enterococcus faecalis*, *E. coli*, *Enterobacter cloacae*, and the *Klebsiella pneumonia* [14]. Unexpectedly, *C. difficile* colonization does not increase in the early neonatal period despite the use of antibiotics such as cephalosporins [15]. Ferraris et al., who investigated the incidence and perinatal determinants of clostridial colonization in premature neonates, have recently confirmed the absence of effect of either antenatal or postnatal antibiotics on the overall clostridial colonization in neonates admitted to intensive care [16]. In the absence of antibiotics to disrupt the microbiota, it is not clear which event precedes *C. difficile* proliferation. It has been speculated that a decrease in *Bacteroides* and an increase in facultative anaerobes might facilitate colonization by *C. difficile* without a need for the action of antibiotics [17]. The influence of antibiotic use may be more relevant in affecting the severity of *Clostridium difficile* infection as suggested by the results of a recent study by Kim et al. showing that the most significant risk factors for severe *C. difficile* disease in children included young age (adjusted odds ratio [95% confidence interval]: 1.12 [1.02; 1.24]) and receipt of 3 antibiotic classes in the 30 days before infection (3.95 [1.19; 13.11]) [18]. In addition to antibiotic use, several other factors may predispose the gut ecology to alterations during the care of critically ill children. The use of gastroprotectant agents favors proliferation of acid sensitive organisms within the upper intestinal tract. Vasoactive agents that severely limit mesenteric perfusion induce profound luminal hypoxia and hypercarbia that are potent activators of bacterial virulence gene expression [19, 20]. The use of opioids as sedative analgesic agents in mechanically ventilated patients create intestinal atonia and bacterial overgrowth [21]. Finally, nutrients delivered to critically ill patients intravenously or as highly processed foods, whose absorption is nearly complete within the small intestine, create nutrient scarcity within the distal intestinal tract, the area where the highest microbial burden exists. Under such circumstances, the proliferation of highly virulent microorganisms creates a state of perturbed host pathogen balance, and an undesired activation of local inflammation. Systemic inflammation can thus be initiated by mucosally derived cytokines, or when microbial products enter the systemic circulation through the disrupted intestinal epithelial barrier [22].

3. Concepts of Bacterial Translocation

Bacterial translocation has been defined as the process by which live bacteria, their products, or both cross the intestinal barrier where they may either directly cause infection or excite the immune system resulting in a massive inflammatory reaction causing diffuse organ damage and eventually organ failure and death [23]. To some extent, bacterial translocation from the gut occurs regularly even in healthy individuals (5–10%) [24], but bacteremia is generally limited by an intact immune system [8]. This may be a normal physiologic process by which animals and humans sample different luminal antigens in order to produce immunocompetent cells [25]. The first line of defense for preventing bacterial translocation is the mucous coat overlying gut epithelia, produced by goblet cells, which includes degraded mucin and antimicrobial peptides. In the neonatal enterocytes, the ability of mucin to inhibit bacterial translocation may be diminished compared to adults [26]. This might help to explain the propensity of the neonatal rats to spontaneous bacterial translocation in the first two weeks of life when intestinal concentrations of gram-negative bacilli and gram-positive cocci are high, and the concentration of lactobacilli is low [27]. Translocation occurs in between cells or through cells of the intestinal epithelium after loss of tight junctions between enterocytes. Several mechanisms of bacterial translocation have been identified from studies of enteropathogenic bacteria, such as a zipper mechanism that utilizes transmembrane cell-adhesion proteins as receptors for the bacteria [28]; a bacterial needle-like probe that injects dedicated bacterial effectors into epithelia and the injected molecules modify the cytoskeleton to facilitate bacterial entry [29]; an increased nitric oxide production during inflammatory states that alters expression and localization of the tight junction proteins that surround the upper part and lateral surfaces of enterocytes leading to intestinal hyperpermeability [30]; toll-like receptors that are present on the luminal surface of enterocytes to sense danger and activate host defenses and that can also be harmful by mediating phagocytosis and translocation of bacteria across the intestinal barrier [31]. Once pathogens pass the mucus and epithelial barriers, they are ingested by submucosal macrophages. This process occurs without initiation of an inflammatory response [32]. If intestinal macrophages are dysfunctional as in very low-birth-weight infants [33, 34], this dysfunction may contribute to diffusion of bacteria in the systemic circulation.

Several factors may enhance bacterial translocation in critically ill children. Bacterial overgrowth and breakdown of a tight junction in the setting of intestinal obstruction has been shown to promote bacterial translocation in animal models [35] and in humans [36]. There is evidence in *in vitro* and animal studies that obstructive jaundice impairs reticuloendothelial function [37], interferes with macrophage activation [38], alters Kupffer cell function [39], promotes disruption of desmosomes and formation of lateral spaces between enterocytes [40], and, therefore, alters epithelial barrier permeability. Another possible mechanism for increased translocation associated with obstructive jaundice is considered to be the inhibitory effect of bile on bacterial invasion of enterocytes shown *in vitro* [41]. Intra-abdominal hypertension and abdominal compartment syndrome may also cause gut barrier dysfunction [42]. Another factor that promotes bacterial translocation and predisposes to development of SIRS and organ dysfunction is intestinal ischemia/reperfusion injury. The gut is an organ extremely sensitive to systemic cardiovascular and pulmonary disturbances. The physiological response to hypoperfusion is the shunting of blood away from splanchnic circulation toward more vital organs. The consequence is ischemia/reperfusion injury of villi, release of proinflammatory factors, mucosal disruption, increased intestinal permeability, and bacterial translocation. In addition, secondary ileus seen after ischemia/reperfusion injury seems to promote bacterial overgrowth and proximal gut colonization which are linked with the development of septic complication [43].

Among the factors that influence bacterial translocation, postnatal age and prematurity appear to play a significant predisposing role. Prematurity reduces mucosal barrier function and consequently foster gut permeability [20]. Moy gave definitive experimental evidence that spontaneous bacterial translocation occurs in the neonate by demonstrating that transformed *E. coli* K1 fed to healthy rabbit pups spontaneously translocated from the intestinal lumen and subsequently disseminated to the mesenteric lymph nodes, spleen, and liver [44]. A high proportion of bacterial translocation in neonates results not only from immaturity of host defense functions, but also from the dominant colonization of aerobic bacteria in the intestine. Bacterial colonization develops differently in breast-fed, formula-fed, premature, and full-term infants. In a model of newborn rats, Yajima showed that breastfeeding inhibited systemic bacterial translocation in the suckling period of the rat, even though this phenomenon is not necessarily correlated to modification of the colonizing flora [24].

There are precise criteria to define that bacterial translocation has occurred in a subject including: gut-origin bacteria found in mesenteric lymph nodes (the first organ encountered by the organism undergoing translocation) or portal venous blood; endotoxin found in mesenteric lymph nodes or portal venous blood; bacterial DNA or proteins found in mesenteric lymph nodes, portal venous blood, or the systemic circulation; development of infectious complications with organisms that presumably originated from the gut; increased levels of circulating and tissue cytokines and inflammatory mediators; increased permeability of the gut to large molecules. Increased intestinal permeability as measured by the lactulose-to-mannitol ratio may be a permissive factor for bacterial translocation, but finding an increased lactulose-to-mannitol ratio does not prove that translocation has occurred [8]. In humans in whom direct culture of mesenteric lymph nodes or portal blood is not routinely possible, we often use indirect ways to confirm or monitor bacterial translocation, thus extreme caution should be practiced when drawing conclusions.

4. Bacterial Translocation in Critically Ill Patients

Few studies have investigated the pathogenic potential of bacterial translocation to septic morbidity in critically ill children. Pathan et al. examined the role of intestinal injury and subsequent endotoxemia in the pathogenesis of organ dysfunction after surgery for congenital heart disease [45]. They analyzed blood levels of endotoxin alongside global transcriptomic profiling and monocyte endotoxin receptor expression in children undergoing surgery for congenital heart disease and found that these infants present an increased risk of intestinal mucosal injury and endotoxemia which may contribute to inflammatory activation and organ dysfunction postoperatively. Cicalese et al. evaluated the correlation between bacterial translocation and preservation injury or acute rejection in 50 pediatric small bowel transplant immunosuppressed recipients [46]. Bacterial translocation episodes were considered when microorganisms were found simultaneously in blood or liver biopsy and stool. In this study, a substantial percentage of bacterial translocation was associated with acute rejection, the presence of a colon allograft, and a long cold ischemia time. In a prospective observational cohort study in 94 neonates and infants who underwent surgical procedures and required parenteral nutrition because of gastrointestinal abnormalities, Pierro et al. explored the relevance of septicemias due to microbial translocation in relation to long-term parenteral nutrition [47]. Microbial translocation was diagnosed when the microorganisms that were isolated from the blood sample were also carried in the throat and/or rectum. Sepsis associated with microbial translocation was found in 15 cases on 6 infants. *Escherichia coli*, *Klebsiella*, *Candida* species, and enterococci were more commonly isolated. The authors concluded that in neonates and infants who are receiving parenteral nutrition, septicemia may be a gut-related phenomenon. From the results of these studies, it appears that bacterial translocation could indeed be a critical component to the development of SIRS; however, the numerous methodological problems that plague these studies make a cause/effect relationship questionable.

5. Prebiotics

The introduction of foods that sustain the growth of intestinal microorganisms might prevent the process by which translocation of potentially pathogen bacteria occurs [48]. In 1995, Gibson and Roberfroid [49, 50] introduced the concept of prebiotic as a nondigestible food ingredient that beneficially affects the host by selectively stimulating the growth and/or activity of one or a limited number of bacteria in the colon, and thus improves host health. Prebiotics, which are not digested in the small intestine, enter the colon as intact large carbohydrates that are then fermented by the resident bacteria to produce short-chain fatty acids. The nature of this fermentation and the resulting pH of the intestinal contents dictate proliferation of specific resident bacteria. For example, infants fed-breast milk containing prebiotics support increased proliferation of bifidobacteria and lactobacilli (probiotic), whereas formula-fed infants produce more enterococci and enterobacteria. Clear criteria have been established for classifying a food ingredient as a prebiotic. These criteria are (1) resistance to gastric acidity, to hydrolysis by mammalian enzymes, and to gastrointestinal absorption; (2) fermentation by intestinal microflora; (3) selective stimulation of the growth and/or activity of those intestinal bacteria that contribute to health and well-being. Presently there are only 2 food ingredients that fulfill these criteria, that is, inulin and transgalactooligosaccharides (TOS) [51]. The current most popular targets for prebiotic use are lactobacilli and bifidobacteria. Bifidobacteria are able to break down and utilize inulin-type fructans by the action of the b-fructofuranosidase enzyme. The mechanisms of action of prebiotics are complex and so far not yet well known or understood. These mechanisms are summarized below [52–60].

(1) Sources of carbon and energy for bacteria growing in the large bowel [precursors of short-chain fatty acids (SCFA): acetate, propionate, and butyrate] by saccharolytic fermentation [53, 55, 61]. (2) Immunological effects: (a) activate leukocytes in the gut-associated lymphoid tissue (GALT) system [56–58]; (b) increase cell numbers in Peyer's patches [57]; (c) enhance production of bacteriocins [59]; (d) enhance IgA levels in the small intestine and caecum [60]. (3) Improve gut-barrier function [57, 58]. (4) Acidify intestinal contents. (5) Improve bioavailability of calcium and magnesium [61, 62]. (6) Foster absorption of water and sodium [54]. (7) Promote gut transit.

6. Probiotics

Probiotics are commercially available microorganisms which, when ingested as individual strains or in combinations, offer potential health benefits to the host. These agents are often concurrently administered with substances that promote bacterial colonization and growth (prebiotics): in this instance, they are referred to as synbiotics [63]. The beneficial effects of probiotics in critically ill patients can be summarized as follows: (1) immunological effects, such as activation of leukocytes in the gut-associated lymphoid tissue system, increased cell number in Peyer's patches, enhanced immunoglobulin A levels [64], production of bacteriocins, inhibitory effect on proinflammatory cytokines tumor necrosis factor-alpha, interleukins IL-1 and IL-6, stimulation of anti-inflammatory cytokines (IL-10); (2) improved gut barrier function; (3) acidifying intestinal contents inhibition of pathogenic bacteria; (4) improved bioavailability of calcium and magnesium; (5) facilitated absorption of water and sodium; (6) increased intestinal motility. The effectiveness of probiotics is related to their ability to survive in the acidic and alkaline environment of stomach and duodenum, respectively, as well as their ability to adhere to the colonic mucosa and to colonize the colon [11]. Some probiotics, for example, *Lactobacillus* GG and *L. plantarum* 299v are better able to colonize the colon than others [11]. *Saccharomyces boulardii* are non-LA yeast and secret a protease causing proteolysis of Toxin A and Toxin B of *C. difficile* responsible for antibiotic-associated diarrhea (AAD) and, therefore, should be avoided [11]. Bifidobacteria are gram-positive anaerobic lactic acid bacilli (LAB), colonize the colon within days of birth and its population remains stable until advanced age. Lactobacilli are gram-positive, facultative anaerobic LAB, and are normal inhabitant of human gut. *L. plantarum* 299v adheres to the intestinal mucosa to reinforce its barrier function, thus prevents attachment of pathogens to the intestinal wall [65]. *Lactobacilllus* GG was found to eradicate *C. difficile* in patients with relapsing colitis. *L. plantarum* ST31 produces bacteriocins to limit the growth of potential pathogens. *L. casei* increases the level of circulating IgA. *L. acidophilus* and *B. bifidum* appear to enhance the phagocyte activity of circulating granulocytes. *Bacillus clausii* are gram-positive spore-forming strictly aerobic non-LAB and constitute less than 1% of gut microflora. *B. clausii* stimulates CD4 proliferation and lymphocytic activity in Peyer's patches. It also leads to increase in IgA-positive lymphocytes and HLA-DR positive T lymphocytes [66]. Multistrain probiotics seem to be better than single-strain ones, as individual probiotics have different functions and show synergistic effects when administered together [67]. Various single-strain and multistrain probiotics are commercially available for clinical use, majority being LABs.

7. Pre- and Pro Biotics in Critically Ill Children: Available Evidence

Circumstantial evidence suggests that probiotics alone or in combination with prebiotics are effective in preventing necrotizing enterocolitis (NEC), fungal colonization, and in improving feeding tolerance. They have also proved to be microbiologically safe and clinically well tolerated. To this regard, Manzoni reported a 6-year, two-center experience of routinary *Lactobacillus rhamnosus* GG (LGG) use in very low-birth-weight infants (3×10^9 CFU/day, in single oral dose, since 4th day of life, for 4-to 6-week courses) [68]. No adverse effects or intolerances nor clinical sepsis attributable to LGG occurred. In a randomized controlled trial on 231 infants weighing 750–1499 g at birth, Braga et al. found that oral supplementation of human milk with *Bacillus breve* and *Lactobacillus casei* reduced the occurrence of NEC [69]. It was considered that an improvement in intestinal motility might have contributed to this result. Similarly, Bin-Nun showed in a randomized controlled trial that a probiotic mixture of *Bifidobacteria infantis*, *Streptococcus thermophilus*, and *Bifidobacteria bifidus* reduced the incidence and severity of NEC in very low-birth-weight neonates [70]. In a larger randomized, multicenter controlled trial, Lin et al. showed that *Bifidobacterium bifidum* and *Lactobacillus acidophilus*, added to breast milk or mixed feeding, twice daily for 6 weeks, reduced the incidence of death or NEC [71]. Taken together, these data favor the use of oral probiotics for the prevention of NEC in preterm infants. An improved outcome after treatment with *Lactobacillus reuteri* or *L. rhamnosus* was also observed in a group of 249 preterm infants by Romeo et al. who showed a reduced colonization by *Candida*, protection from late-onset sepsis and reduced abnormal neurological outcome [72]. In contrast with these positive results, Sari and Dani found no benefit in administering *Lactobacillus sporogenes* or LGG in reducing the incidence of NEC [73, 74]. Type of probiotics used, as well as the timing and dosage, may explain such differences. As compared to neonates, fewer studies have been published in older children admitted to intensive care, the majority of which focusing on safety and beneficial changes of intestinal flora. Simakachorn et al. randomized 94 patients between 1 and 3 years old who were requiring mechanical ventilation to receive either a test formula containing a synbiotic blend (composed of 2 probiotic strains (*Lactobacillus paracasei* NCC 2461 and *Bifidobacterium longum* NCC 3001), fructooligosaccharides, inulin, and Acacia gum, or a control formula. Infants in the test group tolerated well pre- and pro biotics [75]. Faecal bifidobacteria and total lactobacilli were higher in the test group, whereas enterobacteria levels diminished. The authors concluded that the enteral formula supplemented with synbiotics was well tolerated by children in intensive care units; it was safe and produced an increase in faecal bacterial groups. Honeycutt et al. evaluated the efficacy of probiotics in reducing the rates of nosocomial infection in 61 patients admitted to a pediatric intensive care unit [76]. Children were randomized to receive either one capsule of LGG or placebo once a day until discharge from the hospital. In this study, LGG was not shown to be effective in reducing the incidence of nosocomial infections. Similarly, a systematic review on 999 critically ill adults revealed no beneficial effect of probiotics/synbiotics in term of clinical outcomes, length of intensive care unit stay, incidence of nosocomial infection, pneumonia and hospital mortality [77]. In conclusion, it appears that beyond the neonatal age no clear evidence supports the use of probiotics in critically ill patients. Well-designed, large-scale, clinical trials are therefore needed to define optimal probiotic species, doses, and whether combination therapy is superior to single-agent therapy.

8. Conclusions

Bacterial translocation as a direct cause of sepsis is an attractive hypothesis that presupposes that in specific situations bacteria cross the intestinal barrier, enter the systemic circulation, and cause SIRS. Critically ill children are at increased risk for bacterial translocation, particularly in the early postnatal age. Predisposing factors include intestinal obstruction, obstructive jaundice, intra-abdominal hypertension, intestinal ischemia/reperfusion injury and secondary ileus, and immaturity of the intestinal barrier per se. Despite good evidence from experimental studies to support the theory of bacterial translocation as a cause of sepsis, there is no definite evidence in human studies to confirm that translocation is directly correlated to bloodstream infections in critically ill children. Besides, attempts at the use of pre- or pro-biotics have not always translated into clinical benefits to patient care, except for prevention of NEC in the neonatal population. Therefore, further research in this field is needed to help clinicians to make correct decisions concerning protection of the gut in the intensive care unit and to decide for possible therapeutic use of pre- and probiotics.

Conflict of Interests

The authors declare that there are no conflicts of interests.

References

[1] S. L. Bateman and P. C. Seed, "Procession to pediatric bacteremia and sepsis: covert operations and failures in diplomacy," *Pediatrics*, vol. 126, no. 1, pp. 137–150, 2010.

[2] J. C. Marshall, N. V. Christou, and J. L. Meakins, "The gastrointestinal tract: the "undrained abscess" of multiple organ failure," *Annals of Surgery*, vol. 218, no. 2, pp. 111–119, 1993.

[3] R. E. Sarginson, N. Taylor, N. Reilly, P. B. Baines, and H. K. F. Van Saene, "Infection in prolonged pediatric critical illness: a prospective four-year study based on knowledge of the carrier state," *Critical Care Medicine*, vol. 32, no. 3, pp. 839–847, 2004.

[4] G. Iapichino, M. L. Callegari, S. Marzorati et al., "Impact of antibiotics on the gut microbiota of critically ill patients," *Journal of Medical Microbiology*, vol. 57, no. 8, pp. 1007–1014, 2008.

[5] C. J. Doig, L. R. Sutherland, J. D. Sandham, G. H. Fick, M. Verhoef, and J. B. Meddings, "Increased intestinal permeability is associated with the development of multiple organ

dysfunction syndrome in critically ill ICU patients," *American Journal of Respiratory and Critical Care Medicine*, vol. 158, no. 2, pp. 444–451, 1998.

[6] M. A. de La Cal, E. Cerdá, P. García-Hierro et al., "Survival benefit in critically ill burned patients receiving selective decontamination of the digestive tract: a randomized, placebo-controlled, double-blind trial," *Annals of Surgery*, vol. 241, no. 3, pp. 424–430, 2005.

[7] V. Artinian, H. Krayem, and B. DiGiovine, "Effects of early enteral feeding on the outcome of critically ill mechanically ventilated medical patients," *Chest*, vol. 129, no. 4, pp. 960–967, 2006.

[8] S. M. Steinberg, "Bacterial translocation: what it is and what it is not," *American Journal of Surgery*, vol. 186, no. 3, pp. 301–305, 2003.

[9] F. Guarner and J. R. Malagelada, "Gut flora in health and disease," *The Lancet*, vol. 361, no. 9356, pp. 512–519, 2003.

[10] M. B. Roberfroid, F. Bornet, C. Bouley, and J. H. Cummings, "Colonic microflora: nutrition and health—summary and conclusions of an International Life Sciences Institute (ILSI) [Europe] Workshop held in Barcelona, Spain," *Nutrition Reviews*, vol. 53, no. 5, pp. 127–130, 1995.

[11] S. C. Singhi and A. Baranwal, "Probiotic use in the critically ILL," *Indian Journal of Pediatrics*, vol. 75, no. 6, pp. 621–627, 2008.

[12] M. J. G. Farthing, "Bugs and the gut: an unstable marriage," *Best Practice and Research*, vol. 18, no. 2, pp. 233–239, 2004.

[13] J. C. Marshall, "Gastrointestinal flora and its alterations in critical illness1999," *Current Opinion in Clinical Nutrition & Metabolic Care*, vol. 2, no. 5, pp. 405–411.

[14] M. Duman, H. Abacioglu, M. Karaman, N. Duman, and H. Özkan, "β-lactam antibiotic resistance in aerobic commensal fecal flora of newborns," *Pediatrics International*, vol. 47, no. 3, pp. 267–273, 2005.

[15] A. F. Holton, M. A. Hall, and J. A. Lowes, "Antibiotic exposure delays intestinal colonization by Clostridium difficile in the newborn," *Journal of Antimicrobial Chemotherapy*, vol. 24, no. 5, pp. 811–817, 1989.

[16] L. Ferraris, M. J. Butel, F. Campeotto, M. Vodovar, J. C. Roze, and J. Aires, "Clostridia in premature neonates' gut: incidence, antibiotic susceptibility, and perinatal determinants influencing colonization," *PLoS ONE*, vol. 7, no. 1, Article ID e30594, 2012.

[17] C. Rousseau, F. Levenez, C. Fouqueray, J. Doré, A. Collignon, and P. Lepage, "Clostridium difficile colonization in early infancy is accompanied by changes in intestinal microbiota composition," *Journal of Clinical Microbiology*, vol. 49, no. 3, pp. 858–865, 2011.

[18] J. Kim, J. F. Shaklee, S. Smathers et al., "Risk factors and outcomes associated with severe clostridium difficile infection in children," *The Pediatric Infectious Disease Journal*, vol. 31, no. 2, pp. 134–138, 2012.

[19] J. E. Kohler, O. Zaborina, L. Wu et al., "Components of intestinal epithelial hypoxia activate the virulence circuitry of Pseudomonas," *American Journal of Physiology*, vol. 288, no. 5, pp. G1048–G1054, 2005.

[20] J. Alverdy, O. Zaborina, and L. Wu, "The impact of stress and nutrition on bacterial-host interactions at the intestinal epithelial surface," *Current Opinion in Clinical Nutrition and Metabolic Care*, vol. 8, no. 2, pp. 205–209, 2005.

[21] V. B. Nieuwenhuijs, A. Verheem, H. van Duijvenbode-Beumer et al., "The role of interdigestive small bowel motility in the regulation of gut microflora, bacterial overgrowth, and bacterial translocation in rats," *Annals of Surgery*, vol. 228, no. 2, pp. 188–193, 1998.

[22] J. Alverdy, O. Zaborina, and L. Wu, "The impact of stress and nutrition on bacterial-host interactions at the intestinal epithelial surface," *Current Opinion in Clinical Nutrition and Metabolic Care*, vol. 8, no. 2, pp. 205–209, 2005.

[23] J. W. Alexander, S. T. Boyce, G. F. Babcock et al., "The process of microbial translocation," *Annals of Surgery*, vol. 212, no. 4, pp. 496–512, 1990.

[24] P. C. Sedman, J. Macfie, P. Sagar et al., "The prevalence of gut translocation in humans," *Gastroenterology*, vol. 107, no. 3, pp. 643–649, 1994.

[25] C. L. Wells, M. A. Maddaus, and R. L. Simmons, "Proposed mechanisms for the translocation of intestinal bacteria," *Reviews of Infectious Diseases*, vol. 10, no. 5, pp. 958–979, 1988.

[26] A. S. Gork, N. Usui, E. Ceriati et al., "The effect of mucin on bacterial translocation in I-407 fetal and Caco-2 adult enterocyte cultured cell lines," *Pediatric Surgery International*, vol. 15, no. 3-4, pp. 155–159, 1999.

[27] H. H. Wenzl, G. Schimpl, G. Feierl, and G. Steinwender, "Time course of spontaneous bacterial translocation from gastrointestinal tract and its relationship to intestinal microflora in conventionally reared infant rats," *Digestive Diseases and Sciences*, vol. 46, no. 5, pp. 1120–1126, 2001.

[28] P. Cossart and P. J. Sansonetti, "Bacterial invasion: the paradigms of enteroinvasive pathogens," *Science*, vol. 304, no. 5668, pp. 242–248, 2004.

[29] A. R. Hauser, "The type III secretion system of Pseudomonas aeruginosa: infection by injection," *Nature Reviews Microbiology*, vol. 7, no. 9, pp. 654–665, 2009.

[30] R. J. Anand, C. L. Leaphart, K. P. Mollen, and D. J. Hackam, "The role of the intestinal barrier in the pathogenesis of necrotizing enterocolitis," *Shock*, vol. 27, no. 2, pp. 124–133, 2007.

[31] M. D. Neal, C. Leaphart, R. Levy et al., "Enterocyte TLR4 mediates phagocytosis and translocation of bacteria across the intestinal barrier," *Journal of Immunology*, vol. 176, no. 5, pp. 3070–3079, 2006.

[32] B. Weber, L. Saurer, and C. Mueller, "Intestinal macrophages: differentiation and involvement in intestinal immunopathologies," *Seminars in Immunopathology*, vol. 31, no. 2, pp. 171–184, 2009.

[33] L. C. Duffy, "Interactions mediating bacterial translocation in the immature intestine," *Journal of Nutrition*, vol. 130, no. 2, supplement, pp. 432S–436S, 2000.

[34] A. Yachie, N. Takano, K. Ohta et al., "Defective production of interleukin-6 in very small premature infants in response to bacterial pathogens," *Infection and Immunity*, vol. 60, no. 3, pp. 749–753, 1992.

[35] E. A. Deitch, W. M. Bridges, J. W. Ma, R. D. Berg, and R. D. Specian, "Obstructed intestine as a reservoir for systemic infection," *American Journal of Surgery*, vol. 159, no. 4, pp. 394–401, 1990.

[36] E. A. Deitch, "Simple intestinal obstruction causes bacterial translocation in man," *Archives of Surgery*, vol. 124, no. 6, pp. 699–701, 1989.

[37] J. W. Ding, R. Andersson, V. Soltesz, R. Willén, and S. Bengmark, "Obstructive jaundice impairs reticuloendothelial function and promotes bacterial translocation in the rat," *Journal of Surgical Research*, vol. 57, no. 2, pp. 238–245, 1994.

[38] J. V. Reynolds, P. Murchan, H. P. Redmond et al., "Failure of macrophage activation in experimental obstructive jaundice:

association with bacterial translocation," *British Journal of Surgery*, vol. 82, no. 4, pp. 534–538, 1995.

[39] S. M. Sheen-Chen, P. Chau, and H. W. Harris, "Obstructive jaundice alters kupffer cell function independent of bacterial translocation," *Journal of Surgical Research*, vol. 80, no. 2, pp. 205–209, 1998.

[40] R. W. Parks, C. H. Stuart Cameron, C. D. Gannon, C. Pope, T. Diamond, and B. J. Rowlands, "Changes in gastrointestinal morphology associated with obstructive jaundice," *The Journal of Pathology*, vol. 192, no. 4, pp. 526–532, 2000.

[41] C. L. Wells, R. P. Jechorek, and S. L. Erlandsen, "Inhibitory effect of bile on bacterial invasion of enterocytes: possible mechanism for increased translocation associated with obstructive jaundice," *Critical Care Medicine*, vol. 23, no. 2, pp. 301–307, 1995.

[42] H. J. Sugerman, G. L. Bloomfield, and B. W. Saggi, "Multisystem organ failure secondary to increased intraabdominal pressure," *Infection*, vol. 27, no. 1, pp. 61–66, 1999.

[43] F. A. Moore, "The role of the gastrointestinal tract in postinjury multiple organ failure," *American Journal of Surgery*, vol. 178, no. 6, pp. 449–453, 1999.

[44] J. Moy, D. J. Lee, C. M. Harmon, R. A. Drongowski, and A. G. Coran, "Confirmation of translocated gastrointestinal bacteria in a neonatal model," *Journal of Surgical Research*, vol. 87, no. 1, pp. 85–89, 1999.

[45] N. Pathan, M. Burmester, T. Adamovic et al., "Intestinal injury and endotoxemia in children undergoing surgery for congenital heart disease," *American Journal of Respiratory and Critical Care Medicine*, vol. 184, no. 11, pp. 1261–1269, 2011.

[46] L. Cicalese, P. Sileri, M. Green, K. Abu-Elmagd, S. Kocoshis, and J. Reyes, "Bacterial translocation in clinical intestinal transplantation," *Transplantation*, vol. 71, no. 10, pp. 1414–1417, 2001.

[47] A. Pierro, H. K. F. van Saene, S. C. Donnell et al., "Microbial translocation in neonates and infants receiving long-term parenteral nutrition," *Archives of Surgery*, vol. 131, no. 2, pp. 176–179, 1996.

[48] W. Manzanares and G. Hardy, "The role of prebiotics and synbiotics in critically ill patients," *Current Opinion in Clinical Nutrition and Metabolic Care*, vol. 11, no. 6, pp. 782–789, 2008.

[49] G. R. Gibson and M. B. Roberfroid, "Dietary modulation of the human colonic microbiota: Introducing the concept of prebiotics," *Journal of Nutrition*, vol. 125, no. 6, pp. 1401–1412, 1995.

[50] G. R. Gibson, H. M. Probert, J. Van Loo, R. A. Rastall, and M. B. Roberfroid, "Dietary modulation of the human colonic microbiota: updating the concept of prebiotics," *Nutrition Research Reviews*, vol. 17, no. 2, pp. 259–275, 2004.

[51] S. Kolida, K. Tuohy, and G. R. Gibson, "Prebiotic effects of inulin and oligofructose," *British Journal of Nutrition*, vol. 87, supplement 2, pp. S193–S197, 2002.

[52] S. Kolida and G. R. Gibson, "Prebiotic capacity of inulin-type fructans," *Journal of Nutrition*, vol. 137, no. 11, pp. 2503S–2506S, 2007.

[53] J. Van Loo, "The specificity of the interaction with intestinal bacterial fermentation by prebiotics determines their physiological efficacy," *Nutrition Research Reviews*, vol. 17, no. 1, pp. 89–98, 2004.

[54] C. C. Roy, C. L. Kien, L. Bouthillier, and E. Levy, "Short-chain fatty acids: ready for prime time?" *Nutrition in Clinical Practice*, vol. 21, no. 4, pp. 351–366, 2006.

[55] D. J. Morrison, W. G. Mackay, C. A. Edwards, T. Preston, B. Dodson, and L. T. Weaver, "Butyrate production from oligofructose fermentation by the human faecal flora: what is the contribution of extracellular acetate and lactate?" *British Journal of Nutrition*, vol. 96, no. 3, pp. 570–577, 2006.

[56] M. L. Forchielli and W. A. Walker, "The role of gut-associated lymphoid tissues and mucosal defence," *British Journal of Nutrition*, vol. 93, supplement 1, pp. S41–S48, 2005.

[57] B. Watzl, S. Girrbach, and M. Roller, "Inulin, oligofructose and immunomodulation," *British Journal of Nutrition*, vol. 93, supplement 1, pp. S49–S55, 2005.

[58] F. Guarner, "Inulin and oligofructose: impact on intestinal diseases and disorders," *British Journal of Nutrition*, vol. 93, supplement 1, pp. S61–S65, 2005.

[59] Y. S. Chen, S. Srionnual, T. Onda, and F. Yanagida, "Effects of prebiotic oligosaccharides and trehalose on growth and production of bacteriocins by lactic acid bacteria," *Letters in Applied Microbiology*, vol. 45, no. 2, pp. 190–193, 2007.

[60] M. Roller, G. Rechkemmer, and B. Watzl, "Prebiotic inulin enriched with oligofructose in combination with the probiotics Lactobacillus rhamnosus and Bifidobacterium lactis modulates intestinal immune functions in rats," *Journal of Nutrition*, vol. 134, no. 1, pp. 153–156, 2004.

[61] E. Bruzzese, M. Volpicelli, M. Squaglia, A. Tartaglione, and A. Guarino, "Impact of prebiotics on human health," *Digestive and Liver Disease*, vol. 38, 2, pp. S283–S287, 2006.

[62] S. Macfarlane, G. T. Macfarlane, and J. H. Cummings, "Review article: prebiotics in the gastrointestinal tract," *Alimentary Pharmacology and Therapeutics*, vol. 24, no. 5, pp. 701–714, 2006.

[63] S. Bengmark, "Synbiotics and the mucosal barrier in critically ill patients," *Current Opinion in Gastroenterology*, vol. 21, no. 6, pp. 712–716, 2005.

[64] C. Alberda, L. Gramlich, J. Meddings et al., "Effects of probiotic therapy in critically ill patients: a randomized, double-blind, placebo-controlled trial," *American Journal of Clinical Nutrition*, vol. 85, no. 3, pp. 816–823, 2007.

[65] B. Klarin, M. L. Johansson, G. Molin, A. Larsson, and B. Jeppsson, "Adhesion of the probiotic bacterium Lactobacillus plantarum 299v onto the gut mucosa in critically ill patients: a randomised open trial," *Critical Care*, vol. 9, no. 3, pp. R285–293, 2005.

[66] M. C. Urdaci, P. Bressollier, and I. Pinchuk, "Bacillus clausii probiotic strains: antimicrobial and immunomodulatory activities," *Journal of Clinical Gastroenterology*, vol. 38, no. 6, supplement, pp. S86–S90, 2004.

[67] C. M. C. Chapman, G. R. Gibson, and I. Rowland, "Health benefits of probiotics: are mixtures more effective than single strains?" *European Journal of Nutrition*, vol. 50, no. 1, pp. 1–17, 2011.

[68] P. Manzoni, G. Lista, E. Gallo et al., "Routine Lactobacillus rhamnosus GG administration in VLBW infants: a retrospective, 6-year cohort study," *Early Human Development*, vol. 87, supplement 1, pp. S35–S38, 2011.

[69] T. D. Braga, G. A. P. da Silva, P. I. C. de Lira, and M. de Carvalho Lima, "Efficacy of Bifidobacterium breve and Lactobacillus casei oral supplementation on necrotizing enterocolitis in very-low-birth-weight preterm infants: a double-blind, randomized, controlled trial," *American Journal of Clinical Nutrition*, vol. 93, no. 1, pp. 81–86, 2011.

[70] A. Bin-Nun, R. Bromiker, M. Wilschanski et al., "Oral probiotics prevent necrotizing enterocolitis in very low birth weight neonates," *Journal of Pediatrics*, vol. 147, no. 2, pp. 192–196, 2005.

[71] H. C. Lin, C. H. Hsu, H. L. Chen et al., "Oral probiotics prevent necrotizing enterocolitis in very low birth weight preterm infants: a Multicenter, Randomized, Controlled trial," *Pediatrics*, vol. 122, no. 4, pp. 693–700, 2008.

[72] M. G. Romeo, D. M. Romeo, L. Trovato et al., "Role of probiotics in the prevention of the enteric colonization by Candida in preterm newborns: incidence of late-onset sepsis and neurological outcome," *Journal of Perinatology*, vol. 31, no. 1, pp. 63–69, 2011.

[73] C. Dani, R. Biadaioli, G. Bertini, E. Martelli, and F. F. Rubaltelli, "Probiotics feeding in prevention of urinary tract infection, bacterial sepsis and necrotizing enterocolitis in preterm infants: a prospective double-blind study," *Biology of the Neonate*, vol. 82, no. 2, pp. 103–108, 2002.

[74] F. N. Sari, E. A. Dizdar, S. Oguz, O. Erdeve, N. Uras, and U. Dilmen, "Oral probiotics: lactobacillus sporogenes for prevention of necrotizing enterocolitis in very low-birth weight infants: a randomized, controlled trial," *European Journal of Clinical Nutrition*, vol. 65, no. 4, pp. 434–439, 2011.

[75] N. Simakachorn, R. Bibiloni, P. Yimyaem et al., "Tolerance, safety, and effect on the faecal microbiota of an enteral formula supplemented with pre- and probiotics in critically ill children," *Journal of Pediatric Gastroenterology and Nutrition*, vol. 53, no. 2, pp. 174–181, 2011.

[76] T. C. B. Honeycutt, M. El Khashab, R. M. Wardrop III et al., "Probiotic administration and the incidence of nosocomial infection in pediatric intensive care: a randomized placebo-controlled trial," *Pediatric Critical Care Medicine*, vol. 8, no. 5, pp. 452–458, 2007.

[77] P. J. Watkinson, V. S. Barber, P. Dark, and J. D. Young, "The use of pre- pro- and synbiotics in adult intensive care unit patients: systematic review," *Clinical Nutrition*, vol. 26, no. 2, pp. 182–192, 2007.

5

Molecular Analysis of the Bacterial Communities in Crude Oil Samples from Two Brazilian Offshore Petroleum Platforms

Elisa Korenblum,[1] Diogo Bastos Souza,[1] Monica Penna,[2] and Lucy Seldin[1]

[1] *Laboratório de Genética Microbiana, Instituto de Microbiologia Prof. Paulo de Góes, Universidade Federal do Rio de Janeiro, Centro de Ciências da Saúde, Bloco I, Ilha do Fundão, 21941-590 Rio de Janeiro, RJ, Brazil*
[2] *Gerência de Biotecnologia e Tratamentos Ambientais, CENPES-PETROBRAS, Ilha do Fundão, 21949-900 Rio de Janeiro, RJ, Brazil*

Correspondence should be addressed to Lucy Seldin, lseldin@micro.ufrj.br

Academic Editor: J. Wiegel

Crude oil samples with high- and low-water content from two offshore platforms (PA and PB) in Campos Basin, Brazil, were assessed for bacterial communities by 16S rRNA gene-based clone libraries. RDP Classifier was used to analyze a total of 156 clones within four libraries obtained from two platforms. The clone sequences were mainly affiliated with *Gammaproteobacteria* (78.2% of the total clones); however, clones associated with *Betaproteobacteria* (10.9%), *Alphaproteobacteria* (9%), and Firmicutes (1.9%) were also identified. *Pseudomonadaceae* was the most common family affiliated with these clone sequences. The sequences were further analyzed by MOTHUR, yielding 81 operational taxonomic units (OTUs) grouped at 97% stringency. Richness estimators also calculated by MOTHUR indicated that oil samples with high-water content were the most diverse. Comparison of bacterial communities present in these four samples using LIBSHUFF and Principal Component Analysis (PCA) indicated that the water content significantly influenced the community structure only of crude oil obtained from PA. Differences between PA and PB libraries were observed, suggesting the importance of the oil field as a driver of community composition in this habitat.

1. Introduction

Biodegraded oils resulting from the action of microorganisms that destroy hydrocarbons and other oil components have been a problem for the petroleum industry. Biodegradation is responsible for the increase in viscosity and acidity of the oil and the reduction of its API (American Petroleum Institute) grade [1]. Different studies have already demonstrated the existence of large and diverse populations of microbes with different metabolic activities in petroleum systems [2–5], but due to difficulties in sampling and in the efficiency of DNA extraction from oily samples, little is known about the bacterial diversity in crude oil [6–8]. It is also challenging to work with samples with low indigenous biomass as well as a high, oily, and viscous emulsion.

During the aging of oil fields, industries make use of secondary oil recovery (SOR), which consists of water injection inside the reservoir to maintain the formation pressure, resulting in an increase of water content in crude oil (oil : water ratio of the production fluids). SOR and high-water content in the petroleum reservoir may reduce internal temperature and allow the biodegradation of crude oil by autochtone or allochtone bacterial populations [9]. To control the bacterial contamination in the reservoir and subsequently in the production line, petroleum industries use physical and chemical treatments of water injected into the reservoir during SOR [10]. In Brazil, seawater is used in SOR at offshore platforms. The seawater is treated by filtration, chlorination, deoxygenation, and/or biocide addition to reduce bacterial contamination prior to water injection. However, these treatments do not completely eliminate the bacterial contamination from the injection water [8]. Besides waterflooding during SOR, other sources of contamination to the system are drilling operation, well equipment, and damaged tubing or casings. Hence, exogenous bacteria can penetrate the reservoir, form biofilms inside the rock and in the oil-producing line and be recovered at producing wells [2, 11]. Therefore, microbiology in crude oil samples might reflect the indigenous organisms in the respective oil formations and also, to a significant extent, allochthonous bacterial populations.

Despite the recent use of molecular techniques for a broader survey of microbial communities in oil fields, our knowledge of the nature and diversity of bacteria present in these ecosystems is still scarce, especially in waterflooded oil reservoirs. Therefore, in this study, the bacterial communities from crude oil samples containing high- and low-water content from two platforms in Campos Basin were analyzed using the 16S rRNA gene sequencing approach in order to determine whether the different water content could influence the bacterial communities from crude oils and, consequently, pose a risk of decreasing oil quality.

2. Materials and Methods

2.1. Samples. The Caratinga and Barracuda oil fields are located in the south-central portion of the Campos Basin approximately 90 km offshore of Rio de Janeiro State, Brazil, at water depths ranging between 600 and 1200 m. The Caratinga and Barracuda oil fields are denoted as platform A (PA) and platform B (PB), respectively. These two oil fields cover a total area of approximately 234 km^2. The oil densities from these fields range from 20 to 26 API degrees. The temperatures of the Caratinga and Barracuda fields at 2,800 m below sea level are 78 and 79.5°C, respectively. Seawater was injected into these platforms in 2005 for SOR. Two wellheads were selected within each platform, one with high-water content and another with low-water content. The wellheads were denoted as PAH, PAL, PBH, or PBL, corresponding to platform PA or PB and high- (H) or low- (L) water content (PAH, 60% water content; PAL, 5%; PBH, 40%; and PBL, 1%). The crude oil samples were collected in sterile 1 L flasks directly from each platform wellhead after enough crude oil was drained off to clean the wellheads. Samples were stored at 4°C until their arrival at the laboratory.

2.2. DNA Extraction. Crude oil samples (10 mL) were mixed with an equal volume of Winogradsky buffer [7] and incubated for 10 min at 80°C. The aqueous phase was recovered with a sterile pipette. This procedure was repeated five times, collecting a total volume of 50 mL of aqueous phase for each sample. The total volume of aqueous phase for each sample was pelleted by centrifugation at 12,000 ×g for 20 min at 4°C. An aliquot (2 mL) of TE buffer (10 mM Tris; 1 mM EDTA) was added to the pellet, and DNA extraction was further performed as described previously [12]. In order to exclude the possibility of bacterial contamination from the reagents, buffers and/or enzymes used, DNA extraction of a blank tube without sample was additionally carried out.

2.3. Polymerase Chain Reaction Conditions for Bacterial 16S rRNA Gene Amplification. 16S rRNA gene sequences were amplified by polymerase chain reaction (PCR) using the primers and the PCR conditions described previously [13]. Primers U968 and L1401 were used to amplify the V6–V8 variable regions in the *Escherichia coli* small subunit rRNA genes. The 50 μL PCR reaction contained 50 mM KCl, 2.5 mM $MgCl_2$, 2 mM dNTPs, 0.2 μM of each primer (U968: 5′AACGCGAAGAACCTTAC3′ and L1401: 5′GCGTGTGTACAAGACCC3′), 2.5 U of *Taq* DNA polymerase (Promega, Madison, WI), and 1 μL (20–50 ng) of the DNA extract. The amplification conditions included a denaturing step at 94°C for 2 min followed by 35 cycles at 94°C for 1 min, an annealing step at 48°C for 1 min and 30 s, and an extension step at 72°C for 1 min and 30 s; the final extension step was performed at 72°C for 10 min. Negative controls (without DNA) were also included in all sets of PCR reactions. PCR products were then purified using the Wizard PCR Clean-Up System (Promega) and eluted with 50 μL of distilled water. The presence of PCR products was confirmed by 1.4% agarose gel electrophoresis.

2.4. Molecular Analyses. The partial 433-bp 16S rRNA gene sequences obtained by PCR were cloned using the pGEM T-easy vector according to instructions from the manufacturer (Promega). The resulting ligation mixtures were transformed into *Escherichia coli* JM109 competent cells, and clones containing inserts were sequenced. All sequencing reactions were performed by Macrogen Inc. (South Korea). The sequence data set was screened for potential chimeric structures by using Bellerophon [14] (available at the greengenes website (version 3, http://greengenes.lbl.gov/cgi-bin/nph-bel3_interface.cgi)). A total of 156 valid 16S rRNA gene sequences were analyzed for taxonomic affiliation by the RDP Classifier [15] and for the closest match to sequences in the GenBank database by BLASTN [16]. The 16S rRNA gene sequences were submitted to GenBank with the accession numbers: HQ341990–HQ3411999, HQ342010, HQ342021, HQ342032, HQ342043, HQ342054, HQ342064–HQ342083, HQ342084–HQ342121, HQ342122–HQ342146, HQ342000–HQ342009, HQ342011–HQ342017, HQ342018–HQ342031, HQ342033–HQ342039, HQ342041–HQ342053, and HQ-342055–HQ342063.

16S rRNA gene sequences were then clustered as OTUs at an overlap identity cut-off of 97% by MOTHUR software [17]. Richness and diversity statistics including the non-parametric richness estimators ACE, Chao1, and Shannon diversity index were calculated also using MOTHUR. The diversity of OTUs and community overlap were examined using rarefaction analysis and Venn diagram. A phylogenetic tree was constructed with representatives of each OTU found within the four libraries (at a distance level of 3%) and with closely related sequences that were recovered from the GenBank database. Sequence alignment was done by Clustal-X software [18], and the aligned sequences were then used to construct the phylogenetic tree with the neighbor-joining method by using the MEGA5 software [19]. Bootstrap analyses were performed with 1,000 repetitions, and only values higher than 50% are shown in the phylogenetic tree.

2.5. Statistical Analyses. Based on the sequence alignment mentioned above, a distance matrix was constructed using DNAdist from PHYLIP (version 3.6) [20], and pairwise comparisons of each clone library were performed using

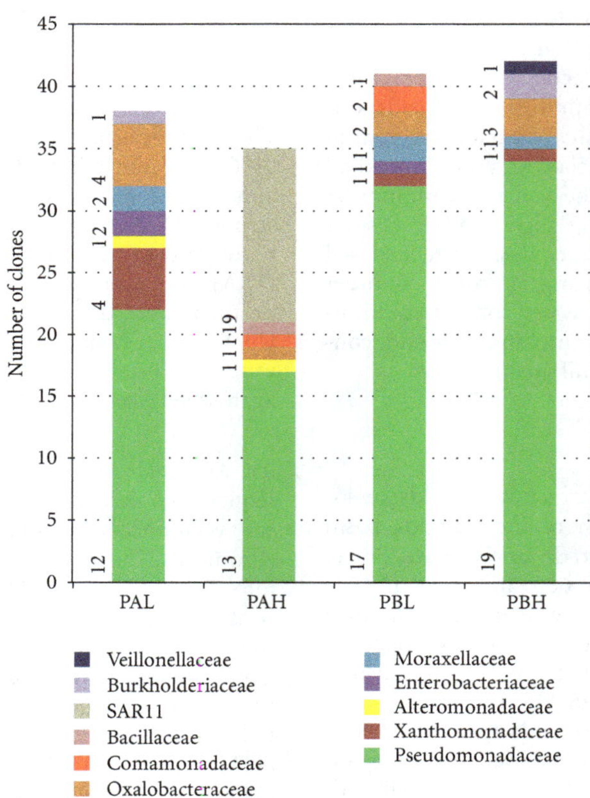

Figure 1: Frequency of clones affiliated with different bacterial families found in each oil sample from platforms A and B with low- (L) and high- (H) water contents. A total of 156 16S rRNA gene clones were classified by the RDP Classifier tool. The number of OTUs corresponding to each family is indicated close to the graph bar. OTUs were defined using a distance level of 3% by using the furthest neighbor algorithm in MOTHUR. PAL—crude oil from Platform A with low-water content (5%); PAH—crude oil from Platform A with high-water content (60%); PBL—crude oil from Platform B with low water content (1%); PBH—crude oil from Platform B with high water content (40%).

LIBSHUFF (version 0.96; http://www.mothur.org/wiki/Libshuff) [21]. Additionally, the relationship among bacterial community structures was evaluated using the Principal Component Analysis (PCA). The matrix used for the PCA was a quantitative matrix of abundance of all OTUs detected from each clone library.

3. Results

3.1. 16S rRNA Gene Sequence Analysis. After recovering the microbial fraction, DNA was successfully extracted from the crude oil samples. However, a low amplicon yield was obtained with the bacterial 16S rRNA gene primers used, probably due to the inefficiency of DNA extraction from oily samples. As expected, no amplicons were obtained from the blank tube. The PCR products were cloned, and a total of 156 valid 16S rRNA gene sequences were analyzed for taxonomic affiliation by the RDP Classifier. Figure 1 shows the bacterial clone frequencies obtained in the PAH (35 clones), PAL (38 clones), PBH (42 clones), and PBL (41 clones) samples and the clone affiliation. The number of OTUs corresponding to each family is indicated close to the bars on the graph (Figure 1).

Among the clones, 122 (78.2%) were *Gammaproteobacteria,* and most were associated with the family *Pseudomonadaceae* (105 clones). The remaining clones (17) were affiliated with *Moraxellaceae, Enterobacteriaceae, Alteromonadaceae,* and *Xanthomonadaceae.* Also, 17 (10.9%) of the total clones were *Betaproteobacteria,* and most were associated with the families *Comamonadaceae* (3 clones from samples PAH and PBL), *Burkholderiaceae* (3 clones from PAL and PBH), and *Oxalobacteraceae* (11 clones from samples PAH, PAL, PBH, and PBL). A few clones (14) from the PBH sample, corresponding to 9% of the total amount of clones analyzed, were affiliated with SAR11 clade, a lineage of bacteria from the *Alphaproteobacteria* class, which is common in the ocean [22]. To a lesser extent, clones from *Bacillaceae* (two clones, one of each sample PAH and PBL) and *Veillonellaceae* (one clone from sample PBH) were present, representing the Firmicutes.

All 16S rRNA gene sequences were then clustered as OTUs, and using the 81 resulting OTUs, a phylogenetic tree was constructed with the closely related sequences that were recovered from the GenBank database (Figure 2). The phylogenetic tree showed that the majority of OTUs from Platforms A and B fall within *Gammaproteobacteria.*

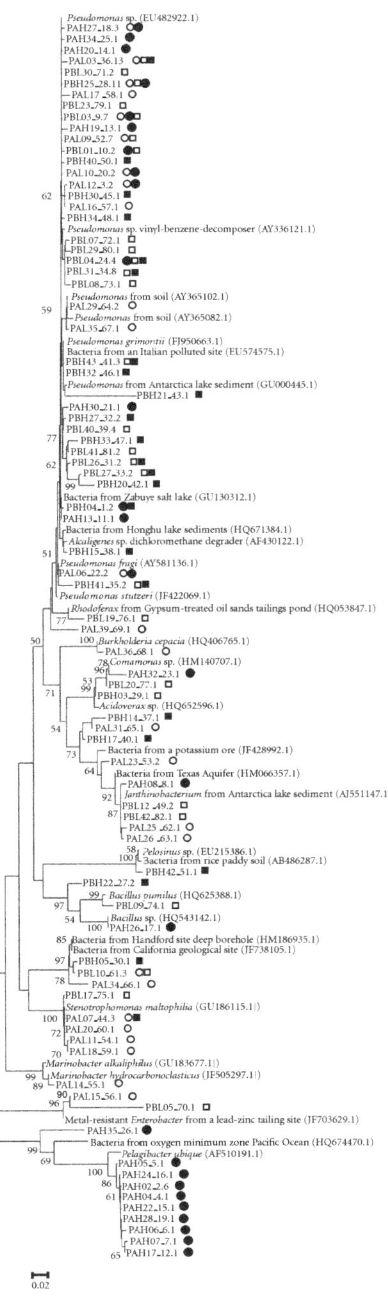

FIGURE 2: Phylogenetic tree of the 16S rRNA gene-based libraries obtained from platforms A and B. OTUs were defined by MOTHUR using the furthest neighbor algorithm with a distance level of 3%. One representative clone of each OTU (81 OTUs) was used for phylogenetic analysis. Reference sequences from GenBank are highlighted in bold. The tree was constructed based on the neighbor-joining method. Bootstrap analyses were performed with 1,000 repetitions and only values higher than 50% are shown. The scale bar indicates the distance in substitutions per nucleotide. Clones indicated with prefixes PAH and PAL originated from platform A with high- and low-water content, respectively. Clones PBH and PBL were obtained from platform B with high- and low-water content, respectively. The access name of each sequence is formed by numbers that correspond to the following: a representative clone number, followed by the OTU and the number of clones in that OTU (e.g., sequence PAH27_18, 3 = name of the clone_OTU, number of clones grouped in this OTU). The symbols after each access name indicate that the OTU was found in oil samples from PAH (●), PAL (○), PBH (■), and/or PBL (□).

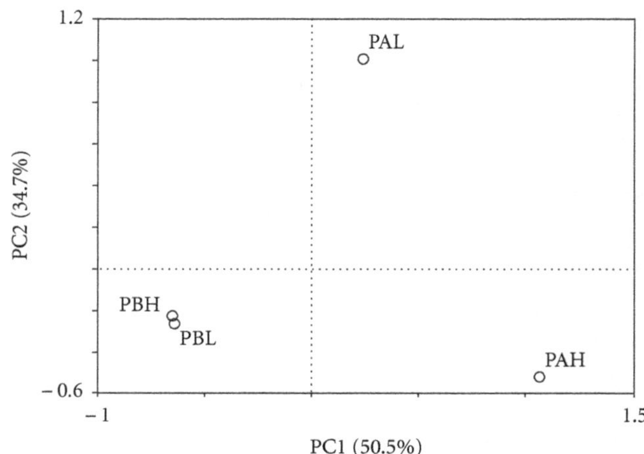

FIGURE 3: Comparison of bacterial communities in samples PAH, PAL, PBH, and PBL. Principal coordinates plots (PCA) were generated using the presence of each OTU (at a distance level of 3%) found in each clone library. PAL—crude oil from Platform A with low-water content (5%); PAH—crude oil from Platform A with high-water content (60%); PBL—crude oil from Platform B with low-water content (1%); PBH—crude oil from Platform B with high-water content (40%).

3.2. Diversity Analyses. Sequences obtained from the different sampling sites were evaluated by pairwise analysis with LIBSHUFF. Crude oil samples with high water content values (PAH and PBH) were statistically different ($P = 0.001$). PAL and PAH libraries also were statistically different. However, LIBSHUFF did not show statistical differences between PB libraries (PBL and PBH) nor between samples with low water content (PAL and PBL). This latter finding was similar to the PCA result (Figure 3). The PCA grouped together PBH and PBL samples, whereas the PAH and PAL samples diverged in this analysis. In addition, the first component determined 50.5% of the total variation, showing a dichotomic separation of the crude oil sample from PAH and the three other samples (PAL, PBL and PBH).

The number of OTUs from each sampling site as well as richness and diversity indexes are shown in Table 1. Total coverage of bacterial richness was almost achieved in all libraries (data not shown). Libraries from oil samples with high water content (PAH and PBH) had a higher richness based on ACE and Chao1 than PAL and PBL. While the Shannon diversity index showed that bacterial communities are 97% similar, the Venn diagram showed that no OTUs are shared between all four samples, indicating that the bacterial communities are different in these two platforms (Figure 4).

TABLE 1: Species richness estimates and diversity of 16S rRNA gene clones calculated by MOTHUR.

	Samples[a]			
	PAL	PAH	PBL	PBH
OTUs[b]	26	26	25	27
ACE[c]	50	122	40	306
Chao1	34	100	38	88
H′[d]	2.94	2.94	2.88	2.87

[a]PAL—crude oil from Platform A with low water content (5%); PAH—crude oil from Platform A with high-water content (60%); PBL—crude oil from Platform B with low-water content (1%); PBH—crude oil from Platform B with high-water content.
[b]Number of unique OTUs defined by using the furthest neighbor algorithm by MOTHUR at 97% stringency.
[c]ACE (Abundance-based coverage estimator).
[d]H′ (Shannon-weaver index of diversity).

4. Discussion

Barracuda and Caratinga are giant oil fields from the deep water Campos Basin, which is one of the most economically important petroleum basins in Brazil. The bacterial communities in crude oil samples from these two platforms were assessed for the first time in this study. Moreover, the difference in the water content of the oil samples was considered in each platform. Water content may influence the bacterial diversity and biodegradation of oil in the reservoir. The latter process requires not only water but also nutrients and hydrocarbons for microbial growth [4]. Biodegraded oil reservoirs usually produce oil-water emulsified fluids, which may complicate the recovery of the totality of the microorganisms present in those samples. Oil from the Caratinga and Barracuda fields ranges from 20 to 26 API degrees, values corresponding to low/medium petroleum degradation. The results obtained in this study indicate that these fields may have a potential for biodegradation because of the bacteria found in the oil samples.

An alternative to microorganism isolation and cell counts, one of the main strategies to study the microbial diversity, is the PCR-based approach, which also may provide an understanding of the uncultured microbial community. However, crude oil samples contain low amounts of biomass that may result in a low DNA yield, which affects the efficiency of the molecular methods [23]. These PCR-based techniques are also highly prone to DNA contamination and can result in false positive amplifications. To overcome these problems in this study, the oil samples were washed several times with Winogradsky buffer to recover bacterial cells and

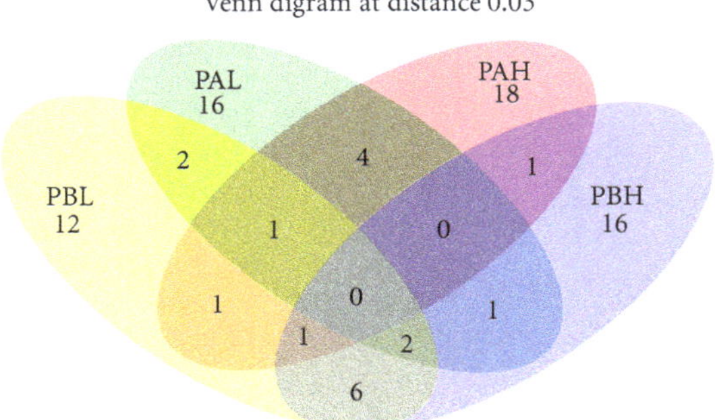

FIGURE 4: Venn diagram of bacterial OTUs clustered with a 3% distance threshold, showing the number of OTUs shared by the four crude oil samples. PAL—crude oil from Platform A with low-water content (5%); PAH—crude oil from Platform A with high-water content (60%); PBL—crude oil from Platform B with low-water content (1%); PBH—crude oil from Platform B with high-water content (40%).

to obtain higher amounts of biomass prior to DNA extraction [7]. Low but sufficient amounts of bacterial DNA for PCR amplification in these oil samples were achieved. Negative controls were also included during the DNA extraction and PCR amplification procedures, and no amplicons were obtained in these controls.

Data obtained in this study show a spatial heterogeneity of bacterial community composition in Campos Basin, comparing Caratinga (PA) and Barracuda (PB) fields. In addition, water injection might have stimulated bacterial growth by supplying nutrients and decreasing downhole temperature, as high-water content samples (PAH, and PBH) showed higher richness index values (Table 1). Most libraries were not statistically different from each other, except for PAH. The difference in the amount of water in PBL (1% water content) and PBH (40% water content) seemed not to be large enough to significantly affect the community structure. When PAH (which had the highest water content—60%) was compared with PAL (5% water content), as well as when PAH was matched up to PBH, a shift of bacterial diversity was observed. These data were also observed by PCA, where PC1 represents the OTUs spatial distribution, separating PA and PB, while PC2 represents the water content, which influenced only PAL and PAH. PAH divergence may be due to an abundant group that was observed only in this library. Almost half of PAH clones were related to *Pelagibacter*. This genus belongs to the SAR11 clade, which is a very small, heterotrophic marine *Alphaproteobacteria* found throughout the oceans [22]. Therefore, we believe that this bacterium may have been introduced with the seawater during secondary oil recovery, not being indigenous of the reservoir.

An abundant group belonging to the family *Pseudomonadaceae* was observed in the four 16S rRNA gene libraries. The presence of this group may be a potential risk for petroleum biodegradation inside the reservoir, because members of this family have been isolated and also detected by molecular techniques in deep ocean crust and in high temperature oil reservoirs [24, 25]. Moreover, the genus *Pseudomonas* has been shown to degrade alkanes and other polyaromatic hydrocarbons as well as to emulsify and degrade resins from Arabian light crude oil [26]. Nelson et al. [27] demonstrated that the species *P. putida* is a metabolically versatile bacterium that is able to degrade a wide variety of xenobiotic compounds. Also, Chayabutra and Ju [28] showed that another member of this genus, *P. aeruginosa*, degrades *n*-hexadecane under anaerobic denitrifying conditions.

In addition to *Pseudomonadaceae*, clone sequences affiliated with *Moraxellaceae*, *Enterobacteriaceae*, *Alteromonadaceae*, and *Xanthomonadaceae* families were also found in the oil samples studied. Some members of these families have been shown to be involved in hydrocarbon degradation [29–32]. *Betaproteobacteria*-related clones belonging to the *Burkholderiales* order were observed in crude oil samples from Barracuda and Caratinga fields. Some genera of this group (*Burkholderia* and *Comamonas*) have been previously found to grow under anaerobic conditions in petroleum reservoirs or in contaminated soils [1, 33]. Therefore, the presence of these groups in the oil reservoirs studied may suggest marine bacterial contamination after the water injection used for SOR. Korenblum et al. [8] have also detected *Marinobacter*, *Burkholderia*, and *Pseudomonas* in water used in SOR.

Few clones related to the Firmicutes phylum were detected in samples PAH, PBH, and PBL. Within this phylum, *Bacillus* strains have been previously isolated from oil reservoirs in Brazil [8, 34], where different strains have been shown to degrade petroleum hydrocarbons [34]. However, this is the first time the presence of another genus of Firmicutes, a *Pelosinus*-related clone, was observed in a petroleum environment. *Bacillus* and *Pelosinus* species may also survive in a petroleum environment due to spore formation or use of nitrate or iron for respiration.

In conclusion, we report the bacterial composition present in crude oil samples with high- and low-water content from two Brazilian oil fields, Caratinga and Barracuda. This information on bacterial diversity in crude oil increases the current knowledge of the microbial ecology in this environment, which may help to predict the potential for biodegradation in these fields. Based on their 16S rRNA sequences, most of the clones obtained were related to different bacterial families, with a predominance of *Pseudomonadaceae*, which was observed in all crude oil samples. Moreover, few clones were related to genera that have not been described before in this environment. However, we are aware that a limited number of clones were evaluated in this study. Bacterial isolation is still necessary to determine the industrial and ecological significance of these different bacteria.

Acknowledgment

This study was performed as a part of a project supported by PETROBRAS.

References

[1] Y. J. Liu, Y. P. Chen, P. K. Jin, and X. C. Wang, "Bacterial communities in a crude oil gathering and transferring system (China)," *Anaerobe*, vol. 15, no. 5, pp. 214–218, 2009.

[2] M. Magot, B. Ollivier, and B. K. C. Patel, "Microbiology of petroleum reservoirs," *Antonie van Leeuwenhoek*, vol. 77, no. 2, pp. 103–116, 2000.

[3] J. D. Van Hamme, A. Singh, and O. P. Ward, "Recent advances in petroleum microbiology," *Microbiology and Molecular Biology Reviews*, vol. 67, no. 4, pp. 503–549, 2003.

[4] C. M. Aitken, D. M. Jones, and S. R. Larter, "Anaerobic hydrocarbon biodegradation in deep subsurface oil reservoirs," *Nature*, vol. 431, no. 7006, pp. 291–294, 2004.

[5] J. C. Philp, A. S. Whiteley, L. Ciric, and M. J. Bailey, "Monitoring bioremediation," in *Bioremediation: Applied Microbial Solutions for Real-World Environmental Cleanup*, R. M. Atlas and J. C. Philp, Eds., pp. 237–268, ASM Press, Washington, DC, USA, 2005.

[6] V. M. Oliveira, L. D. Sette, K. C. M. Simioni, and E. V. dos Santos Neto, "Bacterial diversity characterization in petroleum samples from Brazilian reservoirs," *Brazilian Journal of Microbiology*, vol. 39, no. 3, pp. 445–452, 2008.

[7] I. von der Weid, E. Korenblum, D. Jurelevicius et al., "Molecular diversity of bacterial communities from subseafloor rock samples in a deep-water production basin in Brazil," *Journal of Microbiology and Biotechnology*, vol. 18, no. 1, pp. 5–14, 2008.

[8] E. Korenblum, É. Valoni, M. Penna, and L. Seldin, "Bacterial diversity in water injection systems of Brazilian offshore oil platforms," *Applied Microbiology and Biotechnology*, vol. 85, no. 3, pp. 791–800, 2010.

[9] J. Connan, "Biodegradation of crude oils in reservoirs," in *Advances in Petroleum Geochemistry*, J. Brooks and D. H. Welte, Eds., vol. 1, pp. 299–335, Academic Press, London, UK, 1984.

[10] M. J. McInerney and K. L. Sublette, "Oil field microbiology," in *Manual of Environmental Microbiology*, C. Hurst, R. L. Crawford, G. R. Knudsen, M. J. McInerney, and L. D. Stetzenbach, Eds., pp. 777–787, ASM Press, Washington, DC, USA, 2002.

[11] H. Dahle, F. Garshol, M. Madsen, and N. Birkeland, "Microbial community structure analysis of produced water from a high-temperature North Sea oil-field," *Antonie van Leeuwenhoek*, vol. 93, no. 1-2, pp. 37–49, 2008.

[12] L. Seldin and D. Dubnau, "DNA homology among *Bacillus polymyxa*, *Bacillus azotofixans* and other nitrogen fixing *Bacillus* strains," *International Journal of Systematic Bacteriology*, vol. 35, no. 2, pp. 151–154, 1985.

[13] H. Heuer and K. Smalla, "Application of denaturing gradient gel electrophoresis (DGGE) and temperature gradient gel electrophoresis (TGGE) for studying soil microbial communities," in *Modern Soil Microbiology*, J. D. van Elsas, J. T. Trevors, and E. M. H. Wellington, Eds., pp. 353–370, Marcel Dekker, New York, NY, USA, 1997.

[14] T. Huber, G. Faulkner, and P. Hugenholtz, "Bellerophon: a program to detect chimeric sequences in multiple sequence alignments," *Bioinformatics*, vol. 20, no. 14, pp. 2317–2319, 2004.

[15] Q. Wang, G. M. Garrity, J. M. Tiedje, and J. R. Cole, "Naïve Bayesian classifier for rapid assignment of rRNA sequences into the new bacterial taxonomy," *Applied and Environmental Microbiology*, vol. 73, no. 16, pp. 5261–5267, 2007.

[16] S. F. Altschul, T. L. Madden, A. A. Schäffer et al., "Gapped BLAST and PSI-BLAST: a new generation of protein database search programs," *Nucleic Acids Research*, vol. 25, no. 17, pp. 3389–3402, 1997.

[17] P. D. Schloss, S. L. Westcott, T. Ryabin et al., "Introducing mothur: open-source, platform-independent, community-supported software for describing and comparing microbial communities," *Applied and Environmental Microbiology*, vol. 75, no. 23, pp. 7537–7541, 2009.

[18] J. D. Thompson, T. J. Gibson, F. Plewniak, F. Jeanmougin, and D. G. Higgins, "The CLUSTAL X windows interface: flexible strategies for multiple sequence alignment aided by quality analysis tools," *Nucleic Acids Research*, vol. 25, no. 24, pp. 4876–4882, 1997.

[19] K. Tamura, D. Peterson, N. Peterson, G. Stecher, M. Nei, and S. Kumar, "MEGA5: molecular evolutionary genetics analysis using maximum likelihood, evolutionary distance, and maximum parsimony methods," *Molecular Biology and Evolution*, vol. 28, no. 10, pp. 2731–2739, 2011.

[20] J. Felsenstein, *PHYLIP (Phylogeny Inference Package) Version 3.6*, Department of Genome Science, University of Washington, Seattle, Wash, USA, 2005.

[21] D. R. Singleton, M. A. Furlong, S. L. Rathbun, and W. B. Whitman, "Quantitative comparisons of 16S rDNA sequence libraries from environmental samples," *Applied and Environmental Microbiology*, vol. 67, no. 9, pp. 4374–4376, 2001.

[22] R. M. Morris, M. S. Rappé, S. A. Connon et al., "SAR11 clade dominates ocean surface bacterioplankton communities," *Nature*, vol. 420, no. 6917, pp. 806–810, 2002.

[23] T. B. P. Oldenburg, S. R. Larter, J. J. Adams et al., "Methods for recovery of microorganisms and intact microbial polar lipids from oil-water mixtures: laboratory experiments and natural well-head fluids," *Analytical Chemistry*, vol. 81, no. 10, pp. 4130–4136, 2009.

[24] O. U. Mason, T. Nakagawa, M. Rosner et al., "First investigation of the microbiology of the deepest layer of ocean crust," *PLoS One*, vol. 5, no. 11, Article ID e15399, 2010.

[25] V. J. Orphan, L. T. Taylor, D. Hafenbradl, and E. F. Delong, "Culture-dependent and culture-independent characterization of microbial assemblages associated with high-temperature petroleum reservoirs," *Applied and Environmental Microbiology*, vol. 66, no. 2, pp. 700–711, 2000.

[26] F. F. Evans, L. Seldin, G. V. Sebastian, S. Kjelleberg, C. Holmström, and A. S. Rosado, "Influence of petroleum contamination and biostimulation treatment on the diversity of *Pseudomonas* spp. in soil microcosms as evaluated by 16S rRNA based-PCR and DGGE," *Letters in Applied Microbiology*, vol. 38, no. 2, pp. 93–98, 2004.

[27] K. E. Nelson, C. Weinel, I. T. Paulsen et al., "Complete genome sequence and comparative analysis of the metabolically versatile *Pseudomonas putida* KT2440," *Environmental Microbiology*, vol. 4, no. 12, pp. 799–808, 2002.

[28] C. Chayabutra and L. K. Ju, "Degradation of n-hexadecane and its metabolites by *Pseudomonas aeruginosa* under microaerobic and anaerobic denitrifying conditions," *Applied and Environmental Microbiology*, vol. 66, no. 2, pp. 493–498, 2000.

[29] N. M. Gorshkova, E. P. Ivanova, A. F. Sergeev et al., "*Marinobacter excellens* sp. nov., isolated from sediments of the Sea of Japan," *International Journal of Systematic and Evolutionary Microbiology*, vol. 53, no. 6, pp. 2073–2078, 2003.
[30] X. Hua, J. Wang, Z. Wu et al., "A salt tolerant *Enterobacter cloacae* mutant for bioaugmentation of petroleum- and salt-contaminated soil," *Biochemical Engineering Journal*, vol. 49, no. 2, pp. 201–206, 2010.
[31] S. Phrommanich, S. Suanjit, S. Upatham et al., "Quantitative detection of the oil-degrading bacterium *Acinetobacter* sp. strain MUB1 by hybridization probe based real-time PCR," *Microbiological Research*, vol. 164, no. 4, pp. 486–492, 2009.
[32] L. Yu, Y. Liu, and G. Wang, "Identification of novel denitrifying bacteria *Stenotrophomonas* sp. ZZ15 and *Oceanimonas* sp. YC13 and application for removal of nitrate from industrial wastewater," *Biodegradation*, vol. 20, no. 3, pp. 391–400, 2009.
[33] X. Y. Zhu, J. Lubeck, and J. J. Kilbane, "Characterization of microbial communities in gas industry pipelines," *Applied and Environmental Microbiology*, vol. 69, no. 9, pp. 5354–5363, 2003.
[34] C. D. Cunha, A. S. Rosado, G. V. Sebastián, L. Seldin, and I. von der Weid, "Oil biodegradation by *Bacillus* strains isolated from the rock of an oil reservoir located in a deep-water production basin in Brazil," *Applied Microbiology and Biotechnology*, vol. 73, no. 4, pp. 949–959, 2006.

6

Extreme Heat Resistance of Food Borne Pathogens *Campylobacter jejuni*, *Escherichia coli*, and *Salmonella typhimurium* on Chicken Breast Fillet during Cooking

Aarieke E.I. de Jong,[1,2] Esther D. van Asselt,[1,3] Marcel H. Zwietering,[4] Maarten J. Nauta,[1,5] and Rob de Jonge[1]

[1] *Laboratory for Zoonoses and Environmental Microbiology, National Institute for Public Health and the Environment (RIVM), 3720 BA Bilthoven, The Netherlands*
[2] *Division Consumer and Safety, New Food and Consumer Product Safety Authority (nVWA), 1018 BK Amsterdam, The Netherlands*
[3] *Rikilt, Institute of Food Safety, 6700 AE Wageningen, The Netherlands*
[4] *Laboratory of Food Microbiology, Wageningen University, 6700 EV Wageningen, The Netherlands*
[5] *National Food Institute, Technical University of Denmark, 1790 Copenhagen V, Denmark*

Correspondence should be addressed to Aarieke E.I. de Jong, aarieke.de.jong@vwa.nl

Academic Editor: Giuseppe Comi

The aim of this research was to determine the decimal reduction times of bacteria present on chicken fillet in boiling water. The experiments were conducted with *Campylobacter jejuni*, *Salmonella*, and *Escherichia coli*. Whole chicken breast fillets were inoculated with the pathogens, stored overnight (4°C), and subsequently cooked. The surface temperature reached 70°C within 30 sec and 85°C within one minute. Extremely high decimal reduction times of 1.90, 1.97, and 2.20 min were obtained for *C. jejuni*, *E. coli*, and *S. typhimurium*, respectively. Chicken meat and refrigerated storage before cooking enlarged the heat resistance of the food borne pathogens. Additionally, a high challenge temperature or fast heating rate contributed to the level of heat resistance. The data were used to assess the probability of illness (campylobacteriosis) due to consumption of chicken fillet as a function of cooking time. The data revealed that cooking time may be far more critical than previously assumed.

1. Introduction

Improper cooking is one of the main factors causing food borne illness [1–5], and a large part, 40–60%, of the cases of food borne illness are expected to originate from private households [6–9]. This is partly caused by the consumption of undercooked meat. Most consumers do not use a meat thermometer [10, 11] but determine the doneness of meat most often by cutting the meat to evaluate changes in color and texture, or by other subjective techniques. Especially for chicken breast fillet these techniques frequently result in undercooked meat [12].

Chicken breast fillet is, apart from minced meat, the most popular type of meat in The Netherlands [13] and a pathogen associated with it is *Campylobacter jejuni*, a microorganism responsible for 50% of confirmed cases of bacterial gastroenteritis in Western Europe (Austria, Belgium, Finland, France, Germany, Ireland, Italy, The Netherlands, Norway, Portugal, Sweden, Switzerland, and the UK) and the USA [1, 14–18]. Furthermore, a predominant risk factor for *C. jejuni* infection is consumption of undercooked chicken meat [19–22].

In a consumer study on food handling practices in the Netherlands by our research group [23], focusing on the preparation of a chicken breast fillet salad artificially contaminated with nonmotile, nonpathogenic *Lactobacillus casei*, it was shown that for consumers who apparently made no mistakes regarding cross-contamination, heating time of chicken meat (boiled in chicken stock) was negatively correlated with the bacterial contamination level of the prepared salad. Although this correlation may be expected, it is surprising that the bacteria present were not immediately killed upon contact with boiling water, as the bacteria were only present on the surface of the meat and not, as with ground meat, on the inside as well.

In addition, it was shown by Bergsma et al. [24] that consumer-style cooking (pan frying) of chicken breast inoculated with *C. jejuni* resulted in high bacterial heat resistance levels of this pathogen. Ample research has been dedicated to heat inactivation of bacteria in meat, but most of these studies use thin-layered ground meat as a model system and normal cooking temperatures, that is, temperatures >70°C are hardly ever applied [25–27] (also see Figure 3). As these studies do not suffice to explain the observed heat resistance by Bergsma et al. [24], the aim of this research was to further elaborate on this phenomenon. Therefore, in the current study, the chicken fillets were boiled instead of pan fried to obtain better reproducible temperatures in the experiments and decimal reduction times of bacteria present on each sample were determined. Heating profiles of chicken breast fillets during cooking were obtained both by measurement and by calculation.

Apart from *C. jejuni, Salmonella* and *Escherichia coli* were also studied as these bacteria are as well associated with food borne illness caused by consumption of raw or undercooked meat. In addition, *L. casei* was used, as this bacterium was used as indicator organism for *C. jejuni* to study real-life consumer practices by our research group [23]. Furthermore, the effect of food matrix, matrix size, and refrigerated storage on the survival of bacteria during cooking was studied as well as the effect of refrigerated storage on cell attachment. In addition, the obtained decimal reduction times were used to assess the relationship between the probability of illness (campylobacteriosis) due to consumption of chicken breast fillet and cooking time.

2. Materials and Methods

2.1. Bacterial Strains and Growth Conditions. Five *C. jejuni* strains (NCTC 11168, NCTC 11828, B258, LB99hu, and 82/69), a *Lactobacillus casei* strain (isolated from a food), and *Escherichia coli* WG5 (ATCC 700078) were used as described by De Jong et al. [28]. *C. jejuni, L. casei,* and *E. coli* were cultured as described by De Jong et al. [28]. *S. typhimurium* DT104 was grown in Brain Heart Infusion broth (37°C; BHI, Oxoid, Basingstoke, UK).

2.2. Study Design. The heat resistance of bacteria present on the surface of chicken breast fillets during cooking of the meat in boiling water was expressed as a decimal reduction time during boiling (D_{boil}): time needed to reduce the number of bacteria present on the outside of a large matrix exposed to boiling water by 10-fold (min). This decimal reduction time is comparable to the formal, frequently applied *D*-value. However, to determine a D_T-value of a bacterium on a matrix, the temperature of the matrix must be constant during the heating experiment. Therefore, *D*-values of bacteria on food items are determined using small sized food samples, mainly with a 1 mm thickness. As these small-sized samples are not representative for sizes of meats prepared by consumers during cooking, the heating experiments in this study were conducted using a normal sized meat matrix at cooking temperature. Thus, the term D_{boil} was introduced, with "boil" as index as the temperature could not be given explicitly, since the product temperature was dynamically changing (see Figure 2).

Furthermore, the effect of food matrix and refrigerated storage on survival of *C. jejuni* during cooking was studied. In addition, cell attachments studies were conducted. To study the matrix-effect on heat resistance of *C. jejuni*, cold-stored (4°C, 24 h) inoculated chicken breast fillets and carrots were used as matrix and survival of *C. jejuni* on these matrices after a 5 minute cooking time was compared. The effect of refrigerated storage of chicken meat on heat resistance of *C. jejuni* present on the meat surface was studied by comparing survival of *C. jejuni* present after 5 min cooking using chicken breast fillets refrigerated (4°C) for 0, 1, or 24 h. The effect of refrigerated storage on cell attachment of *C. jejuni* to the matrix was studied for both chicken breast fillet and carrot. The inoculated matrices were stored in the refrigerator (chicken meat: 0, 1, and 24 h at 4°C; carrots: 0 and 1 h at 4°C) and subsequently manually rinsed for 10 s under cold running water and the number of *C. jejuni* cells attached was determined. This was done by comparing the number of cells present on the food item after washing to the number of cells applied to the food item.

2.3. Inoculation Method. Food items were inoculated with a multiple-species cocktail. A multiple-species cocktail of *C. jejuni/L. casei* was obtained by mixing a multiple-strain cocktail (5 strains mixed in equal volumes) of *C. jejuni* with a single strain of *L. casei* in order to compare the behavior of these different bacterial species under equal test conditions, similar to what was done by the authors in a study on cross-contamination during consumer cooking practices [28]. In addition, a multiple-species cocktail containing one strain of *E. coli* and one strain *S. typhimurium* was used. The cocktails were prepared by combining equal volumes of each single-species culture [28]. An inoculum level of $10^{8\text{-}9}$ CFU mL^{-1} was used to inoculate the food items with. Cell counts of bacterial suspensions were determined (N_0) by spread plating appropriate dilutions on appropriate media.

Heating experiments were conducted using chicken fillet or carrot. Whole chicken breast fillets (98–218 g; maximal thickness: 3.5–4 cm) purchased in different batches at a local supermarket were used fresh or were stored frozen and defrosted before use. Thawed or fresh fillets were dabbed with paper tissue to remove superficial moisture, after which each fillet was contaminated with 1 mL of inoculation culture (each side 0.5 mL), which was evenly spread using a plastic, sterile rod. Each contaminated fillet was stored overnight in a separate plastic bag at 4°C to mimic retail storage.

The (large) carrot was longitudinally sliced to obtain slices of similar weight (77–198 g) and similar thickness as the chicken breast fillets. Both sides of these slices were allowed to dry in a laminar flow cabinet (30 min each side) to remove moist, after which the carrot slices were similarly inoculated and stored as the chicken fillets.

2.4. Heat Treatment. For the heating experiments, one single inoculated refrigerated food item was placed in a pan (∅ 24 cm) with boiling water (4-5 L) at a time (each data point is one single heating experiment), thus limiting the decrease of the water temperature due to the addition of the food item. The water was constantly heated and boiling. At time zero the food item was placed in the boiling water and then heated for times ranging from 0 (not heated) to 15 min. After heat treatment, the food item was immediately transferred into a sterile Waring commercial blender, weighed, and cooled with 200 mL peptone ($1\,g\,L^{-1}$) physiological salt ($9\,g\,L^{-1}$) solution (PPS) of 4°C and blended for 1 min resulting in blended chicken slurry or blended carrot slurry. The pan was washed up after cooking for every single food item and clean tap water was brought to the boil again. Experiments were conducted at least in duplicate, on different days.

2.5. Thermal Heat Profiles. For the heat-inactivation test, a cooking pan (∅ 24 cm) with water (4-5 L) was brought to the boil. The water was constantly heated and as the weight of the added matrices was small (<200 g) compared to that of the water (>4000 g), the temperature profile of the water was considered to be constant at 100°C. However, the temperature of refrigerated chicken meat has a come-up time and the temperature profile of the meat can be estimated using (1):

$$u_{\text{slab}} = u_{\text{chicken fillet}} = \frac{T_{x,t} - T_0}{T_\infty - T_0} = 1 - \text{erf}\left(\frac{x}{2\sqrt{at}}\right), \quad (1)$$

in which u is the fraction of nontransferred heat ($0 \le u \le 1$), $T_{x,t}$ is the temperature of the product at place coordinate x at time t, T_0 is the temperature of the product at $t = 0$, T_∞ is the temperature of the surroundings, erf is the standard error function from Excel, x [m] is the place coordinate, and a [$m^2\,s^{-1}$] is the product's thermal diffusion coefficient, which is given by:

$$a = \frac{\lambda}{\rho c_p}, \quad (2)$$

where λ [$W\,m^{-1}\,K^{-1}$] is the product's thermal conductivity, ρ [$kg\,m^{-3}$] is the product's specific density, and c_p [$Ws\,kg^{-1}\,K^{-1}$] is the specific heat capacity of the product, and in which only conductive heat transfer effects via the determining dimension of the chicken fillet, the height, are taken into account (the chicken fillet is thus considered to be an infinite slab, since the width and length of the fillet are more than 2.5 and 5 times as long as the height, resp.). In reality, the temperature of the chicken will increase more rapidly, so this is a fail safe assumption. Equation (1) is applicable when the external heat transfer resistance is negligible (boundary condition 1) and when the temperature of the centre of the product has not changed yet (boundary condition 2) [29].

To determine whether boundary condition 1 is met, the ratio between the transport resistance of the product and its surroundings, the Biot number (Bi) [29], can be calculated by:

Table 1: Parameter values for estimating the temperature profile of a chicken fillet in boiling water.

Parameter	Value
T_0	4°C
T_∞	100°C
$\alpha_{\text{still water}}$	350–580 $W\,m^{-2}\,K^{-1}$
$L_{\text{chicken meat}}$	0.02 m[a]
$\lambda_{\text{chicken meat}}$	0.6 $W\,m^{-1}\,K^{-1}$
$\rho_{\text{chicken meat}}$	1192 $kg\,m^{-3}$
$c_{p,\text{chicken meat}}$	4080 $Ws\,kg^{-1}\,K^{-1}$

[a]Maximal measured half of thickness of 10 chicken fillets.

$$Bi = \frac{\alpha L}{\lambda} = \frac{\text{transport resistance product}}{\text{transport resistance surroundings}}, \quad (3)$$

where α [$W\,m^{-2}\,K^{-1}$] is the convective-heat-transfer coefficient of the surroundings, L [m] is half height of the product, and λ [$W\,m^{-1}\,K^{-1}$] the product's thermal conductivity.

When the value of Bi exceeds 10, the transport resistance of the surroundings is considered to be negligible. With:

$$Fo = \frac{at}{L^2}, \quad (4)$$

where Fo (Fourier number) [29] is the dimensionless time, one can calculate until which heating time, the short times boundary, boundary condition 2 is met. For an infinite slab, the short times boundary is set by $Fo = 0.20$.

The temperature profile of the chicken fillet (at different distances from the surface of the meat) in boiling water was estimated using (1) and the parameter values given in Table 1.

A PT100 temperature probe attached to an Applikon ADI-1030 biocontroller (Applikon, Schiedam, The Netherlands) was used to measure the water and surface temperature of the meat during cooking. The temperature probe was calibrated between 0°C (melting ice) and 100°C (boiling water) prior to use.

According to the manufacturer, the actual sensor is in the last two centimetres of the temperature probe. Therefore, to measure the temperature at the surface of the meat, a 20 cm long stainless steel probe was bent in the shape of an ice-hockey stick, allowing the actual sensor (last two cm) to be pressed firmly to the surface of the meat.

The temperature profiles of the water and meat surface were generated using noninoculated chicken fillets (weight: 142–172 g). The meat was placed on a plastic test tube rack so that only the last 7 cm of the temperature probe was under water. After adding the meat in the pan with boiling water, the temperature probe was immediately pressed on the meat surface or held in the water and temperature recording was started. Each experiment was repeated three times.

2.6. Sampling and Microbiological Enumeration. Culturability of the inoculated bacteria on the food items after heat-treatment was determined by use of the Most Probable

Number method (MPN, see De Man [30]) in combination with spread-plating suitable dilutions on agar plates. Chicken breast fillets were sampled and analyzed as described by De Jong et al. [28]; the carrots were similarly treated. The lower level of detection used was 1.4 CFU per food item.

Media used for the MPN method and plate counts, respectively, were as follows, as described by De Jong et al. [28]: Preston broth and Karmali agar for *C. jejuni*, MRS broth and agar for *L. casei*, and modified-Tryptic Soy Broth and Tryptic Soy Agar both supplemented with 1 g/L nalidixic acid for *E. coli*. For *S. typhimurium* Buffered Peptone Water (NVI) and Brilliant Green Agar (Oxoid) were used. MPN samples were checked for growth of the respective organisms after incubation (time/temperature details see [28]) by streak plating on abovementioned agar plates. Suspected colonies of *C. jejuni* and *L. casei* were confirmed by phase contrast microscopy.

C. jejuni media were microaerobically incubated (broth: 48 h, agar: 72 h, at 37°C (to allow the recovery of any sublethally injured *C. jejuni*) either in a three-gas incubator (5% CO_2, 10% O_2, 85% N_2) or in jars with BBL Campypak (Becton Dickinson, Sparks, USA). For reasons of comparison [28], media for *L. casei* were aerobically incubated at 30°C (broth 48 h; agar: 72 h), those for *E. coli* and *S. typhimurium* overnight at 37°C.

The contamination levels of the food items after heat treatment were calculated, taking into account the exact weights used for enumeration, using an Excel spreadsheet based on the MPN method described by De Man [30].

2.7. Data and Statistical Analysis. Data were represented as count data (log CFU and MPN per fillet) plotted versus heating time (min), to which the linear and Weibull inactivation models were fitted. Levels below the detection limit were taken into account as censored data by using maximum likelihood estimation assuming Poisson-distributed data [31]. The best fitting model was determined by applying an F-test using the RSS of the Weibull and linear model. Analyses were performed using Mathematica 5.2 (Wolfram Research Inc, Champaign, USA). According to the F-test, the linear model fitted best and therefore decimal reduction times during boiling (D_{boil}) were calculated using the slope of the graphs, using:

$$\log N_t = \log N_0 - \frac{t}{D_{\text{boil}}} \tag{5}$$

with log as the 10-based logarithm, N_t as number of viable microorganisms at a given time, N_0 as the initial number of microorganisms, t as time in min, and D_{boil} as time needed to reduce the number of bacteria present on the meat surface by 10-fold when exposed to boiling water.

The effect of different heat treatment scenarios was tested for its significance with ANOVA on the log transformed data in SPSS (SPSS, Chicago, USA). A significance level of 0.05 was used.

2.8. Comparison with Literature Data. For comparison of currently obtained data to literature data, D-values were collected from literature for different temperatures, strains, and products (mainly meat) or media following the approach of Van Asselt and Zwietering [32]. Microorganisms studied were *Campylobacter jejuni* ($n = 176$), *Lactobacilli* ($n = 6$), *Escherichia coli* ($n = 79$), and *Salmonella* ($n = 287$). The relation between heating temperature and log D is linear:

$$\log D_{\text{ref}} = \text{intercept } (\log D, T) - \frac{T_{\text{ref}}}{z}. \tag{6}$$

$\log D_{\text{ref}}$ is the logarithm of the D-value (log min) at temperature T_{ref}, T_{ref} is the reference temperature (°C), and z is the temperature increase (°C) needed to reduce the D-value with a factor of 10. The log D versus temperature plot was used to compare our decimal reduction times with data (D-values) published in literature.

2.9. Risk Assessment. Decimal reduction times can be used to assess the human health risk of pathogen exposure consequential to undercooking. As an example, we studied *Campylobacter* on chicken breast fillets using an empirical distribution of concentrations on fillets after cutting at the cutting plant, obtained from a Dutch risk assessment model [21, 33]. It showed that most contaminated fillets carried between 0 between 4 log cfu/fillet and less than 5% carried higher numbers, results which are similar to *Campylobacter* contamination data reported for German retail products [34]. After incorporation of the survival after storage [21, 33], yielding a decrease in concentration described by a BetaPert distribution with minimum 0.1, most likely value 0.9 and maximum 2.1 log CFU/fillet, these results can be used as a distribution for input (N_0) of (5). The exposure distribution after an inactivation time t is then given by the resulting distribution of N_t. Using the same approach as Nauta et al. [35] the associated health risk, expressed as the probability of illness per fillet, can be assessed by implementing this distribution of doses into the "classic" dose response relationship for *Campylobacter* [36]:

$$P_{\text{ill}}(N_t) = 0.33 \times \left(1 - \frac{\Gamma(\alpha+\beta)\Gamma(\beta+N_t)}{\Gamma(\beta)\Gamma(\alpha+\beta+N_t)}\right) \tag{7}$$

with dose response parameters α and β.

3. Results

The effect of cooking on bacterial survival on whole chicken breast fillets was studied for a cocktail of *C. jejuni.* During cooking, cell numbers declined, following a straight line (Figure 1). Using (5), we calculated that the decimal reduction time for *C. jejuni* when present on meat during cooking of the meat in boiling water was 1.90 min.

In Figure 2, the estimated temperature profiles at the surface of a chicken fillet in boiling water and at 0.25 mm, 0.5 mm, and 0.75 mm depth are shown. To estimate the temperature profile of the chicken fillet in boiling water, two boundary conditions must be met. The first condition states that the external heat transfer resistance is negligible, thus that $Bi > 10$ (3). To verify whether this boundary condition

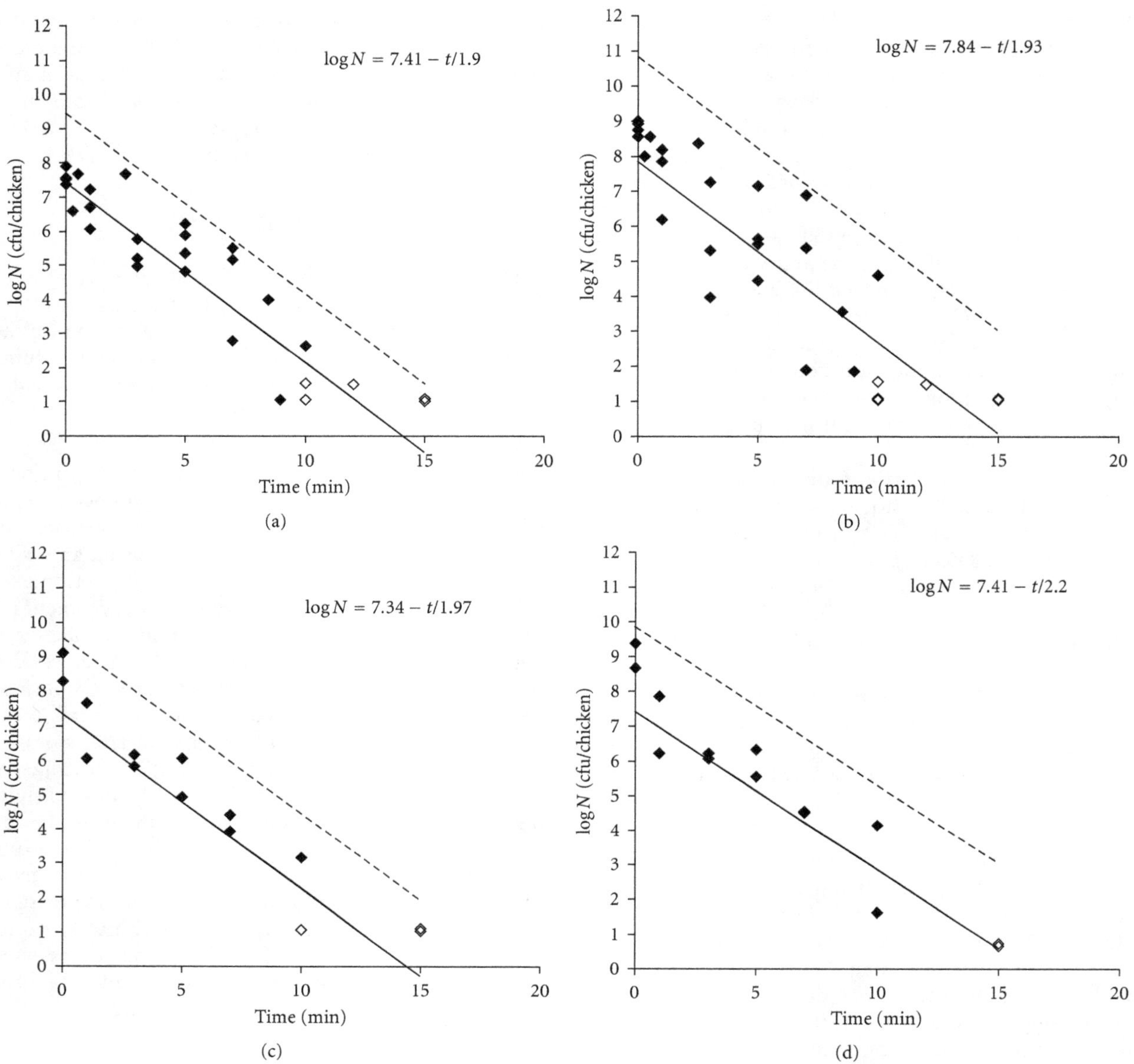

FIGURE 1: Survival of (a) *C. jejuni*, (b) *L. casei*, (c) *E. coli*, and (d) *S. typhimurium* DT104 on whole chicken breast fillets during cooking in boiling water. Straight black line: predicted model; grey line: 95% upper confidence limit; open symbols: count below detection limit.

is met, *Bi* was calculated for still water with the parameter values given in Table 1. For still water, *Bi* ranges between 11.67 and 19.33; thus for boiling water *Bi* will be higher and boundary condition 1 is met. To verify until which heating time boundary condition 2 was met, the short times boundary was calculated, using (4). We calculated a short time boundary of 10.8 min. This implies that for heating times >10.8 min the temperature according to the heating profile estimated by (1) is overestimated.

We also measured the temperature profile of the surface of the chicken fillets. After 30 sec, the average measured surface temperature of the chicken fillets approximated the estimated profile at 0.5 mm depth in the meat, and measured data ranged between the estimated profiles at 0.25 mm and 0.75 mm depth. Within 1 min, the surface temperature of chicken fillets (4°C at time zero) reached a value of at least 85°C (Figure 2). The water temperature was only slightly affected by the addition of the refrigerated meat to the water and did not drop below 99.3°C.

Ample research has been dedicated to heat resistance levels of food borne pathogens, and in Figure 3 an overview is given from a selection of heat inactivation data not only of *Campylobacter* but also of Lactobacilli, *E. coli*, and *Salmonella*. Data are both from fluid and solid (i.e., meat) matrices (see also Van Asselt and Zwietering [32]. This graph shows the extremity of our obtained bacterial heat resistance levels (encircled data).

The heat resistance of bacteria depends on many factors. To investigate some specificities of the observed heat resistance, different heating experiments were conducted with

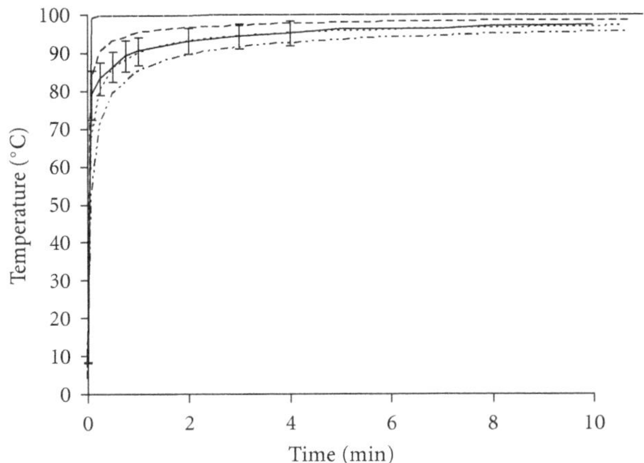

FIGURE 2: (Estimated) Temperature profiles of the surface of a chicken fillet during boiling in water. Estimated: thin —: at the surface; – –: at 0.25 mm depth; - - -: at 0.5 mm depth; –··– : at 0.75 mm depth. Measured: thick —: surface of chicken fillet including error bars for stdev.

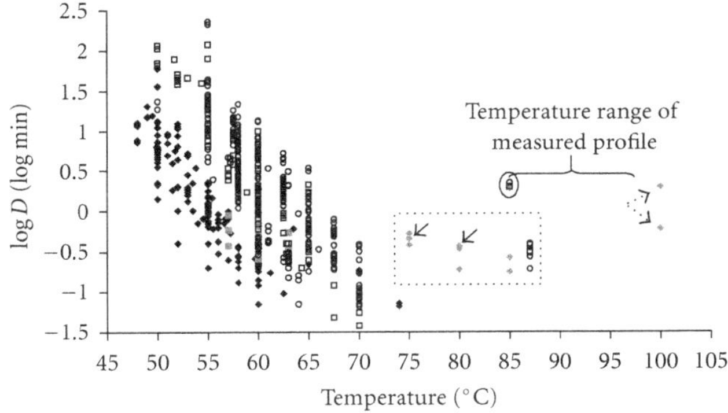

FIGURE 3: log*D*-values plotted against temperature for *Campylobacter* (black diamond), Lactobacilli (grey square), *Escherichia coli* (hollow square), and *Salmonella* (hollow circle) as reported in literature. Campylobacter data for which a decimal reduction time was plotted against heating temperature, instead of *D*-value against matrix temperature: grey diamond. Encircled solid line: current D_{boil} data plotted against lowest surface temperature measured; }: range of the increasing temperature profile the meat was exposed to; boxed dotted line: *D*-values or decimal reduction times based on heating times ≤60 s, $T > 70$°C. Dotted arrows indicate data from Bergsma et al. [24]; solid arrows indicate data from Purnell et al. [37].

C. jejuni. In addition, the same experiment was done, but now with different bacterial species.

The heat resistance of *C. jejuni* was also determined when inoculated onto a slice of carrot of similar size as a chicken fillet to investigate whether the observed heat resistance was related to the matrix type. No bacteria were recovered from cold stored inoculated carrots (24 h, 4°C) after 5 min of boiling (data not shown).

The effect of refrigerated storage time on the heat resistance of *C. jejuni* on chicken meat was investigated. We observed that after a 24 h storage period, a 5 min cooking time reduced cell numbers by 3.0 (stdev 0.5) log units, for 1 h stored fillets this was 4.3 (stdev 0.8) log units, and for noncold stored fillets a 4.7 (stdev 0.5) log reduction was observed. The heat resistances after 0 h and 1 h storage time were not significantly different ($P = 0.392$), but those observed after a 24 h storage time were significantly higher than the other two storage times (24 h versus 1 h storage time: $P = 0.037$; 24 h versus 0 h storage time: $P = 0.002$). Cell attachment to a solid surface increases heat survival of cells [38]. Refrigerated storage for various periods of time did have no effect on the level of attachment of cells to chicken meat, as rinsing for 10 s under cold running water resulted in a 1.4–1.9 log CFU removal of Campylobacter from chicken meat, independent of the storage time (0, 1, and 24 h).

The heat resistance of *L. casei, E. coli, and S. typhimurium* was also tested under the same conditions as used for *Campylobacter*. For all bacterial species tested cells could still be recovered from meat after a 10 minute heating time in boiling water. Again, cell numbers declined, following a straight line. We calculated (5) decimal reduction times of 1.93, 1.97, and 2.20 min, respectively for *L. casei*, *E. coli*, and *S. typhimurium*.

3.1. Impact for Food Safety. We finally determined the effect of the observed heat resistance levels on food safety health risks. For this purpose we applied a Dutch risk assessment study on *Campylobacter* in broiler meat, as explained in the Methods section. Equation (1) was used to calculate the expected ingested doses N_t after boiling the meat for t minutes, given the value $D_{boil} = 1.90$ min, as found in our experiments. After implementation of the dose response relationship, this resulted in Figure 4, showing the decreasing probability of illness per fillet bought at retail in The Netherlands, as a function of cooking time. Consumption of chicken breast fillets cooked for 10 minutes resulted in an estimated probability of illness of 5.5×10^{-6}.

4. Discussion

This research studied the effect of consumer cooking practices on survival of various bacterial species applied to chicken fillet meat. Frying is the most frequently used method for the preparation of chicken meat by Dutch consumers [24]. The temperature at the surface of chicken meat during frying is on average 127°C, but as the temperature is difficult to control (stdev: 18°C) [24], we decided to study survival in boiling water. All tested species died off following a straight line from which a *D*-value of approximately 2 minutes was calculated. The advantage of the study design was that it immediately revealed the fate of bacteria present on meat during nonisothermal heating to high temperatures, mimicking a consumer's style of cooking. Thus, no extrapolation of data obtained from isothermal *D*- and *z*-value experiments with small-sized matrices at temperatures <70°C was needed. The disadvantage was the difficulty of determining the actual surface temperature of

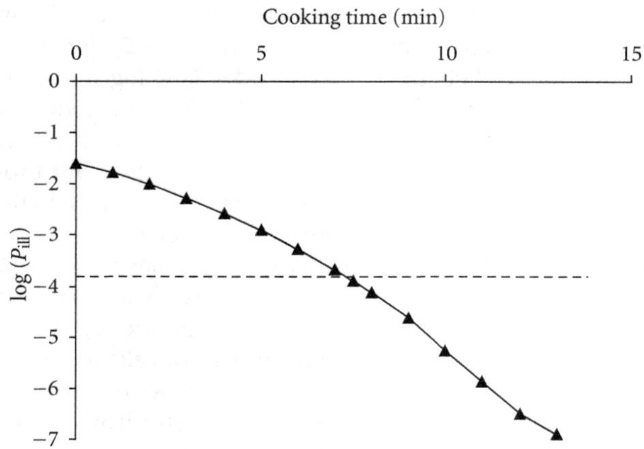

FIGURE 4: The probability of illness per consumed chicken breast fillet as a function of the cooking time. Results are obtained by a Monte Carlo simulation (400.000 runs for each cooking time t, given as a triangle) of (4), using the distribution of Campylobacters per fillet at retail level in The Netherlands, as found by Nauta et al. [33]. The dashed line shows the mean probability of illness per meal due to cross-contamination as found by these authors.

the meat and thus the exact temperature the bacteria were exposed to.

We measured the surface temperature and we used a mathematical approach to determine the surface temperature during cooking. With a temperature probe, we measured that the surface temperature of chicken fillets reached 85°C within 1 min. The mathematical approach can be applied when the external heat transfer resistance is negligible and when the temperature of the centre of the product has not changed. We calculated that during the first 10 minutes of cooking, both conditions are met. The calculated temperature at 0.5 mm depth in the meat approximated the average measured surface temperature of the chicken fillets. The shape of the calculated heating profile was similar to that of the measured profile, and the results of our calculations are comparable to those of Houben and Eckenhausen [39], who calculated the temperature at certain surface depths during pasteurization of a model product in a water bath of 96°C. Together this shows that both our measured and calculated temperature reliably reflect the temperature bacteria experience at the surface of a chicken breast fillet during cooking in boiling water.

Heat resistance studies are generally conducted in solid (thin patties of mainly ground meat) or in liquid matrices (broth, milk, etc.), with higher heat resistance levels obtained in solid matrices [25, 40–46]. Temperatures during such studies range from 55 to 72°C, allowing accurate determination of D- and z-values. At higher temperatures, D-values can only be calculated. In our experimental set-up the surface temperature of chicken fillets reached 85°C within 1 minute. According to the literature (after extrapolation) the D-value of bacterial cells at 85°C is less than one second [32, 41, 47], yet in our experiments it took 2 minutes to obtain a 1 log reduction of bacteria. When inoculated on another solid matrix of approximately the same size (slice of carrot) bacteria showed different behaviour. No bacteria could be detected after a heating time of 5 min. The study of Van Asselt and Zwietering [32] revealed that heat resistance can be significantly increased for certain specific matrix (high fat and low Aw)/bacterium combinations. But as both carrot and chicken meat are low fat/high Aw products, our results indicate that chicken meat itself (e.g., the presence of some specific component like iron, or an amino acid) affects the heat resistance of bacteria.

To determine the D-value of a bacterial species at a certain temperature, a small volume of a bacterial culture is transferred to a relatively large volume of a preheated menstruum. In this way, the temperature come-up time is as short as possible and bacteria experience a constant temperature. As the temperature of a large and solid matrix does not come up as fast as the temperature of a small or liquid matrix, the dimension of a product can affect the heating experiment. A weak size effect on heat resistance is indeed demonstrated in a similar study conducted in our laboratory by Bergsma et al. [24], who pan fried chicken meat inoculated with *C. jejuni* using whole and diced fillets. In literature, some other heat resistance data have been published based on large size meat samples. Purnell et al. [37] observed that elevated heat resistance levels of naturally *C. jejuni* contaminated whole chicken carcasses during water immersion treatments at temperatures ≥70°C (20–30 s). The other *Campylobacter* data points in Figure 3 boxed by the dotted line were calculated using data from Whyte et al. [27]. These authors conducted hot water immersion treatments (75–80°C; 0, 10, and 20 s) with naturally and artificially contaminated chicken thighs. Again high survival levels of *Campylobacter* cells were obtained after heat treatments. In the study by Purnell et al. [37] not only was the size of the test products larger than that used in most other studies, but also the challenge temperatures in this studies were higher than normally used in heat resistance testing. This high challenge temperature might also have its effect. Interestingly, in small sized meat discs artificially inoculated with *E. coli* and *Salmonella* (see Figure 3 in dotted box) and pasteurized with hot steam (87°C), also high survival levels of bacteria were measured after a 60 s heat treatment [48]. Although the calculated D-values or decimal reduction times for data from Purnell, Whyte, and McCann are based on very short heating times, ≤60 s, and may therefore be less accurate, these D-values or decimal reduction times and our current data show that bacteria on meat are not as easily killed by temperatures >70°C as predicted based on D-values obtained at temperatures <70°C. Both our data and those discussed previously indicate that the combination of products size and challenge temperature might affect the level of heat resistance of bacteria.

Another difference of our study design compared to heating experiments described in literature is the overnight refrigerated storage of the contaminated fillets to mimic consumer/retail storage conditions. This could allow for physiological adaptation and attachment. Although overnight cold storage did not affect the number of bacteria attached to the meat, we observed that storing the meat in a refrigerator increased the number of surviving cells. At a low temperature (no growth, stress) physiological properties may have changed which may have affected the heat resistance of the bacteria, a phenomenon known as cross-protection [49, 50].

Although the pathogens used were all motile and thus can move from the surface to more inner parts of fillets, the observed high level of heat resistance cannot be explained by such movement, as the heat survival of the non-motile *L. casei* was equal to that of the motile pathogens tested.

So chicken meat, challenge temperature, or heating rate and cold storage have their effect on the heat resistance of *C. jejuni*, *S. typhimurium*, *E. coli*, and *L. casei*. They survive for longer periods of time than expected during cooking. Friedman et al. [51] concluded that "...because bacteria on the surface of poultry would be destroyed by limited cooking, the recurrent association of illness with undercooked poultry suggests either that the poultry is regularly re-contaminated after cooking or that *Campylobacter* is somehow present deep in the tissues of a single poultry carcass, where it survives limited cooking", which is also in concordance with FAO/WHO conclusions [52]. However, data presented in our paper reveal that limited cooking does not necessarily eliminate all bacteria present on the surface of poultry meat. Furthermore, our data add to the deeper understanding of the frequent association of food borne illness with consumption of undercooked poultry. As a consequence, our data show that research conclusions based on the assumption that cooked chicken meat does not substantially contribute to the risk of food borne illness [53–55] must be interpreted with more consideration. Consumption of chicken meat cooked for 10 min cooking still results in a probability of illness of 5.5×10^{-6}.

The estimated probability of illness per meal containing chicken breast fillet and a salad cross-contaminated with *C. jejuni* from the chicken fillet to the salad, as assessed in the risk assessment of Nauta et al. [21], is 1.6×10^{-4}. This not only confirms that the risk of acquiring campylobacteriosis consequential to undercooking is much smaller than that consequential to cross-contamination [56] but also shows that the cooking time may be far more critical than previously assumed: with a cooking time of about 7.5 min the risk of undercooking is comparable to that calculated for cross-contamination (see Figure 4). Taking consumer behavior into account, using data of the observational study of Fischer et al. [23], it becomes clear that undercooking of chicken meat by consumers is certainly not negligible, as 33% of the participants (Dutch consumers) in their study applied heating times during cooking of chicken breast fillets of less than 7.5 min.

When inoculated on chicken breast fillets, the heat resistance of bacteria increased to unexpected high levels. Chicken meat, the challenge temperature, or heating rate and cold storage affected the level of resistance. It can be concluded from our study that, until now, the effect of cooking on the survival of bacteria present on the outside of chicken meat has been overestimated. It is therefore recommended to reconsider all statements made based on meat heating trials that do not use consumer-style meat types, sizes, and cooking techniques.

Acknowledgments

The authors would like to acknowledge Ellen Delfgou and John Dufrenne (Laboratory for Zoonoses and Environmental Microbiology, National Institute for Public Health and the Environment) for assisting them with part of the heating experiments. This research was funded by the ZonMw (Grant 014-12-33).

References

[1] Anonymous, WHO surveillance programme for control of foodborne infections and intoxications in Europe. 8th report 1999-2000, 2001, http://www.bfr.bund.de/cd/2352.

[2] N. H. Bean, J. S. Goulding, C. Lao, and F. J. Angulo, "Surveillance for foodborne-disease outbreaks—United States, 1988–1992," *MMWR*, vol. 45, no. 5, pp. 1–66, 1996.

[3] N. H. Bean and P. M. Griffin, "Foodborne disease outbreaks in the United States, 1973–1987: pathogens, vehicles, and trends," *Journal of Food Protection*, vol. 53, pp. 804–817, 1990.

[4] S. J. Olsen, L. C. MacKinnon, J. S. Goulding, N. H. Bean, and L. Slutsker, "Surveillance for foodborne-disease outbreaks—United States, 1993–1997," *MMWR*, vol. 49, no. 1, pp. 1–62, 2000.

[5] P. J. Panisello, R. Rooney, P. C. Quantick, and R. Stanwell-Smith, "Application of foodborne disease outbreak data in the development and maintenance of HACCP systems," *International Journal of Food Microbiology*, vol. 59, no. 3, pp. 221–234, 2000.

[6] T. A. Cogan, J. Slader, S. F. Bloomfield, and T. J. Humphrey, "Achieving hygiene in the domestic kitchen: the effectiveness of commonly used cleaning procedures," *Journal of Applied Microbiology*, vol. 92, no. 5, pp. 885–892, 2002.

[7] S. B. Duff, E. A. Scott, M. S. Mafilios et al., "Cost-effectiveness of a targeted disinfection program in household kitchens to prevent foodborne illnesses in the United States, Canada, and the United Kingdom," *Journal of Food Protection*, vol. 66, no. 11, pp. 2103–2115, 2003.

[8] S. B. Fein, C. T. Jordan Lin, and A. S. Levy, "Foodborne illness: perceptions, experience, and preventive behaviors in the United States," *Journal of Food Protection*, vol. 58, no. 12, pp. 1405–1411, 1995.

[9] T. J. Humphrey, K. W. Martin, J. Slader, and K. Durham, "*Campylobacter* spp. in the kitchen: spread and persistence," *Journal of Applied Microbiology Symposium Supplement*, vol. 90, no. 30, pp. 115s–150s, 2001.

[10] Anonymous, Fight BAC! Keep food safe from bacteria, 2010, http://www.fightbac.org/safe-food-handling/cook/.

[11] A. R. H. Fischer, L. J. Frewer, and M. J. Nauta, "Toward improving food safety in the domestic environment: a multi-item rasch scale for the measurement of the safety efficacy of domestic food-handling practices," *Risk Analysis*, vol. 26, no. 5, pp. 1323–1338, 2006.

[12] J. B. Anderson, T. A. Shuster, K. E. Hansen, A. S. Levy, and A. Volk, "A Camera's view of consumer food-handling behaviors," *Journal of the American Dietetic Association*, vol. 104, no. 2, pp. 186–191, 2004.

[13] Anonymous, Vlees, cijfers en trends 2004—marktverkenning over het consumptiegedrag in een dynamische samenleving. Voorlichtingsbureau Vlees, Zoetermeer.18, 2005.

[14] Centers for Disease Control and Prevention, "Preliminary FoodNet data on the incidence of infection with pathogens

transmitted commonly through food—10 States, 2009," *Morbidity and Mortality Weekly Report*, vol. 9, pp. 418–422, 2010.

[15] T. A. Cogan, S. F. Bloomfield, and T. J. Humphrey, "The effectiveness of hygiene procedures for prevention of cross-contamination from chicken carcases in the domestic kitchen," *Letters in Applied Microbiology*, vol. 29, no. 5, pp. 354–358, 1999.

[16] A. J. Lawson, J. M. J. Logan, G. L. O'Neill, M. Desai, and J. Stanley, "Large-scale survey of *Campylobacter* species in human gastroenteritis by PCR and PCR-enzyme-linked immunosorbent assay," *Journal of Clinical Microbiology*, vol. 37, no. 12, pp. 3860–3864, 1999.

[17] S. F. Park, "The physiology of *Campylobacter* species and its relevance to their role as foodborne pathogens," *International Journal of Food Microbiology*, vol. 74, no. 3, pp. 177–188, 2002.

[18] European Food Safety Authority, "The EU summary report on trends and sources of zoonoses, zoonotic agents and food-borne outbreaks in 2009," *EFSA Journal*, vol. 9, article 2090, 2011.

[19] J. P. Butzler, "*Campylobacter*, from obscurity to celebrity," *Clinical Microbiology and Infection*, vol. 10, no. 10, pp. 868–876, 2004.

[20] E. O. Göksoy, C. James, J. E. L. Ćorry, and S. J. James, "The effect of hot-water immersions on the appearance and microbiological quality of skin-on chicken-breast pieces," *International Journal of Food Science and Technology*, vol. 36, no. 1, pp. 61–69, 2001.

[21] M. J. Nauta, W. F. Jacobs-Reitsma, E. G. Evers, W. van Pelt, and A. H. Havelaar, "Risk assessment of *Campylobacter* in the Netherlands via broiler meat and other routes," RIVM Report 250911006:128, RIVM, Bilthoven, The Netherlands, 2005.

[22] P. Padungton and J. B. Kaneene, "*Campylobacter* spp. in human, chickens, pigs and their antimicrobial resistance," *Journal of Veterinary Medical Science*, vol. 65, no. 2, pp. 161–170, 2003.

[23] A. R. H. Fischer, A. E. I. De Jong, E. D. Van Asselt, R. De Jonge, L. J. Frewer, and M. J. Nauta, "Food safety in the domestic environment: an interdisciplinary investigation of microbial hazards during food preparation," *Risk Analysis*, vol. 27, no. 4, pp. 1065–1082, 2007.

[24] N. J. Bergsma, A. R. H. Fischer, E. D. van Asselt, M. H. Zwietering, and A. E. I. de Jong, "Consumer food preparation and its implication for survival of *Campylobacter jejuni* on chicken," *British Food Journal*, vol. 109, no. 7, pp. 548–561, 2007.

[25] L. E. Blankenship and S. E. Craven, "*Campylobacter jejuni* survival in chicken meat as a function of temperature," *Applied and Environmental Microbiology*, vol. 44, no. 1, pp. 88–92, 1982.

[26] V. K. Juneja, "A comparative heat inactivation study of indigenous microflora in beef with that of *Listeria monocytogenes*, *Salmonella* serotypes and *Escherichia coli* O157:H7," *Letters in Applied Microbiology*, vol. 37, no. 4, pp. 292–298, 2003.

[27] P. Whyte, K. McGill, and J. D. Collins, "An assessment of steam pasteurization and hot water immersion treatments for the microbiological decontamination of broiler carcasses," *Food Microbiology*, vol. 20, no. 1, pp. 111–117, 2003.

[28] A. E. I. De Jong, L. Verhoeff-Bakkenes, M. J. Nauta, and R. De Jonge, "Cross-contamination in the kitchen: effect of hygiene measures," *Journal of Applied Microbiology*, vol. 105, no. 2, pp. 615–624, 2008.

[29] D. P. DeWitt, T. A. Bergman, and A. S. Lavine, "Textbook incropera," in *Fundamentals of Heat and Mass Transfer*, P. Frank, Ed., pp. 260–261, John Wiley & Sons, New York, NY, USA, 6th edition, 2006.

[30] J. C. De Man, "The probability of most probable numbers," *European Journal of Applied Microbiology*, vol. 1, no. 1, pp. 67–78, 1975.

[31] M. F. Lorimer and A. Kiermeier, "Analysing microbiological data: Tobit or not Tobit?" *International Journal of Food Microbiology*, vol. 116, no. 3, pp. 313–318, 2007.

[32] E. D. Van Asselt and M. H. Zwietering, "A systematic approach to determine global thermal inactivation parameters for various food pathogens," *International Journal of Food Microbiology*, vol. 107, no. 1, pp. 73–82, 2006.

[33] M. J. Nauta, W. F. Jacobs-Reitsma, and A. H. Havelaar, "A risk assessment model for *Campylobacter* in broiler meat," *Risk Analysis*, vol. 27, no. 4, pp. 845–861, 2007.

[34] P. Luber and E. Bartelt, "Enumeration of *Campylobacter* spp. on the surface and within chicken breast fillets," *Journal of Applied Microbiology*, vol. 102, no. 2, pp. 313–318, 2007.

[35] M. J. Nauta, A. R. H. Fischer, E. D. Van Asselt, A. E. I. De Jong, L. J. Frewer, and R. De Jonge, "Food safety in the domestic environment: the effect of consumer risk information on human disease risks," *Risk Analysis*, vol. 28, no. 1, pp. 179–192, 2008.

[36] P. F. M. Teunis and A. H. Havelaar, "The Beta Poisson dose-response model is not a single-hit model," *Risk Analysis*, vol. 20, no. 4, pp. 513–520, 2000.

[37] G. Purnell, K. Mattick, and T. Humphrey, "The use of 'hot wash' treatments to reduce the number of pathogenic and spoilage bacteria on raw retail poultry," *Journal of Food Engineering*, vol. 62, no. 1, pp. 29–36, 2004.

[38] T. J. Humphrey, S. J. Wilde, and R. J. Rowbury, "Heat tolerance of *Salmonella* typhimurium DT104 isolates attached to muscle tissue," *Letters in Applied Microbiology*, vol. 25, no. 4, pp. 265–268, 1997.

[39] J. H. Houben and F. Eckenhausen, "Surface pasteurization of vacuum-sealed precooked ready-to-eat meat products," *Journal of Food Protection*, vol. 69, no. 2, pp. 459–468, 2006.

[40] G. R. Acuff, C. Vanderzant, M. O. Hanna, J. G. Ehlers, and F. A. Gardner, "Effects of handling and preparation of turkey products on the survival of *Campylobacter jejuni*," *Journal of Food Protection*, vol. 49, pp. 627–631, 1986.

[41] M. E. Doyle and A. S. Mazzotta, "Review of studies on the thermal resistance of Salmonellae," *Journal of Food Protection*, vol. 63, no. 6, pp. 779–795, 2000.

[42] M. P. Doyle and D. J. Roman, "Growth and survival of *Campylobacter fetus* subsp. *jejuni* as a function of temperature and pH," *Journal of Food Protection*, vol. 44, pp. 596–601, 1981.

[43] P. Koidis and M. P. Doyle, "Survival of *Campylobacter jejuni* in fresh and heated red meat," *Journal of Food Protection*, vol. 46, pp. 771–774, 1983.

[44] R. Y. Murphy, B. P. Marks, E. R. Johnson, and M. G. Johnson, "Inactivation of *Salmonella* and Listeria in ground chicken breast meat during thermal processing," *Journal of Food Protection*, vol. 62, no. 9, pp. 980–985, 1999.

[45] S. Quintavalla, S. Larini, P. Mutti, and S. Barbuti, "Evaluation of the thermal resistance of different *Salmonella* serotypes in pork meat containing curing additives," *International Journal of Food Microbiology*, vol. 67, no. 1-2, pp. 107–114, 2001.

[46] S. C. Waterman, "The heat-sensitivity of *Campylobacte jejuni* in milk," *Journal of Hygiene*, vol. 88, no. 3, pp. 529–533, 1982.

[47] S. Sörqvist, "Heat resistance in liquids of *Enterococcus* spp., *Listeria* spp., *Escherichia coli*, *Yersinia enterocolitica*, *Salmonella*

spp. and *Campylobacter* spp," *Acta Veterinaria Scandinavica*, vol. 44, no. 1-2, pp. 1–19, 2003.

[48] M. S. McCann, J. J. Sheridan, D. A. McDowell, and I. S. Blair, "Effects of steam pasteurisation on *Salmonella* Typhimurium DT104 and *Escherichia coli* O157:H7 surface inoculated onto beef, pork and chicken," *Journal of Food Engineering*, vol. 76, no. 1, pp. 32–40, 2006.

[49] S. García, J. C. Limón, and N. L. Heredia, "Cross protection by heat and cold shock to lethal temperatures in Clostridium perfringens," *Brazilian Journal of Microbiology*, vol. 32, no. 2, pp. 110–112, 2001.

[50] H. H. Wemekamp-Kamphuis, A. K. Karatzas, J. A. Wouters, and T. Abee, "Enhanced levels of cold shock proteins in Listeria monocytogenes LO28 upon exposure to low temperature and high hydrostatic pressure," *Applied and Environmental Microbiology*, vol. 68, no. 2, pp. 456–463, 2002.

[51] C. R. Friedman, J. Neimann, H. C. Wegener, and R. V. Tauxe, "Epidemiology of *Campylobacter jejuni* infections in the United States and other industrialized nations," in *Campylobacter*, I. Nachamkin and M. J. Blaser, Eds., pp. 121–138, American Society for Microbiology Press, Washington, DC, USA, 2nd edition, 2000.

[52] FAO/WHO Expert Consultation, *Risk Assessment of* Campylobacter *spp. in Broiler Chickens and Vibrio spp. in Seafood*, FAO/WHO, Bangkok, Thailand, 2002.

[53] E. G. Evers, H. J. V. D. Fels, M. H. Nauta, J. F. Schijven, and A. H. Havelaar, "Het relatieve belang van *Campylobacter* transmissieroutes op basis van blootstellingsschatting," RIVM Report 250911003:62, RIVM, Bilthoven, The Netherlands, 2004.

[54] H. Rosenquist, N. L. Nielsen, H. M. Sommer, B. Nørrung, and B. B. Christensen, "Quantitative risk assessment of human campylobacteriosis associated with thermophilic *Campylobacter* species in chickens," *International Journal of Food Microbiology*, vol. 83, no. 1, pp. 87–103, 2003.

[55] J. M. Straver, A. F. W. Janssen, A. R. Linnemann, M. A. J. S. Van Boekel, R. R. Beumer, and M. H. Zwietering, "Number of *Salmonella* on chicken breast filet at retail level and its implications for public health risk," *Journal of Food Protection*, vol. 70, no. 9, pp. 2045–2055, 2007.

[56] S. Brynestad, L. Braute, P. Luber, and E. Bartelt, "Quantitative microbiological risk assessment of campylobacteriosis cases in the German population due to consumption of chicken prepared in homes," *International Journal of Risk Assessment and Management*, vol. 8, no. 3, pp. 194–213, 2008.

7

Diversity across Seasons of Culturable *Pseudomonas* from a Desiccation Lagoon in Cuatro Cienegas, Mexico

Alejandra Rodríguez-Verdugo,[1,2] Valeria Souza,[1] Luis E. Eguiarte,[1] and Ana E. Escalante[1,3]

[1] *Departamento de Ecología Evolutiva, Instituto de Ecología, Universidad Nacional Autónoma de México, Apartado Postal 70-275, 04510 México, DF, Mexico*
[2] *Department of Ecology and Evolutionary Biology, University of California, Irvine, CA 92091, USA*
[3] *Departamento de Ecología de la Biodiversidad, Instituto de Ecología, Universidad Nacional Autónoma de México, Apartado Postal 70-275, 04510 México, DF, Mexico*

Correspondence should be addressed to Ana E. Escalante, anaelena.escalante@gmail.com

Academic Editor: Isabel Sá-Correia

Cuatro Cienegas basin (CCB) is a biodiversity reservoir within the Chihuahuan desert that includes several water systems subject to marked seasonality. While several studies have focused on biodiversity inventories, this is the first study that describes seasonal changes in diversity within the basin. We sampled *Pseudomonas* populations from a seasonally variable water system at four different sampling dates (August 2003, January 2004, January 2005, and August 2005). A total of 70 *Pseudomonas* isolates across seasons were obtained, genotyped by fingerprinting (BOX-PCR), and taxonomically characterized by 16S rDNA sequencing. We found 35 unique genotypes, and two numerically dominant lineages (16S rDNA sequences) that made up 64% of the sample: *P. cuatrocienegasensis* and *P. otitidis*. We did not recover genotypes across seasons, but lineages reoccurred across seasons; *P. cuatrocienegasensis* was isolated exclusively in winter, while *P. otitidis* was only recovered in summer. We statistically show that taxonomic identity of isolates is not independent of the sampling season, and that winter and summer populations are different. In addition to the genetic description of populations, we show exploratory measures of growth rates at different temperatures, suggesting physiological differences between populations. Altogether, the results indicate seasonal changes in diversity of free-living aquatic *Pseudomonas* populations from CCB.

1. Introduction

The Cuatro Cienegas basin (CCB), in Mexico, has been described as an important biodiversity reservoir within the Chihuahuan desert. The basin consists of a small (<840 km^2) intermontane valley that contains different water systems. Most of the aquatic habitats are ephemeral, not permanent or subject to marked seasonal fluctuations [1]. Moreover, most aquatic systems in the area are extremely oligotrophic due to the almost negligible phosphorous levels [2]. Despite this, CCB is one of only two North American desert ecosystems characterized by high levels of species endemism including vertebrates, invertebrates [1, 3], and more recently a considerable list of microbes either bentonic, planktonic, or part of stromatolites and microbial mats [4–7]. Using culture-independent approaches, gammaproteobacteria in CCB appears as a dominant group in the aquatic environments [4, 8]. Within proteobacteria, *Pseudomonas* is itself a dominant group, with ample distribution and new endemic lineages or species described within the basin [5, 9], as well as a clear dominance in some microbial mats [7]. The unusual levels of biodiversity and endemism have led to describe CCB as well as either a time machine [10] or a "microbial Galapagos" [1, 4, 11] and have made it priority for conservation efforts by (comisión nacional para el conocimiento y uso de la biodiversidad) CONABIO, the world wildlife fund (wwf), the ramsar convention on wetlands and man, and the biosphere (MAB)/UNESCO.

Previous studies in CCB have sought to describe the unusual levels of microbial diversity across environmental or geographic gradients [6, 12, 13], as well as to understand the evolutionary and ecological origins of the observed diversity [4, 10, 14, 15]. However, nothing has been done to characterize bacterial diversity across seasons, despite (1) the existence of analytical models that indicate that seasonal fluctuation can influence the origin and maintenance of diversity [16] and (2) the marked seasonality in many of the aquatic systems in the basin [1, 17].

To evaluate the changes in microbial diversity associated with seasonality in water systems of CCB, we characterize, for the first time, microbial diversity across seasons in one of the seasonally variable freshwater systems (desiccation lagoon). The studied system is relatively small, and evidence exists of the strong influence that its seasonal environmental changes have in genetic variation of fish [17]. Within this context, we analyzed a fraction of total microbial diversity, the culturable *Pseudomonas* populations, and hypothesize that seasonal variation of microbial populations should track seasonal changes of the desiccation lagoon.

We statistically show that taxonomic identity of isolates is not independent of the sampling season, and that winter and summer populations are different. In addition to the genetic description of populations, we show exploratory measures of growth rates at different temperatures, suggesting physiological differences between populations. Altogether, the results indicate seasonal changes in diversity of free-living aquatic *Pseudomonas* populations from CCB.

2. Materials and Methods

2.1. Study Site. We chose a seasonal aquatic ecosystem within CCB subject to marked fluctuations of chemical and physical parameters across seasons [18] temperature being one of them, as shown in Figure 1(b) (0–38°C range; [17]). The site is locally known as "Laguna Grande" (LG), and is located in the hydrological system of Churince on the western side of CCB (Figure 1(a)). Temperature was measured hourly over approximately two-week intervals at two sites (LG1 and LG3) using iButton temperature sensors (Maxim Integrated, Dallas, TX, USA).

2.2. Sampling and Isolation of Bacterial Strains. We sampled four sites in the desiccation lagoon "Laguna Grande": LG1 (26°50.830′N, 102°09.335′W), LG2 (26°51.199′N, 102°09.009′W), LG3 (26°51.146′N, 102°08.964′W), and LG4 (26°51.222′N, 102°09.040′W). At a single time point, there were not significant temperature differences between sampling sites (Figure 1 and [6]), the multiple site sampling per time point was done to cover as much area as possible. Temperature variation was mostly through seasons, with temperatures reaching lows close to 0°C and highs close to 40°C (Figure 1; [17]). Samples were taken in summer (August 2003 and 2005) and winter (January 2004 and 2005). No further sampling was possible since 2006 because overexploitation of CCB aquifer associated with agricultural practices dried out the aquatic environment of "Laguna Grande."

Triplicate samples of 15 mL of water were taken from surface water (15–20 cm depth) at each of the four samples sites using sterile BD Falcon vials (BD Biosciences, MA, USA). Each replicate sample was plated in triplicate by spreading 200 μL of each vial. Culture plates contained GSP culture media (*Pseudomonas-Aeromonas* selective agar base): 10.0 (g L^{-1}) sodium L(+) glutamate, 20.0 (g L^{-1}) soluble starch, 2.0 (g L^{-1}) potassium dihydrogen phosphate, 0.5 (g L^{-1}) magnesium sulfate, 0.36 (g L^{-1}) phenol red, and 12.0 (g L^{-1}) agar-agar [19]. Strains that belong to the genus *Aeromonas* degrade the starch and produce acid, causing change in color (red to yellow). Strains that belong to the genus *Pseudomonas* did not produce acid; therefore, we selected the colonies that did not decolorize the media into yellow. The plates were incubated according to instructions for enrichments of *Pseudomonas-Aeromonas* [19]. Colonies were purified by subculturing on the same medium and maintained at −80°C in GSP media and 15% (w/v) glycerol.

2.3. DNA Extraction and BOX-PCR Genomic Fingerprint Analysis. DNA was extracted by using DNeasy Blood and Tissue Kit (Qiagen, CA, USA) according to the manufacturer's instructions. Repetitive extragenic palindromic PCR (rep-PCR) genomic fingerprinting of the isolates was carried out with a BOX-A1R primer (5′-CTACGGCAAGGCGAC-GCTGACG-3′) according to the protocol of [20]. The following PCR conditions were used: 7 min at 95°C, followed by 30 cycles of 94°C for 1 min, 53°C for 1 min, 65°C for 8 min, and a final extension at 65°C for 8 min. PCR products were analyzed on 1.5% (w/v) agarose gels containing 0.5X TAE-buffer (200 mM trisacetate, 0.5 mM EDTA, pH 8). The electrophoresis was performed for 5 hours at 180 mV (5 V cm^{-1}). A 1-kb Plus DNA size ladder (INVITROGEN) was run at both sides and in the central lane of each gel. The gels were stained with ethidium bromide.

2.4. Computer-Assisted Analysis of BOX-PCR Genomic Fingerprints. Gel images were digitized with a charge-couple device video camera (Gel Logic 100, Kodak) and stored on disk as TIFF files. These digitized images were converted, normalized with the abovementioned DNA size markers, and analyzed with GelCompar software (version 4.0; Applied Maths, Kortrijk, Belgium). The "rolling disk" background subtraction method was applied. To analyse BOX-PCR patterns, similarity matrices of whole densitometric curves of the gel tracks were calculated by using the pair-wise Pearson's product-moment correlation coefficient (r value of 1 is equivalent to 100% similarity). This approach compares the whole densitometric curves of the fingerprints [21, 22]. Cluster analyses of similarity matrices were performed by the unweighted pair group method using arithmetic averages (UPGMA). We performed a cluster analysis of all DNA ladders to choose a similarity value to define isolates belonging to a same group of genotypes.

2.5. 16S rRNA Gene Sequencing and Analysis. We chose one isolate per genotype (as defined by rep-PCR analysis and determined by having at least 90% similarity in banding patterns) to obtain the 16S rDNA sequence. Previous studies have shown that clones with very similar BOX-PCR fingerprints (r values of more than 0.8) had identical 16S rRNA

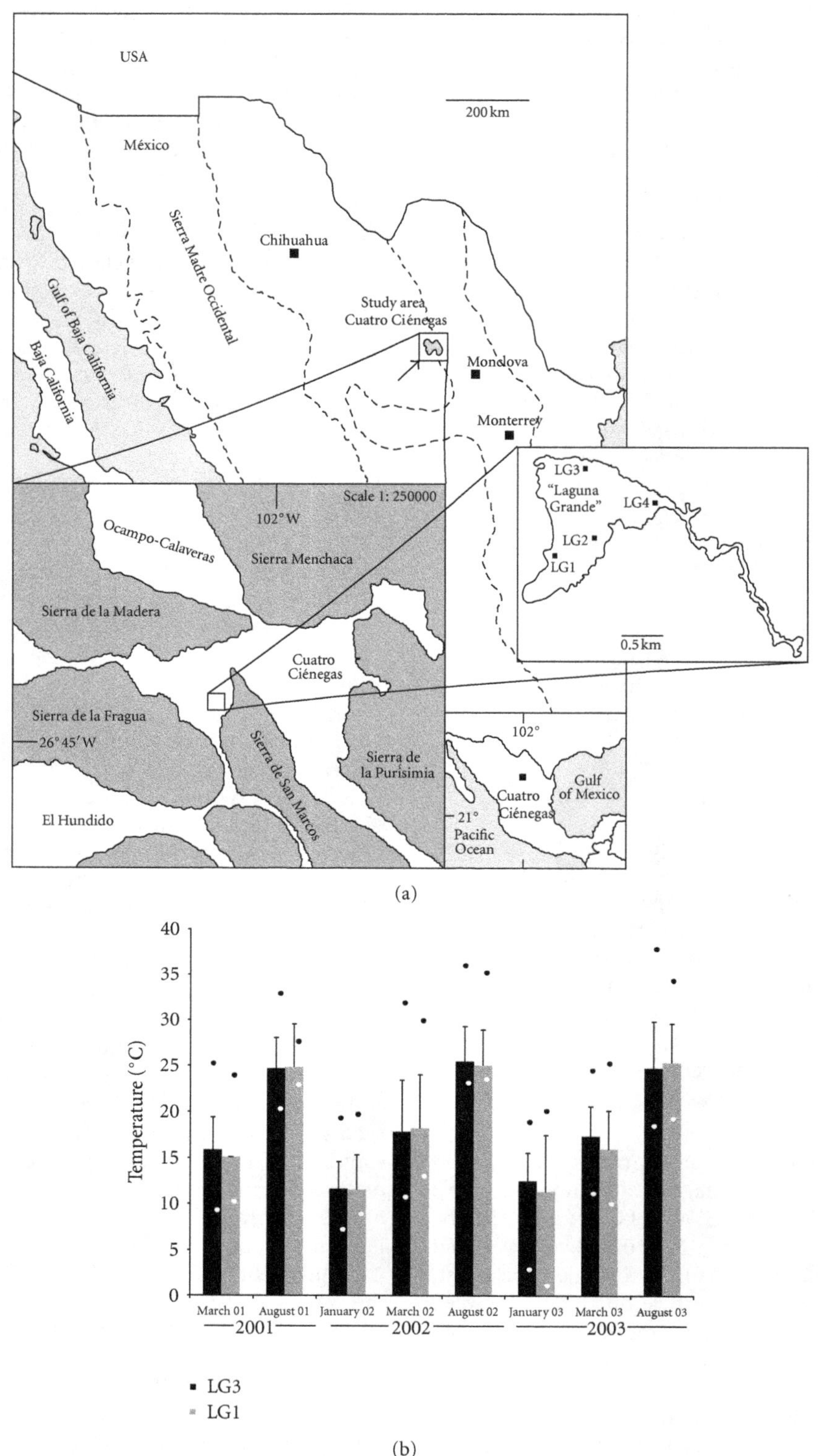

Figure 1: Study site in the Cuatro Cienegas basin. (a) Geographic location of the study site, indicating the sampling sited within "Laguna Grande" (desiccation lagoon) of the Churince system (modified from [8, 11]). (b) Average water temperature in two sites (LG1 and LG3) of "Laguna Grande." Temperature was measured hourly over two weeks; average temperatures are significantly higher during summer than winter [17]. Error bars represent standard deviations, white dots and dark dots represent minimum and maximum temperatures, respectively [16].

gene sequences [23]. The 16S rRNA gene was amplified using the 27F and 1492R primers under conditions described previously [24] in 100 μL final volume. The PCR products were purified using the QIAquick gel extraction kit (Qiagen, Hilden, Germany). For sequencing the 16S rRNA gene (ca. 1450 bp) primers 27F, 357R, 530R, 530F, 790F, 981R, and 1492R were used [25]. The sequencing reaction had a total volume of 15 μL consisting of 2 μL Big Dye Terminator sequencing buffer (Applied Biosystems, Foster City, CA, USA), 1.6 μM primer, and 5 μL-purified amplified product. The amplification conditions were as follows: one cycle of 5 min at 95°C, and 45 cycles of 10 s at 95°C, 10 s at 50°C and 4 min at 60°C. Sequencing was done in a capillary sequencer (ABI-Avant 100). Sequences were assembled and revised using Consed software [26].

2.6. Nucleotide Accession Numbers. The 16S rRNA gene sequences obtained have been submitted to the GenBank database under accession numbers EU791282 and FJ976048-FJ976083.

2.7. Phylogenetic Analysis. The BLAST 2.0.6 algorithm of GenBank and the SIMILARITY_RANK tool of the Ribosomal Database Project II (RDP-II) were employed to search for closest matches found in the RDP-II and GenBank. Sequences were aligned using the CLUSTAL_W program [27]. Model generator (version 0.84, [28]) was used to determine the optimal nucleotide substitution model. Neighbor-joining (NJ) algorithm was used to generate a genealogy as implemented in PAUP (version 4.0, [29]), by using the GTR evolutionary model with gamma correction 0.40 and 1500 bootstrap replicates for all sequences.

2.8. Growth Rates. As an exploratory approach towards potential differences in physiological responses of winter and summer populations, growth curves at different temperatures were constructed, and maximum growth rates determined for a subset of isolates. The subset of isolates from the total sample represented winter and summer populations. The criteria for assembling this subset looked for a fair representation of genotype diversity at the individual level, as well as the inclusion of isolates that were obtained at different sampling dates and belong to the observed dominant lineages (*P. otitidis* and *P. cuatrocienegasensis*). By applying these criteria, the subset resulted in 6 genotypes of winter samples (*P. cuatrocienegasensis*) and 11 genotypes of summer samples (*P. otitidis*). We determined individual maximum growth rates at 5 different temperatures (28, 32, 26, 40, and 44°C), likely experienced in summer time, and ran the experiments in triplicate. A Biotek Synergy Microplate Reader (Synergy 2 Multi-Mode Microplate Reader Model, BioTek) was used to measure optical density of individual cultures every 10 min. Optical density measures were then used to construct growth curves and determine maximum growth rates.

2.9. Statistical Analyses

2.9.1. Diversity Calculations and Genotypes. Genotypic diversity was obtained from the BOX-PCR fingerprinting. We calculated the index G/N, where G is the number of isolates with the same BOX-banding patterns and N the total number of isolates. The Shannon index of diversity was calculated using the formula: $H = -\sum(G/N)\ln(G/N)$ [30]. The abundance of each genotype was calculated as the number of isolates in each genotypic group divided by total number of isolates. To determine how sampling effort affected these estimates, rarefaction curves were constructed comparing the number of isolates versus number of observed genotypes using ECOSIM (version 7.72) [31].

2.9.2. Diversity Calculations and Phylotypes. Given the small sample size, and in order to evaluate diversity differences between summer and winter populations correcting for this, we constructed rarefaction curves [32] for the abundance of phylotypes (lineages) using ECOSIM (version 7.72) [31]. We also estimated the actual number of lineages (phylotypes) that may be present in the sample, by the calculation of a nonparametric Chao1 richness estimator using estimates 8.2.0 [33, 34].

To statistically determine the existence of two populations (summer and winter), we constructed a contingency table with the frequencies of lineages for the different sampling seasons and used a G test to evaluate the significance of our frequency distribution of lineages [35]. Finally, we performed a generalization of Fisher's exact test as using the Fisher test routine as provided in the R statistical package, using the simulate P value = TRUE flag.

2.9.3. Comparisons of Growth Rates. Differences in growth rates at different temperatures were observed between summer and winter populations (*P. cuatrocienegasensis* and *P. otitidis*, resp.). To evaluate the statistical significance of these differences we performed a one-way analysis of variance as implemented in the R statistical package, using the function one-way test.

3. Results

To characterize the diversity of natural *Pseudomonas* isolates and its changes associated with seasonality in a CCB water system, we sampled a desiccation lagoon subject to marked seasonal fluctuations. Cultures were obtained from surface water samples in four sampling events (two summers, two winters). Individual isolates (70) were genotyped and temporal structure of the total sample analyzed.

3.1. Genetic Structure of Populations (Genotypic Diversity). Genotypic diversity was measured through genomic fingerprinting for each isolate using BOX-PCR technique, which permits the identification of individual clones, and each unique pattern was considered a different genotype. We chose a similarity value of 90% or more to indicate strains of the same (or very similar) genotype. Very similar or identical banding patterns have been demonstrated to have the same genotype and identical 16S rRNA gene sequences [23]. Cluster analysis resulted in a total of 35 representative genotypes (Figure 2). We identified 9 genotypes (15 isolates) from August 2003, 7 genotypes (31 isolates) from January

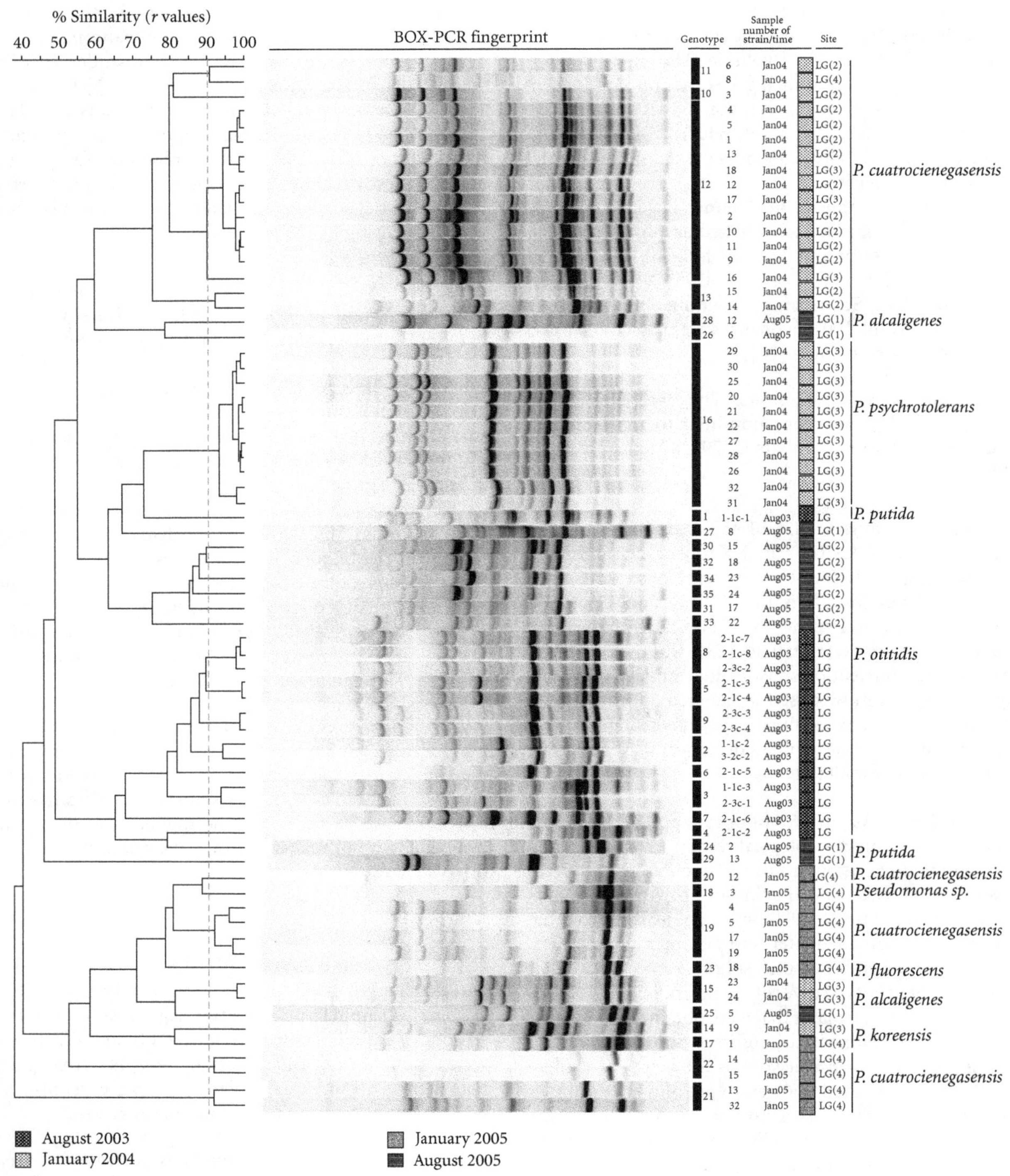

FIGURE 2: *Pseudomonas* isolates cluster analysis of genetic similarity. BOX-PCR genomic fingerprints of individual isolates were analyzed and grouped using product-moment UPGMA algorithm. A similarity value (r) of 90% was used to determine the same genotypes (dashed line).

2004, 7 genotypes (12 isolates) from January 2005, and 12 genotypes (12 isolates) from August 2005.

The genotypic diversity calculated for the total sample (70 *Pseudomonas* isolates) and all estimates derived from this sample should be taken with caution given the small sample size. It has been said that the standard diversity description of the sample indicates that Shannon index (H) is 3.14. Additional analyses include the observation that genotypic diversity was heterogeneously distributed in the different samples. In January 2004, we observed the lowest diversity ($G/N = 0.22$) and the highest number of isolates having the same genotypic pattern (12 strains having the same genotype). While, in August 2005, we observed the highest diversity with 12 isolates out of 12 unique genotypes ($G/N = 1$). All genotypes were found to be unique to one sample occasion (Figure 2). Even when we applied a cutoff value of 80% to define clusters, the majority of genotypes (92.9%) were collected only once, except for two genotypes that included isolates from different sampling occasions. Rarefaction analysis showed that more sampling is needed to gain confidence on the genotype diversity present (data not shown). Thus, these observations are only suggestive of not reoccurrence of genotypes from year to year.

3.2. Seasonal Changes (P. otitidis and P. cuatrocienegasensis). Phylogenetic diversity was defined by the identification of species or lineages as unique 16S rDNA sequences. To determine the seasonal structure of lineages, the 16S rDNA was sequenced from all the unique genotypes as identified by fingerprinting. The Neighbor-joining genealogy of 16S rDNA sequences represents an estimate of the phylogenetic relationship of the 35 genotypes identified by BOX-PCR and is shown in Figure 3. Using a 97% sequence similarity cutoff for the 16S rDNA sequences, the data revealed two numerically dominant clusters. The first cluster (8 sequences representing 24 strains of the total sample) is closely related to *P. cuatrocienegasensis* [5] and was isolated exclusively in winter samples (January 2004 and January 2005), while the second cluster (15 sequences representing 21 strains of the total sample) is closely related to *P. otitidis* and was isolated exclusively in summer samples (August 2003 and August 2005). The seasonal reappearance of phylotypes, identified by 16S rDNA sequences, was not observed at the BOX-PCR fingerprinting level, since all the patterns were different from one sample occasion to the other (Figure 2). These results show that there is seasonal reoccurrence of specific lineages in this site, but the populations that define them have different genotypic composition from one year to the next.

We also analyzed the possibility that the two distinct populations (summer and winter) were not statistically different in terms of the observed diversity, by correcting for sampling size using rarefaction curves. The resulting curves show sampling saturation and that the two populations truly differ in diversity levels (Figure 4). In accordance with rarefaction results, Chao1 richness indices show that the observed number of lineages will not change significantly with more sampling (Table 1).

Additionally, we performed a generalized Fisher's test and a G test of independence. Fisher's test was done to evaluate the statistical significance of a seasonal effect on the distribution of phylotypes, as based upon a contingency table. We observed a strongly statistically significant result ($P = 0.0004998$), indicating that the probability of observing the particular arrangement of lineages/seasons by chance is extremely small. The G test was done to evaluate the association of phylotypes to sampling seasons and indicated that the probability of finding a particular phylotype is highly dependent on the season ($G = 108.92$; df = 24; $P = 8.6\times 10^{-13}$). These results indicate that the observed seasonal distribution of lineages is statistically significant and is not likely due to random events.

Table 1: Diversity estimates of culturable *Pseudomonas* populations. Total and sample occasion diversity are indicated. Observed diversity in terms of total number of different lineages is indicated as S_{obs} and nonparametric richness estimator of the actual number of lineages is indicated as S_{Chao1}.

Sample occasion	Number of isolates	S_{obs}	S_{Chao1} (SD)
Summer	**27**	**2**	**2 (0)**
August 2003	15	1	1 (0)
August 2005	12	2	2 (0.05)
Winter	**43**	**8**	**9.33 (0.92)**
January 2004	31	5	7 (3.74)
January 2005	12	6	9.6 (7.19)
Total	**70**	**8**	**10 (3.74)**

Finally, we explored the possibility that *P. cuatrocienegasenesis* and *P. otitidis* populations may differ in their maximum growth rates at different temperatures that can be experienced during summer time (28, 32, 36, 40, and 44°C). We followed the same approach as [36]. We observed that, on average, differences between populations are statistically significant only at 40°C, where *P. otitidis* "summer lineage" grows faster than *P. cuatrocienegasensis* "winter lineage" (Figure 5).

4. Discussion

In CCB there is an extraordinary microbial biodiversity, and each site seems to be unique [5–7, 14, 37–39]. As in other places, even if the diversity is high, most of it remains unreachable by traditional culture approaches. Some culturable groups such as *Pseudomonas, Bacillus, Exiguobacterium*, and other Firmicutes [6, 13] are an exception. We have found these groups being in high numbers in clone libraries and metagenomes from environmental samples [38, 39] and also have been able to culture them in the laboratory. The microbial diversity information from CCB comes mainly from the study of water systems and ponds, most of which are subject to seasonal fluctuations [1], and nothing is known of the biodiversity changes that occur associated with these environmental cycles. The present study is part of this exploration focusing on the genus *Pseudomonas* and seasonality.

4.1. Genetic Structure of Populations (Genotypic Diversity). BOX-PCR fingerprint analysis and 16S rDNA sequences of all the unique BOX-PCR genotypes were used to determine the temporal structure of the sampled populations. Our results revealed that half of the total number of genotypes were unique ($G/N = 0.5$). This diversity value is relatively low in comparison with reported values for *Escherichia coli* ($G/N = 0.73$; [40]). However, undersampling, shown by rarefaction curves (data not shown), calls for caution in the interpretation of diversity calculations at at the genotype level.

4.2. Seasonal Changes (P. otitidis and P. cuatrocienegasensis). Characterization of the phylogenetic diversity leads to the

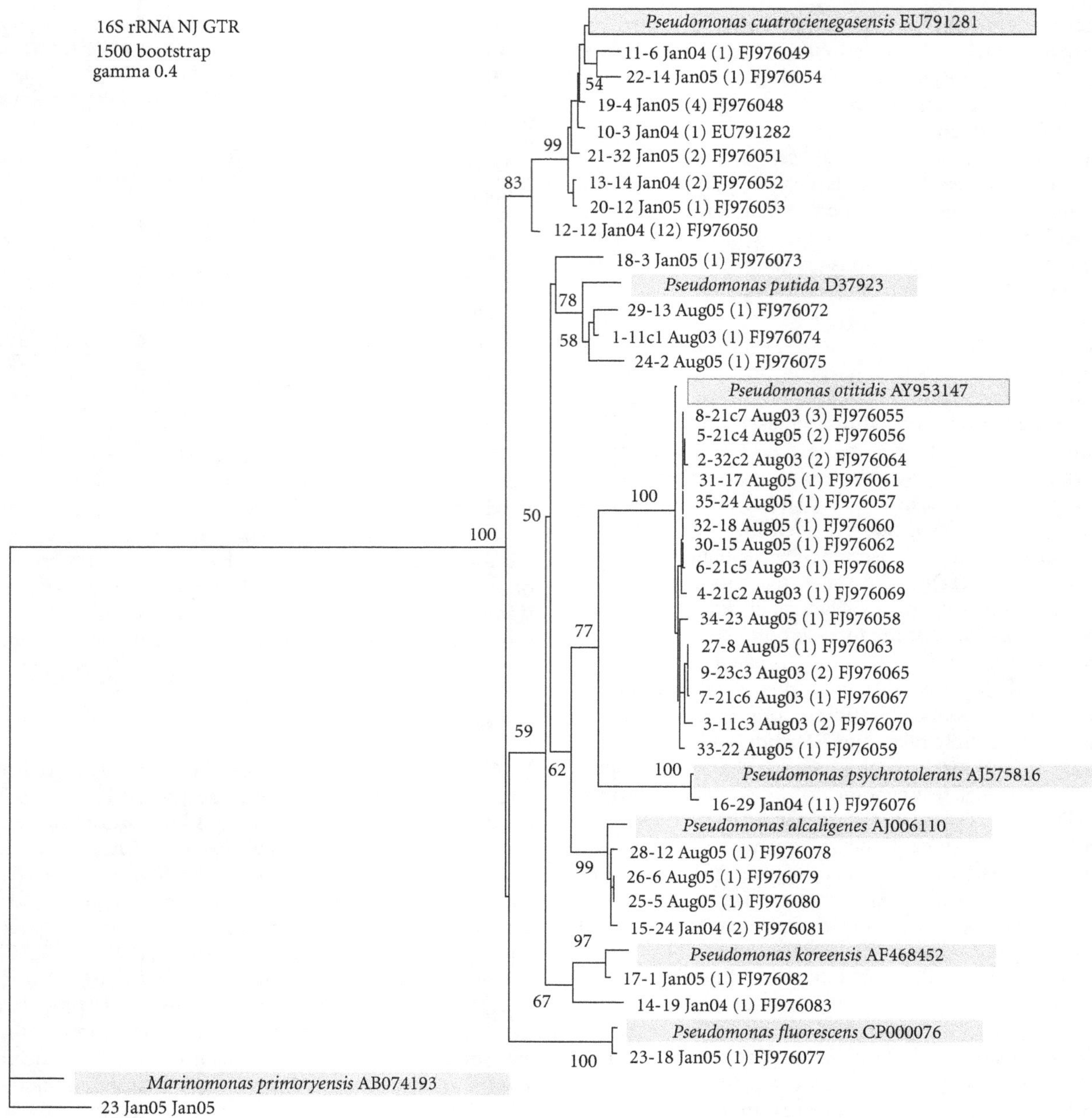

FIGURE 3: Neighbor-joining tree (GTR, gamma correction 0.40) of 16S rDNA sequences. Sequences are from *Pseudomonas* isolates from Laguna Grande water samples taken in August 2003, January 2004, January 2005, and August 2005. The number of strains having identical fingerprint pattern is in parentheses. The numbers at the nodes are bootstrap values based on 1500 resamplings.

finding of seasonal structure of two numerically dominant lineages: *P. cuatrocienegasensis* and *P. otitidis*. Although diversity may be underestimated at the genotype level due to reduced sample size, we were able to test statistically the correlation between genetic structure and seasonality with a G test of independence, a generalized Fisher test, through sampling size correction via rarefaction curves analysis, and by the estimation of the expected richness with nonparametric richness estimator Chao1. G test of independence and generalized Fisher test indicate that phylotype (species or lineage) identity is not independent of sampling season ($G = 108.92$; df $= 24$; $P = 8.6 \times 10^{-13}$) and that probability of observing the particular arrangement of lineages/seasons by chance is extremely small ($P = 0.0004998$). Rarefaction curves of [41] winter and summer populations showed differentiation between the two and a saturation of diversity for summer samples, giving evidence that both populations differ significantly in their diversity levels (Figure 4).

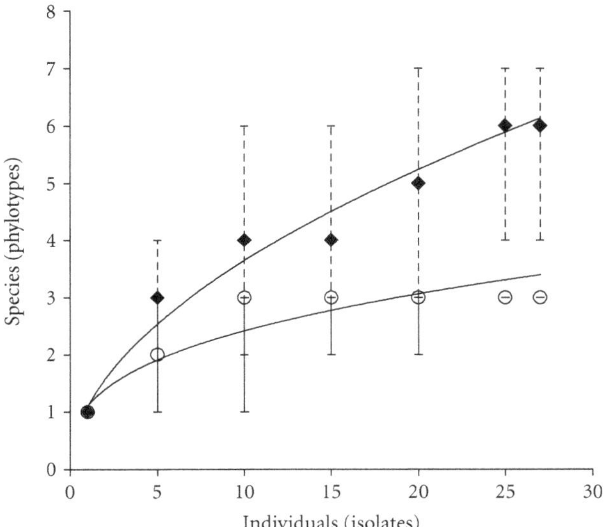

FIGURE 4: Rarefaction curve constructed with the abundances of the different phylotypes (16S rDNA) in winter and summer samples. Error bars represent the confidence limits for the distributions. ♦ Winter samples (January 2004 and 2005); O summer samples (August 2003 and 2005).

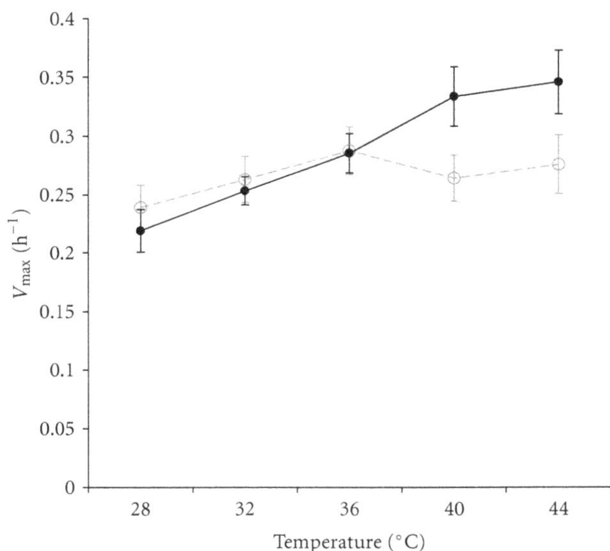

FIGURE 5: Maximum growth rate, V_{max}, of a subset of summer (solid line) and winter (dashed line) populations as a function of temperature. Error bars represent 95% confidence intervals, based on 3-fold replication and number of genotypes (9 summer, 6 winter).

Finally, expected richness indices (Chao1) do not deviate significantly from the observed number of lineages (Table 1). Altogether, these tests indicate that in fact winter and summer populations are statistically different both in their composition and in their diversity levels.

Other studies have found similar patterns in leaf-associated fluorescent pseudomonad populations [41]. Using restriction fragment length polymorphism (RFLP) of leaves samples taken monthly over 3-year period, they found seasonal reappearance of long-term survival ribotypes [41]. In our study, although we were able to discern a seasonal pattern on lineage composition, the factors causing this pattern are more difficult to determine unambiguously. One obvious factor that can be involved in the maintenance of different populations across seasons is temperature. As an attempt to examine this hypothesis, we measured maximum growth rates of the most abundant lineages (*P. otitidis* and *P. cuatrocienegasensis*) at different temperatures. As expected, *P. otitidis* grew faster than *P. cuatrocienegasensis* at high temperatures, but this differential growth was only statistically significant at 40°C. This result provides a clue that temperature can be a relevant environmental factor affecting growth and persistence of isolates, The presented growth rate experiments are far from definitive and must be interpreted with caution, as laboratory conditions invariably differ from the environment in multiple ways beyond that being investigated [42], besides the fact that other environmental parameters that can be associated with temperature changes need to be investigated as well. Nonetheless, these experiments give a good perspective of what can be further done to investigate the factors involved in the observed genetic structure associated with seasonality. We consider that detailed investigation of the physiological responses over a wider temperature range, using more lineages and do measurements with competing isolates, is needed to advance knowledge into the causes of the observed genetic structure of the studied populations.

Another potential explanation for the observed seasonal pattern can be found in the documented transition of certain bacteria into a dormancy state triggered by unfavourable environmental conditions such as oxygen and temperature stress or resource limitation. A recent study by Jones and Lennon [43] demonstrates that only some taxa of the total bacterial community in various lakes were in an active state, and the rest were in a dormant state triggered by environmental stress. Although members of the genus *Pseudomonas* do not form spores, they could enter reversible states of reduced metabolic activity described as viable but nonculturable (VBNC) [44]. Thus, a dormancy/VBNC state could explain the observed seasonal pattern, without excluding other ecological mechanisms (i.e., adaptation). This possibility is one of the limitations that culture-dependent-techniques can have when characterizing microbial diversity. However, several culture-independent techniques have found similar patterns [45] suggesting that the seasonal shifts and reoccurrences of bacterial populations or microbial functional groups occur in the bacterial aquatic communities and, therefore, are not an artefact of the culture-dependent techniques or microbiological procedures [46, 47]. Research in CCB aquatic habitats, including other culturable and nonculturable groups, has recently been published [6] or soon to be [12, 38] that will contribute to determine the generality of the observations here presented.

While we found that lineages or phylotypes (16S rDNA sequences) are seasonally recurrent, genotypes (isolate fingerprints) within each lineage are not, leading to a different genotype composition each year. Despite that undersampling was verified at the genotype level (rarefaction), correcting for sample size at the lineage level, it still gave evidence of differences between summer and winter samples (Figure 4; Table 1). Looking at seasonality on phylotype composition and taking cautiously genotypic composition (fingerprints), we see three possible explanations for our observations: (1) selection associated with seasonality, (2) neutral or stochastic fixation of different genotypes or lineages each season, and (3) artefact due to limited sample size at each sampling date. Given the strong association of phylotypes to sampling season, the selection-mediated possibility is favoured over a purely stochastic explanation. The fact that we do not recover identical fingerprint patterns is debatable due to undersampling and cannot be interpreted as evidence of selective sweeps [47], or simple rapid diversification of bacteria after each seasonal change unless more isolates are analyzed. The third possible explanation relates to the second and implies that genotypes previously "unseen" are present in low numbers; however, this will not necessarily contradict the possibility of seasonal selection acting as an ecological process occurring.

5. Conclusion

We showed that the simultaneous utilization of phylogenetic markers and genomic fingerprinting can be used to characterize diversity changes across seasons, and to formulate hypotheses about the potential mechanisms that structure populations. Future experiments that include more phylogenetic groups, larger samples, over extended periods of time, and in controlled laboratory conditions will be necessary to test these hypotheses and further investigate the role of seasonality in the maintenance of lineage (or species) diversity and bacterial diversification in CCB.

The results presented here are the first temporal characterization of the biological composition and dynamics of microorganisms at the CCB study site. The strong correlation of seasonality with the lineage composition contributes with information to formulate future experiments that test hypothesis on the mechanisms involved in the origins and maintenance of microbial diversity in the area.

Acknowledgments

This research was supported by Grants from SEMARNAT (0237, 23459) and CONACyT SEP (44673 and 57507) to V. Souza and L. E. Eguiarte. A. E. Escalante was supported by CONACyT-UNAM scholarship. The authors thank R. González-Chauvet and Modern American School for help in sample collection; E. Carson, L. Espinosa-Asuar, and M. G. Rosas for technical assistance; P. Vinuesa for help with BOX-PCR pattern analysis; L. Falcón and M. Travisano for ideas and revision of earlier versions of the manuscript. Finally they thank R. Cerritos for his collaboration and helpful discussion.

References

[1] W. Minckley, *Environments of the Bolson of Cuatro Cienegas, Coahuila, Mexico*, vol. 2 of *(El Paso) Science Series*, University of Texas, 1969.

[2] J. J. Elser, J. H. Schampel, F. Garcia-Pichel et al., "Effects of phosphorus enrichment and grazing snails on modern stromatolitic microbial communities," *Freshwater Biology*, vol. 50, no. 11, pp. 1808–1825, 2005.

[3] A. Contreras-Arquieta, "New records of the snail Melanoides tuberculata (Muller, 1774) (Gastropoda: Thiaridae) in the Cuatro Cienegas Basin, and its distribution in the state of Coahuila, Mexico," *Southwestern Naturalist*, vol. 43, no. 2, pp. 283–286, 1998.

[4] V. Souza, L. Espinosa-Asuar, A. E. Escalante et al., "An endangered oasis of aquatic microbial biodiversity in the Chihuahuan desert," *Proceedings of the National Academy of Sciences of the United States of America*, vol. 103, no. 17, pp. 6565–6570, 2006.

[5] A. E. Escalante, J. Caballero-Mellado, L. Martínez-Aguilar et al., "*Pseudomonas cuatrocienegasensis* sp. nov., isolated from an evaporating lagoon in the Cuatro Ciénegas valley in Coahuila, Mexico," *International Journal of Systematic and Evolutionary Microbiology*, vol. 59, no. 6, pp. 1416–1420, 2009.

[6] R. Cerritos, L. E. Eguiarte, M. Avitia et al., "Diversity of culturable thermo-resistant aquatic bacteria along an environmental gradient in Cuatro Ciénegas, Coahuila, México," *Antonie van Leeuwenhoek, International Journal of General and Molecular Microbiology*, vol. 99, no. 2, pp. 303–318, 2011.

[7] G. Bonilla-Rosso, M. Peimbert, G. Olmedo et al., "Microbial mat metagenomes reveal common patterns in microbialite community structure and composition," *Astrobiology*, vol. 12, no. 7, pp. 659–673, 2012.

[8] A. E. Escalante, L. E. Eguiarte, L. Espinosa-Asuar, L. J. Forney, A. M. Noguez, and V. Souza Saldivar, "Diversity of aquatic prokaryotic communities in the Cuatro Cienegas basin," *FEMS Microbiology Ecology*, vol. 65, no. 1, pp. 50–60, 2008.

[9] J. Toribio, A. E. Escalante, J. Caballero-Mellado et al., "Characterization of a novel biosurfactant producing *Pseudomonas koreensis* lineage that is endemic to Cuatro Ciénegas Basin," *Systematic and Applied Microbiology*, vol. 34, no. 7, pp. 531–535, 2011.

[10] A. Moreno-Letelier, G. Olmedo, L. E. Eguiarte, and V. Souza, "Divergence and phylogeny of Firmicutes from the Cuatro Cienegas Basin, Mexico: a window to an ancient ocean," *Astrobiology*, vol. 12, no. 7, pp. 674–684, 2012.

[11] D. M. Olson and E. Dinerstein, "The global 200: priority ecoregions for global conservation," *Annals of the Missouri Botanical Garden*, vol. 89, no. 2, pp. 199–224, 2002.

[12] E. A. Rebollar, M. Avitia, L. E. Eguiarte et al., "Water-sediment niche differentiation in ancient marine lineages of *Exiguobacterium* endemic to the Cuatro Cienegas Basin," *Environmental Microbiology*, vol. 14, no. 9, pp. 2323–2333, 2012.

[13] V. Souza, L. E. Eguiarte, J. Siefert, and J. J. Elser, "Microbial endemism: does phosphorus limitation enhance speciation?" *Nature Reviews Microbiology*, vol. 6, no. 7, pp. 559–564, 2008.

[14] V. Souza, J. Siefert, A. E. Escalante et al., "The Cuatro Cienegas Basin in Coahuila, Mexico: and astrobiological Precambrian park," *Astrobiology*, vol. 12, no. 7, pp. 641–647, 2012.

[15] P. Chesson, "Mechanisms of maintenance of species diversity," *Annual Review of Ecology and Systematics*, vol. 31, pp. 343–366, 2000.

[16] E. Carson, M. Tobler, W. Minckley et al., "Relationships between spatio-temporal environmental and genetic variation

reveal and important influence of exogenous selection in a pupfish hybrid zone," *Molecular Ecology*, vol. 21, pp. 1209–1222, 2012.

[17] E. Carson, *Hybridization between Cyprinodon atrorus and C. bifasciatus: history, patterns and dynamics [Ph.D. thesis]*, Arizona State University, Tempe, Ariz, USA, 2005.

[18] G. Kielwein, "Die isolierung und differenzierung von Pseudomonaden aus lebensmittel," *Archiv für Lebensmittelhygiene*, vol. 22, pp. 29–37, 1971.

[19] J. Versalovic, T. Koeuth, and J. R. Lupski, "Distribution of repetitive DNA sequences in eubacteria and application to fingerprinting of bacterial genomes," *Nucleic Acids Research*, vol. 19, no. 24, pp. 6823–6831, 1991.

[20] B. G. Häne, K. Jäger, and H. G. Drexler, "The Pearson product-moment correlation coefficient is better suited for identification of DNA fingerprint profiles than band matching algorithms," *Electrophoresis*, vol. 14, no. 10, pp. 967–972, 1993.

[21] J. Rademaker and F. de Brujin, "Characterization and classification of microbes byr rep-PCR genomic fingerprinting and computer assited pattern analysis," in *DNA Markers: Protocols, Applications and Overviews*, G. Caetano-Anollés and P. Gresshoff, Eds., pp. 151–171, Wiley & Sons, New York, NY, USA, 1997.

[22] Y. Oda, W. Wanders, L. A. Huisman, W. G. Meijer, J. C. Gottschal, and L. J. Forney, "Genotypic and phenotypic diversity within species of purple nonsulfur bacteria isolated from aquatic sediments," *Applied and Environmental Microbiology*, vol. 68, no. 7, pp. 3467–3477, 2002.

[23] D. J. Lane, "16S/23S rDNA sequencing," in *Nucleic Acid Techniques*, E. Stackebrandt and M. Goodfellow, Eds., pp. 115–175, John Wiley & Sons, New York, NY, USA, 1991.

[24] C. T. Sacchi, A. M. Whitney, L. W. Mayer et al., "Sequencing of 16S rRNA gene: a rapid tool for identification of *Bacillus anthracis*," *Emerging Infectious Diseases*, vol. 8, no. 10, pp. 1117–1123, 2002.

[25] D. Gordon, C. Abajian, and P. Green, "Consed: a graphical tool for sequence finishing," *Genome Research*, vol. 8, no. 3, pp. 195–202, 1998.

[26] J. D. Thompson, D. G. Higgins, and T. J. Gibson, "CLUSTAL W: improving the sensitivity of progressive multiple sequence alignment through sequence weighting, position-specific gap penalties and weight matrix choice," *Nucleic Acids Research*, vol. 22, no. 22, pp. 4673–4680, 1994.

[27] T. M. Keane, C. J. Creevey, M. M. Pentony, T. J. Naughton, and J. O. McInerney, "Assessment of methods for amino acid matrix selection and their use on empirical data shows that ad hoc assumptions for choice of matrix are not justified," *BMC Evolutionary Biology*, vol. 6, article 29, 2006.

[28] D. Swofford, *PAUP: Phylogenetic Analysis Using Parsimony*, 2000.

[29] R. M. Atlas and R. Bartha, *Microbial Ecology: Fundamentals and Applications*, Benjamin/Cummings, 1993.

[30] N. J. Gotelli and G. L. Entsminger, "EcoSim 7. 72," Acquired Intelligence, http://www.uvm.edu/~ngotelli/EcoSim/EcoSim.html.

[31] N. J. Gotelli and R. K. Colwell, "Quantifying biodiversity: Procedures and pitfalls in the measurement and comparison of species richness," *Ecology Letters*, vol. 4, no. 4, pp. 379–391, 2001.

[32] A. Chao, "Non parametric estimation of the number of classes in a population," *Scandinavian Journal of Statistics*, vol. 11, pp. 265–270, 1984.

[33] R. K. Colwell, "EstimateS: statistical estimation of species richness and shares species from samples," 2009, http://purl.oclc.org/estimates.

[34] R. Sokal and F. Rohlf, *Biometry: The Principles and Practice of Statistics in Biological Research*, WH Freeman, New York, NY, USA, 1995.

[35] V. S. Cooper, A. F. Bennett, and R. E. Lenski, "Evolution of thermal dependence of growth rate of *Escherichia coli* populations during 20,000 generations in a constant environment," *Evolution*, vol. 55, no. 5, pp. 889–896, 2001.

[36] C. Desnues, B. Rodriguez-Brito, S. Rayhawk et al., "Biodiversity and biogeography of phages in modern stromatolites and thrombolites," *Nature*, vol. 452, no. 7185, pp. 340–343, 2008.

[37] M. Peimbert, G. Bonilla-Rosso, G. Olmedo et al., "Comparative metagenomics of microbialites of Cuatro Cienegas: a window to Precambrianfunction," *Astrobiology*, vol. 12, no. 7, pp. 648–658, 2012.

[38] N. E. López-Lozano, G. Bonilla-Rosso, F. García-Oliva et al., "Bacterial communities and nitrogen cycle in the gypsum soil in Cuatro Cienegas Basin, Coahuila," *Astrobiology*, vol. 12, no. 7, pp. 699–709, 2012.

[39] L. G. D. A. Borges, V. Dalla Vechia, and G. Corção, "Characterisation and genetic diversity via REP-PCR of *Escherichia coli* isolates from polluted waters in southern Brazil," *FEMS Microbiology Ecology*, vol. 45, no. 2, pp. 173–180, 2003.

[40] R. J. Ellis, I. P. Thompson, and M. J. Bailey, "Temporal fluctuations in the pseudomonad population associated with sugar beet leaves," *FEMS Microbiology Ecology*, vol. 28, no. 4, pp. 345–356, 1999.

[41] C. M. Jessup, R. Kassen, S. E. Forde et al., "Big questions, small worlds: microbial model systems in ecology," *Trends in Ecology and Evolution*, vol. 19, no. 4, pp. 189–197, 2004.

[42] S. E. Jones and J. T. Lennon, "Dormancy contributes to the maintenance of microbial diversity," *Proceedings of the National Academy of Sciences of the United States of America*, vol. 107, no. 13, pp. 5881–5886, 2010.

[43] H. S. Xu, N. Roberts, and F. L. Singleton, "Survival and viability of nonculturable *Escherichia coli* and *Vibrio cholerae* in the estuarine and marine environment," *Microbial Ecology*, vol. 8, no. 4, pp. 313–323, 1982.

[44] M. A. J. Hullar, L. A. Kaplan, and D. A. Stahl, "Recurring seasonal dynamics of microbial communities in stream habitats," *Applied and Environmental Microbiology*, vol. 72, no. 1, pp. 713–722, 2006.

[45] S. D. Sutton and R. H. Findlay, "Sedimentary microbial community dynamics in a regulated stream: East Fork of the Little Miami River, Ohio," *Environmental Microbiology*, vol. 5, no. 4, pp. 256–266, 2003.

[46] J. A. Fuhrman, I. Hewson, M. S. Schwalbach, J. A. Steele, M. V. Brown, and S. Naeem, "Annually reoccurring bacterial communities are predictable from ocean conditions," *Proceedings of the National Academy of Sciences of the United States of America*, vol. 103, no. 35, pp. 13104–13109, 2006.

[47] J. A. G. M. de Visser and D. E. Rozen, "Clonal interference and the periodic selection of new beneficial mutations in *Escherichia coli*," *Genetics*, vol. 172, no. 4, pp. 2093–2100, 2006.

8

Candida albicans versus *Candida dubliniensis*: Why Is *C. albicans* More Pathogenic?

Gary P. Moran, David C. Coleman, and Derek J. Sullivan

Division of Oral Biosciences, Dublin Dental University Hospital, Trinity College Dublin, Dublin 2, Ireland

Correspondence should be addressed to Derek J. Sullivan, derek.sullivan@dental.tcd.ie

Academic Editor: Julian R. Naglik

Candida albicans and *Candida dubliniensis* are highly related pathogenic yeast species. However, *C. albicans* is far more prevalent in human infection and has been shown to be more pathogenic in a wide range of infection models. Comparison of the genomes of the two species has revealed that they are very similar although there are some significant differences, largely due to the expansion of virulence-related gene families (e.g., *ALS* and *SAP*) in *C. albicans*, and increased levels of pseudogenisation in *C. dubliniensis*. Comparative global gene expression analyses have also been used to investigate differences in the ability of the two species to tolerate environmental stress and to produce hyphae, two traits that are likely to play a role in the lower virulence of *C. dubliniensis*. Taken together, these data suggest that *C. dubliniensis* is in the process of undergoing reductive evolution and may have become adapted for growth in a specialized anatomic niche.

1. Introduction

Fungi are an important cause of human infection, and yeast species of the genus *Candida* are the most pathogenic fungi. While most *Candida* species are found in the environment, approximately a dozen or so are associated with colonization and infection of humans [1]. *Candida* species are common commensals of the oral cavity, intestinal tract and vagina, with newborns being colonized soon after birth. While these species are innocuous in most individuals, under certain circumstances they can opportunistically overgrow and cause a variety of diseases [2]. These diseases range from superficial infections of the vaginal and oral mucosae, to life-threatening systemic infections that can spread via the bloodstream to organs throughout the body. The risk factors for candidal vaginitis are poorly understood; however, other candidal infections are largely the result of host-related defects. These include depletion of CD4 T cells in HIV-infected individuals, which predisposes to oropharyngeal candidosis, or neutropenia and intestinal surgery, both of which are significant risk factors for systemic infection [1–3].

Candida albicans is widely recognized as being the most pathogenic yeast species and in the majority of epidemiological studies has been found to be the most common cause of superficial and systemic infections. Other species, such as *Candida glabrata, Candida parapsilosis,* and *Candida tropicalis* have also been associated with most forms of candidiasis and the relative distribution of each species can vary depending on geographic location, patient cohort, and previous exposure to antifungal drugs [2, 4]. In 1995, a new *Candida* species was identified in HIV-infected individuals with oropharyngeal candidosis in Dublin, Ireland [5]. This species, which was subsequently named *Candida dubliniensis*, is very closely related to *C. albicans* with which it shares many phenotypic properties, including the ability to produce hyphae and chlamydospores, traits previously specifically associated only with *C. albicans* [6–8]. Phylogenetic studies indicate that *C. dubliniensis* is the species that is most closely related to *C. albicans*, and it is often quite difficult to discriminate between the two species in clinical samples [9, 10]. Indeed it was only when DNA fingerprinting techniques were applied to the large-scale analysis of *C. albicans* populations in epidemiological studies that the first isolates of *C. dubliniensis* were originally identified [5]. Surprisingly, despite the close phylogenetic relationship of the two species epidemiological data show that *C. albicans* is far more prevalent than *C. dubliniensis*. In particular, in most analyses of systemic infection, *C. albicans* is found in >50% of cases, while if it is identified at all, *C. dubliniensis* has only been found in at most 2-3% of cases [11–13]. This apparent discrepancy between the ability of the two species to cause infection is also reflected in data obtained from comparative studies in a wide range of infection models (e.g., systemic and mucosal) which clearly show that *C. albicans* is significantly more pathogenic than *C. dubliniensis* [10, 14–18].

The identification of virulence-associated factors in *Candida* species is complicated by the fact that they are opportunistic pathogens that usually exist in harmony with the human host as part of the commensal flora and only cause infection when host deficiencies permit. Since it is by far the most pathogenic *Candida* species, *C. albicans* is the best-studied member of the genus in terms of pathogenesis. The most commonly cited *C. albicans* virulence factors include adhesins (e.g., Hwp1 [19] and the Als family [20]), extracellular enzymes (e.g., the secreted aspartyl proteinase (Sap) family [21] and phospholipases [22]), and most importantly of all, the ability to alternate between unicellular yeast and filamentous hyphal forms of growth [23]. Both morphological forms have been shown to be essential for virulence. Hyphae have been proposed to play a major role in adhesion, invasion, and biofilm formation while yeast cells are likely to be important for dissemination and initial colonization of host surfaces [24]. Comparative phenotypic analysis of *C. albicans* and *C. dubliniensis* has suggested that *in vitro* isolates of *C. dubliniensis* exhibit higher levels of proteinase activity, are more adherent to buccal epithelial cells, and undergo phenotypic switching at a higher rate than *C. albicans* [10, 25–27]. In addition, as described earlier, *C. dubliniensis* is the only *Candida* species, other than *C. albicans* that is able to produce hyphae [5, 6]. Given the close relationship between the two species and the fact that they are so alike phenotypically, at first glance, it is difficult to understand why there is such disparity in the capacity of *C. albicans* and *C. dubliniensis* to colonise and cause disease in humans. This short review appraises recent findings that help to clarify this conundrum and explain how *C. albicans* appears to have evolved to be a better commensal and opportunistic pathogen than *C. dubliniensis*.

2. Comparative Genomic Analysis of *C. albicans* and *C. dubliniensis*

The *C. albicans* genome sequence was first published in 2004 [28], with improved annotation and analysis subsequently reported in 2005 [29] and 2007 [30]. In an early attempt to identify genomic differences that might serve to explain the disparity in the virulence of *C. albicans* and *C. dubliniensis*, Moran et al. cohybridized genomic DNA from each species to *C. albicans* whole genome microarrays to identify genes that are only present in *C. albicans* [31]. This relatively crude experiment suggested that there are 247 (approx. 4%) *C. albicans* genes that are either absent or highly (i.e., >60%) divergent in the *C. dubliniensis* genome. Interestingly, several genes strongly associated with *C. albicans* virulence are included in the list of absent/divergent genes. In 2009, in order to further investigate the genetic differences between the two species the Wellcome Trust Sanger Institute sequenced the entire *C. dubliniensis* genome [32]. Comparison of the two genome sequences revealed that, despite major karyotypic differences, the genomes of the two species are remarkably similar with 96.3% of genes exhibiting >80% identity, while 98% of genes are syntenic, thus, confirming the very close phylogenetic relationship and the relatively recent divergence of the two species (estimated to have occurred approx. 20 million years ago [33]). When transposable elements were discounted, comparison of the two genome sequences revealed that there are 29 *C. dubliniensis*-specific genes and 168 *C. albicans*-specific genes. The majority of the differences observed between the two species can be accounted for by the expansion of gene families in *C. albicans*, many of which have been previously associated with virulence. In particular, genes missing from the *C. dubliniensis* genome include those encoding hypha-specific virulence factors, such as the cell surface proteins Hyr1 and Als3 and two members of the secreted aspartyl proteinase family (i.e., Sap5 and Sap6), while the gene encoding the well-characterized epithelial adhesin Hwp1 is highly divergent [31, 32]. Hyr1 has been shown to confer resistance to neutrophil killing activity [34] and, along with Hwp1, has been shown recently to play an important role in oral mucosal biofilm formation [35]. Als3 has been shown to play an important role in adhesion to host cells and has been shown to have invasin-like [36] and iron-sequestering [37] activity, while the Saps are well-known virulence factors [21]. The biggest difference in gene family size between the two species is the TeLOmere-associated (TLO) family which is comprised of 14 genes in *C. albicans,* but only two genes in *C. dubliniensis.* Sequence comparisons suggest that the TLO genes encode transcriptional regulators, and preliminary analysis of the phenotype of *C. dubliniensis* Δ*tlo* mutants suggests that these genes may play a role in the control of hypha formation [32]. In addition to these differences, a range of genes appear to be in the process of being lost by *C. dubliniensis.* There are 78 *C. dubliniensis* pseudogenes with intact positional orthologs in *C. albicans*, including genes identified as filamentous growth regulators (FGR) in haploinsufficiency studies [38]. These findings suggest that *C. dubliniensis* is undergoing a process of reductive evolution leading to the loss of genes that have been associated with *C. albicans* virulence. Interestingly, many of these genes are only expressed by the hyphal form of growth and are likely to play a prominent role in host-pathogen interaction.

One of the most prominent phenotypic differences between *C. albicans* and *C. dubliniensis* is their different capacity to tolerate environmental stress, with the former being far more tolerant of thermal, osmotic, and oxidative stress [5, 14, 39, 40]. Indeed, comparative growth at 45°C is commonly used as a simple diagnostic test to discriminate between the two species [41]. Comparative transcriptional profiling analysis revealed that although the two species express similar core stress responses, *C. dubliniensis* mounts a more robust response to thermal stress and a very poor transcriptional response to oxidative and osmotic stress [39]. Forward genetic screens using a *C. albicans* library to try and identify genes that might increase the tolerance of *C.*

dubliniensis to environmental stress failed to identify any single gene that could complement oxidative and thermal sensitivity, suggesting that these are likely to be polygenic traits. However, the *C. albicans ENA21* gene, which encodes a sodium efflux pump, was found to increase the salt tolerance of *C. dubliniensis* [39]. Since the *C. dubliniensis* ortholog of this gene appears to be functional but not upregulated in response to the presence of salt, it is likely that the differential salt stress susceptibility of the two species is due to differences in stress-related transcriptional regulatory pathways.

3. Comparative Analysis of Hypha Formation by *C. albicans* and *C. dubliniensis*

One of the most important and best-studied virulence factors of *C. albicans* is its ability to switch between yeast and filamentous growth forms (i.e., dimorphism), a trait also shared by *C. dubliniensis* [5]. However, although *C. dubliniensis* is capable of producing germ tubes and true hyphae, it does so far less efficiently than *C. albicans,* both *in vivo* and under a wide range of *in vitro* conditions [16, 42, 43]. Given the perceived importance of dimorphism in *C. albicans* virulence, we have previously suggested that the lower virulence of *C. dubliniensis* may, at least in part, be related to its relatively poor ability to switch between yeast and hyphal forms [16]. Evidence in support of this was obtained from murine systemic infection model studies [14, 15] and the neonatal orogastric infection model [16]. In the latter, stomach and kidney samples in infected animals contained only *C. dubliniensis* yeast cells, while *C. albicans* cells were found in both the yeast and hyphal forms [16].

We have used the RHE model of superficial infection [44] to compare the invasive potential of both species (Figure 1). In particular, in this model, *C. albicans* grows as both yeast and hyphae and invades the tissue causing major damage. In contrast, *C. dubliniensis* grows exclusively in the yeast form in this model, therefore, causing relatively limited tissue invasion and damage [16, 17]. In order to investigate why the two species differ so markedly in virulence in this model and in order to identify novel virulence-associated genes; Spiering et al. compared their global gene expression profiles during the early stages of RHE infection [17]. Both species showed similar expression profiles for ribosomal and general metabolic genes, however, unsurprisingly, *C. albicans* showed increased expression of hypha-specific virulence genes (e.g., *ECE1, HWP1, HYR1, and ALS3*) within 30 minutes of infection. In contrast, *C. dubliniensis* showed a far less robust transcriptional response and no expression of hypha-specific genes. In addition, several genes with unknown function were found to be specifically upregulated in *C. albicans* that are absent from or very divergent in the *C. dubliniensis* genome. One of these genes, named *SFL2* due to its sequence similarity to the transcription factor-encoding gene *SFL1*, encodes a putative DNA-binding heat shock factor protein. When the gene was deleted in *C. albicans*, it resulted in the failure to produce hyphae under a wide range of growth conditions, including the RHE infection model. Interestingly, the Δ*sfl2* mutation had no effect on survival in the murine systemic infection model, although histological analysis revealed that the kidneys of infected mice were infected only with yeast cells, while the kidneys of mice infected with the wild-type parental strains contained both yeast and predominantly hyphal cells [17]. In a subsequent study, it has been shown that the Δ*sfl2* mutant exhibits reduced virulence in a mouse model of gastrointestinal infection, suggesting that Sfl2 is required for the penetration of the gut wall and subsequent dissemination throughout the body [45]. The *C. dublinensis* ortholog of *SFL2* is only 50% identical and is not expressed under the same conditions as the *C. albicans* gene, therefore it is possible that the divergence of this gene and its apparent lack of expression may be partly responsible for its lower virulence.

Recent studies by our group have been directed towards investigating the molecular basis for differences in the signaling pathways responsible for filamentation in the two species. Comparative genome analysis suggests that orthologs of the known components of the major *C. albicans* morphogenetic pathways (e.g., Cph1-mediated MAPK and the Efg1-mediated Ras1-cAMP pathways) are highly conserved in *C. dubliniensis,* so the reduced capacity of *C. dubliniensis* to produce hyphae and express hypha-specific genes such as *SFL2* is unlikely to be due to the absence of regulators involved in these pathways. Forced stimulation of the Ras1-cAMP pathway with a hyperactive $RAS1^{G113V}$ allele did not result in increased true hypha formation in *C. dubliniensis,* suggesting strong repression of the RAS1-cAMP pathway itself or downstream regulators [43]. One of the most important transcriptional regulators involved in the control of morphogenesis in *C. albicans* is Nrg1, which Staib and Morschhäuser. showed that it is differentially expressed by *C. dubliniensis* when grown on media such as Staib agar [46]. In *C. albicans* this protein targets the negative regulator Tup1 to specific sequences in the promoters of genes involved in hypha formation. *NRG1* expression is rapidly downregulated in *C. albicans* cells incubated under hypha-inducing conditions, including when cells are phagocytosed by murine macrophages, which results in germination and escape from the phagocytes. However, in *C. dubliniensis, NRG1* expression remains high under these conditions, preventing hypha formation and causing cells to remain in the yeast phase, which in the murine macrophage model results in failure to escape from the phagocytes and the death of the fungus [43]. Deletion of the *NRG1* gene in *C. dubliniensis* resulted in an increase in the rate of hypha and particularly pseudohypha formation, which in turn led to increased survival when exposed to murine macrophages as well as increased virulence in the reconstituted human epithelial (RHE) cell model of oral candidosis. Surprisingly, the *C. dubliniensis* Δ*nrg1* mutant was no more virulent than its parent strain in the murine systemic infection model and formed mainly pseudohyphae in infected kidneys, suggesting an additional level of repression preventing true hypha formation *in vivo* [43].

Recent investigations by O'Connor et al. have suggested that repression of filamentation in *C. dubliniensis* is mediated by nutrients. In order to improve our understanding of how environmental signals trigger these pathways O'Connor

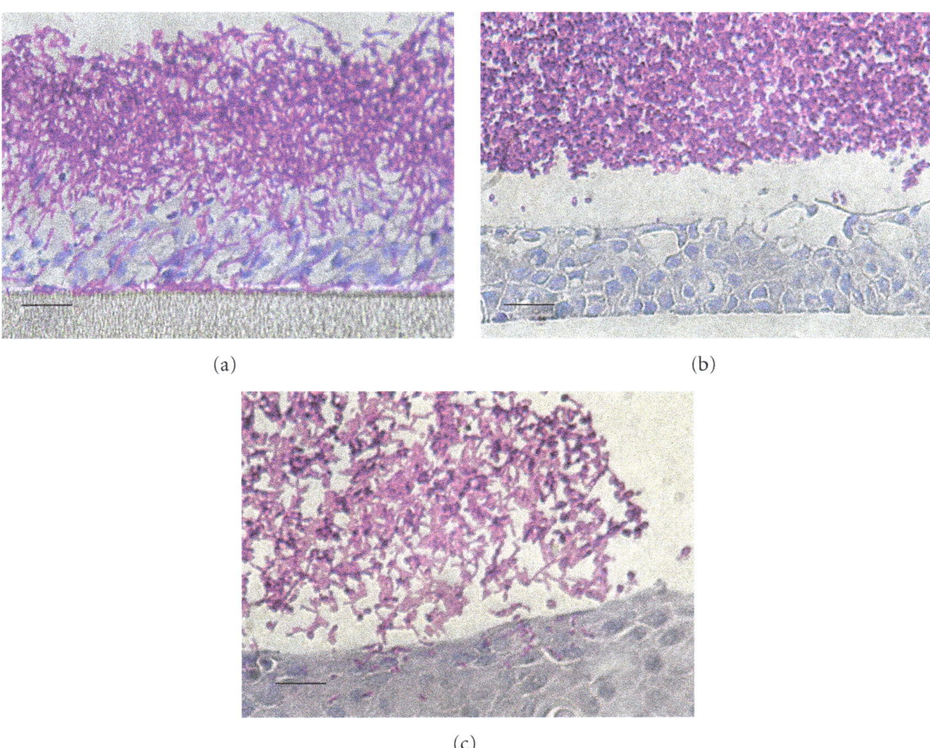

FIGURE 1: Photomicrograph of *C. albicans* SC5314 and *C. dubliniensis* CD36 infecting oral reconstituted human epithelial (RHE) tissue. (a) *C. albicans* originally grown in nutrient-rich YPD, note the presence of hyphae and extensive tissue invasion and damage; (b) *C. dubliniensis* originally grown in YPD, note the absence of hyphae and the limited level of invasion and tissue damage; (c) *C. dubliniensis* originally grown in Lee's medium, note the increased level of filamentation and invasion. Scale bars, approximately 25 μm.

et al. investigated the effects of nutrient availability on the rate of hypha formation [42]. One of the most common incubation conditions for inducing hypha formation in *C. albicans* is incubation in the nutrient-rich medium YPD supplemented with 10% (vol/vol) fetal calf serum at 37°C. Under these conditions >80% of *C. albicans* cells produced germ tubes/filaments within two hours, in contrast only ~20% of *C. dubliniensis* cells were observed to produce hyphae under the same induction conditions. However, when *C. dubliniensis* cells were incubated in water supplemented with 10% (vol/vol) fetal calf serum (WS) at 37°C, the level of hypha producing cells increased to 90% (i.e., similar to the level of hypha formation by *C. albicans*), suggesting that a nutrient-rich environment, in particular the presence of complex mixtures of peptides, suppressed hypha formation in this species. This was confirmed when the addition of peptone and peptone and glucose was found to significantly reduce the levels of hyphae, suggesting that nutrient starvation is a prerequisite for hypha formation by *C. dubliniensis* [42]. These morphological changes were coupled with changes in the expression of genes encoding key transcriptional regulators, such as *NRG1* and *UME6*, which were significantly altered in WS, with the former downregulated by 70% and the latter upregulated 30-fold. Overexpression of the *UME6* gene (which encodes a protein required for hyphal extension), using a doxycycline-inducible promoter led to *C. dubliniensis* cells being able to produce true hyphae, even in nutrient-rich media such as YPD. Similarly, preculture of *C. dubliniensis* cells in nutrient poor media, such as Lee's medium, pH 4.5, prior to the induction of hyphae in YPDS also resulted in a transient ability of *C. dubliniensis* cells to produce hyphae which increased the ability of these cells to adhere to and invade epithelial tissue in the RHE model (see Figure 1) and increased survival in the murine macrophage infection model [42].

These data suggest that factors controlling *UME6* expression in *C. dubliniensis* are repressed by the presence of nutrients, and unlike *C. albicans*, this repression cannot be lifted by a shift to alkaline pH, which occurs when serum is added to the medium. *UME6* is likely to be regulated by Efg1 and Eed1 and is therefore under the control of the Ras1-cAMP pathway. Few studies have investigated how nutrients regulate this pathway in *C. albicans* or *C. dubliniensis*. Preliminary investigations in our laboratory have shown that rapamycin, an inhibitor of the nutrient sensing kinase Tor1 (a kinase that plays a central role in the control of responses to nutrient availability [47]), can stimulate transient hypha formation in *C. dubliniensis* in nutrient-rich YPD serum [48]. This derepression of hypha formation in the presence of nutrients is concomitant with a reduction of *NRG1* and an increase in *UME6* expression. These data suggest that differences in Tor1 activity may play a role in the differential ability of *C. albicans* and *C. dubliniensis* to form hyphae. The

Table 1: Comparison of *C. albicans* and *C. dubliniensis*.

	C. albicans	*C. dubliniensis*	References
Growth and morphology			
Growth at ≥42°C	Yes	No	[39, 41]
Growth in high salt media	Yes	No	[39, 40]
Hypha formation in YPD + serum	Yes	Poor	[16, 42]
Hypha formation in water + serum	Yes	Yes	[42]
RHE infection model	yeasts and hyphae	yeasts only	[17]
Genome			
Chromosome number	8	9–11 chromosome-sized fragments	[7]
No. of species-specific genes	168	29	[32]
ALS3	Present	Absent	[32]
HYR1	Present	Absent	[32]
SAP4, 5 and *6*	All three genes	One gene	[32]
HWP1	Present	Divergent	[32]
TLO family	14 genes	2 genes	[32]

molecular basis for the difference in the activity of Tor1 in the two species is currently under investigation.

4. Conclusions

Candidal pathogenicity involves the complex interplay of a wide range of virulence-associated factors. Comparative analysis of *C. albicans* and *C. dubliniensis* genomic and transcriptomic data has revealed that the reasons for the differences in the capacity of these two species to cause disease are also complex and are not due to a simple defect in *C. dubliniensis*. Instead these studies have revealed genetic differences in the two species, which, at least in part, may explain the differences in their capacity to tolerate stress and to filament. It is clear that the *C. dubliniensis* genome is missing important virulence genes (e.g., *ALS3* and *HYR1*), is in the process of losing others (e.g., the *FGR* genes), has failed to expand certain gene families (e.g., the *SAP* and *TLO* families), and has undergone some degree of transcriptional rewiring (e.g., the Tor pathway and Sfl2). All of these differences suggest that the main discrepancy between these two closely related species relates to differences in hypha formation and the expression of hypha-specific products (summarized in Table 1). We propose that *C. dubliniensis* is in the process of undergoing reductive evolution, whereby its genetic repertoire is diminishing in comparison with *C. albicans* and their common ancestor. One of the main phenotypic manifestations of this is the narrowing of environmental conditions permissive for hypha formation, perhaps as a result of specialization for survival in a specific (as yet unidentified) anatomic niche where hyphae are not required for colonization or growth. By further investigating the molecular basis for the differences between *C. albicans* and *C. dubliniensis*, we hope to improve our understanding of candidal virulence, in particular the relative contribution of hyphae and hypha-specific proteins to the pathogenesis of candidal infections.

Acknowledgments

The authors would like to acknowledge support from the Dublin Dental University Hospital, the Health Research Board (Grant HRA/2009/3), and Science Foundation Ireland (Grant SFI 11 RFP.1/GEN/3042).

References

[1] R. A. Calderone, Ed., *Candida and Candidiasis*, ASM Press, Washington, DC, USA, 2002.

[2] M. Ruhnke, "Epidemiology of *Candida albicans* infections and role of non-*Candida albicans* yeasts," *Current Drug Targets*, vol. 7, no. 4, pp. 495–504, 2006.

[3] D. C. Coleman, D. E. Bennett, D. J. Sullivan et al., "Oral *Candida* in HIV infection and AIDS: new perspectives/new approaches," *Critical Reviews in Microbiology*, vol. 19, no. 2, pp. 61–82, 1993.

[4] V. Krcmery and A. J. Barnes, "Non-*albicans Candida* spp. causing fungaemia: pathogenicity and antifungal resistance," *Journal of Hospital Infection*, vol. 50, no. 4, pp. 243–260, 2002.

[5] D. J. Sullivan, T. J. Westerneng, K. A. Haynes, D. E. Bennett, and D. C. Coleman, "*Candida dubliniensis* sp. nov.: phenotypic and molecular characterization of a novel species associated with oral candidosis in HIV-infected individuals," *Microbiology*, vol. 141, no. 7, pp. 1507–1521, 1995.

[6] D. Sullivan and D. Coleman, "*Candida dubliniensis*: characteristics and identification," *Journal of Clinical Microbiology*, vol. 36, no. 2, pp. 329–334, 1998.

[7] D. J. Sullivan, G. P. Moran, and D. C. Coleman, "*Candida dubliniensis*: ten years on," *FEMS Microbiology Letters*, vol. 253, no. 1, pp. 9–17, 2005.

[8] F. Citiulo, G. P. Moran, D. C. Coleman, and D. J. Sullivan, "Purification and germination of *Candida albicans* and *Candida dubliniensis* chlamydospores cultured in liquid media," *FEMS Yeast Research*, vol. 9, no. 7, pp. 1051–1060, 2009.

[9] B. A. McManus, D. C. Coleman, G. Moran et al., "Multilocus sequence typing reveals that the population structure of *Candida dubliniensis* is significantly less divergent than that of *Candida albicans*," *Journal of Clinical Microbiology*, vol. 46, no. 2, pp. 652–664, 2008.

[10] G. D. Gilfillan, D. J. Sullivan, K. Haynes, T. Parkinson, D. C. Coleman, and N. A. R. Gow, "*Candida dubliniensis*: phylogeny and putative virulence factors," *Microbiology*, vol. 144, no. 4, pp. 829–838, 1998.

[11] C. C. Kibbler, S. Seaton, R. A. Barnes et al., "Management and outcome of bloodstream infections due to Candida species in England and Wales," *Journal of Hospital Infection*, vol. 54, no. 1, pp. 18–24, 2003.

[12] F. C. Odds, M. F. Hanson, A. D. Davidson et al., "One year prospective survey of Candida bloodstream infections in Scotland," *Journal of Medical Microbiology*, vol. 56, no. 8, pp. 1066–1075, 2007.

[13] M. A. Pfaller and D. J. Diekema, "Epidemiology of invasive candidiasis: a persistent public health problem," *Clinical Microbiology Reviews*, vol. 20, no. 1, pp. 133–163, 2007.

[14] M. M. S. Vilela, K. Kamei, A. Sano et al., "Pathogenicity and virulence of *Candida dubliniensis*: comparison with *C. albicans*," *Medical Mycology*, vol. 40, no. 3, pp. 249–257, 2002.

[15] L. R. Ásmundsdóttir, H. Erlendsdóttir, B. A. Agnarsson, and M. Gottfredsson, "The importance of strain variation in virulence of *Candida dubliniensis* and *Candida albicans*: results of a blinded histopathological study of invasive candidiasis," *Clinical Microbiology and Infection*, vol. 15, no. 6, pp. 576–585, 2009.

[16] C. Stokes, G. P. Moran, M. J. Spiering, G. T. Cole, D. C. Coleman, and D. J. Sullivan, "Lower filamentation rates of *Candida dubliniensis* contribute to its lower virulence in comparison with *Candida albicans*," *Fungal Genetics and Biology*, vol. 44, no. 9, pp. 920–931, 2007.

[17] M. J. Spiering, G. P. Moran, M. Chauvel et al., "Comparative transcript profiling of *Candida albicans* and *Candida dubliniensis* identifies SFl2, a *C. albicans* gene required for virulence in a reconstituted epithelial infection model," *Eukaryotic Cell*, vol. 9, no. 2, pp. 251–265, 2010.

[18] C. Y. Koga-Ito, E. Y. Komiyama, C. A. De Paiva Martins et al., "Experimental systemic virulence of oral *Candida dubliniensis* isolates in comparison with *Candida albicans,Candida tropicalis* and *Candida krusei*," *Mycoses*. In press.

[19] J. F. Staab, S. D. Bradway, P. L. Fidel, and P. Sundstrom, "Adhesive and mammalian transglutaminase substrate properties of *Candida albicans* Hwp1," *Science*, vol. 283, no. 5407, pp. 1535–1538, 1999.

[20] L. L. Hoyer, "The ALS gene family of *Candida albicans*," *Trends in Microbiology*, vol. 9, no. 4, pp. 176–180, 2001.

[21] J. R. Naglik, S. J. Challacombe, and B. Hube, "*Candida albicans* secreted aspartyl proteinases in virulence and pathogenesis," *Microbiology and Molecular Biology Reviews*, vol. 67, no. 3, pp. 400–428, 2003.

[22] M. Niewerth and H. C. Korting, "Phospholipases of *Candida albicans*," *Mycoses*, vol. 44, no. 9-10, pp. 361–367, 2001.

[23] P. Sudbery, N. Gow, and J. Berman, "The distinct morphogenic states of *Candida albicans*," *Trends in Microbiology*, vol. 12, no. 7, pp. 317–324, 2004.

[24] J. S. Finkel and A. P. Mitchell, "Genetic control of *Candida albicans* biofilm development," *Nature Reviews Microbiology*, vol. 9, pp. 109–118, 2010.

[25] J. Hannula, M. Saarela, S. Alaluusua, J. Slots, and S. Asikainen, "Phenotypic and genotypic characterization of oral yeasts from Finland and the United States," *Oral Microbiology and Immunology*, vol. 12, no. 6, pp. 358–365, 1997.

[26] L. De Repentigny, F. Aumont, K. Bernard, and P. Belhumeur, "Characterization of binding of *Candida albicans* to small intestinal mucin and its role in adherence to mucosal epithelial cells," *Infection and Immunity*, vol. 68, no. 6, pp. 3172–3179, 2000.

[27] M. McCullough, B. Ross, and P. Reade, "Characterization of genetically distinct subgroup of *Candida albicans* strains isolated from oral cavities of patients infected with human immunodeficiency virus," *Journal of Clinical Microbiology*, vol. 33, no. 3, pp. 696–700, 1995.

[28] T. Jones, N. A. Federspiel, H. Chibana et al., "The diploid genome sequence of *Candida albicans*," *Proceedings of the National Academy of Sciences of the United States of America*, vol. 101, no. 19, pp. 7329–7334, 2004.

[29] B. R. Braun, M. van het Hoog, C. d'Enfert et al., "A human-curated annotation of the *Candida albicans* genome," *PLoS Genetics*, vol. 1, no. 1, pp. 36–57, 2005.

[30] M. Van het Hoog, T. J. Rast, M. Martchenko et al., "Assembly of the *Candida albicans* genome into sixteen supercontigs aligned on the eight chromosomes," *Genome Biology*, vol. 8, no. 4, article no. R52, 2007.

[31] G. Moran, C. Stokes, S. Thewes, B. Hube, D. C. Coleman, and D. Sullivan, "Comparative genomics using *Candida albicans* DNA microarrays reveals absence and divergence of virulence-associated genes in *Candida dubliniensis*," *Microbiology*, vol. 150, no. 10, pp. 3363–3382, 2004.

[32] A. P. Jackson, J. A. Gamble, T. Yeomans et al., "Comparative genomics of the fungal pathogens *Candida dubliniensis* and *Candida albicans*," *Genome Research*, vol. 19, no. 12, pp. 2231–2244, 2009.

[33] P. K. Mishra, M. Baum, and J. Carbon, "Centromere size and position in *Candida albicans* are evolutionarily conserved independent of DNA sequence heterogeneity," *Molecular Genetics and Genomics*, vol. 278, no. 4, pp. 455–465, 2007.

[34] G. Luo, A. S. Ibrahim, B. Spellberg, C. J. Nobile, A. P. Mitchell, and Y. Fu, "*Candida albicans* Hyr1p confers resistance to neutrophil killing and is a potential vaccine target," *Journal of Infectious Diseases*, vol. 201, no. 11, pp. 1718–1728, 2010.

[35] P. Dwivedi, A. Thompson, Z. Xie et al., "Role of Bcr1-activated genes Hwp1 and Hyr1 in *Candida albicans* oral mucosal biofilms and neutrophil evasion," *PLoS One*, vol. 6, no. 1, Article ID e16218, 2011.

[36] Q. T. Phan, C. L. Myers, Y. Fu et al., "Als3 is a *Candida albicans* invasin that binds to cadherins and induces endocytosis by host cells.," *PLoS Biology*, vol. 5, no. 3, article e64, 2007.

[37] R. S. Almeida, S. Brunke, A. Albrecht et al., "The hyphal-associated adhesin and invasin Als3 of *Candida albicans* mediates iron acquisition from host ferritin," *PLoS Pathogens*, vol. 4, no. 11, Article ID e1000217, 2008.

[38] M. A. Uhl, M. Biery, N. Craig, and A. D. Johnson, "Haploinsufficiency-based large-scale forward genetic analysis of filamentous growth in the diploid human fungal pathogen *C.albicans*," *EMBO Journal*, vol. 22, no. 11, pp. 2668–2678, 2003.

[39] B. Enjalbert, G. P. Moran, C. Vaughan et al., "Genome-wide gene expression profiling and a forward genetic screen show that differential expression of the sodium ion transporter Ena21 contributes to the differential tolerance of *Candida albicans* and *Candida dubliniensis* to osmotic stress," *Molecular Microbiology*, vol. 72, no. 1, pp. 216–228, 2009.

[40] S. H. Alves, E. P. Milan, P. De Laet Sant'Ana, L. O. Oliveira, J. M. Santurio, and A. Lopes Colombo, "Hypertonic sabouraud broth as a simple and powerful test for *Candida dubliniensis* screening," *Diagnostic Microbiology and Infectious Disease*, vol. 43, no. 1, pp. 85–86, 2002.

[41] E. Pinjon, D. Sullivan, I. Salkin, D. Shanley, and D. Coleman, "Simple, inexpensive, reliable method for differentiation of

Candida dubliniensis from *Candida albicans*," *Journal of Clinical Microbiology*, vol. 36, no. 7, pp. 2093–2095, 1998.

[42] L. O'Connor, N. Caplice, D. C. Coleman, D. J. Sullivan, and G. P. Moran, "Differential filamentation of *Candida albicans* and *Candida dubliniensis* is governed by nutrient regulation of UME6 expression," *Eukaryotic Cell*, vol. 9, no. 9, pp. 1383–1397, 2010.

[43] G. P. Moran, D. M. MacCallum, M. J. Spiering, D. C. Coleman, and D. J. Sullivan, "Differential regulation of the transcriptional repressor NRG1 accounts for altered host-cell interactions in *Candida albicans* and *Candida dubliniensis*," *Molecular Microbiology*, vol. 66, no. 4, pp. 915–929, 2007.

[44] M. Schaller, K. Zakikhany, J. R. Naglik, G. Weindl, and B. Hube, "Models of oral and vaginal candidiasis based on in vitro reconstituted human epithelia," *Nature Protocols*, vol. 1, no. 6, pp. 2767–2773, 2007.

[45] W. Song, H. Wang, and J. Chen, "*Candida albicans* Sfl2, a temperature-induced transcriptional regulator, is required for virulence in a murine gastrointestinal infection model," *FEMS Yeast Research*, vol. 11, no. 2, pp. 209–222, 2011.

[46] P. Staib and J. Morschhäuser, "Differential expression of the NRG1 repressor controls species-specific regulation of chlamydospore development in *Candida albicans* and *Candida dubliniensis*," *Molecular Microbiology*, vol. 55, no. 2, pp. 637–652, 2005.

[47] J. R. Rohde, R. Bastidas, R. Puria, and M. E. Cardenas, "Nutritional control via Tor signaling in Saccharomyces cerevisiae," *Current Opinion in Microbiology*, vol. 11, no. 2, pp. 153–160, 2008.

[48] D. J. Sullivan and G. P. Moran, "Differential virulence of *Candida albicans* and *C. dubliniensis* a role for tor1 kinase?" *Virulence*, vol. 2, no. 1, pp. 77–81, 2011.

9

Impact of HMGB1/TLR Ligand Complexes on HIV-1 Replication: Possible Role for Flagellin during HIV-1 Infection

Piotr Nowak,[1] Samir Abdurahman,[1] Annica Lindkvist,[2] Marius Troseid,[3] and Anders Sönnerborg[1,2]

[1] *Department of Infectious Diseases, Institution of Medicine, Karolinska University Hospital and Karolinska Institutet, 14186 Stockholm, Sweden*
[2] *Department of Clinical Microbiology, Institution of Laboratory Medicine, Karolinska University Hospital and Karolinska Institutet, 14186 Stockholm, Sweden*
[3] *Department of Infectious Diseases, Oslo University Hospital, Ullevål, 0424 Oslo, Norway*

Correspondence should be addressed to Piotr Nowak, piotr.nowak@ki.se

Academic Editor: Giancarlo Ceccarelli

Objective. We hypothesized that HMGB1 in complex with bacterial components, such as flagellin, CpG-ODN, and LPS, promotes HIV-1 replication. Furthermore, we studied the levels of antiflagellin antibodies during HIV-1-infection. *Methods.* Chronically HIV-1-infected U1 cells were stimulated with necrotic extract/recombinant HMGB1 in complex with TLR ligands or alone. HIV-1 replication was estimated by p24 antigen in culture supernatants 48–72 hours after stimulation. The presence of systemic anti-flagellin IgG was determined in 51 HIV-1-infected patients and 19 controls by immunoblotting or in-house ELISA. *Results.* Flagellin, LPS, and CpG-ODN induced stronger HIV-1 replication when incubated together with necrotic extract or recombinant HMGB1 than activation by any of the compounds alone. Moreover, the stimulatory effect of necrotic extract was inhibited by depletion of HMGB1. Elevated levels of anti-flagellin antibodies were present in plasma from HIV-1-infected patients and significantly decreased during 2 years of antiretroviral therapy. *Conclusions.* Our findings implicate a possible role of HGMB1-bacterial complexes, as a consequence of microbial translocation and cell necrosis, for immune activation in HIV-1 pathogenesis. We propose that flagellin is an important microbial product, that modulates viral replication and induces adaptive immune responses *in vivo.*

1. Introduction

Antiretroviral therapy (ART) suppresses efficiently the replication of human immunodeficiency virus type 1 (HIV-1) to undetectable levels with standard techniques in most treated patients, but there is still an ongoing low-grade replication in most or all patients [1]. Also, immune activation is a central feature of progressive HIV-1 infection [2–4], and although the degree of immune activation is decreased during ART, it is not normalized [5]. The pathogenic mechanisms for the persistent immune activation remain further to be determined. This is especially important since studies suggest that the remaining immune activation may cause organ damage, for example, an increased risk for cardiovascular diseases and possibly neurocognitive dysfunction [5, 6].

The gastrointestinal (GI) immune system seems to play a central role in the pathogenesis of immune activation [7]. The early dramatic depletion of CD4+ T cells from the gut mucosa may drive immune activation, as this mucosal immune damage impairs the normal barrier function and allows increased translocation of bacterial products from the gut lumen into the circulation [8]. We and others have shown that microbial translocation is present in HIV-1 infection through increased plasma LPS levels in subjects with progressive disease and that the levels are decreased by ART [9–11].

Furthermore, we and others have implied that the alarmin high-mobility group binding 1 protein (HMGB1) modulates HIV-1 replication *in vitro* and contributes to the activation of immune system [12–14]. Thus, plasma HMGB1 levels are elevated in HIV-1-infected patients and reduced with effective ART [11, 15]. HMGB1 is released from damaged or necrotic cells to the extracellular milieu, in which it may act as a potent proinflammatory marker by stimulating cytokine expression in monocytes and endothelial cells [16, 17]. HMGB1 per se does not seem to have a proinflammatory activity [18, 19] but has a high affinity to form complexes with other molecules such as LPS and CpG-DNA [20]. These complexes are likely to bind to various receptors, including TLR4 and TLR9, and promote a large variety of inflammatory and immunological responses [20, 21].

The aim of our study was to explore whether complexes of HMGB1 and TLR ligands, such as flagellin, could synergistically induce HIV-1 replication in a promonocytic cell line.

2. Methods

2.1. Ethic Statement. All research involving human participants has been conducted according to the principles expressed in the Declaration of Helsinki. Patients gave their informed written consent and the study protocol was approved by the Regional Ethics Committee in Stockholm, Sweden (Dnr 2005/3:10).

2.2. Reagents. Lipopolysaccharide (LPS) and (phorbol-12-myristate-13-acetate) PMA were obtained from Sigma (St. Louise, MO, USA), IL-1β from R&D Systems (Minneapolis, MN, USA), and CpG-ODN type B (ODN2006), purified flagellin (*S.typhimurium*), and anti-flagellin (FliC) antibodies from InvivoGen (San Diego, CA, USA and Abcam, Cambridge, UK). Recombinant HMGB1 (HM-116) was purchased from HMGBiotech (Milan, Italy) or from R&D systems (Minneapolis, MN, USA). We also used recombinant HMGB1 [19] that was a kind gift from Professor Helena Erlandsson-Harris CMM/KI, Stockholm.

2.3. Cell Cultures. U1 cells, a subclone derived from U937 cells, were obtained through the AIDS Research and Reference Reagent Program (NIAID, NIH). The U1 cells are chronically infected with HIV-1 and are characterized by low constitutive levels of virus expression that can be upregulated by several cytokines and phorbol esters. The cells were maintained in RPMI medium (Gibco) supplemented with 10% fetal calf serum, glutamine, and antibiotics. Cells were seeded at 200 000 cells/mL in 96-well plates and complexes/TLR-ligands/controls were added and incubated for 48 or 72 hours.

2.4. Patients. Patients ($n = 51$) given ART, followed at the Department of Infectious Diseases, Karolinska University Hospital, Stockholm, and 19 healthy controls were included. Patients' recruitment was based on sample availability as well as virologic response after 2 years of ART. Thirty-three individuals had undetectable viral load and 18 had detectable viraemia (nonresponders) after 2 years of treatment. This cohort (42 of 51 patients) has been described previously [11]. The age and sex distribution of the patients and controls was similar (median age 38 years, 52% women).

2.5. Preparation of Necrotic Cell Extracts. Necrotic extracts were obtained as previously described [22]. Briefly, necrosis was induced in peripheral blood mononuclear cells (PBMCs) from healthy donors (30×10^6 cells/mL) by exposing the cells to six cycles of freezing and thawing. Cell debris was removed by centrifugation and the supernatant was passed through a 0.2 μm membrane and collected. The concentration of HMGB1 in necrotic extracts was 40 μg/mL, as estimated by immunoblot (data not shown). Furthermore, 250 μL of necrotic extract was incubated with polyclonal anti-HMGB-1 antibodies (Abs) from ABCAM (Cambridge, UK). Samples were incubated at 4°C for 16 hours. As control, nonspecific Abs (Rabbit polyclonal IgG) was utilized. Immune complexes were removed by adding 25 μL of Sepharose A/G to the extract, incubated for 1.5 hours at 4°C, and centrifuged. The supernatant was collected and the procedure was repeated again with 25 μL sepharose for 1 hour at 4°C.

2.6. Preparation of HMGB1 Complexes. Necrotic extract or HMGB1was mixed with the TLR-ligands, LPS, CpG-ODN, IL-1β, and flagellin in PBS in different concentrations and incubated for 16 hours at 4°C. Concentrations are given in Figures 1–3. The suboptimal stimulatory concentrations (capable to trigger HIV replication from U1 cells) of necrotic extract as well as LPS, flagellin, CpG-ODN, and Il-1β were estimated in a series of experiments (data not included). Complexes were also mixed and denatured by heating at 95°C for five minutes to verify the stimulatory effect of complex formation on U1 cells.

2.7. Characterization of HMGB1 with Immunoblotting. Equal volumes of necrotic cell extracts, HMGB1-depleted necrotic cell extracts, as well as recombinant HMGB1 proteins were resolved on 10–20% Tris/glycine gel and transferred onto nitrocellulose membrane (Invitrogen, Carlsbad, USA). The membranes were then incubated overnight with anti-HMGB1 Abs at 1 : 2000 dilution. The following day, the membranes were incubated 1 h with horseradish-peroxidase (HRP-) conjugated secondary antibody (GE, Healthcare), raised against rabbit IgG at 1 : 10,000 dilution. The proteins were finally visualized using ECL reagents (GE, Healthcare).

2.8. Immunoblotting of Antiflagellin Antibodies. Approximately 1.88 μg of recombinant flagellin was twofold serially diluted (4 series) and subjected to gel electrophoresis on 10–20% precasted SDS-PAGE gel (Invitrogen) in Tris/glycine/SDS buffer. Similarly, bacterial extracts of flagellated *E. coli* strain O126:H2 and aflagellate *E. coli* O21:H- (CCUG catalog number 11425 and 11326, kind gift from Professor Andrej Weintraub, Clinical Microbiology/KI, Stockholm) that were prepared freshly from an overnight inoculation were also resolved on SDS-PAGE gel as described previously. The proteins were then electroblotted onto iBlot

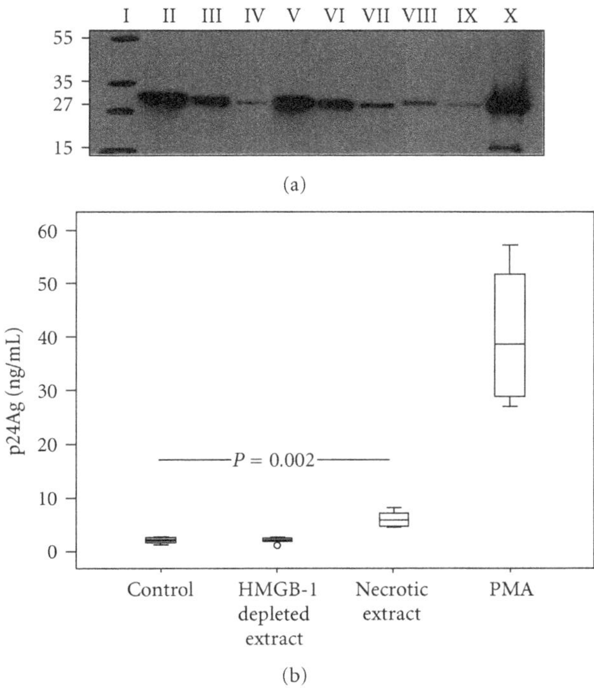

Figure 1: HMGB1 present in necrotic extract induces HIV-1 replication in U1 cells. (a) Western blot of cell supernatants (necrotic extracts) obtained after freeze-thawing cycles of peripheral blood mononuclear cells (PBMC) (30×10^6 cells/mL) from healthy donors: Molecular weight marker (I); supernatants after immune depletion of HMGB1 with nonspecific rabbit polyclonal antibody (II); depletion with anti-HMGB1 antibody $-5\,\mu$g (III) and $10\,\mu$g (IV); necrotic extract loaded $20\,\mu$L (V), $10\,\mu$L (VI), and $5\,\mu$L (VII); 100 ng (VIII) and 75 ng (IX) of recombinant HMGB1; cell debris (X). Numbers to the left depict positions of molecular mass markers (in kDa). (b) Levels of HIV p24 protein in cell culture supernatants after 72 h incubation of U1 cells with necrotic extract (HMGB1 concentration $1\,\mu$g/mL): HMGB1-depleted necrotic extract and mock cells. PMA served as a positive control (20 nM). The levels of viral replication were approximately 2-fold higher after stimulation by necrotic extract compared to the mock cells ($P = 0.002$). Results from three independent experiments in duplicates are presented.

gel transfer stacks nitrocellulose membranes, using the iBlot dry blotting system (Invitrogen), as recommended by the manufacturer. After blocking the nitrocellulose membranes for 1 hour in blocking buffer (PBS supplemented with 0.05% Tween and containing 10% nonfat milk), the blots were probed with primary antibodies overnight at 4°C with a slow agitation. As primary antibody, serum from HIV-1-infected or control subjects diluted 1 : 1000, monoclonal, or polyclonal anti-flagellin Abs was used. The following day, the membranes were washed with PBS containing 0.05% (vol/vol) Tween and bound antibodies were then detected by using HRP-conjugated secondary Abs (Pierce) against human IgG in 1 : 10,000 dilution. Protein bands were visualized by chemiluminescence (Thermo Scientific). To confirm the protein bands, two immunoblotted membranes from HIV-1-infected patients and control subjects were stripped off and reprobed with mouse monoclonal antibody directed against flagellin. Bound antibodies were then detected by using an HRP-conjugated secondary antibody, raised against mouse (DAKO; 1 : 4.000).

2.9. Specific and Total Antibodies Measurement. Antibody titers against flagellin, measles, and total IgG levels were assessed by ELISA. An in-house anti-flagellin-specific IgG ELISA was developed using purified flagellin monomers from *S. typhimurium* (InvivoGen). It has been previously shown that human sera have a similar recognition pattern of flagellin monomers whether isolated from flagellated *E. coli* or *S. typhimurium* [23]. Briefly, microwell plates (MWP) were coated overnight with purified flagellin from *S. typhimurium* (25 ng/well). The following day, plasma samples from HIV-1-infected and control subjects diluted 1 : 1000 were applied to wells coated with flagellin. After incubation and washing, the MWPs were incubated with HRP-conjuggated anti-human IgG. For total IgG ELISA, the manufacturer's procedure was followed (MABTECH, Nacka, Sweden). The Enzygnost Measeles virus IgG ELISA kit (Behring, Germany) was utilized for quantification of antimeasles antibodies.

2.10. Plasma HIV-1 RNA Quantification and CD4+/CD8+ T-Cell Counts. Plasma HIV-1 RNA levels (COBAS Amplicor test Roche Molecular Systems; USA; detection limit 40 copies/mL) and T-cell counts (flow cytometry) were evaluated as part of clinical routine.

2.11. HIV-1 Replication Assay. Supernatants were collected at indicated time points and tested for the presence of HIV p24 antigen with Architect i2000 HIV-1 Ag/Ab combo detection system (Abbott Diagnostics, Abbott Park, IL, USA). The p24 concentration was calculated based on the several standard dilutions of p24 protein included in each run.

2.12. Statistics. Data are presented as median, interquartile range, and total range. Differences between groups were analysed with the Mann-Whitney U-test, and intragroup changes from baseline to the end of the study were evaluated by Wilcoxon test. Jonckheere-Terpstra test was used for trend analyses and correlation analyses were performed using the Spearman method. A two-tailed significance level of 0.05 was used. The statistical analyses were performed with SPSS software, version 15.0 (SPSS Inc, Chicago, USA).

3. Results

3.1. Necrotic Cellular Extract Upregulates HIV-1 Replication in U1 Cells. To determine the impact of an endogenous signal, associated with cell injury, on HIV-1 replication, we generated soluble necrotic extracts from healthy donors PBMCs. Western blot confirmed the presence of large amounts of HMGB1 in the extracts. Additionally HMGB1-depleted extracts were obtained by immune depletion utilizing specific anti-HMGB1 antibodies (Figure 1(a)). In the initial experiment, the U1 cells were exposed to necrotic

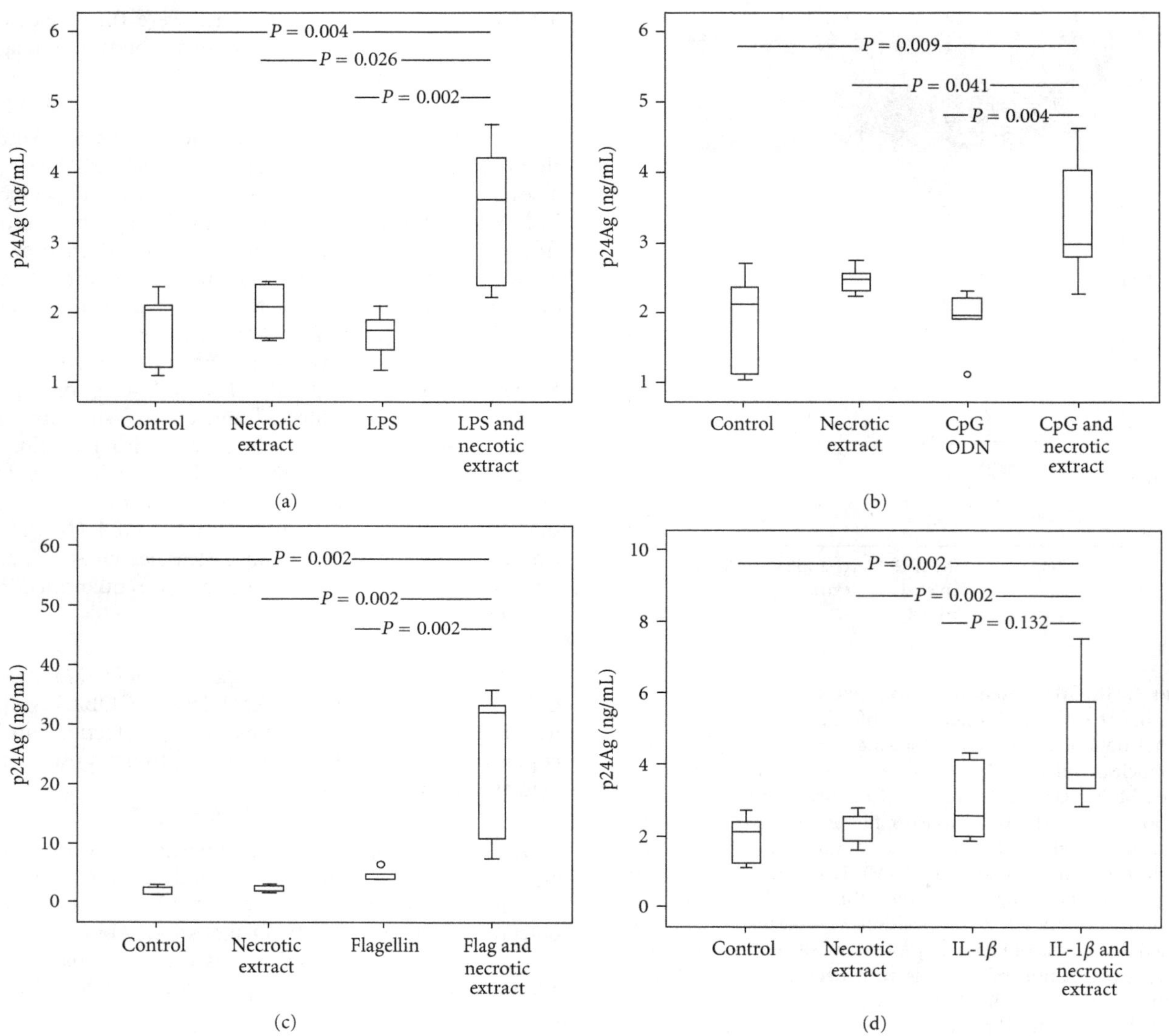

Figure 2: Necrotic extract and TLR-ligands in complexes upregulate viral replication in U1 cells. U1 cell cultures were stimulated with necrotic extract (HMGB1 concentration 1 μg/mL) and Toll-like receptor ligands: LPS 10 ng/mL (a), CpG-ODN 1 μg/mL (b), flagellin 50 ng/mL (c), and IL-1β 0.25 ug/mL (d) alone or in complexes. Supernatants from mock cells served as controls. Supernatants were collected from cell cultures after 72 hours. Results from three independent experiments in duplicates are presented.

extract, HMGB1-depleted extract, and PMA, respectively. The HIV p24 antigen concentration in the cell supernatants was measured after 72 hours (Figure 1(b)). The levels of viral replication were approximately 2-fold higher after stimulation by necrotic extract compared to the mock cells ($P = 0.002$). The stimulation with PMA gave a 10-fold higher viral replication than stimulation with necrotic extract. Notably, addition of necrotic extract depleted of HMGB1 did not result in an increase of viral replication, as compared to the controls, suggesting that HMGB1 crucially contributes to the stimulatory effect of the necrotic extract.

3.2. Interacting Effect of TLR Ligands and Necrotic Extract on HIV-1 Replication in U1 Cells. Thereafter, we stimulated the U1 cells with necrotic extract, TLR ligands (LPS, flagellin, CpG-ODN), and IL-1β alone or with the complexes of necrotic extract and the TLR ligands or IL-1β. Notably, stimulation with all the TLR ligands, in combination with necrotic extract, resulted in a higher viral replication than stimulation with necrotic extract or TLR ligands alone (Figure 2). Hence, stimulation with LPS, CpG-ODN and IL-1β in complexes with necrotic extract resulted in a 1.5–2-fold-increased viral replication compared to each component alone, whereas flagellin in combination with necrotic extract resulted in a 7-fold increased replication compared to flagellin alone and a 13-fold-increased replication compared to necrotic extract alone. The preheating of complexes prior incubation with cells resulted in abrogation of stimulatory signal, implying that the active compound relies on intact protein structure (data not included).

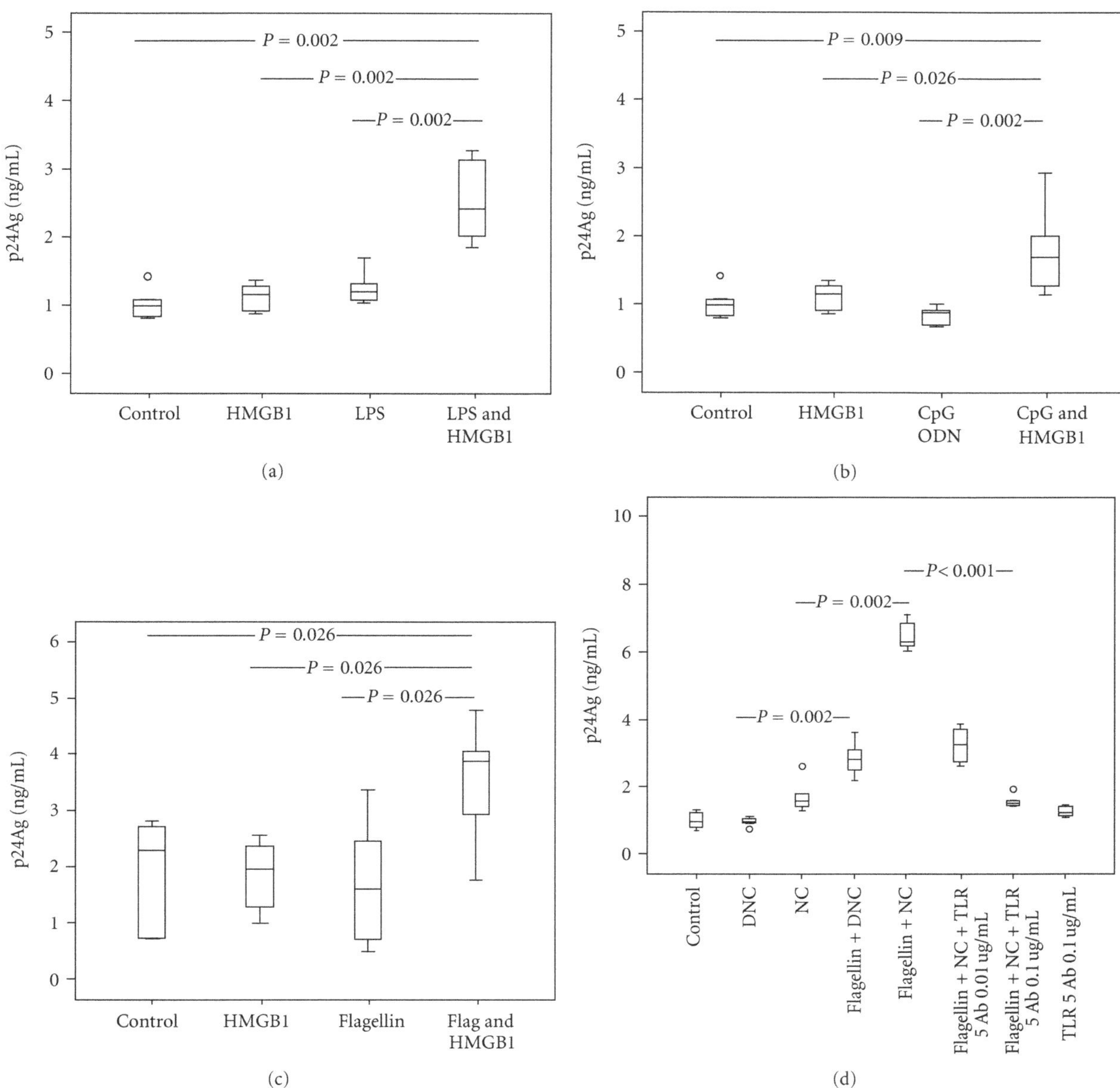

FIGURE 3: Interacting effect of recombinant HMGB1 and TLR-ligand complexes in U1 cells. Inhibition of flagellin complexes induced HIV-1 replication by anti-TLR5 antibodies. U1 cells were stimulated with recombinant HMGB1 (1 μg/mL) and TLR ligands: LPS 10 ng/mL (a), CpG-ODN 1 μg/mL (b) and flagellin 10 ng/mL (c) alone or in complexes. (d) U1 cells were incubated with 0.1 and 0.01 ug/mL anti-TLR5 antibodies (TLR5 Ab) for 1 hour and then exposed to necrotic extract (NC) and HMGB1-depleted necrotic extract (DNC), alone or in complexes with flagellin (10 ng/mL). HIV-1 replication was estimated after 48 hours of incubation. A dose-dependent inhibition of flagellin-necrotic extract complexes stimulatory effect is present in wells pretreated with anti-TLR5 antibodies (P for trend < 0.001). Results from three independent experiments in duplicates are shown.

3.3. Interacting Effect of TLR Ligands and HMGB1 on HIV-1 Replication in U1 Cells. In order to explore if HMGB1 could mimic the synergistic effects of necrotic extract-TLR ligands, we challenged the U1 cells with complexes consisting of HMGB1 and bacterial substances. Indeed, stimulation with microbial products (LPS, flagellin, or CpG-ODN) in combination with HMGB1 resulted in a higher viral replication than stimulation with HMGB1 or TLR ligands alone (Figures 3(a)–3(c)). Stimulation with LPS, flagellin, and CpG-ODN in combination with HMGB1 resulted in a 1.5–2-fold-increased viral replication compared to each component alone, although the stimulatory effect was not as prominent as with TLR ligands in combination with necrotic extract.

3.4. Dose-Dependent Inhibition of Flagellin by Anti-TLR5. It is known that immune response to flagellin is mediated by TLR5. To investigate whether anti-TLR5 antibodies could block the inducing effects of flagellin, we first preincubated U1 cells with anti-TLR5 antibodies and subsequently added the necrotic extract complexed with flagellin. Flagellin in combination with HMGB1-depleted extract gave a 3-fold-increased viral replication compared to depleted extract alone, whereas flagellin in combination with necrotic extract gave a 4-fold increase compared to necrotic extract alone (Figure 3(d)). Preincubation of U1 cells with anti-TLR-5 antibodies before addition of necrotic extract-flagellin complexes resulted in a dose-dependent inhibition of this stimulatory effect (P for trend < 0.001). Addition of anti-TLR5 antibodies alone did not affect viral replication.

3.5. Detection of Antiflagellin Antibodies in HIV-1-Infected Patients. Encouraged by the *in vitro* data we aimed to evaluate whether flagellin is a potentially important antigen *in vivo* during HIV-1 infection. Therefore, serum samples from HIV-1-infected patients and control subjects were used to measure the level of flagellin-specific antibodies by Western blot analysis. When diluted 1 : 1000, all of the sera samples from the HIV-1-infected patients analyzed exhibited easily detectable bands that recognized the first two dilutions of flagellin derived from *S. typhimurium* (Figure 4(a), upper panels). A relative increase in flagellin-specific IgG in HIV patients was observed if serum samples were diluted 1 : 500. In contrast, in only one control subject (CS#2) flagellin was detected faintly at the highest dilution (Figure 4(a), lower panels). Although semiquantitative analysis of detected bands was not performed, the levels of flagellin-specific IgG observed were in all cases strikingly elevated in HIV-1-infected patients. Similar pattern of anti-flagellin IgG was observed when plasma instead of sera was used. In order to address the specificity of the flagellin IgG, we subjected the bacterial lysates from flagellated and aflagellate *E. coli* to protein separation on the SDS-PAGE gel. The Western blotting with HIV-1 serum (as a primary antibody) showed similar pattern as when the polyclonal anti-flagellin antibody was used confirming that the specificity of the antibodies was not limited to the recombinant protein (Figure 4(b)).

Furthermore, we used the anti-flagellin ELISA to evaluate the levels of flagellin IgG in plasma of HIV-1-infected patients before and after two years of ART. At baseline significantly elevated levels of flagellin antibodies were found in HIV-1-infected patients as compared to controls ($P < 0.001$) (Figure 5(a)). This difference persisted ($P < 0.001$) when the flagellin antibodies were adjusted to the total IgG (Figure 5(b)), suggesting that the elevation of flagellin antibodies was not due to hypergammaglobulinemia. Moreover, analysis of antimeasles antibodies in 10 patients with severe immune deficiency (CD4+ T-cell counts ≤ 200) supported that the elevation of the flagellin antibodies was not caused by polyclonal activation Supplementary Table 1 (see supplementary material available online at doi:10.1155/2012/263836). The levels of flagellin IgG, total IgG, and the ratio flagellin IgG/total IgG were significantly reduced after two years of ART for the whole group ($P < 0.001$, $P = 0.03$, and $P < 0.001$, resp. Figures 5(a)–5(c)). Additionally a significant reduction of flagellin IgG levels was observed also when the patients were subdivided into those with successful ART and the nonresponders who had remaining low levels of viral replication two years after initiating the ART ($P = 0.009$; $P = 0.001$, resp.). The total IgG levels after ART did not decrease in nonresponders as they did in successfully treated patients ($P < 0.001$) (data not shown).

We found no correlation between the levels of flagellin IgG and the viral load nor the CD4/CD8 T-cell counts. In contrast, among the subgroup of 42 patients in whom we had earlier analysed HGMB1 and LPS in plasma, significant correlations were found between the levels of flagellin IgG and LPS ($r = 0.32$; $P = 0.02$) as well as between the flagellin IgG/total IgG ratio and LPS ($r = 0.25$; $P = 0.007$) (data not shown).

4. Discussion

Microbial translocation has been described in different conditions like inflammatory bowel disease, neutropenia, and chronic viral infections [6, 24]. In HIV-1 infection, the proof for the translocation of bacterial products is based mainly on LPS data [9, 11, 25, 26]. However, the original observation that increased LPS levels were associated with both activated memory CD8+ T cells and enhanced IFN-α levels implies the involvement of other factors [9].

We therefore hypothesized that HMGB1 could be such a link between the microbial products and hyperinflammation [27]. Mounting evidence shows that HMGB1 does not act alone but forms stable potent proinflammatory complexes with other molecules, such as bacterial products or single stranded DNA [18–20, 28]. Since we have earlier shown that HMGB1 alone activates latent HIV-1 replication *in vitro* [12], we decided to expand our analysis to the effect of HGMB1 in complexes with bacterial products. Here, we present that HMGB1 in complex with the TLR ligands (LPS, CpG-ODN, flagellin) and IL-1β induce viral replication in a promonocytic cell line, U1 cells. The data obtained with both the HMGB1 derived from necrotic extract as well as recombinant protein yielded similar results, although the stimulatory signals associated with necrotic HMGB1 were more potent. This is not surprising as other endogenous danger signals should be anticipated in this process [20]. The reduction of the stimulatory effect by depletion of HGMB1 also supports our hypothesis that HMGB1 is an important component of these complexes.

These *in vitro* findings brought our attention to flagellin as a potent activator of HIV-1 replication alone or in complexes with HMGB1. Bacterial flagellins are present in all motile bacteria and play an important role in mediating gut inflammation associated with infection by enteric pathogens or in inflammatory bowel diseases [29]. Their proinflammatory activity is exerted mainly through TLR5 [30, 31]. It has been recently demonstrated that flagellin is the major antigen activating innate and adaptive immune response in intestinal inflammation observed in Crohn's disease [32, 33]. Disruption of the intestinal barrier promotes translocation of flagellated commensal bacteria across the epithelium driving the activation of innate immune cells residing in the lamina propria. This phenomenon results also in abnormal exposure

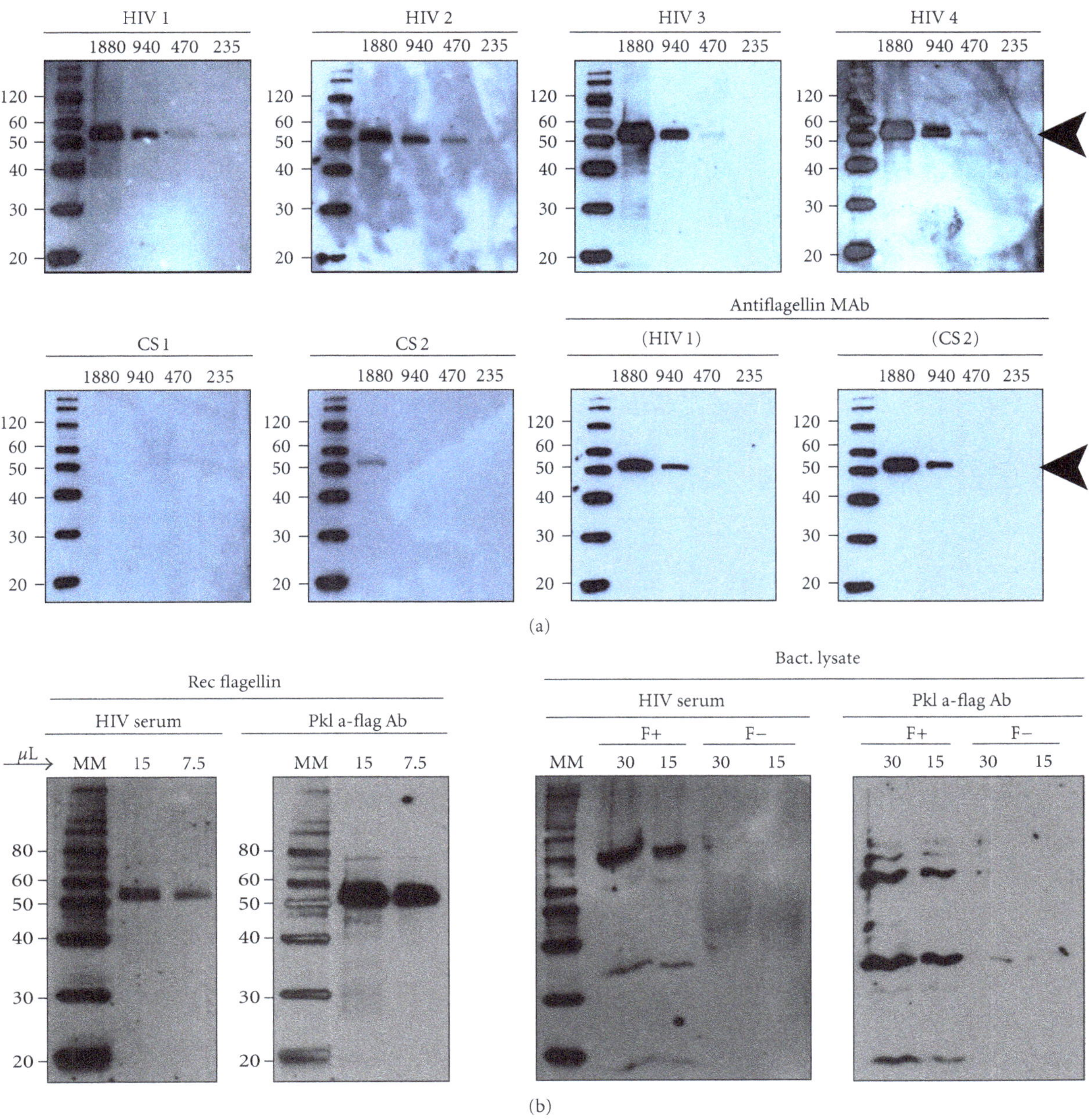

Figure 4: HIV-1-infected patients exhibit elevated levels of flagellin-specific antibodies. (a) Approximately 1.8 μg of recombinant flagellin was twofold serially diluted (4 series) and resolved on 10–20% SDS-PAGE gel, transferred onto a nitrocellulose membrane, and detected with immunoblot assay. Serum from HIV-1-infected or control subjects diluted 1 : 1000 was used as a primary antibody. Each panel is a separate immunoblot of the same recombinant flagellin detected with serum from HIV-1 patients (HIV 1, HIV 2, HIV 3, HIV 4) or control subjects (CS 1, CS 2). To confirm equal sample loading and protein transfer, immunoblotted membranes from HIV 1 and CS 2 were stripped off and reprobed with monoclonal antibody directed against flagellin (blots in the lower panel). The experiments were performed using sera from ten HIV-1-infected patients and four control subjects. These presented data are representative for all immunoblots. The position of recombinant flagellin protein is indicated with an arrow to the right. Numbers to the left depict positions of molecular mass markers (in kDa). (b) Recombinant flagellin (column 1) and lysates of flagellated or aflagellate *E. coli* (columns 3 and 4) were subjected to SDS-PAGE and immunoblotted using serum from HIV infected patient (column 1 and 3, diluted 1 : 1000) as primary antibody or polyclonal anti-flagellin antibody (columns 2 and 4, diluted 1 : 1000). The membrane in column 1 was stripped off and reprobed with polyclonal anti-flagellin antibody (column 2). The numbers above the figures indicate the amount of sample loaded in each well (in μL). MM, magic marker (Invitrogen); F+: flagellated; F−: aflagellate whole bacterial lysate; Pkl: polyclonal; Ab: antibody. Numbers to the left depict positions of molecular mass markers (in kDa).

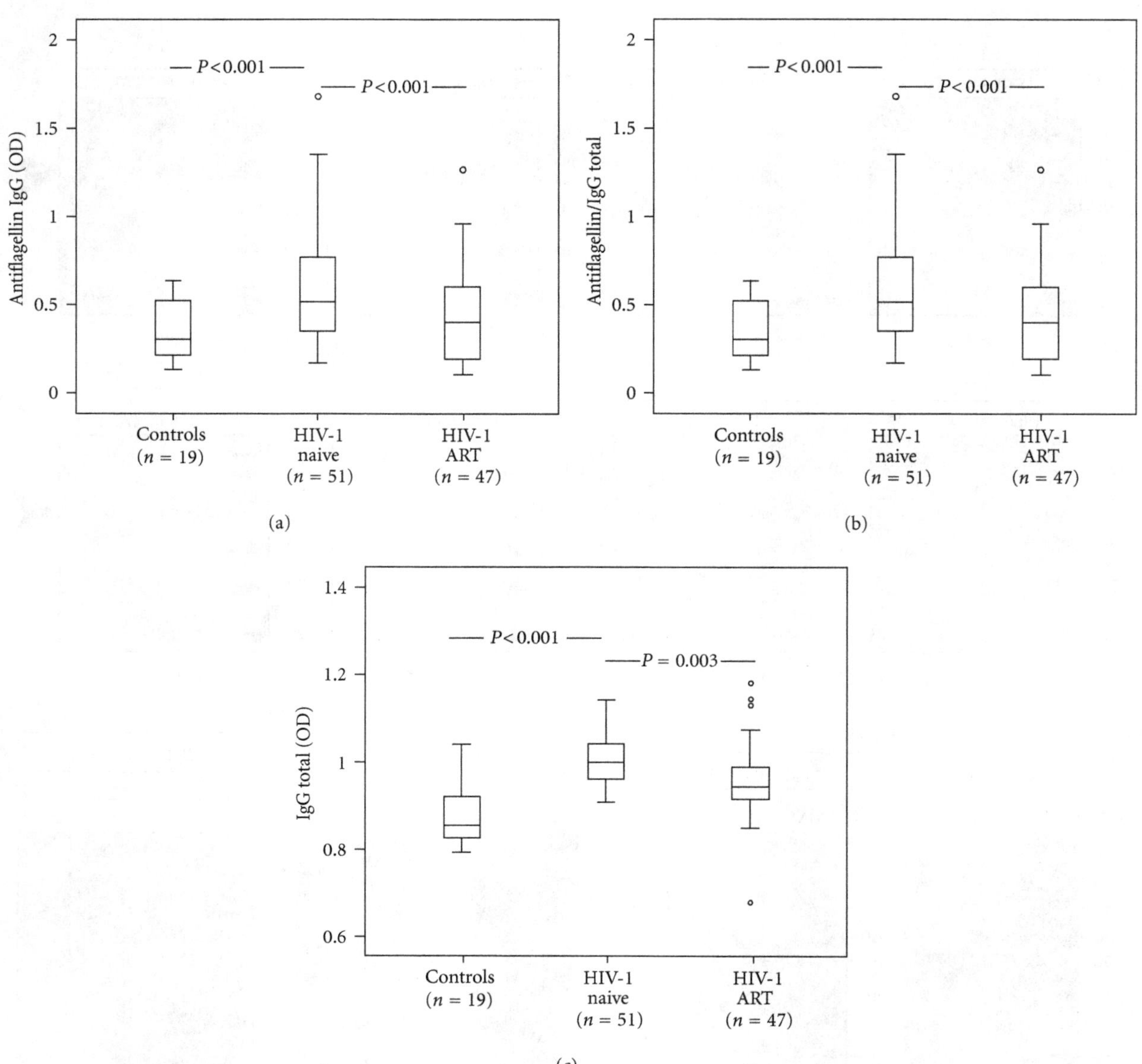

Figure 5: Elevated levels of anti-flagellin IgG are reduced during ART. Plasma levels (OD) of anti-flagellin IgG (a), ratio anti-flagellin IgG/total IgG (b), and total IgG (c) in healthy controls, HIV-1-infected individuals before (= naïve) and after two years of ART (antiretroviral therapy). The plasma samples were diluted 1:1000. The data are presented as median 25–75 interquartile range and total range. P values refer to intergroup differences.

of immune cells to flagellin, a process that may influence the balance and function of the different T-cell subsets present in the gut-associated immune system (GALT) and promote inflammation [29, 34].

The contribution of flagellin to immune activation during HIV-1 infection has been anticipated [7], but the data are scarce. Thus, exposure of PBMCs to flagellin resulted in activation of T cells predominantly of central memory and effector memory phenotype [35]. Moreover, flagellin is able to induce HIV-1 gene expression in resting memory CD4+ T cells that are considered as a key cellular HIV-1 reservoir in infected individuals [36].

Our *in vitro* and *in vivo* findings are clearly in line with the hypothesis of an important role of flagellin in HIV-1 pathogenesis. Thus, we demonstrated that flagellin complexes are able to significantly stimulate the HIV-1 replication, at least from cells of monocytic origin. Furthermore, our finding that elevated levels of anti-flagellin IgG are present in HIV-1-infected individuals anticipates that this observation has *in vivo* implications. Hence, we show not only presence of elevated levels of flagellin antibodies before the initiation of ART but also a reduction after two years of ART suggesting decreased exposure to the antigen probably due to partial restoration of gut-blood barrier [25, 37].

The hypergammaglobulinemia present during the HIV-1 infection cannot be solely responsible for the elevation of flagellin IgG levels as the normalisation to the IgG did not influence the results.

An elevated adaptive immune response to flagellin has been previously observed in conditions associated with gut barrier dysfunction such as Crohn's disease and short bowel syndrome [23, 24] and relates to the severity of Crohn's disease [38]. Also, Kamat et al. [39] reported recently the presence of a subgroup of anti-flagellin antibodies (anti-CBir1) in 4/26 HIV-1-infected patients with CD4+ T-cells counts <300 cells/ul and high LPS levels. The CBir1 flagellin has been identified as an immune dominant flagellin in Crohn's disease and linked to Clostridia species. Interestingly a recently presented work has shown alterations in bacterial composition of microbiota during HIV-1 infection with significantly lower ratio of Clostridia taxa in faeces obtained from HIV-1-infected patients as compared to controls [40]. Although our assay is based on the recognition of *Salmonella typhimurium* flagellin which spans the two well conserved N- and C-terminal domains of flagellin [41] enabling us the detection of broader range of flagellin antibodies, additional studies are needed to determine the antibody specificity. Furthermore, our results deserve future studies of adaptive flagellin immune response in a larger cohort of HIV-1-infected individuals.

In summary, the novelty of our findings is that the bacterial products and HMGB1 form active complexes which can efficiently not only create a proinflammatory milieu but also directly trigger viral replication in infected cells. This synergistic effect may require lower levels of the interacting substances when present in complexes, as suggested by others [19]. We also report that flagellin has to be considered as a microbial product that can contribute to the immune activation during the HIV-1 infection. The formation of HMGB1/TLR ligand complexes has direct implications on immune activation, particularly in late stage of disease, where cell destruction and necrosis are dominant phenomena due to CD4+ T-cell loss, opportunistic infections, and other pathological conditions [27].

Acknowledgments

The study was supported by the Swedish Medical Research Council, Stockholm County Council, Swedish Physicians against AIDS, Swedish Medical Society (SLS), and Swedish Society for Medical Research (SSMF for Piotr Nowak). Data included in the paper were presented in part at the CROI meeting 2010 and 2011.

References

[1] J. B. Dinoso, S. Y. Kim, A. M. Wiegand et al., "Treatment intensification does not reduce residual HIV-1 viremia in patients on highly active antiretroviral therapy," *Proceedings of the National Academy of Sciences of the United States of America*, vol. 106, no. 23, pp. 9403–9408, 2009.

[2] M. D. Hazenberg, S. A. Otto, B. H. B. Van Benthem et al., "Persistent immune activation in HIV-1 infection is associated with progression to AIDS," *AIDS*, vol. 17, no. 13, pp. 1881–1888, 2003.

[3] M. Hellerstein, M. B. Hanley, D. Cesar et al., "Directly measured kinetics of circulating T lymphocytes in normal and HIV-1-infected humans," *Nature Medicine*, vol. 5, no. 1, pp. 83–89, 1999.

[4] J. V. Giorgi, L. E. Hultin, J. A. McKeating et al., "Shorter survival in advanced human immunodeficiency virus type 1 infection is more closely associated with T lymphocyte activation than with plasma virus burden or virus chemokine coreceptor usage," *Journal of Infectious Diseases*, vol. 179, no. 4, pp. 859–870, 1999.

[5] S. G. Deeks and A. N. Phillips, "HIV infection, antiretroviral treatment, ageing, and non-AIDS related morbidity," *Bratislava Medical Journal*, vol. 338, article a3172, 2009.

[6] J. M. Brenchley, D. A. Price, and D. C. Douek, "HIV disease: fallout from a mucosal catastrophe?" *Nature Immunology*, vol. 7, no. 3, pp. 235–239, 2006.

[7] J. M. Brenchley and D. C. Douek, "The mucosal barrier and immune activation in HIV pathogenesis," *Current Opinion in HIV and AIDS*, vol. 3, no. 3, pp. 356–361, 2008.

[8] J. M. Brenchley and D. C. Douek, "HIV infection and the gastrointestinal immune system," *Mucosal Immunology*, vol. 1, no. 1, pp. 23–30, 2008.

[9] J. M. Brenchley, D. A. Price, T. W. Schacker et al., "Microbial translocation is a cause of systemic immune activation in chronic HIV infection," *Nature Medicine*, vol. 12, no. 12, pp. 1365–1371, 2006.

[10] S. Baroncelli, C. M. Galluzzo, M. F. Pirillo et al., "Microbial translocation is associated with residual viral replication in HAART-treated HIV+ subjects with <50 copies/ml HIV-1 RNA," *Journal of Clinical Virology*, vol. 46, no. 4, pp. 367–370, 2009.

[11] M. Trøseid, P. Nowak, J. Nyström, A. Lindkvist, S. Abdurahman, and A. Sönnerborg, "Elevated plasma levels of lipopolysaccharide and high mobility group box-1 protein are associated with high viral load in HIV-1 infection: reduction by 2-year antiretroviral therapy," *AIDS*, vol. 24, no. 11, pp. 1733–1737, 2010.

[12] P. Nowak, B. Barqasho, C. J. Treutiger et al., "HMGB1 activates replication of latent HIV-1 in a monocytic cell-line, but inhibits HIV-1 replication in primary macrophages," *Cytokine*, vol. 34, no. 1-2, pp. 17–23, 2006.

[13] S. Thierry, J. Gozlan, A. Jaulmes et al., "High-mobility group box 1 protein induces HIV-1 expression from persistently infected cells," *AIDS*, vol. 21, no. 3, pp. 283–292, 2007.

[14] L. Cassetta, O. Fortunato, L. Adduce et al., "Extracellular high mobility group box-1 inhibits R5 and X4 HIV-1 strains replication in mononuclear phagocytes without induction of chemokines and cytokines," *AIDS*, vol. 23, no. 5, pp. 567–577, 2009.

[15] P. Nowak, B. Barqasho, and A. Sönnerborg, "Elevated plasma levels of high mobility group box protein 1 in patients with HIV-1 infection," *AIDS*, vol. 21, no. 7, pp. 869–871, 2007.

[16] U. Andersson, H. Wang, K. Palmblad et al., "High mobility group 1 protein (HMG-1) stimulates proinflammatory cytokine synthesis in human monocytes," *Journal of Experimental Medicine*, vol. 192, no. 4, pp. 565–570, 2000.

[17] C. Fiuza, M. Bustin, S. Talwar et al., "Inflammation-promoting activity of HMGB1 on human microvascular endothelial cells," *Blood*, vol. 101, no. 7, pp. 2652–2660, 2003.

[18] A. Rouhiainen, S. Tumova, L. Valmu, N. Kalkkinen, and H.

Rauvala, "Pivotal advance: analysis of proinflammatory activity of highly purified eukaryotic recombinant HMGB1 (amphoterin)," *Journal of Leukocyte Biology*, vol. 81, no. 1, pp. 49–58, 2007.

[19] H. S. Hreggvidsdottir, T. Östberg, H. Wähämaa et al., "The alarmin HMGB1 acts in synergy with endogenous and exogenous danger signals to promote inflammation," *Journal of Leukocyte Biology*, vol. 86, no. 3, pp. 655–662, 2009.

[20] M. E. Bianchi, "HMGB1 loves company," *Journal of Leukocyte Biology*, vol. 86, no. 3, pp. 573–576, 2009.

[21] S. Ivanov, A. M. Dragoi, X. Wang et al., "A novel role for HMGB1 in TLR9-mediated inflammatory responses to CpG-DNA," *Blood*, vol. 110, no. 6, pp. 1970–1981, 2007.

[22] B. Barqasho, P. Nowak, S. Abdurahman, L. Walther-Jallow, and A. Sönnerborg, "Implications of the release of high-mobility group box 1 protein from dying cells during human immunodeficiency virus type 1 infection in vitro," *Journal of General Virology*, vol. 91, no. 7, pp. 1800–1809, 2010.

[23] S. V. Sitaraman, J. M. Klapproth, D. A. Moore et al., "Elevated flagellin-specific immunoglobulins in Crohn's disease," *American Journal of Physiology*, vol. 288, no. 2, pp. G403–G406, 2005.

[24] T. R. Ziegler, M. Luo, C. F. Estívariz et al., "Detectable serum flagellin and lipopolysaccharide and upregulated anti-flagellin and lipopolysaccharide immunoglobulins in human short bowel syndrome," *American Journal of Physiology*, vol. 294, no. 2, pp. R402–R410, 2008.

[25] W. Jiang, M. M. Lederman, P. Hunt et al., "Plasma levels of bacterial DNA correlate with immune activation and the magnitude of immune restoration in persons with antiretroviral-treated HIV infection," *Journal of Infectious Diseases*, vol. 199, no. 8, pp. 1177–1185, 2009.

[26] N. G. Sandler, H. Wand, A. Roque et al., "Plasma levels of soluble CD14 independently predict mortality in HIV infection," *Journal of Infectious Diseases*, vol. 203, no. 6, pp. 780–790, 2011.

[27] M. Troseid, A. Sönnerborg, and P. Nowak, "High mobility group box protein-1 in HIV-1 infection: connecting microbial translocation, cell death and immune activation," *Current HIV Research*, vol. 9, no. 1, pp. 6–10, 2011.

[28] J. Tian, A. M. Avalos, S. Y. Mao et al., "Toll-like receptor 9-dependent activation by DNA-containing immune complexes is mediated by HMGB1 and RAGE," *Nature Immunology*, vol. 8, no. 5, pp. 487–496, 2007.

[29] M. Vijay-Kumar and A. T. Gewirtz, "Flagellin: key target of mucosal innate immunity," *Mucosal Immunology*, vol. 2, no. 3, pp. 197–205, 2009.

[30] F. Hayashi, K. D. Smith, A. Ozinsky et al., "The innate immune response to bacterial flagellin is mediated by Toll-like receptor 5," *Nature*, vol. 410, no. 6832, pp. 1099–1103, 2001.

[31] M. Vijay-Kumar, J. D. Aitken, and A. T. Gewirtz, "Toll like receptor-5: protecting the gut from enteric microbes," *Seminars in Immunopathology*, vol. 30, no. 1, pp. 11–21, 2008.

[32] M. J. Lodes, Y. Cong, C. O. Elson et al., "Bacterial flagellin is a dominant antigen in Crohn disease," *Journal of Clinical Investigation*, vol. 113, no. 9, pp. 1296–1306, 2004.

[33] M. Vijay-Kumar and A. T. Gewirtz, "Role of flagellin in Crohn's disease: emblematic of the progress and enigmas in understanding inflammatory bowel disease," *Inflammatory Bowel Diseases*, vol. 15, no. 5, pp. 789–795, 2009.

[34] A. T. Gewirtz, T. A. Navas, S. Lyons, P. J. Godowski, and J. L. Madara, "Cutting edge: bacterial flagellin activates basolaterally expressed TLR5 to induce epithelial proinflammatory gene expression," *Journal of Immunology*, vol. 167, no. 4, pp. 1882–1885, 2001.

[35] N. Funderburg, A. A. Luciano, W. Jiang, B. Rodriguez, S. F. Sieg, and M. M. Lederman, "Toll-like receptor ligands induce human T cell activation and death, a model for HIV pathogenesis," *PLoS ONE*, vol. 3, no. 4, Article ID e1915, 2008.

[36] S. Thibault, M. Imbeault, M. R. Tardif, and M. J. Tremblay, "TLR5 stimulation is sufficient to trigger reactivation of latent HIV-1 provirus in T lymphoid cells and activate virus gene expression in central memory CD4+ T cells," *Virology*, vol. 389, no. 1-2, pp. 20–25, 2009.

[37] H. J. Epple, T. Schneider, H. Troeger et al., "Impairment of the intestinal barrier is evident in untreated but absent in suppressively treated HIV-infected patients," *Gut*, vol. 58, no. 2, pp. 220–227, 2009.

[38] S. R. Targan, C. J. Landers, H. Yang et al., "Antibodies to CBir1 flagellin define a unique response that is associated independently with complicated Crohn's disease," *Gastroenterology*, vol. 128, no. 7, pp. 2020–2028, 2005.

[39] A. Kamat, P. Ancuta, R. S. Blumberg, and D. Gabuzda, "Serological markers for inflammatory bowel disease in aids patients with evidence of microbial translocation," *PLoS ONE*, vol. 5, no. 11, Article ID e15533, 2010.

[40] E. Collin, C.-S. Li, Z.-M. Ma, H. Overman, and T. Knight, "Shifts in proportions of higher-taxa bacterial orders measured by 16S rDNA PCR in the stools of HIV patients following ART and correlations with HIV immunopathogenesis," in *Proceedings of the 18th Conference on Retroviruses and Opportunistic Infections*, Boston, Mass, USA, 2011.

[41] R. M. Macnab, "How bacteria assemble flagella," *Annual Review of Microbiology*, vol. 57, pp. 77–100, 2003.

10

Histoplasma Virulence and Host Responses

Mircea Radu Mihu[1] and Joshua Daniel Nosanchuk[2]

[1] *Department of Medicine, Sound Shore Medical Center of Westchester, New Rochelle, NY 10802, USA*
[2] *Division of Infectious Diseases, Departments of Medicine and Microbiology and Immunology, Albert Einstein College of Medicine, Bronx, NY 10461, USA*

Correspondence should be addressed to Joshua Daniel Nosanchuk, josh.nosanchuk@einstein.yu.edu

Academic Editor: Julian R. Naglik

Histoplasma capsulatum is the most prevalent cause of fungal respiratory disease. The disease extent and outcomes are the result of the complex interaction between the pathogen and a host's immune system. The focus of our paper consists in presenting the current knowledge regarding the multiple facets of the dynamic host-pathogen relationship in the context of the virulence arsenal displayed by the fungus and the innate and adaptive immune responses of the host.

1. Introduction

Histoplasmosis was first described in 1906 by Darling among the workers of the Panama Canal [1], and it is currently the most common cause of fungal respiratory disease with almost 500,000 individuals acquiring the fungus each year [2]. The etiologic agent responsible for histoplasmosis is *Histoplasma capsulatum*, a thermally dimorphic fungus with worldwide distribution. The fungus is primarily found in soil, where it exists in a mycelia form. In the United States, highly endemic areas include regions along the Mississippi and Ohio River valleys, where seroprevalence studies have shown that up to 80% of individuals are skin test positive for histoplasmin [3].

The entry portal of *H. capsulatum* is through inhalation of aerosolized of 2–4 μm diameter microconidia [4]. Morphogenesis is initiated after infection with the conidia developing into a 2–4 μm oval yeast form. The fungus is rapidly ingested by macrophages and neutrophils, but manages to avoid intracellular destruction. Intracellular yeast can be transported diffusely via the lymphatics and into the bloodstream. Nevertheless, initial infection is typically contained by innate and adaptive host responses. In immunocompetent individuals, the pulmonary disease is usually subclinical to limited, typically with flu-like symptoms, including fever, cough, headaches, and myalgias. However, lethal disease can occur in otherwise healthy individuals who acquire a large inoculum infection. Additionally, severe primary infection is more common and reactivation of latent infection occurs in immunocompromised persons, particularly in HIV-infected population and transplant recipients. Disseminated disease occurs in a small fraction of infected individuals, but this form of histoplasmosis continues to carry a high fatality rate even in patients receiving appropriate medical treatment [5]. More recently, treatments with inhibitors of tumor necrosis factor-α have been shown to place patients at high risk for developing histoplasmosis [6].

2. *H. capsulatum* Virulence Factors

The characterized virulence determinants of *H. capsulatum* are mainly surface expressed molecules that mediate the interaction between the fungus and the host's immune cells allowing the pathogen to evade destruction by innate immune response and facilitate the replication of the yeast in its new environment.

Heat shock protein 60 (HSP60), which has important roles in chaperoning intracellular proteins and supervising adequate protein folding, has also recently been described as an essential surface molecule, mediating the recognition and phagocytosis of the yeast by macrophages [7]. It acts as a ligand for CD11/CD18 macrophage receptor and, despite low number of antigenic sites, the coupling with the CR3 receptor is followed by rapid ingestion of the yeast. The interaction between *Histoplasma* and macrophage through HSP60 binding to CR3 results only in a mild host immune reaction, as it does not lead to a significant activation of phagocytes in the absence of other costimulatory signals [8]. This process allows the microorganism to survive and replicate inside the host cells [9, 10].

Heat shock protein 82 (HSP82) is another important molecule in normal development of *H. capsulatum* that also participates in the response to cellular stresses; it binds to a variety of cellular proteins, keeping them inactive until they have reached their proper intracellular location or have received the proper activation signal [11]. The role of HSP82 is further complicated by the thermal dimorphism displayed by *Histoplasma*, since changes in temperature represent both a stress inducer and a signal for yeast phase transformation. Recently, Edwards et al. demonstrated that a reduction in HSP82 decreases *Histoplasma* virulence in macrophages and severely impairs the fungus' ability to infect lungs in a murine infection model [12]. This suggests that a low basal level of HSP82 expression is sufficient only to preserve cellular functions at mammalian body temperature, but not to withstand other stresses encountered during infection. The temperature values during febrile episodes in the host, for example, represent one stress that requires HSP82 function, and defective mutant strains show a decreased ability to recover from transient in vitro incubation at 40°C. However, even at 37°C these defective mutants showed decreased virulence within macrophages, despite having identical growth in culture at similar temperature; this implies that HSP82 extends its role to enduring additional, nonthermal stresses during host infection. One example is the ability to survive oxidative stress, as shown by peroxide challenge studies [12].

YPS3 is a yeast phase-specific gene encountered in a subset of *H. capsulatum* strains. The encoded protein is found both as a fungal cell wall constituent and as a secreted molecule [13]. The exact function of this gene remains to be determined, but its importance in virulence is a certainty, since YPS3 mutants are attenuated in vivo [14].

The production of cell wall melanin is associated with virulence for diverse fungi; *H. capsulatum* conidia and yeast produce melanin or melanin-like pigments in vitro and yeast cells are melanized during mammalian infections [15]. The melanization process decreases the susceptibility of the fungus to amphotericin B and caspofungin and melanin can abrogate the potency of certain host defense mechanisms, such as free radicals and microbicidal peptides [16, 17].

Calcium-binding protein (CBP) represents another important factor in *Histoplasma* pathogenicity. CBP is secreted by the fungal cells during the yeast-phase of intracellular growth within the macrophage [18], and its importance for the virulence of the fungus was demonstrated both in vitro and in vivo. For example, CBP1 gene deletion yeast cells were rapidly cleared from the lungs of infected mice. Additionally, *H. capsulatum* growth is inhibited in limiting calcium conditions [19]. One of the hypotheses is that calcium acquisition represents an important factor for intracellular survival of the microorganism; another hypothesis targets the modulating effect of CBP in binding calcium to facilitate optimal phagolysosomal conditions for yeast growth.

Many *H. capsulatum* strains express α-(1,3)-glucan on their yeast cell surface. This polysaccharide forms a layer that conceals cell surface β glucans, which have antigenic properties, eluding the identification by the host phagocytic cells. The β glucan found in the cell wall of *Histoplasma* and other fungi is recognized by the Dectin-1 receptor on macrophages resulting in the triggered formation of reactive oxygen species and secretion of proinflammatory cytokines [20]. Confirmatory evidence for the role of the α-glucan was obtained using *Ags-1*-deficient *Histoplasma* mutant strains, where yeasts lacking the cell wall α-(1,3)-glucan were attenuated for virulence [21].

Histone 2B (H2B) has also been found to play a role in pathogenesis [22]. Histones are mainly intracellular components, but a study investigating passive immunity through administration of monoclonal antibodies from mice immunized with *Histoplasma* revealed antibody recognition of H2B present on the cell wall. The means by which historically intracellular-based molecules, as HSP60 or H2B, reach the yeast cell wall where they can interact with host cells was unclear until recently when macromolecular transport to the extracellular space was demonstrated to occur via vesicular secretion and active vesicular transport [23, 24].

Hydroxamate siderophores production by *Histoplasma* is another newly characterized virulence factor. Strains defective for the gene coding for siderophore production display impaired intracellular growth in both human and murine macrophages, which can be reversed by either exogenous iron addition or restoration of SID1 expression [25, 26].

3. Host Defense Mechanisms

After being exposed to *Histoplasma*, the host relies on both innate and adaptive immune response mechanisms to neutralize the pathogen and withstand infection. Macrophages and dendritic cells have major roles in the activation of cellular pathways, and the numerous cytokines, especially IFN-γ and TNF-α, significantly impact host responses.

Macrophages have a central role in the interaction between the fungus and the host, although their contribution has a dual nature. They represent the first line of defense during infection with *H. capsulatum*, as they rapidly phagocytose the inhaled conidia and transforming yeast cells, and the infected macrophage subsequently activate effector T cells and enhance the release of Th1-associated proinflammatory cytokines (IL-12, IFN-γ, and TNF-α) [27, 28]. Deprivation of zinc and iron is amongst the means used by macrophages to neutralize the intruding pathogen [29], along with production of superoxide, nitric oxide, and lysosomal hydrolysis. However, the fungus displays various mechanisms to elude destruction after phagocytosis. For example, *H. capsulatum* yeast cells are able to regulate the pH of the phagolysosomes at a neutral pH (approximately pH 6.5), where lysosomal hydrolases have decreased activity. Hence, the *H. capsulatum* yeast cells manage to survive and even replicate inside macrophages [30].

Dendritic cells are also an important effector of the innate immunity. They are able to phagocytose and degrade the fungal cells with higher efficacy than macrophages, which might be due to recognition of the pathogen via a different type of receptor (fibronectin receptor on dendritic cells versus CD18 on macrophages) [31]. Dendritic cells also are extremely efficient at processing and presenting antigens to specific CD8 T cells, either following ingestion of the yeast, or through "cross presentation" of fungal antigens engulfed from infected apoptotic macrophages [32]. In a recent study, the addition of antigen-presenting dendritic cells was found to suppress excessive production of IL-4 by CD4 T cells in lungs of CCR2-deficient mice infected with *H. capsulatum*, demonstrating the importance of these cells in the regulation of immune responses [33].

Cellular immunity is crucial in the host defense against intracellular pathogens; therefore, T cells, as the central effectors of the cellular immunity, have a substantial role in neutralizing *H. capsulatum* yeast cells. Mice depleted of both CD4 and CD8 T cells have accelerated time to death after challenge with *H. capsulatum* yeast cells, especially in a primary histoplasmosis model, which underlines the importance of the interaction between the two cell subsets in withstanding *Histoplasma* infection by eliciting a Th1 response [34]. CD4 cell depletion is associated with survival during primary infection, as a result of impaired IFN-γ production. The elimination of CD8 T cells results in decreased clearance of yeast cells in primary but not secondary infection. One particular subpopulation of T cells, Vβ4$^+$ T cells, is preferentially expended during infection with *H. capsulatum*, and elimination of these cells from mice impairs their ability to resolve infection [35]. Th17 T cells and their interaction with regulatory T cells have recently been linked via the chemoattractant mediator CCR5 to the host's ability to effectively combat *H. capsulatum* infection; increases in Th17 cytokines and reductions in the number of regulatory T cells were associated with accelerated fungal clearance in CCR5-deficient animals [36].

Although cytokine responses are complex in histoplasmosis and alter over the course of disease, the main cytokines involved in *Histoplasma* clearance from the host are IL-12, IFN-γ, and TNF-α [34]. IL-12 through its ability to regulate IFN-γ production is critical in inducing a protective immune response in primary infection with the pathogen. IFN-γ is pivotal for the host's innate resistance to systemic infection with *H. capsulatum*. Survival of mice is significantly reduced in IFN-γ-deficient mice as well as in wild-type mice treated with neutralizing antibody to IFN-γ [37]. Patients with impaired IFN-γ signaling due to genetic defects are at increased risk for severe disease forms and administration of the cytokine can be therapeutic. For example, a report of recurrent disseminated *H. capsulatum* osteomyelitis in a patient with genetic IFN-γ receptor 1 deficiency describes progressive clearing of all bone lesions and normalization of inflammatory markers following subcutaneous therapy with IFN-γ [38]. Although IFN-γ is critical in primary infection, survival in secondary infection can be achieved in the absence of IFN-γ, as immunization of IFN-γ-deficient mice with an initial sublethal inoculum can prolong the survival of these mice when subsequently challenged with a high concentration of *H. capsulatum* yeast cells [39]. The major mechanism by which these mice were able to control secondary infection was through increased production of TNF-α.

TNF-α is a key modulator of disease in both primary and secondary histoplasmosis, though different protective mechanisms are involved in these conditions [34, 40]. In primary infection, decreased survival of TNF-α-deficient mice has been attributed to an impaired ability to generate reactive nitrogen intermediates in the alveolar macrophages, although inducible nitric oxide synthase expression in lung tissue is preserved. During secondary infection, the increased mortality is largely due to a biased host reaction to a Th2-type response that is associated with elevated levels of both IL-4 and IL-10. These findings parallel the clinical data that clearly demonstrates that therapy with TNF-α inhibitors poses a significant increased risk for reactivation of latent histoplasmosis with a greater likelihood of severe, disseminated disease [8].

Humoral immune responses generally have a limited role in the clearance of intracellular pathogens; however, the protective role of antibodies against surface molecules of *H. capsulatum* has been described. Administration of monoclonal antibodies to *Histoplasma* H2B reduces fungal burden, decreases pulmonary inflammation, and prolongs survival in murine infection models [22]. The protective response was associated with increased levels of IL-4, IL-6, and IFN-γ. Similarly, antibodies to *H. capsulatum* HSP60 prolong the survival of the lethally infected animals [41, 42].

4. Discussion

Histoplasmosis is the most common endemic dimorphic fungal pathogen of man. The continuously expanding population of immunocompromised patients, secondary to the ongoing HIV epidemic, the increasing use of immunosuppressant therapies and rising number of transplant recipients, represents a high risk cohort for histoplasmosis. The mortality rate associated with invasive histoplasmosis is still unacceptably high, despite the use of broad spectrum antifungal agents, which emphasizes the need for developing novel therapies and effective preventive strategies.

As outlined in this paper, targeting virulence determinants of *H. capsulatum* and attempts to modify the capacity of the host to respond to the fungal invader are actively being pursued by researchers. Recent studies investigating the capacity of *H. capsulatum* to release a large number of proteins and other immunologically active compounds [23, 24, 43, 44] demonstrate the breadth of the fungus' ability to modify host responses. The high frequency of disease in the endemic areas and the increasing prevalence of the disseminated disease forms justify the development of adequate immunization strategies. Most recently, data has shown that vaccine-induced fungus-specific Th17 cells can confer protection against pulmonary histoplasmosis by recruiting and activating neutrophils and macrophages to the alveolar space [45]. Harnessing the host's existing armamentarium or supplementing the host's capacity, such as with the administration of cytokines or antibody to *H. capsulatum*, will be rich areas of study for the future.

References

[1] S. T. Darling, "The morphology of the parasite (*Histoplasma capsulatum*) and the lesions of histoplasmosis, a fatal disease of tropical America," *Journal of Experimental Medicine*, vol. 11, no. 4, pp. 515–531, 1909.

[2] L. J. Wheat and C. A. Kauffman, "Histoplasmosis," *Infectious Disease Clinics of North America*, vol. 17, no. 1, pp. 1–19, 2003.

[3] http://www.cdc.gov/nczved/divisions/dfbmd/diseases/histoplasmosis .

[4] R. A. Goodwin, J. E. Loyd, and R. M. Des Prez, "Histoplasmosis in normal hosts," *Medicine*, vol. 60, no. 4, pp. 231–266, 1981.

[5] G. S. Deepe Jr., "The immune response to *Histoplasma capsulatum*: unearthing its secrets," *Journal of Laboratory and Clinical Medicine*, vol. 123, no. 2, pp. 201–205, 1994.

[6] G. S. Deepe Jr., "Modulation of infection with *Histoplasma capsulatum* by inhibition of tumor necrosis factor-α activity," *Clinical Infectious Diseases*, vol. 41, no. 3, pp. S204–S207, 2005.

[7] K. H. Long, F. J. Gomez, R. E. Morris, and S. L. Newman, "Identification of heat shock protein 60 as the ligand on *Histoplasma capsulatum* that mediates binding to CD18 receptors on human macrophages," *Journal of Immunology*, vol. 170, no. 1, pp. 487–494, 2003.

[8] M. R. W. Ehlers, "CR3: a general purpose adhesion-recognition receptor essential for innate immunity," *Microbes and Infection*, vol. 2, no. 3, pp. 289–294, 2000.

[9] L. G. Eissenberg and W. E. Goldman, "*Histoplasma capsulatum* fails to trigger release of superoxide from macrophages," *Infection and Immunity*, vol. 55, no. 1, pp. 29–34, 1987.

[10] J. E. Wolf, V. Kerchberger, G. S. Kobayashi, and J. R. Little, "Modulation of the macrophage oxidative burst by *Histoplasma capsulatum*," *Journal of Immunology*, vol. 138, no. 2, pp. 582–586, 1987.

[11] K. A. Borkovich, F. W. Farrelly, D. B. Finkelstein, J. Taulien, and S. Lindquist, "Hsp82 Is an essential protein that is required in higher concentrations for growth of cells at higher temperatures," *Molecular and Cellular Biology*, vol. 9, no. 9, pp. 3919–3930, 1989.

[12] J. A. Edwards, O. Zemska, and C. A. Rappleye, "Discovery of a role for Hsp82 in Histoplasma virulence through a quantitative screen for macrophage lethality," *Infection and Immunity*, vol. 79, no. 8, pp. 3348–3357, 2011.

[13] C. H. Weaver, K. C. F. Sheehan, and E. J. Keath, "Localization of a yeast-phase-specific gene product to the cell wall in *Histoplasma capsulatum*," *Infection and Immunity*, vol. 64, no. 8, pp. 3048–3054, 1996.

[14] M. L. Bohse and J. P. Woods, "RNA interference-mediated silencing of the YPS3 gene of *Histoplasma capsulatum* reveals virulence defects," *Infection and Immunity*, vol. 75, no. 6, pp. 2811–2817, 2007.

[15] J. D. Nosanchuk, B. L. Gómez, S. Youngchim et al., "*Histoplasma capsulatum* synthesizes melanin-like pigments in vitro and during mammalian infection," *Infection and Immunity*, vol. 70, no. 9, pp. 5124–5131, 2002.

[16] D. Van Duin, A. Casadevall, and J. D. Nosanchuk, "Melanization of *Cryptococcus neoformans* and *Histoplasma capsulatum* reduces their susceptibilities to amphotericin B and caspofungin," *Antimicrobial Agents and Chemotherapy*, vol. 46, no. 11, pp. 3394–3400, 2002.

[17] J. D. Nosanchuk and A. Casadevall, "Impact of melanin on microbial virulence and clinical resistance to antimicrobial compounds," *Antimicrobial Agents and Chemotherapy*, vol. 50, no. 11, pp. 3519–3528, 2006.

[18] J. W. Batanghari, G. S. Deepe Jr., E. Di Cera, and W. E. Goldman, "*Histoplasma* acquisition of calcium and expression of CBP1 during intracellular parasitism," *Molecular Microbiology*, vol. 27, no. 3, pp. 531–539, 1998.

[19] T. S. Sebghati, J. T. Engle, and W. E. Goldman, "Intracellular parasitism by *Histoplasma capsulatum*: fungal virulence and calcium dependence," *Science*, vol. 290, no. 5495, pp. 1368–1372, 2000.

[20] C. A. Rappleye, L. G. Eissenberg, and W. E. Goldman, "*Histoplasma capsulatum* α-(1,3)-glucan blocks innate immune recognition by the β-glucan receptor," *Proceedings of the National Academy of Sciences of the United States of America*, vol. 104, no. 4, pp. 1366–1370, 2007.

[21] C. A. Rappleye, J. T. Engle, and W. E. Goldman, "RNA interference in *Histoplasma capsulatum* demonstrates a role for α-(1,3)-glucan in virulence," *Molecular Microbiology*, vol. 53, no. 1, pp. 153–165, 2004.

[22] J. D. Nosanchuk, J. N. Steenbergen, L. Shi, G. S. Deepe Jr., and A. Casadevall, "Antibodies to a cell surface histone-like protein protect against *Histoplasma capsulatum*," *Journal of Clinical Investigation*, vol. 112, no. 8, pp. 1164–1175, 2003.

[23] J. D. Nosanchuk, L. Nimrichter, A. Casadevall, and M. L. Rodrigues, "A role for vesicular transport of macromolecules across cell walls in fungal pathogenesis," *Communicative and Integrative Biology*, vol. 1, no. 1, pp. 37–39, 2008.

[24] P. C. Albuquerque, E. S. Nakayasu, M. L. Rodrigues et al., "Vesicular transport in *Histoplasma capsulatum*: an effective mechanism for trans-cell wall transfer of proteins and lipids in ascomycetes," *Cellular Microbiology*, vol. 10, no. 8, pp. 1695–1710, 2008.

[25] J. Hilty, A. George Smulian, and S. L. Newman, "*Histoplasma capsulatum* utilizes siderophores for intracellular iron acquisition in macrophages," *Medical Mycology*, vol. 49, no. 6, pp. 633–642, 2011.

[26] L. H. Hwang, J. A. Mayfield, J. Rine, and A. Sil, "Histoplasma requires SID1, a member of an iron-regulated siderophore gene cluster, for host colonization," *PLoS Pathogens*, vol. 4, no. 4, Article ID e1000044, 2008.

[27] E. Lázár-Molnár, A. Gácser, G. J. Freeman, S. C. Almo, S. G. Nathenson, and J. D. Nosanchuk, "The PD-1/PD-L costimulatory pathway critically affects host resistance to the pathogenic fungus *Histoplasma capsulatum*," *Proceedings of the National Academy of Sciences of the United States of America*, vol. 105, no. 7, pp. 2658–2663, 2008.

[28] P. Zhou, M. C. Sieve, J. Bennett et al., "IL-12 prevents mortality in mice infected with *Histoplasma capsulatum* through induction of IFN-γ," *Journal of Immunology*, vol. 155, no. 2, pp. 785–795, 1995.

[29] M. S. Winters, Q. Chan, J. A. Caruso, and G. S. Deepe Jr., "Metallomic analysis of macrophages infected with *Histoplasma capsulatum* reveals a fundamental role for zinc in host defenses," *Journal of Infectious Diseases*, vol. 202, no. 7, pp. 1136–1145, 2010.

[30] K. Seider, A. Heyken, A. Lüttich, P. Miramón, and B. Hube, "Interaction of pathogenic yeasts with phagocytes: survival, persistence and escape," *Current Opinion in Microbiology*, vol. 13, no. 4, pp. 392–400, 2010.

[31] L. A. Gildea, R. E. Morris, and S. L. Newman, "*Histoplasma capsulatum* yeasts are phagocytosed via very late antigen-5, killed, and processed for antigen presentation by human dendritic cells," *Journal of Immunology*, vol. 166, no. 2, pp. 1049–1056, 2001.

[32] J. S. Lin, C. W. Yang, D. W. Wang, and B. A. Wu-Hsieh, "Dendritic cells cross-present exogenous fungal antigens to stimulate a protective CD8 T cell response in infection by *Histoplasma capsulatum*," *Journal of Immunology*, vol. 174, no. 10, pp. 6282–6291, 2005.

[33] W. A. Szymczak and G. S. Deepe Jr., "Antigen-presenting dendritic cells rescue CD4-depleted CCR2-/- mice from lethal *Histoplasma capsulatum* infection," *Infection and Immunity*, vol. 78, no. 5, pp. 2125–2137, 2010.

[34] R. Allendörfer, G. D. Brunner, and G. S. Deepe Jr., "Complex requirements for nascent and memory immunity in pulmonary histoplasmosis," *Journal of Immunology*, vol. 162, no. 12, pp. 7389–7396, 1999.

[35] F. J. Gomez, J. A. Cain, R. Gibbons, R. Allendoerfer, and G. S. Deepe Jr., "Vβ4+ T cells promote clearance of infection in murine pulmonary histoplasmosis," *Journal of Clinical Investigation*, vol. 102, no. 5, pp. 984–995, 1998.

[36] D. N. Kroetz and G. S. Deepe Jr., "CCR5 dictates the equilibrium of proinflammatory IL-17+and regulatory Foxp3+ T cells in fungal infection," *Journal of Immunology*, vol. 184, no. 9, pp. 5224–5231, 2010.

[37] K. V. Clemons, W. C. Darbonne, J. T. Curnutte, R. A. Sobel, and D. A. Stevens, "Experimental histoplasmosis in mice treated with anti-murine interferon- γ antibody and in interferon-γ gene knockout mice," *Microbes and Infection*, vol. 2, no. 9, pp. 997–1001, 2000.

[38] C. S. Zerbe and S. M. Holland, "Disseminated histoplasmosis in persons with interferon-gamma receptor 1 deficiency," *Clinical Infectious Diseases*, vol. 41, no. 4, pp. e38–41, 2005.

[39] P. Zhou, G. Miller, and R. A. Seder, "Factors involved in regulating primary and secondary immunity to infection with *Histoplasma capsulatum*: TNF-α plays a critical role in maintaining secondary immunity in the absence of IFN-γ," *Journal of Immunology*, vol. 160, no. 3, pp. 1359–1368, 1998.

[40] R. Allendoerfer and G. S. Deepe Jr., "Blockade of endogenous TNF-α exacerbates primary and secondary pulmonary histoplasmosis by differential mechanisms," *Journal of Immunology*, vol. 160, no. 12, pp. 6072–6082, 1998.

[41] A. J. Guimaraes, S. Frases, F. J. Gomez, R. M. Zancope-Oliveira, and J. D. Nosanchuk, "Monoclonal antibodies to heat shock protein 60 alter the pathogenesis of *Histoplasma capsulatum*," *Infection and Immunity*, vol. 77, no. 4, pp. 1357–1367, 2009.

[42] A. J. Guimarães, S. Frases, B. Pontes et al., "Agglutination of *Histoplasma capsulatum* by IgG monoclonal antibodies against Hsp60 impacts macrophage effector functions," *Infection and Immunity*, vol. 79, no. 2, pp. 918–927, 2011.

[43] A. J. Guimarães, E. S. Nakayasu, T. J. P. Sobreira et al., "*Histoplasma capsulatum* heat-shock 60 orchestrates the adaptation of the fungus to temperature stress," *PLoS ONE*, vol. 6, no. 2, article e14660, 2011.

[44] E. D. Holbrook, J. A. Edwards, B. H. Youseff, and C. A. Rappleye, "Definition of the extracellular proteome of pathogenic-phase *Histoplasma capsulatum*," *Journal of Proteome Research*, vol. 10, no. 4, pp. 1929–1943, 2011.

[45] M. Wüthrich, B. Gern, C. Y. Hung et al., "Vaccine-induced protection against 3 systemic mycoses endemic to North America requires Th17 cells in mice," *Journal of Clinical Investigation*, vol. 121, no. 2, pp. 554–568, 2011.

11

Abiotic and Biotic Factors Affecting Resting Spore Formation in the Mite Pathogen *Neozygites floridana*

Vanessa da Silveira Duarte,[1] Karin Westrum,[2] Ana Elizabete Lopes Ribeiro,[3] Manoel Guedes Corrêa Gondim Junior,[3] Ingeborg Klingen,[2] and Italo Delalibera Júnior[1]

[1] *Department of Entomology and Acarology, ESALQ/University of São Paulo, 13418-900 Piracicaba, SP, Brazil*
[2] *Plant Health and Plant Protection Division, Norwegian Institute for Agricultural and Environmental Research (Bioforsk), 1432 Ås, Norway*
[3] *Department of Agronomy, Federal Rural University of Pernambuco, 52171-900 Recife, PE, Brazil*

Correspondence should be addressed to Italo Delalibera Júnior; delalibera@usp.br

Academic Editor: Carla Pruzzo

Neozygites floridana is an obligate mite pathogenic fungus in the Entomophthoromycota. It has been suggested that resting spores of this fungus are produced as a strategy to survive adverse conditions. In the present study, possible mechanisms involved in the regulation of resting spore formation were investigated in the hosts *Tetranychus urticae* and *Tetranychus evansi*. Abiotic and biotic factors mimicking conditions that we, based on earlier field studies, thought might induce resting spores in temperate and tropical regions were tested with isolates from Norway and Brazil. A total of 42 combinations of conditions were tested, but only one induced the formation of a high number of resting spores in only one isolate. The Brazilian isolate ESALQ1420 produced a large number of resting spores (51.5%) in *T. urticae* at a temperature of 11°C, photoperiod of 10L:14D, and light intensity of 42–46 (μmol m^{-2} s^{-1}) on nonsenescent plants (nondiapausing females). Resting spores of the Brazilian *N. floridana* isolate ESALQ1421 were found at very low levels (up to 1.0%). Small percentages of *T. urticae* with resting spores (0–5.0%) were found for the Norwegian isolate NCRI271/04 under the conditions tested. The percentages of resting spores found for the Norwegian isolate in our laboratory studies are similar to the prevalence reported in earlier field studies.

1. Introduction

The entomopathogenic fungal genus *Neozygites* belongs to the order Neozygitales in the class Neozygitomycetes in the phylum Entomophthoromycota [1]. Fungi in this genus attack small arthropods such as mealybugs, aphids, thrips, and mites [2]. *Neozygites floridana* (Weiser and Muma) Remaudière and Keller is pathogenic to several species of plant-feeding spider mites [3], and it is an important natural enemy of the two-spotted spider mite, *Tetranychus urticae* Koch, and the red tomato spider mite, *Tetranychus evansi* Baker and Pritchard (Acari: Tetranychidae) [4–6].

For many of the fungal species within the Entomophthoromycota, zygospores and azygospores are important for fungal survival during periods of adverse conditions (e.g., winter, dry season, or host absence), and they are therefore called resting spores [7]. *N. floridana* is an obligate pathogen, and this fungal species may also form resting spores to survive adverse conditions [7–10]. Resting spores of *N. floridana* have been reported in the field in temperate regions in *T. urticae* populations in late summer, fall, and winter [6, 8, 11], and *N. floridana* resting spore prevalences of up to 13.8% were found in *T. urticae* in Norway [8]. Carner [12] suggested that *Neozygites* resting spores were restricted to northern/temperate regions, where the weather is often below freezing during the fall and winter. No field studies on the prevalence of resting spores of *N. floridana* under tropical conditions have been performed, but field studies with *Neozygites tanajoae* Delalibera Jr., Humber, Hajek showed that resting spores of *N. tanajoae* in *Mononychellus tanajoa*

Bondar populations were found under tropical conditions in Brazil. Low prevalences of resting spores of *N. tanajoae* in *M. tanajoa* (up to 3.8%) were detected in Brazil by Delalibera Jr. et al. [10], whereas higher prevalences (34–38%) were found by Elliot et al. [9]. However, resting spores of *Neozygites* have not been found in other studies in tropical regions [13–15].

Several factors, such as photoperiod, temperature, host age, inoculum density, and the fungal isolate, may be important for the induction of resting spores in fungi in the Entomophthoromycota [16–21]. For *Zoophthora radicans* (Brefeld) Batko, the resting spore production was negatively correlated with temperature and positively correlated with the relative humidity (RH) and inoculum density [21]. Hajek and Shimazu [20] tested the effects of temperature, photoperiod, and host molting status on resting spore formation by *Entomophaga maimaiga* Humber, Shimazu, and Soper in *Lymantria dispar* (L.); they found that the factor with the greatest impact on the type of spore produced was host age. Resting spore formation was negatively associated with larval molting status; the cadavers of those larvae that molted or exhibited premolt characteristics during the period between infection and death contained fewer resting spores. High levels of fungal inoculum also increased resting spore formation. In a field study, Thomsen and Eilenberg [19] found that *Entomophthora muscae* (Cohn) Fresenius forms resting spores only in female *Delia radicum* (L.) and that the proportion of females with resting spores was negatively correlated with day length. Further, Huang and Feng [18] hypothesized that the resting spore formation of the aphid pathogenic fungus *Pandora nouryi* (Remaudière and Hennebert) Humber depends on the inoculum concentration. Later, Zhou and Feng [17] tested the effects of three parameters on the resting spore formation of *P. nouryi*. Their results suggest that the most important factor for resting spore production is spore density but that temperature and photoperiod are also important. In an even more recent study, Zhou et al. [16] suggested that temperature is the most important factor for the resting spore production of *P. nouryi* in *Myzus persicae* Sulzer under winter field conditions. To the best of our knowledge, no controlled experiments have been conducted to determine which factors are most important for the induction of resting spores in *N. floridana* isolates from temperate or tropical regions. One laboratory study with a Brazilian strain of *N. tanajoae* reported resting spores in 24.2% of *M. tanajoa* individuals under conditions mimicking field conditions at which high prevalences of resting spores were found [9].

Therefore, in the present study, we conducted controlled experiments to identify factors that might be important for the induction of resting spores in *N. floridana* isolates from spider mites from temperate (Norway) and tropical (Brazil) regions. The conditions tested mimic the field conditions under which resting spores have been observed in temperate and tropical regions. Thus, we tested conditions found at the beginning of the dry season in tropical regions and conditions found during the fall and winter in temperate regions.

T. urticae females are known to hibernate during winter [22], and the diapause is induced by short day length [23], but temperature and a lack of nutrition from the host plant may also contribute to the induction of this stage [22]. We hypothesized that infection in diapausing mites might induce resting spore production in *N. floridana*; therefore, we tested diapausing and nondiapausing *T. urticae* as one of the variables in the temperate region treatments.

In the experiment mimicking tropical region conditions, the temperature and RH used were similar to the conditions under which resting spores were found in the field in northeast Brazil, as reported by Delalibera Jr. et al. [10] and Elliot et al. [9]. We also included an experiment where mites were coinfected with two strains of the fungus to test the effect of heterothallism on resting spore production. The nature of *N. floridana* resting spores is still unknown; Humber [24] affirms that there is evidence of heterothallism within the Entomophthoromycota, but Keller [3] suggested that there are indications that *Neozygites fresenii* might be heterothallic.

2. Materials and Methods

2.1. Experiments Mimicking Temperate Region Conditions

2.1.1. T. urticae Culture Reared on Nonsenescent and Senescent Plants. The *T. urticae* used in this culture was collected on the strawberry *Fragaria × ananassa* in Ås, Akershus, in southeastern Norway (59° 42″ N, 10° 44″ E) in 2003. *T. urticae* were reared on nonsenescent bean plants, *Phaseolus vulgaris* L., in an acclimatized room at 21°C, 60% RH, and L16 : D8. The plants were watered three times per week. Old and weak plants were replaced as needed, usually once a week.

Diapausing *T. urticae* was obtained from old bean plants by maintaining the old plants in a Plexiglas cage in the acclimatized room as described above, but these plants were watered only once per week to stress them and accelerate the process of plant senescence.

2.1.2. N. floridana Isolate. Norwegian and Brazilian *N. floridana* isolates were used in the experiments mimicking temperate region conditions. The Norwegian isolate (NCRI271/04) was collected in August 2004, in the same location at which the *T.urticae* was collected, and the Brazilian isolate (ESALQ1420) was collected from *T. urticae* on the jack bean, *Canavalia ensiformis*, in Piracicaba, SP, Brazil (22° 42′ 30″ S, 47° 38′ 00″ W).

2.1.3. N. floridana Cadaver Production. Leaf discs (1.5 cm diameter) from bean plants were placed underside up on 1.5% water agar in a Petri dish (5 cm in diameter and 2 cm high), and three *N. floridana*-killed *T. urticae* cadavers were placed with their dorsal sides up on the leaf disc. Petri dishes with cadavers on leaf discs were then placed in a plastic box (22 × 16 × 7 cm), covered with aluminum foil to ensure darkness, and incubated at 20°C and 90% RH in a climatic chamber. Cadavers were checked under a compound microscope (80x) after 24 h of incubation, and only the leaf discs with cadavers with good capilliconidia production were used. Thirty uninfected adult *T. urticae* females were then placed on each leaf disc with cadavers for *N. floridana* inoculation. Water was added to the water agar surrounding the leaf disk in the Petri dish to prevent the mites from escaping from the leaf disc. The leaf discs with *T. urticae* were then incubated for 24 h under the conditions described above. Leaf discs containing *N. floridana*-inoculated *T. urticae* were then transferred to uninfested bean plants after 24 h. The mites then walked from the leaf disc onto the bean plant and remained there until they died and mummified. Pods and tendrils were removed to prevent the plant from dangling and allowing the *T. urticae* to crawl off the plant. Leaves that overlapped or grew close together were also cut off to ensure a dry microclimate, keeping the newly mummified cadavers dry and preventing them from sporulating. Plants with *N. floridana*-inoculated *T. urticae* were kept under ambient laboratory conditions at 22–25°C, 20–30% RH, and 24 h light. The dry, nonsporulating cadavers produced on the plant were collected after 7–10 days and kept in small, unbleached cotton cloth pieces in 1.8 mL NUNC Cryo Tubes and stored at 5°C until used in experiments.

2.1.4. Experimental Setup for Abiotic (Light, Temperature) and Biotic (T. urticae "Diapause" Condition) Factors. To infect adult *T. urticae* females with the fungus *N. floridana*, we used the protocol described above. Inoculated mites were then transferred with a fine paintbrush onto a bean leaf disc (1.5 cm diameter) placed underside up on 1.5% water agar in 30 mL vials with lids. Twelve holes were made in the lids of the vials with a number 2 insect pin for aeration. At least 60 individual mites were included in each treatment for each isolate. Vials with adult *T. urticae* females were kept under the treatment conditions described in Table 1 until they died of *N. floridana* infection. The two light qualities tested were provided by (1) warm white fluorescent lamps (Philips-Master TL-D 90, referred to as "light quality 1" in this paper) and (2) cool white fluorescent lamps (Mitsubishi-40SW (Ra61), referred to as "light quality 2" in this paper). The effects of a short decrease in temperature were also tested and these treatments were maintained for 4 h at −10, −5, 0, or 5°C during the light period.

Adult *T. urticae* females were evaluated daily during the light period, and dead mites were then mounted in 0.075% Cotton Blue in 50% lactic acid to permit the observation of hyphal bodies and resting spores under a compound microscope (400x). The time of infection lethality (the time from infection to mite death) was calculated for mites with hyphal bodies and for mites with resting spores.

2.2. Experiments Mimicking Tropical Region Conditions

2.2.1. T. urticae and T. evansi Stock Cultures. Both spider mite species (*T. evansi* and *T. urticae*) were collected in Piracicaba, SP, Brazil (22° 42′ 30″ S, 47° 38′ 00″ W), and reared on plants maintained in the greenhouse. *T. evansi* was reared on tomato (*Solanum lycopersicum* L.), and *T. urticae* was reared on jack bean (*Canavalia ensiformis* L. (DC)).

2.2.2. N. floridana Isolate. The *N. floridana* isolates used in this experiment were collected from *T. urticae* (isolate ESALQ1420) and *T. evansi* (ESALQ1419) on jack bean and tomato, respectively, in Piracicaba, SP, Brazil (22° 42′ 30″ S, 47° 38′ 00″ W). A third *N. floridana* isolate (ESALQ1421) was collected from *T. evansi* on tomato, in Recife, PE, Brazil (8° 04′ 03″ S, 34° 55′ 00″ W).

2.2.3. N. floridana Cadaver Production. Leaf discs (1.2 cm diameter) from jack bean and tomato plants were placed underside up on top of a moist sponge in closed Petri dishes (9 cm diameter). One fungus-killed *T. urticae* or *T. evansi* cadaver was placed on the leaf disc. Petri dishes with cadavers on leaf discs were then placed in a paper box (20 × 20 × 10 cm) to create dark conditions and incubated at 25 ± 2°C and 100% RH in a climatic chamber. Cadavers were checked under a compound microscope (80x) after 24 h of incubation for the production of capilliconidia, and only leaf discs with cadavers with good sporulation and capilliconidia production were used. Twenty uninfected adult *T. urticae* or *T. evansi* females were then transferred with a fine paintbrush to the leaf disc with cadavers for *N. floridana* inoculation. Both spider mites were then incubated for 24 h under the conditions described above. Mites exposed to the sporulating cadavers were then transferred to leaf discs (20 mm diameter) from tomato or jack bean placed underside up on top of moist cotton pads in closed vials (30 mm diameter × 20 mm high) with lids and maintained in a climatic chamber at 25 ± 2°C, 50% RH, and 24 h light. Dry (nonsporulating) cadavers were collected 3–7 days later and kept at −10°C in vials containing silica gel until they were used in the experiment.

2.2.4. Experimental Setup to Test Abiotic (Temperature, RH) and Biotic (Coinfection, Plant Quality, and Mite Age) Factors. To infect spider mites with *N. floridana* isolates (ESALQ1419 or ESALQ1421 to *T. evansi* and ESALQ1420 to *T. urticae*), we used the same protocols described in Section 2.2. (*N. floridana* cadaver production). Vials with *T. urticae* or *T. evansi* were kept under the treatment conditions described in Tables 2 and 3 until they died of *N. floridana* infection.

Abiotic Factors: Temperature and RH. Inoculated mites were transferred with a fine paintbrush onto a tomato leaf disc (20 mm diameter) placed underside up on moist cotton pads in vials (30 mm diameter and 20 mm high) closed with lids.

The vials for each treatment were placed on metal supports inside chambers containing saturated salt solutions to achieve the desired humidity. The RH conditions inside the chambers were 50%, 70%, 80%, and 90%, obtained using saturated salt solutions of $Mg(NO_3)_2 6H_2O$, NaCl, KCl, and K_2SO_4, respectively, according to Winston and Bates [25]. The RH was measured by a hygrometer at the beginning of the experiment. The chambers were closed with Parafilm to maintain the same RH until the end of the experiment. The chambers were placed in incubators at 32 ± 2°C and 35 ± 2°C and a photoperiod of L12 : D12, and each chamber represented one treatment. After ten days, the mites were checked, and dead and live mites were mounted in Aman Blue for the observation of hyphal bodies and resting spores under a compound microscope (400x). Each treatment and isolate

TABLE 1: Effect of different combinations of photoperiod, mean temperature, temperature drop, light intensity, and light quality on resting spores produced in *N. floridana*-killed *T. urticae* by one isolate from Norway (NCRI271/04) and one from Brazil (ESALQ1420).

Photoperiod	Mean temperature (temperature drop[1]) °C	Host plant conditions	Light intensity (light quality)[2]	Isolate ESALQ1420			Isolate NCRI271/04		
				No. of mites	Hyphal bodies (%)	Resting spores (%)	No. of mites	Hyphal bodies (%)	Resting spores (%)
12L : 12D	25 (−10)	Nonsenescent	165–243 (2)	40	22.5	0	47	17.0	0
	25 (−5)			42	35.7	0	47	25.5	0
	25 (0)			40	30.0	0	46	26.1	0
	25 (5)			40	20.0	0	46	15.2	0
	15		42–46 (1)	60	78.3	0	59	79.7	0
10L : 14D	15 (−10)	Nonsenescent	165–243 (2)	72	50.0	0	69	73.9	0
			247–280 (1)	72	55.6	0	69	66.7	0
			30–35 (2)	72	54.2	0	72	38.9	0
			42–46 (1)	69	68.1	0	72	56.9	**1.4**
		Senescent	165–243 (2)	72	55.6	0	72	59.7	0
			247–280 (1)	71	73.2	0	72	52.8	**2.8**
			30–35 (2)	69	68.1	0	60	56.7	**5.0**
			42–46 (1)	60	81.7	0	63	68.3	**3.2**
	15	Nonsenescent	165–243 (2)	72	51.4	0	69	65.2	0
			247–280 (1)	72	68.1	**1.4**	64	75.0	0
			30–35 (2)	72	75.0	0	72	63.9	0
			42–46 (1)	72	55.6	0	72	50.0	**4.2**
		Senescent	165–243 (2)	72	86.1	0	72	58.3	**1.4**
			247–280 (1)	72	66.7	0	72	69.4	0
			30–35 (2)	69	78.3	0	60	70.0	**1.7**
			42–46 (1)	72	68.1	0	60	58.3	0
	13	Nonsenescent	42–46 (1)	60	91.7	0	60	70.0	0
	11			111	73.0	**51.5**	102	93.6	**4.7**
	6			60	83.3	0	58	62.1	0
14L : 10D	15	Nonsenescent	42–46 (1)	59	67.8	0	60	78.3	0
16L : 08D				58	93.1	0	59	75.7	**1.7**

[1]Temperature drop: fall of the temperature for 4 h during the light period. [2]Light intensity (μmol m^{-2} s^{-1}) and light quality (1 = warm white fluorescent lamps Philips-Master TL-D 90 and 2 = cool white fluorescent lamps Mitsubishi—40SW (Ra61)).

included at least 20 mites, and the experiment was repeated ten times.

Biotic Factors: Coinfection of Isolates. In this study, it was investigated whether coinfection with different *N. floridana* isolates would yield mating between individuals of opposite mating types and thus zygospores. This study was conducted by coinfecting *T. evansi* and *T. urticae* hosts with three different *N. floridana* isolates in the following different combinations: *T. evansi* isolate ESALQ1419 × *T. urticae* isolate ESALQ1420 and *T. evansi* isolate ESALQ1419 × *T. evansi* isolate ESALQ1421.

One *N. floridana*-killed *T. evansi* isolate ESALQ1419 cadaver and another *N. floridana*-killed *T. urticae* isolate ESALQ1420 were placed side by side in the centers of the same jack bean or tomato leaf disc (2.0 cm diameter). In the same way, one mummified mite from the *T. evansi* isolate ESALQ1419 and another mummified mite from the *T. evansi* isolate ESALQ1421 were placed side by side in the centers of a tomato leaf disc (2.0 cm diameter). Each leaf disc was placed with the underside up on a moist sponge in a closed Petri dish (9 cm diameter). The closed Petri dishes with cadavers on leaf discs were then placed in a paper box (20 × 20 × 10 cm) to create dark conditions and incubated at 25 ± 2°C and 100% RH. Cadavers were checked under a compound microscope (80x) after 24 h of incubation to select only the leaf discs with cadavers with good capilliconidia production. *Neozygites* conidia were forcibly discharged from the surface of the host forming a halo conidia on the leaf discs around the cadaver. Only leaf discs with at least 300 capilliconidia were used. Twenty adult *T. urticae* or *T. evansi* females were then transferred, with a paintbrush, onto a jack bean or tomato leaf disc, respectively. The Petri dishes were kept at 25 ± 2°C, L12 : D12, and 70% RH for 24 h, and the mites were then transferred onto new leaf discs. After seven days, the mites were checked, and both dead and live mites were mounted in Aman Blue for the observation of hyphal bodies and resting spores under a compound microscope (400x). At least 20 mites were included in each treatment and isolate, and the experiment was repeated ten times.

Biotic Factors: Host Plant Quality: Leaf Chlorosis and Senescence. Leaf discs (1.2 cm diameter) from jack bean and tomato with or without chlorosis were placed underside up on a moist sponge in closed Petri dishes (9 cm diameter). Chlorosis was induced by infestation with high densities of *T. urticae* until more than 50% of the green color of the leaves was lost. One *N. floridana*-killed *T. urticae* (ESALQ1420) or *T. evansi* (ESALQ1421) cadaver was placed on jack bean or tomato leaf disc, respectively. Petri dishes were then placed in a paper box to create dark conditions and incubated at $25 \pm 2^{\circ}$C and 100% RH. Cadavers were checked under a compound microscope (80x) after 24 h of incubation to ensure good sporulation and production of capilliconidia. Only leaf discs with over 300 capilliconidia were used. Twenty uninfected adult *T. urticae* or *T. evansi* females were then transferred, with a paintbrush, onto each leaf disc with cadavers for *N. floridana* inoculation and incubated for 24 h under the conditions described above. *N. floridana*-inoculated *T. evansi* and *T. urticae* were then transferred to vials (3.0 cm diameter × 2.0 cm high) closed with lids. The vials were incubated at $25 \pm 2^{\circ}$C during the light period and at $15 \pm 2^{\circ}$C during the dark period (L11 : D13) with 70% RH. When the mites died, they were mounted in Aman Blue to enable the observation of hyphal bodies and resting spores under a compound microscope (400x). Each treatment and isolate included 20 mites, and the experiment was repeated ten times. To test the effect of leaf senescence on *N. floridana* resting spore production, 60 senescent leaves of the nightshade *Solanum americanum* Mill. were collected. These leaves were kept in small cages (11 × 11 × 3.5 cm) with a moist sponge. *N. floridana*-infected adult *T. evansi* females were then placed on these leaves and incubated at $25 \pm 2^{\circ}$C, L12 : D12, and 70% RH. Approximately 680 fungus-killed mites were mounted in Aman Blue for the observation of hyphal bodies and resting spores under a compound microscope (400x).

Biotic Factors—Mite Age. To test the effect of mite age on *N. floridana* resting spore production, newly hatched larvae and adult *T. urticae* females were inoculated with *N. floridana* isolate ESALQ1420 using the protocol described above. Inoculated *T. urticae* larvae and adults females were transferred with a fine paintbrush onto a jack bean leaf disc (2.0 cm diameter) placed underside up on moist cotton in vials (3.0 cm diameter and 2.0 cm high) closed with lids. The vials were incubated at $25 \pm 2^{\circ}$C, 70% RH, and L12 : D12. After seven days, the mites were checked, and dead and live mites were mounted in Aman Blue for observation of hyphal bodies and resting spores under a compound microscope (400x). At least 20 mites were included in each treatment, and the experiment was repeated ten times.

2.3. Statistical Analysis. The effects of different abiotic and biotic factors on the percentage of *T. urticae* with resting spores were analyzed with ANOVA after the arcsine transformation of the data. When significant effects were found, post hoc comparisons using Tukey's HSD test were conducted to evaluate the pairwise differences between means ($P < 0.05$). All statistical analyses were carried out in the SAS package (SAS Institute Inc., Cary North Carolina).

3. Results

3.1. Experiments Mimicking Temperate Region Conditions. A total of 3,106 mites (not including a series of pilot experiments) were tested at 26 different combinations of conditions. However, a significantly higher rate of resting spores, 51.5% ($F = 20.5$, $P < 0.0001$), was found for only one condition: the Brazilian *N. floridana* isolate at 11°C (no temperature drop) with a photoperiod of 10L : 14D, light intensity of 42–46 (μmol m^{-2} s^{-1}), and light quality of 1 in nondiapausing *T. urticae* females from nonsenescent plants (Table 1). No significant difference in resting spore production was observed for any of the other combinations of conditions for any of the isolates tested. One combination of conditions resulted in a low level of resting spore production (1.4%) for the Brazilian *N. floridana* isolate in *T. urticae* females; several combinations of conditions also resulted in resting spore production for the Norwegian *N. floridana* isolate in *T. urticae* females, but only at low levels (1.4–5.0%). The majority of the spore-forming conditions (8 out of 9 combinations) included a 10L : 14D light regime.

The time to lethality in *T. urticae* females varied from 10.0 to 18.0 days for the Norwegian isolate (NCRI271/04) and from 20.0 to 21.9 days for the Brazilian isolate (ESALQ1420) at the temperatures tested. The mites containing resting spores survived longer than the mites with hyphal bodies.

T. urticae cadavers containing resting spores from the Norwegian *N. floridana* isolate (NCRI271/04) were quite different from *T. urticae* cadavers containing resting spores from the Brazilian *N. floridana* isolate (ESALQ1420). Swollen fungus-killed cadavers filled with hyphal bodies, referred to as mummies, were opaque orange/light brown for the Brazilian isolate (Figure 1(B1)) but dark brown/black for the Norwegian isolate (Figure 1(A1)). When *N. floridana* produces resting spore cadavers, *T. urticae* first turns gray/light brown and then shiny dark brown/black and slightly swollen (Figures 1(A2) and 1(B2)). When the resting spores reach maturity, the cuticle of the mite becomes fragile. *T. urticae* cadavers with immature Norwegian *N. floridana* resting spores were of equal size and shape, whereas *T. urticae* cadavers with immature Brazilian *N. floridana* resting spores varied in size and shape (Figures 1(A3) and 1(B3)). The majority of the *T. urticae* cadavers with resting spores also contained hyphal bodies (Figures 1(A3) and 1(B3)).

3.2. Experiments Mimicking Tropical Region Conditions. Even though 13,516 *T. urticae* and *T. evansi* (including a pilot experiment, data not shown) were tested under 13 different conditions, no *T. urticae* and a very low percentage of

FIGURE 1: *T. urticae* killed by the fungus *N. floridana*. A(1–3) Norwegian isolate (NCRI271/04), (A1) dark brown cadavers with hyphal bodies, (A2) black/dark brown cadaver with resting spore, (A3) mature resting spores in squash mount, B(1–3) Brazilian isolate (ESALQ1420), (B1) light brown/orange cadavers with hyphal bodies, (B2) black/dark brown cadaver with resting spores, (B3) almost mature resting spores in squash mount.

N. floridana (ESALQ1421)-killed *T. evansi* adult females (up to 1.0%) produced resting spores under the following conditions: 32°C, RH: 70%, 12L : 12D, and young leaves. Further, 0.5% of *N. floridana* (ESALQ1421)-killed *T. evansi* adult females produced resting spores under the following conditions: 35°C, RH: 60%, 12L : 12D, uninfested leaves. A third condition also resulted in 0.5% resting spore production in *N. floridana* (ESALQ1421)-killed *T. evansi* adult females: 25°C (light period) and 15°C (dark period), RH: 60%, 11L : 13D, and leaves with chlorosis (Tables 2 and 3).

T. evansi cadavers containing *N. floridana* isolate ESALQ1421 resting spores were shiny dark brown/black and retained their original mite shape. When the resting spores were mature, the *T. evansi* cuticle became fragile. Further, the mature resting spores of *N. floridana* (ESALQ1421)-killed *T. evansi* were equal in size and shape. *N. floridana* (ESALQ1421)-killed *T. evansi* cadavers with hyphal bodies were distinct from cadavers with resting spores and became swollen and light brown/orange in color.

4. Discussion

In this study a small percentage (1.4–5.0%) of *T. urticae* with resting spores was found for the Norwegian *N. floridana* isolate NCRI271/04 under certain temperate region-mimicking conditions. Most of the resting spores produced by the Norwegian isolate (8 out of 9 conditions) were produced under a 10L : 14D light regime and a 16L : 08D light regime. At Ås, in the Southeastern part of Norway, 10 h of light occurs in the fall (17 October) and winter (24 February), and days with 16 h of light occur at the end of the summer (10 August) and in the spring (1 May) (http://www.timeanddate.no/).

Table 2: Effect of different combinations of abiotic factors (temperature and RH) on resting spores produced in *N. floridana* killed *T. evansi* by two isolates from Brazil (ESALQ1419 and ESALQ1421).

Temperature (°C)	RH (%)	Resting spores (%)	
		ESALQ1421	ESALQ1419
32	60	0	0
	70	1	0
	80	0	0
	90	0	0
35	60	0.5	0
	70	0	0
	80	0	0
	90	0	0

Table 3: Effect of different combinations of biotic factors, coinfection, host plant quality, and mite age, on resting spores produced in *N. floridana*-killed *T. urticae* (Brazilian isolate: ESALQ1420) and *T. evansi* (Brazilian isolates: ESALQ1419 and ESALQ1421) isolates from Brazil.

Biotic factors	Treatments	Host	Resting spores (%)
Coinfection	ESALQ1419 × ESALQ1420	*T. evansi*	0
		T. urticae	0
	ESALQ1419 × ESALQ1421	*T. evansi*	0
Host plant quality	Chlorosis	ESALQ1420 *T. urticae*	0
		ESALQ1421 *T. evansi*	0.5
	Senescence	ESALQ1421 *T. evansi*	0
Mite age	Larvae	ESALQ1420 *T. urticae*	0
	Adult	ESALQ1420 *T. urticae*	0

Our results for the Norwegian *N. floridana* isolate are therefore consistent with earlier field studies in temperate regions that indicated that resting spores of local *N. floridana* isolates in *T. urticae* seem to be induced in fall when the hibernation of *T. urticae* females is also induced [8]. In São Paulo, Brazil, 14 h of darkness never occurs; the shortest day (10 h 40 min) occurs on the winter solstice (21 June). In the experiments mimicking tropical conditions, resting spores were found at very low levels (up to 1.0%) and only in *T. evansi* infected by the Brazilian *N. floridana* isolate ESALQ1421 at high temperatures (32 and 35°C) and a 12L : 12D light regime. In São Paulo, Brazil, days with 12 h of light occur during spring (17 September) and fall (24 March).

Between-strain differences in the ability to form resting spores have been observed for *Z. radicans* [21] and *E. maimaiga* [26], but this phenomenon has never been investigated in species of *Neozygites* affecting tetranychid mites. The low percentages of resting spores found for the Norwegian isolate in our laboratory studies are similar to the prevalences found in earlier field studies [8] in Norway where resting spore levels in hibernating *T. urticae* females ranged from 2.5 to 13.8%. In these field studies, hibernating *T. urticae* females with hyphal bodies were found at much higher levels, however, and peaked at 54.4%. Our laboratory studies resulting in resting spore infection levels in the Norwegian *N. floridana* isolate of no more than 5.0% at any of the conditions tested may further indicate, as suggested by Klingen et al. [8], that the major overwintering strategy of *N. floridana* in temperate regions is to exist as hyphal bodies inside live hibernating *T. urticae* females and that resting spores are produced mainly for sexual recombination. Other reports have described *N. floridana* resting spores in temperate regions during the autumn and winter, but most of these spores are found at low levels. Klubertanz et al. [6] found resting spores of *Neozygites* sp. in overwintering *T. urticae* in soybeans at a level of approximately 8% of mites sampled. Brandenburg and Kennedy [27] investigated the overwintering strategy of *Entomophthora floridana* (syn. *N. floridana*) in *T. urticae* for two years and observed resting spores in only one sample, collected in autumn (28.0% of mites with resting spores). *T. urticae* with resting spores of *Entomophthora* sp. (syn. *N. floridana*) was observed at some locations in the USA (Clemson, Alabama, Blackville), but no resting spores were found [12]. In temperate regions, *T. urticae* hibernates as adult females [22, 28, 29]. *T. urticae* hibernation is induced by short day length, low temperature, and a lack of nutrition from its host plant [22].

In our study, a high percentage (51.5%) of *T. urticae* with *N. floridana* resting spores was found only for the Brazilian isolate ESALQ1420 in the experiments mimicking conditions of temperate regions (11°C, 10L : 14D, and a light intensity of 42–46 μmol m^{-2} s^{-1}). These conditions are common in temperate regions but rare in most tropical sites. In tropical regions, it has never been observed that spider mites survive adverse conditions (e.g., drought and lack of host plants) as hibernating females as one may see in temperate regions (e.g., winter) [8]. Further, it is unclear whether resting spore formation is the major strategy for survival under adverse conditions in tropical climates considering that the conditions that best induced resting spore formation for the Brazilian isolate in our experiment are not common in the tropics.

Resting spores of *Neozygites* sp. have rarely been found in tropical regions. During the nearly 20-year duration of the cassava green mite project investigating *N. tanajoae*, resting spores were found only occasionally in laboratory and field studies. Resting spores of *Neozygites* sp. (= *N. tanajoae*) were observed in northeastern Brazil during the winter [9, 10]. In a field study of *M. tanajoa* and its natural enemy, *N. tanajoae*, Houtondji et al. [30] found only four mites with resting spores in an examination of over 460,000 mites. In more recent studies, however, higher levels of resting spores (34–38%) of *N. tanajoae* in *M. tanajoa* were found by Elliot et al. [9] in northeast Brazil during the winter. The same conditions were tested out in laboratory experiments and also gave high (24.2%) resting spore infection levels. Further, resting spores of *N. floridana* in *T. urticae* were found in southeastern and southern Brazil (Duarte et al. unpublished data; Roggia et al. unpublished data) [4]. These regions have colder winter conditions than those found in northeast Brazil, but the mites with resting spores were found in summer, when the plants become senescent, not in winter when temperatures were as low as in our laboratory experiment.

The *T. urticae* containing resting spores normally did not die as quickly as *T. urticae* with hyphal bodies. The time to lethality was negatively correlated with temperature. This finding is in accordance with Smitley et al. [31], who found that the mean time to lethality of *T. urticae* infected with *N. floridana* was 15, 5, 4, and 7 days after inoculation when maintained at 10, 20, 30, and 37°C, respectively. Normally, hosts infected by the Brazilian isolate ESALQ1420 die five days after inoculation at 25°C, and those infected with the Norwegian isolate die seven days after inoculation at 20°C (Delalibera Jr. personal communication; Klingen personal communication).

According to Humber [32] Neozygitales specifically represents the largest and most important "black box" of the new phylum Entomophthoromycota for which needed data remains unavailable. Basic information such as the nature and role of sexual part of the life cycle (resting spore) is still not well characterized. Although we were not able to answer many of our initial questions about resting spore formation and the role of this type of spore, we identified a set of conditions that can consistently produce resting spores, which will be useful for further investigations.

Conflict of Interests

The authors declare that they have no conflict of interests.

Acknowledgments

This research was funded by the Norwegian Foundation for Research Levy on Agricultural Products (FFL) and the Agricultural Agreement Research Funds (JA) through BERRYSYS (Project no. 190407/199). The first author received a scholarship from The National Council for Scientific and Technological Development (CNPq).

References

[1] A. P. . Gryganskyi, R. A. Humber, M. E. Smith et al., "Molecular phylogeny of the *Entomophthoromycota*," *Molecular Phylogenetics and Evolution*, vol. 65, pp. 682–694, 2012.

[2] S. Keller, "Arthropod-patogenic entomophthorales of Switzerland. II. *Erynia, Eyniopsis, Neozygites, Zoophthora* and *Tarichium*," *Sydowia*, vol. 43, pp. 39–122, 1991.

[3] S. Keller, "The genus *Neozygites* (Zygomycetes, Entomophthorales) with special reference to species found in tropical regions," *Sydowia*, vol. 49, no. 2, pp. 118–146, 1997.

[4] V. S. Duarte, R. A. Silva, V. W. Wekesa, F. B. Rizzato, C. T. S. Dias, and I. Delalibera Jr., "Impact of natural epizootics of the fungal pathogen *Neozygites floridana* (Zygomycetes: Entomophthorales) on population dynamics of *Tetranychus evansi* (Acari: Tetranychidae) in tomato and nightshade," *Biological Control*, vol. 51, no. 1, pp. 81–90, 2009.

[5] R. A. Humber, G. J. Moraes, and J. M. dos Santos, "Natural infection of *Tetranychus evansi* [Acarina: Tetranychidae] by a *Triplosporium* sp. [Zygomycetes: Entomophthorales] in northeastern Brazil," *Entomophaga*, vol. 26, no. 4, pp. 421–425, 1981.

[6] T. H. Klubertanz, L. P. Pedigo, and R. E. Carlson, "Impact of fungal epizootics on the biology and management of the two spotted spider mite (Acari: Tetranychidae) in soybean," *Environmental Entomology*, vol. 20, pp. 731–735, 1991.

[7] A. E. Hajek, "Ecology of terrestrial fungal entomopathogens," *Advances in Microbial Ecology*, vol. 15, no. 1, pp. 193–249, 1999.

[8] I. Klingen, G. Wærsted, and K. Westrum, "Overwintering and prevalence of *Neozygites floridana* (Zygomycetes: Neozygitaceae) in hibernating females of *Tetranychus urticae* (Acari: Tetranychidae) under cold climatic conditions in strawberries," *Experimental and Applied Acarology*, vol. 46, no. 1–4, pp. 231–245, 2008.

[9] S. L. Elliot, J. D. Mumford, and G. J. De Moraes, "The role of resting spores in the survival of the mite-pathogenic fungus *Neozygites floridana* from *Mononychellus tanajoa* during dry periods in Brazil," *Journal of Invertebrate Pathology*, vol. 81, no. 3, pp. 148–157, 2002.

[10] I. Delalibera Jr., G. J. Moraes, S. T. Lapointe, C. A. D. Silva, and M. A. Tamai, "Temporal variability and progression of *Neozygites* sp. (Zygomycetes: Entomophthorales) in populations of *Mononychellus tanajoa* (Bondar) (Acari: Tetranychidae)," *Anais Sociedade Entomologica Do Brasil*, vol. 29, pp. 523–535, 2000.

[11] R. Mietkiewski, S. Balazy, and L. P. S. van der Geest, "Observations on a mycosis of spider mites (Acari: Teranychidae) caused by *Neozygites floridana* in Poland," *Journal of Invertebrate Pathology*, vol. 61, no. 3, pp. 317–319, 1993.

[12] G. R. Carner, "A description of the life cycle of *Entomophthora* sp. in the two-spotted spider mite," *Journal of Invertebrate Pathology*, vol. 28, no. 2, pp. 245–254, 1976.

[13] A. E. L. Ribeiro, M. G. C. Gondim Jr., J. W. S. Melo, and I. Delalibera Jr., "Solanum americanum as a reservoir of natural enemies of the tomato red spider mite, *Tetranychus evansi* (Acari: Tetranychidae)," *International Journal of Acarology*, vol. 38, no. 8, pp. 692–698, 2012.

[14] J. S. Yaninek, S. Saizonou, A. Onzo, I. Zannou, and D. Gnanvossou, "Seasonal and habitat variability in the fungal pathogens, *Neozygites* cf. *floridana* and *Hirsutella thompsonii*, associated with cassava mites in Benin, West Africa," *Biocontrol Science and Technology*, vol. 6, no. 1, pp. 23–33, 1996.

[15] J. M. Alvarez, A. Acosta, A. C. Bellotti, and A. R. Braun, "Estudios de patogenicidad de un hongo asociado a *Mononychellus tanajoa* (Bondar), ácaro plaga de la yuca (Manihot esculenta Crantz)," *Revista Colombiana de Entomologia*, vol. 19, pp. 3–5, 1993.

[16] X. Zhou, M. Feng, and L. Zhang, "The role of temperature on *in vivo* resting spore formation of the aphid-specific pathogen *Pandora nouryi* (Zygomycota: Entomophthorales) under winter field conditions," *Biocontrol Science and Technology*, vol. 22, pp. 93–100, 2012.

[17] X. Zhou and M.-G. Feng, "Biotic and abiotic regulation of resting spore formation *in vivo* of obligate aphid pathogen *Pandora nouryi*: modeling analysis and biological implication," *Journal of Invertebrate Pathology*, vol. 103, no. 2, pp. 83–88, 2010.

[18] Z.-H. Huang and M.-G. Feng, "Resting spore formation of aphid-pathogenic fungus *Pandora nouryi* depends on the concentration of infective inoculum," *Environmental Microbiology*, vol. 10, no. 7, pp. 1912–1916, 2008.

[19] L. Thomsen and J. Eilenberg, "Entomophthora muscae resting spore formation *in vivo* in the host *Delia radicum*," *Journal of Invertebrate Pathology*, vol. 76, no. 2, pp. 127–130, 2000.

[20] A. E. Hajek and M. Shimazu, "Types of spores produced by *Entomophaga maimaiga* infecting the gypsy moth *Lymantria dispar*," *Canadian Journal of Botany*, vol. 74, no. 5, pp. 708–715, 1996.

[21] T. R. Glare, R. J. Milner, and G. A. Chilvers, "Factors affecting the production of resting spores by *Zoophthora radicans* in the spotted alfalfa aphid, *Therioaphis trifolii* f. *Maculata*," *Canadian Journal of Botany*, vol. 67, pp. 848–855, 1989.

[22] A. Veerman, ""Diapause". Spider mites their biology," *Natural Enemies and Control A*, vol. 1, pp. 279–316, 1985.

[23] A. Veerman, "Aspects of the induction of diapause in a laboratory strain of the mite *Tetranychus urticae*," *Journal of Insect Physiology*, vol. 23, no. 6, pp. 703–711, 1977.

[24] R. A. Humber, "An alternative view of certain taxonomic criteria used in the Entomophthorales (Zygomycetes)," in *Mycotaxon*, vol. 13, pp. 191–240, 1981.

[25] P. W. Winston and D. H. Bates, "Saturated soultions for the control of humidity in biological research," in *Ecology*, vol. 41, pp. 232–237, 1960.

[26] P. H. Kogan and A. E. Hajek, "*In vitro* formation of resting spores by the insect pathogenic fungus *Entomophaga maimaiga*," *Journal of Invertebrate Pathology*, vol. 75, no. 3, pp. 193–201, 2000.

[27] R. L. Brandenburg and G. G. Kennedy, "Overwintering of the pathogen *Entomophthora floridana* and its host, the two-spotted spider mite," *Journal of Economic Entomology*, vol. 74, pp. 428–431, 1981.

[28] W. Helle, "Genetic of resistance to orgaphosphorus compounds and its relation to diapause in *Tetranychus urticae* Koch (Acari)," *Tijdschrift Over Plantenziekten*, vol. 68, pp. 155–195, 1962.

[29] C. F. Van de Bund and W. Helle, "Investigations on the *Tetranychus urticae* complex in north west Europe (Acari: Tetranychidae)," *Entomologia Experimentalis et Applicata*, vol. 3, no. 2, pp. 142–156, 1960.

[30] F. C. C. Hountondji, J. S. Yaninek, G. J. De Moraes, and G. I. Oduor, "Host specificity of the cassava green mite pathogen *Neozygites floridana*," *BioControl*, vol. 47, no. 1, pp. 61–66, 2002.

[31] D. R. Smitley, G. G. Kennedy, and W. M. Brooks, "Role of the entomogenous fungus, *Neozygites floridana*, in population declines of the twospotted spider mite, *Tetranychus urticae*, on field corn," *Entomologia Experimentalis et Applicata*, vol. 41, no. 3, pp. 255–264, 1986.

[32] R. A. Humber, "*Entomophthoromycota*: a new phylum and reclassification for entomophthoroid fungi," *Mycotaxon*, vol. 120, pp. 477–492, 2012.

12

Antioxidant Functions of Nitric Oxide Synthase in a Methicillin Sensitive *Staphylococcus aureus*

Manisha Vaish and Vineet K. Singh

Microbiology and Immunology, Kirksville College of Osteopathic Medicine, A.T. Still University of Health Sciences, 800 West Jefferson Street, Kirksville, MO 63501, USA

Correspondence should be addressed to Vineet K. Singh; vsingh@atsu.edu

Academic Editor: John Tagg

Nitric oxide and its derivative peroxynitrites are generated by host defense system to control bacterial infection. However certain Gram positive bacteria including *Staphylococcus aureus* possess a gene encoding nitric oxide synthase (SaNOS) in their chromosome. In this study it was determined that under normal growth conditions, expression of *SaNOS* was highest during early exponential phase of the bacterial growth. In oxidative stress studies, deletion of *SaNOS* led to increased susceptibility of the mutant cells compared to wild-type *S. aureus*. While inhibition of *SaNOS* activity by the addition of L-NAME increased sensitivity of the wild-type *S. aureus* to oxidative stress, the addition of a nitric oxide donor, sodium nitroprusside, restored oxidative stress tolerance of the *SaNOS* mutant. The *SaNOS* mutant also showed reduced survival after phagocytosis by PMN cells with respect to wild-type *S. aureus*.

1. Introduction

Staphylococcus aureus is a Gram-positive bacterial pathogen that colonizes anterior nares and mucosal surfaces in humans and is responsible for causing a wide array of diseases from mild skin infections to life-threatening conditions such as bacteremia, pneumonia, and endocarditis [1–4]. The emerging resistant strains of *S. aureus* exacerbate efforts to control or properly treat staphylococcal infections [5].

The host immune system responds to bacterial infections in a concerted manner to eliminate this pathogen. This involves recruitment of polymorphonuclear leukocytes and macrophages to the site of infection and ingestion of invading bacteria. Uptake of bacteria triggers oxygen-dependent and oxygen-independent microbicidal pathways in the phagocytic cells. The oxygen-dependent pathway generates superoxide anion (O_2^-) that serves as a precursor for additional reactive oxygen species (ROS) such as hydrogen peroxide (H_2O_2), hydroxyl radical, singlet oxygen, hypochlorous acid (HOCl), and peroxynitrite [6–9].

S. aureus utilizes various strategies to defend itself against host immune attack. It produces antioxidant enzymes such as superoxide dismutase that converts superoxide anion to H_2O_2, catalase that converts H_2O_2 to water and oxygen, and alkyl hydroperoxide reductases that detoxify H_2O_2, peroxynitrites and hydroperoxides [10, 11]. In addition to their ability to protect from host's oxidants, *S. aureus* infections impose oxidative stress in a host [12]. During infection with a methicillin resistant *S. aureus* strain, host neutrophils respond by an increase in nitric oxide production [12]. Nitric oxide (NO) is a free radical synthesized by nitric oxide synthase.

Certain Gram-positive bacteria express homologs of nitric oxide synthases (NOS) that have been extensively studied in eukaryotic species. In these species, NOS-derived nitric oxide (NO) is involved in vasodilation, neurotransmission, and host defense [7, 13, 14], but the functions of bacterial NOS are still being defined. Recent genome sequencing has revealed that NOS-like protein exists in many bacteria including *Streptomyces* (StNOS), *Deinococcus* (DrNOS), *Staphylococcus* (SaNOS), and *Bacillus* (BsNOS) species [15]. Bacterial NOS enzymes are homologous with the mammalian NOS, but lack an associated NOS reductase and N-terminal β-hairpin hook that binds Zn^{2+}, the dihydroxypropyl side chain of H_4B, and the adjacent subunit of the oxygenase dimer [15–18].

It has also been reported that in *Bacillus subtilis*, NO protects bacterial cells from reactive oxygen species [19]. In addition, the *in vivo* survival of *Bacillus anthracis* was dependent on its own NOS activity [20]. NOS activity was also shown to protect from oxidative stress, and deletion of the gene encoding NOS reduced the virulence of a methicillin resistant *S. aureus* [21]. In this study, SaNOS-derived NO was seen to be protective in a methicillin sensitive *S. aureus* from lethal oxidative stress conditions, suggesting its moderate role in stress tolerance.

2. Materials and Methods

2.1. Bacterial Strains and Growth Conditions. All experiments were carried out using the methicillin sensitive *S. aureus* strain SH1000 (wild-type) [22], its isogenic *SaNOS* deletion mutant, and the mutant complemented with *SaNOS in trans*. Bacterial cultures were grown in tryptic soy broth/agar (TSB/TSA; Becton Dickinson) at 37°C in a shaking (220 rpm) or static incubator. When needed, tetracycline (10 μg mL^{-1}) and chloramphenicol (10 μg mL^{-1}) were added to the growth medium.

2.2. DNA Manipulations and Analysis. Plasmid DNA was isolated using the Qiaprep kit (Qiagen Inc.); chromosomal DNA was isolated using a DNAzol kit (Molecular Research Center) from lysostaphin-treated *S. aureus* cells as per the manufacturer's instructions. All restriction and modification enzymes were purchased from Promega. DNA manipulations were carried out using standard procedures. PCR was performed with the PTC-200 Peltier Thermal Cycler (MJ Research). Oligonucleotide primers were obtained from Sigma Genosys.

2.3. Construction of SaNOS Mutant. To construct a mutation in the *SaNOS* gene, primers P1 (5′-ACGAATTCTGCTAGCCTTTGTTG-3′) and P2 (5′-GGATCCCAAAATAAACGACCAATGC-3′) were used to amplify an 831 bp DNA fragment using genomic DNA from *S. aureus* strain SH1000 as the template. This amplicon represents *SaNOS* left flanking fragment (starting 207 nt downstream of the *SaNOS* start codon and going upstream). Another set of primers, P3 (5′-GGATCCATTATCTCCAACATTG-3′) and P4 (5′-TCTAGAATCAGCCTGAACGAAAAATCG-3′), was used to amplify an 850 bp DNA fragment representing *SaNOS* right flanking fragment (starting 120 nt upstream of the *SaNOS* stop codon and going downstream). These two fragments were ligated together into vector pTZ18R [23] and a unique *Bam*HI site was engineered between the ligated fragments. To the *Bam*HI site of this fragment (lacking most of the *SaNOS* gene; 750 nt out of a total of 1074 nt of the *SaNOS* gene), a 2.2 kb tetracycline resistance cassette was cloned. The resulting construct was used as a suicidal plasmid to transform *S. aureus* RN4220 cells by electroporation. Transformants were selected on TSA plates containing 10 μg mL^{-1} tetracycline that led to a single crossover event where the mutated *SaNOS* from the plasmid was integrated into the bacterial genome leaving the wild-type *SaNOS* intact. These merodiploids were used to resolve the mutation in the *SaNOS* gene using a phage 80α transduction procedure as described previously [24, 25]. Mutation in the *SaNOS* was verified by PCR. For genetic complementation of the *SaNOS* mutant, a 2.4 kb DNA fragment was PCR amplified using primers P1 and P4 and *S. aureus* SH1000 genomic DNA as template. The amplicon represents a fragment starting from 624 nt upstream and spanning 730 nt downstream of the *SaNOS* gene that was cloned into the shuttle plasmid pCU1 [26] and subsequently transferred to the *SaNOS* mutant of *S. aureus* strain SH1000.

2.4. Quantitative Real-Time RT-PCR (qRT-PCR) Assays. qRT-PCR assays were carried out as described [27] using primers P5 (ATGGTGCTAAAATGGCTTGGC) and P6 (GCTTCGTCAGTAACATCTCTTG) to determine optimum expression of *SaNOS* during different stages of *S. aureus* growth in TSB. Bacterial cells were harvested from early- (OD_{600} = 0.6), mid- (OD_{600} = 1.8), late-exponential (OD_{600} = 3.0), and stationary (OD_{600} = 4.2) phase cultures. Total RNA extracted from these cells was used in qRT-PCR assays as described [27].

2.5. Determination of Nitric Oxide Synthase Activity. Total protein was extracted from lysostaphin treated *S. aureus* cells grown to OD_{600} = 0.6 as described previously [28]. The NOS activity was determined using NOS activity assay kit (Cayman Chemical Company) and radioactive ^{3}H arginine monohydrochloride as substrate (Amersham Biosciences).

2.6. Determination of H_2O_2 Susceptibility. For these studies, *S. aureus* cells from early exponential phase cultures OD_{600} = 0.6 were treated with 350 mM H_2O_2 for 30 min. The surviving bacteria were enumerated by serial dilution and plating on TSA agar plates. L-arginine serves as a substrate for the nitric oxide synthase in the production of NO. Wild-type *S. aureus* cultures in TSB were added with L-arginine (1 mM final concentration) at OD_{600} = 0.5 and subsequently at OD_{600} = 0.6 were stressed with 350 mM H_2O_2 to determine if the addition of L-arginine affected NO production and the oxidative stress tolerance. Additionally, the wild-type *S. aureus* cells were collected from cultures grown to OD_{600} = 0.3 and were resuspended in similar volume of TSB containing 5 mM L-NAME (Tocris Bioscience), an inhibitor of NOS activity. At an OD_{600} = 0.6, these NOS-inhibited cells were stressed with 350 mM H_2O_2 for 30 min and the surviving bacteria were counted. To further ascertain the role of nitric oxide in the protection of *S. aureus* cells, the *SaNOS* mutant cells at OD_{600} = 0.5 were treated with 2.5 mM concentration of an NO donor, sodium nitroprusside (SNP) (Sigma). At OD_{600} = 0.6, these SNP-treated cells were stressed with 350 mM H_2O_2 for 30 min, and the surviving bacteria were counted.

2.7. Phagocytic Killing of S. aureus SaNOS Mutant. The promyelocytic HL-60 cells (ATCC) were grown in Iscove's Modified Dulbecco's Medium (IMDM) (ATCC) with 20% fetal bovine serum (Fisher) and were treated with 1.3%

TABLE 1: Expression of *SaNOS* in *S. aureus* during different phases of growth.

Growth stage	*SaNOS* expression*
Early-exponential	100%
Mid-exponential	19.48%
Late-exponential	10.73%
Stationary	4.90%

*Expression of *SaNOS* is shown relative to its transcript level during early-exponential phase of growth.

TABLE 2: Nitric oxide synthase activity in different *S. aureus* strains.

Strain	NOS activity (%)*
SH1000	3.95 ± 1.61
SH1000Δ*SaNOS*	0
Complemented strain	26.47 ± 3.95

*%Citrulline formed in relation to total L-arginine used in the assay. Citrulline conversion in the mutant strain was below the background level (control reaction with no protein extract). Values represent average of three independent experiments ± standard deviation.

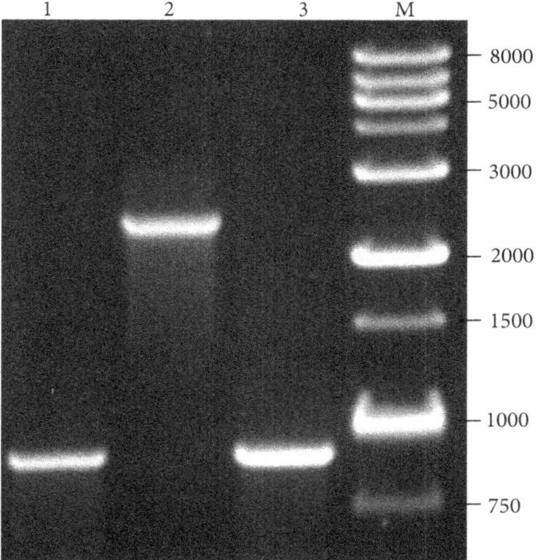

FIGURE 1: PCR verification of a mutation in the *SaNOS* gene in *S. aureus*. Primers P7 (5′-ATACAGAAGAAGAACTTATTTATGG-3′) and P8 (5′- CACCTCTACTAACTTAATGATGG-3′) were used in the PCR that allowed amplification of a 963 bp product (lane 1) when genomic DNA from wild-type *S. aureus* strain SH1000 was used. These primers amplified a ~2.4 kb fragment when genomic DNA from the *SaNOS* mutant of *S. aureus* strains SH1000 was used as template (lanes 2). Lane 3: PCR product when genomic DNA from the *SaNOS* mutants of *S. aureus* strains SH1000 complemented *in trans* with *SaNOS* was used as template. The larger PCR product is not seen because of complementation with wild-type *SaNOS* gene on a high copy plasmid pCU1. Lane M: DNA ladder.

DMSO (Fisher) for 5 days to induce their differentiation into neutrophil-like cells [29, 30]. Morphology of differentiated cells was confirmed by Giemsa staining under inverted microscope. The oxidative burst inside neutrophil cells was determined by the reduction of nitroblue tetrazolium. The differentiated neutrophils were used for phagocytic killing using a method described previously [9] with slight modification. In brief, the neutrophils (1×10^6) were added with *S. aureus* cells (2.5×10^6) (MOI 1 : 2.5) in a 24-well plate. The plate was centrifuged at 4000 rpm for 10 min and incubated in a CO_2 incubator at 37°C for 1 h. The supernatant was gently aspirated and the neutrophils were lysed by the addition of IMDM containing 0.025% Triton X-100. The number of surviving bacteria was enumerated by making serial dilutions and plating of this lysate on TSA plate.

2.8. Statistical Analysis. All results are reported as the mean ± SD of at least three independent experiments. Data were analyzed with Dunnett's Method in one-way analysis of variance or with Student-Newman-Keuls Method in two-way analysis of variance using statistical analysis computer programs (SigmaPlot for Windows, version 12.0, Systat Software, Inc.). Statistical significance was set at $P < 0.05$.

3. Results and Discussion

3.1. Construction of SaNOS Deletion Mutant in S. aureus. To investigate the role of the *S. aureus* nitric oxide synthase and NO produced by this enzyme, the *SaNOS* gene was deleted and replaced with a tetracycline cassette by site-directed mutagenesis. The deletion of *SaNOS* gene was confirmed by PCR (Figure 1).

3.2. Expression of SaNOS and NOS Enzymatic Activity in S. aureus. In qRT-PCR assays, maximum expression of *SaNOS* in strain SH1000 was determined during the early stage of the bacterial growth (Table 1). The expression of *SaNOS* declined dramatically during the late stages of the bacterial growth and was least during the stationary phase (Table 1). A higher bacterial NO production was also noted during the early stages of macrophage infection by *B. anthracis* [19]. The determination of NOS activity, based on the conversion of L-arginine to citrulline, indicated that SaNOS was functional and was able to use L-arginine as the substrate (Table 2). The level of citrulline in the *SaNOS* mutant was similar or below the background level; a reaction mixture that contained only the L-arginine substrate and no protein extract was added to this reaction mixture (Table 2). The complementation of the *SaNOS* mutant with *SaNOS* gene on a high copy plasmid led to a significant increase in the NOS activity in this complemented strain (Table 2). Similar NOS activities in these strains were also verified by measuring the nitrite and nitrate levels using Griess reagent (data not shown).

3.3. Lack of SaNOS in S. aureus Reduces Its Survival under Oxidative Stress. The impact of the deletion of *SaNOS* was investigated for its growth in TSB. There was no change in the growth of the mutant strain and it was comparable to the growth of the wild-type *S. aureus* (data not shown). Under stress conditions such as salt (1.5 mM NaCl) and pH (6.0 or 8.5), the growth rate of the *SaNOS* was comparable to the growth rate of the wild-type *S. aureus* (data not shown). Also,

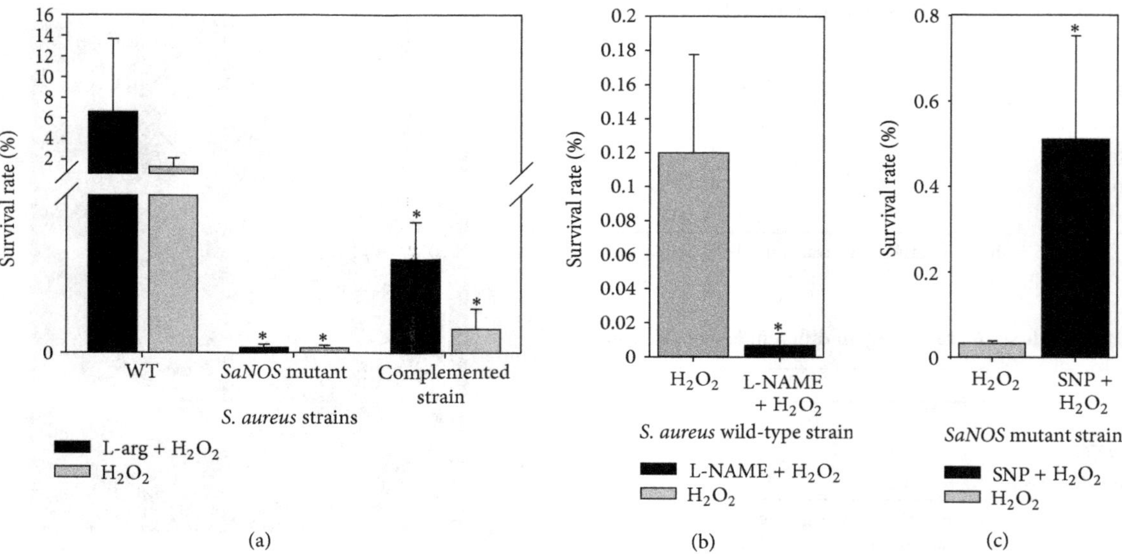

FIGURE 2: (a) Survival of *S. aureus* SH1000, its isogenic *SaNOS* mutant, and the mutant complemented with *SaNOS* gene *in trans* from a lethal dose (350 mM) of H_2O_2 with and without supplementation with 1 mM L-arginine. (b) Survival of wild-type *S. aureus* SH1000 pretreated with 5 mM L-NAME from 350 mM H_2O_2. (c) Survival of *SaNOS* mutant of *S. aureus* SH1000 pre-treated with 2.5 mM sodium nitroprusside from 350 mM H_2O_2. *Significant at $P < 0.05$.

in the presence of 1.1 mM H_2O_2, the growth of the *SaNOS* mutant of *S. aureus* SH1000 was comparable to the wild-type strain (data not shown). However, it has been shown that the priming of the *B. subtilis* cells with nitric oxide for 5 sec leads to a significant increase in their resistance to the exposure of a much higher H_2O_2 concentration (370 mM) [19].

In qRT-PCR assays, maximum expression of *SaNOS* was determined in the cells from the early exponential phase (OD_{600} = 0.6). Thus, cultures at this density were used in H_2O_2 susceptibility assays. When wild-type and the *SaNOS* mutant cells were treated with a lethal dose of 350 mM H_2O_2, there were significantly more surviving wild-type bacteria (>1000-fold) compared to the *SaNOS* mutant bacteria under identical experimental conditions (Figure 2(a)). Addition of L-arginine is expected to increase the production of nitric oxide and thus is expected to also increase the resistance of *S. aureus* cells grown in the presence of L-arginine. Addition of L-arginine indeed increased the resistance of the wild-type *S. aureus* cells but caused no increase in the survival of the *SaNOS* mutant (Figure 2(a)). Complementation of *SaNOS* mutant with the *SaNOS* gene on a plasmid partially restored the ability of these bacteria to survive H_2O_2 stress when it was grown with or without L-arginine (Figure 2(a)). When the NOS activity was inhibited in the wild-type *S. aureus* by the addition of L-NAME, a competitive inhibitor of the NOS enzymatic activity, it dramatically reduced the bacterial survival (Figure 2(b)) under oxidative stress. In addition, when sodium nitroprusside (an NO donor) was added to the *SaNOS* mutant cells, there was significant increase (>300-fold) in the survival of the mutant bacteria when they were exposed to H_2O_2 (Figure 2(c)). These results, collectively, suggest the role of a functional nitric oxide synthase in the protection of *S. aureus* cells from oxidative stress conditions.

3.4. Phagocytic Killing of the SaNOS Mutant. Neutrophils are a critical component of innate immunity and are essential in controlling bacterial infections in a host. Experiments were carried out to determine if the lack of a functional NOS decreased the survival of the *S. aureus* bacteria when it was allowed to interact with neutrophils. In these experiments, the *SaNOS* mutant showed significantly reduced survival compared to the wild-type *S. aureus* (Figure 3). These *SaNOS* mutant bacteria were also used to determine their survival compared to wild-type *S. aureus* in a murine intraperitoneal model as described previously [24, 25]. However, there was no decrease in the survival of the *SaNOS* mutant when compared to the wild-type *S. aureus* bacteria (data not shown). The ability of the *SaNOS* mutant cells to make biofilms was also comparable to the wild-type *S. aureus* cells (data not shown).

In recent years, the presence of NOS has been viewed with great interest for its role in bacterial physiology and virulence. Presence of NOS was determined to be a key factor in the defense of *B. subtilis* and *B. anthracis* from reactive oxygen species generated by the neutrophils and macrophages [19, 20]. It was shown that exposure to nitric oxide enhanced catalase activity in *B. subtilis* [19]. We observed a slight reduction in catalase activity in the *SaNOS* mutant relative to its level in the wild-type *S. aureus* (data not shown). *S. aureus* bacteria are known to produce a very high level of catalase activity. A lower level of superoxide dismutase activity was determined in the *SaNOS* mutant of a methicillin resistant *S. aureus* [21]. The reduced catalase and superoxide dismutase activity levels might be the reasons of the reduced survival of the *SaNOS* mutant under oxidative stress. Lack of the ability of the *S. aureus* cells to produce NO increased the susceptibility to reactive oxygen species and host antimicrobial peptides [21]. The level of the expression of the staphylococcal NOS

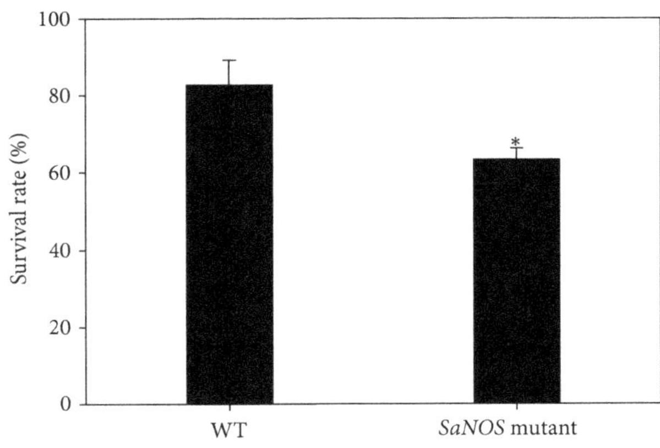

FIGURE 3: *S. aureus* survival in neutrophil cells. Neutrophil cells were infected (MOI 1 : 2.5) with wild-type *S. aureus* SH1000 and its isogenic *SaNOS* mutant for 1 h at 37°C and then plated on TSA plate. *Significant at $P < 0.05$.

was induced by exposure to cell wall-active antibiotics and it was also determined to be a factor in conferring resistance to these antibiotics in a methicillin resistant *S. aureus* [21]. Surprisingly, in that study, the lack of a functional NOS increased the resistance of *S. aureus* to aminoglycosides [21].

Studies utilizing a methicillin resistant *S. aureus* showed reduced virulence subsequent to NOS deletion [21]. Infection with the mutant cells resulted in smaller abscess formation compared to the *S. aureus* cell with a functional NOS suggesting its role in staphylococcal virulence [21]. In our studies that utilized a methicillin sensitive *S. aureus*, there was no difference in the survival of the *SaNOS* mutant in a mouse. There was also no appreciable difference in the survival or growth of the *SaNOS* mutant of *S. aureus* SH1000 under mild stress conditions. The difference in the survival was only detected when the *SaNOS* mutant and the wild-type bacteria were exposed to a lethal dose of H_2O_2. The reduction in virulence of *S. aureus* subsequent to *SaNOS* deletion in the recent report [21] can be attributed to strain differences (methicillin-resistant versus methicillin-sensitive *S. aureus*) and to a difference in the type of animal model used to study the virulence. These strain differences are significant as host neutrophils respond differently when they are exposed to methicillin-resistant *S. aureus* compared to during infection with methicillin-sensitive *S. aureus* [12]. NO production decreased in neutrophils in mice infected with vancomycin sensitive *S. aureus* and exposed to vancomycin but the decrease in neutrophilic NO production was insignificant when the mice were infected with vancomycin resistant *S. aureus* and exposed to vancomycin [12].

During the phagocytic process to control bacterial infections, the respiratory burst generates two very potent toxic substances, H_2O_2 and superoxide anions (O_2^-). A model has been proposed describing how bacterial NO might be protective from the toxic action of these reactive oxygen species [19, 20]. It is suggested that the O_2^- fails to cross the bacterial cell wall and membrane and limits the production of peroxynitrites inside the bacterial cell from a reaction between bacterial NO and phagocytic O_2^-. Although H_2O_2 can diffuse inside the bacterial cell, a higher bacterial catalase should degrade it to protect the bacterial cells from any damage.

Considering the fact that the SaNOS was seen to be significant only during extreme conditions of stress and has a varied role in antibiotic stress tolerance and virulence, more studies need to be carried out to determine the significance of this enzyme in *S. aureus*.

Conflict of Interests

The authors do not have any conflict of interests with the content of the paper.

Acknowledgments

The authors thank Deborah Hudman for her valuable assistance with statistical analysis. This work was supported by Grant 1R15AI090680-01 from the National Institutes of Health to V. K. Singh.

References

[1] J. R. Mediavilla, L. Chen, B. Mathema, and B. N. Kreiswirth, "Global epidemiology of community-associated methicillin resistant *Staphylococcus aureus* (CA-MRSA)," *Current Opinion in Microbiology*, vol. 15, no. 5, pp. 588–595, 2012.

[2] E. Stenehjem and D. Rimland, "MRSA nasal colonization burden and risk of MRSA infection," *American Journal of Infection Control*, 2012.

[3] R. R. Watkins, M. Z. David, and R. A. Salata, "Current concepts on the virulence mechanisms of meticillin-resistant *Staphylococcus aureus*," *Journal of Medical Microbiology*, vol. 61, part 9, pp. 1179–1193, 2012.

[4] F. D. Lowy, "Medical progress: *Staphylococcus aureus* infections," *The New England Journal of Medicine*, vol. 339, no. 8, pp. 520–532, 1998.

[5] A. M. Rivera and H. W. Boucher, "Current concepts in antimicrobial therapy against select gram-positive organisms: methicillin-resistant *Staphylococcus aureus*, penicillin-resistant pneumococci, and vancomycin-resistant enterococci," *Mayo Clinic Proceedings*, vol. 86, no. 12, pp. 1230–1243, 2011.

[6] F. C. Fang, "Antimicrobial reactive oxygen and nitrogen species: concepts and controversies," *Nature Reviews Microbiology*, vol. 2, no. 10, pp. 820–832, 2004.

[7] C. K. Ferrari, P. C. Souto, E. L. França, and A. C. Honorio-França, "Oxidative and nitrosative stress on phagocytes' function: from effective defense to immunity evasion mechanisms," *Archivum Immunologiae et Therapia Experimentalis*, vol. 59, no. 6, pp. 441–448, 2011.

[8] J. MacMicking, Q. W. Xie, and C. Nathan, "Nitric oxide and macrophage function," *Annual Review of Immunology*, vol. 15, pp. 323–350, 1997.

[9] J. M. Voyich, K. R. Braughton, D. E. Sturdevant et al., "Insights into mechanisms used by *Staphylococcus aureus* to avoid destruction by human neutrophils," *Journal of Immunology*, vol. 175, no. 6, pp. 3907–3919, 2005.

[10] F. R. DeLeo, B. A. Diep, and M. Otto, "Host defense and pathogenesis in *Staphylococcus aureus* infections," *Infectious Disease Clinics of North America*, vol. 23, no. 1, pp. 17–34, 2009.

[11] G. Y. Liu, "Molecular pathogenesis of *Staphylococcus aureus* infection," *Pediatric Research*, vol. 65, no. 5, part 2, pp. 71R–77R, 2009.

[12] S. P. Chakraborty, P. Pramanik, and S. Roy, "*Staphylococcus aureus* infection induced oxidative imbalance in neutrophils: possible protective role of nanoconjugated vancomycin," *ISRN Pharmacol*, vol. 2012, Article ID 435214, 2012.

[13] J. Kopincova, A. Puzserova, and I. Bernatova, "Biochemical aspects of nitric oxide synthase feedback regulation by nitric oxide," *Interdisciplinary Toxicology*, vol. 4, no. 2, pp. 63–68, 2011.

[14] S. Mariotto, M. Menegazzi, and H. Suzuki, "Biochemical aspects of nitric oxide," *Current Pharmaceutical Design*, vol. 10, no. 14, pp. 1627–1645, 2004.

[15] B. R. Crane, "The enzymology of nitric oxide in bacterial pathogenesis and resistance," *Biochemical Society Transactions*, vol. 36, part 6, pp. 1149–1154, 2008.

[16] L. E. Bird, J. Ren, J. Zhang et al., "Crystal structure of SANOS, a bacterial nitric oxide synthase oxygenase protein from *Staphylococcus aureus*," *Structure*, vol. 10, no. 12, pp. 1687–1696, 2002.

[17] A. Brunel, J. Santolini, and P. Dorlet, "Electron paramagnetic resonance characterization of tetrahydrobiopterin radical formation in bacterial nitric oxide synthase compared to mammalian nitric oxide synthase," *Biophysical Journal*, vol. 103, no. 1, pp. 109–117, 2012.

[18] K. Pant, A. M. Bilwes, S. Adak, D. J. Stuehr, and B. R. Crane, "Structure of a nitric oxide synthase heme protein from *Bacillus subtilis*," *Biochemistry*, vol. 41, no. 37, pp. 11071–11079, 2002.

[19] I. Gusarov and E. Nudler, "NO-mediated cytoprotection: instant adaptation to oxidative stress in bacteria," *Proceedings of the National Academy of Sciences of the United States of America*, vol. 102, no. 39, pp. 13855–13860, 2005.

[20] K. Shatalin, I. Gusarov, E. Avetissova et al., "Bacillus anthracis-derived nitric oxide is essential for pathogen virulence and survival in macrophages," *Proceedings of the National Academy of Sciences of the United States of America*, vol. 105, no. 3, pp. 1009–1013, 2008.

[21] N. M. van Sorge, F. C. Beasley, I. Gusarov et al., "Methicillin-resistant *Staphylococcus aureus* bacterial nitric oxide synthase affects antibiotic sensitivity and skin abscess development," *The Journal of Biological Chemistry*, 2013.

[22] M. J. Horsburgh, J. L. Aish, I. J. White, L. Shaw, J. K. Lithgow, and S. J. Foster, "δb modulates virulence determinant expression and stress resistance: characterization of a functional rsbU strain derived from *Staphylococcus aureus* 8325-4," *Journal of Bacteriology*, vol. 184, no. 19, pp. 5457–5467, 2002.

[23] D. A. Mead, E. Szczesna-Skorupa, and B. Kemper, "Single-stranded DNA "blue" t7 promoter plasmids: a versatile tandem promoter system for cloning and protein engineering," *Protein Engineering, Design and Selection*, vol. 1, no. 1, pp. 67–74, 1986.

[24] V. K. Singh, D. S. Hattangady, E. S. Giotis et al., "Insertional inactivation of branched-chain α-keto acid dehydrogenase in *Staphylococcus aureus* leads to decreased branched-chain membrane fatty acid content and increased susceptibility to certain stresses," *Applied and Environmental Microbiology*, vol. 74, no. 19, pp. 5882–5890, 2008.

[25] V. K. Singh, S. Utaida, L. S. Jackson, R. K. Jayaswal, B. J. Wilkinson, and N. R. Chamberlain, "Role for dnaK locus in tolerance of multiple stresses in *Staphylococcus aureus*," *Microbiology*, vol. 153, no. 9, pp. 3162–3173, 2007.

[26] J. Augustin, R. Rosenstein, B. Wieland et al., "Genetic analysis of epidermin biosynthetic genes and epidermin-negative mutants of Staphylococcus epidermidis," *European Journal of Biochemistry*, vol. 204, no. 3, pp. 1149–1154, 1992.

[27] V. K. Singh, M. Syring, A. Singh, K. Singhal, A. Dalecki, and T. Johansson, "An insight into the significance of the DnaK heat shock system in *Staphylococcus aureus*," *International Journal of Medical Microbiology*, vol. 302, no. 6, pp. 242–252, 2012.

[28] V. K. Singh, R. K. Jayaswal, and B. J. Wilkinson, "Cell wall-active antibiotic induced proteins of *Staphylococcus aureus* identified using a proteomic approach," *FEMS Microbiology Letters*, vol. 199, no. 1, pp. 79–84, 2001.

[29] S. J. Collins, "The HL-60 promyelocytic leukemia cell line: proliferation, differentiation, and cellular oncogene expression," *Blood*, vol. 70, no. 5, pp. 1233–1244, 1987.

[30] C. Tarella, D. Ferrero, and E. Gallo, "Induction of differentiation of HL-60 cells by dimethyl sulfoxide: evidence for a stochastic model not linked to the cell division cycle," *Cancer Research*, vol. 42, no. 2, pp. 445–449, 1982.

13

Pseudomonas sp. as a Source of Medium Chain Length Polyhydroxyalkanoates for Controlled Drug Delivery: Perspective

Sujatha Kabilan, Mahalakshmi Ayyasamy, Sridhar Jayavel, and Gunasekaran Paramasamy

UGC-Networking Resource Centre in Biological Sciences, School of Biological Sciences, Madurai Kamaraj University, Madurai 625021, India

Correspondence should be addressed to Sridhar Jayavel, srimicro2002@gmail.com

Academic Editor: Barbara H. Iglewski

Controlled drug delivery technology represents one of the most rapidly advancing areas of science. They offer numerous advantages compared to conventional dosage forms including improved efficacy, reduced toxicity, improved patient compliance and convenience. Over the past several decades, many delivery tools or methods were developed such as viral vector, liposome-based delivery system, polymer-based delivery system, and intelligent delivery system. Recently, nonviral vectors, especially those based on biodegradable polymers, have been widely investigated as vectors. Unlike the other polymers tested, polyhydroxyalkanoates (PHAs) have been intensively investigated as a family of biodegradable and biocompatible materials for *in vivo* applications as implantable tissue engineering material as well as release vectors for various drugs. On the other hand, the direct use of these polyesters has been hampered by their hydrophobic character and some physical shortcomings, while its random copolymers fulfilled the expectation of biomedical researchers by exhibiting significant mechanical and thermal properties. This paper reviews the strategies adapted to make functional polymer to be utilized as delivery system.

1. Introduction

For the last decades, drug delivery systems have enormously increased their performances, moving from simple pills to sustained/controlled release and sophisticated programmable delivery systems. Currently, drug delivery has also become more specific from systemic to organ and cellular targeting [1]. In general, the action of a drug molecule is dependent on its inherent therapeutic activity and the efficiency with which it is delivered to the site of action. An increasing appreciation of the latter has led to the evolution and development of novel drug delivery systems (NDDSs) [2], whereas traditional delivery systems (TDSs) are characterized by immediate and uncontrolled drug release kinetics [3]. Accordingly, drug absorption is essentially controlled by the body's ability to assimilate the therapeutic molecule and thus, drug concentration in different body tissues, such as the blood, typically undergoes an abrupt increase followed by a similar decrease. As a consequence, increasing attention has been focused on drug delivery methods which continually delivers drugs for prolonged time periods and in a controlled fashion. The primary method of accomplishing this controlled release has been through incorporating the existing drugs into new drug delivery systems such as polymers. This novel approach considerably improves drug performance in terms of efficacy, safety and patient compliance. A large number of both natural and synthetic polymers have been studied for possible application in an outstanding range of extended/controlled release properties for a wide variety of dosage forms and processing methods. Among the polymer tested, two promising synthetic polymers which have been developed for biomedical applications are polyvinylpyrrolidone and polyethylene glycol acrylate-based hydrogels [3]. Both of them are biodegradable and form copolymers with natural macromolecules. On the other hand, natural polymers have the advantage of high biocompatibility and less immunogenicity [4]. Among the natural polymers studied a special mention has to be made to polyhydroxyalkanoates (PHAs). Other natural polymers are chitosan, alginate, starch, pectin, casein and cellulose

Table 1: Classes of PHA synthases and varieties of P (3HA)s (adapted from Rehm 2007) [10].

Substrate specificity	Class of PHA synthase	Subunit(s) (PHA synthase subunit)	Microorganism	Polymers produced[a]
SCL-3HA-CoA (C_3–C_5)	I III IV	PhaC PhaC, PhaE PhaC,PhaR	*Ralstonia eutropha* *Allochromatium vinosum* *Bacillus megaterium*	P (3HB), P (3HB-*co*-3HV)
MCL-3HA-CoA (C_6–C_{14})	II	PhaC	*Pseudomonas oleovorans* *Pseudomonas putida* *Pseudomonas aeruginosa*	P (3HA)
SCL-MCL-P (3HA)-CoA (C_3–C_{14})	I II	PhaC PhaC	*Aeromonas caviae* FA440 *Pseudomonas* sp.61-3	P (3HB-*co*-3HA)

[a]P (3HB), poly-3-hydroxybutyrate, P (3HB-*co*-3HV), poly-3-hydroxybutyrate-*co*-3-hydroxyvalerate, P (3HA), poly-3-hydroxyalkanoate, P (3HB-*co*-3HA), poly-3-hydroxybutyrate-*co*-3-hydroxyalkanoate.

derivatives. However, in recent years additional polymers designed primarily for medical applications have entered the arena of controlled release because of its biodegradability within the body; among them Polylactic acid (PLA), Polyglycolic acid (PGA), Poly (lactic-co-glycolic acid) (PLGA), Polycaprolactone (PCL), especially PHA (polyhydroxyalkanoate), and PHB (polyhydroxybutyrate) have attracted researcher's attention. Polyhydroxyalkanoates (PHAs) are bacterial polymers that are formed as naturally occurring storage polyesters by a wide range of microorganisms usually under unbalanced growth conditions. PHAs are composed of β-hydroxy fatty acids, where the R group changes from methyl to tridecyl. Poly (3-hydroxy butyrate) is the most investigated PHA [5]. Initially interests on PHAs were focused as replacements for petrochemical plastics such as polyethylene and polypropylene, due to their degradable nature (degradable to carbon dioxide and water through natural microbiological mineralization), but currently due to their biocompatibility, processability and degradability, PHAs have been investigated as matrices for drug delivery and tissue engineering applications [6]. Microorganisms are able to incorporate up to 60 different types of monomer into their storage polymer and a series of PHAs with different monomeric compositions (i.e., different physical and chemical properties (Table 1) can be produced [7]. Over the past years, PHAs, particularly PHB, have been used to develop devices including sutures, repair devices, repair patches, slings, cardiovascular patches, orthopedic pins, adhesion barriers, stents, guided tissue repair/regeneration devices, articular cartilage repair devices, nerve guides, tendon repair devices, bone marrow scaffolds, and wound dressings [8]. Furthermore, the *in vitro* and *in vivo* biocompatibility experiments demonstrated that PHB can be exploited for the purpose of encapsulation and controlled release of different drugs [8, 9]. Formulation performance can be optimized by variations in molecular weights and chemical substitutions. Each range has fundamentally different hydrophilicity, swelling and erosion characteristics which provide flexibility in controlling the release mechanisms.

2. Advantages of Using Natural Polymer

For the past decades, polymeric materials have been used for a variety of applications ranging from food industries, textile and biomedical industries. Biopolymeric materials may be utilized for the encapsulation, delivery of various functional food ingredients, drugs such as bioactive lipids, minerals, enzymes and peptides [11–13]. Most of the polymers initially used for drug delivery applications were hydrophobic and nondegradable in nature for example, poly(dimethylsiloxane) (PS), polyurethanes (PUs) and poly(ethylene-*co*-vinyl acetate) (EVA) [6]. Among the synthetic and natural polymers tested, biopolymers accumulated by microbes such as Polyhydroxyalkanoates (PHAs) and polyhydroxybutyrates (PHB) are attractive carrier matrices for drugs where the drugs can be released by bioerosion. At the same time, materials, which degrade or denature soon after processing, pose a significant threat for mass production and industrial usage leading to limited research interests beyond academia [12]. In order to make biopolymers having broad industrial/medical relevance, it would be better to form composite by incorporating with cross-linking agents. The modification of biopolymers with the addition of functional groups is a common yet elegant mechanism to create durable and industrially relevant biopolymers. Natural polymeric drug delivery systems have the following advantages over other controlled release formulations such as good biocompatibility, flexible drug release profile which could be adjusted through the cross-linking strategies, degradability of the by-products of the polymer and possibility of quick elimination by the excretory system to overcome accumulation in the body.

3. Biosynthesis, Structure, and Properties of PHA

Polyhydroxyalkanoates (PHAs) are biological polyesters accumulated by microorganisms as energy reserve material

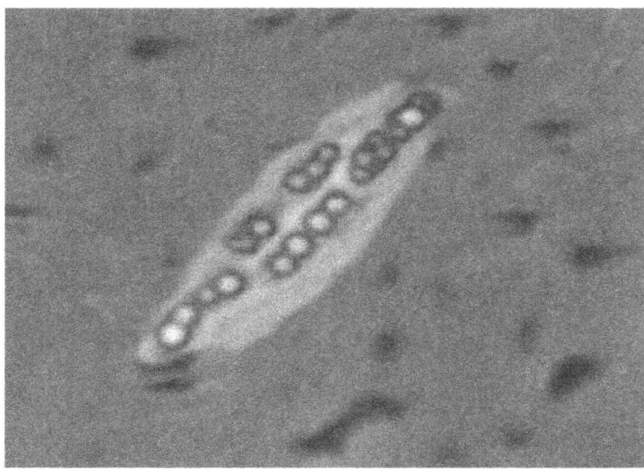

FIGURE 1: Phase Contrast Microscopic view of *Pseudomonas* sp. LDC-5 cells with accumulated PHA granules [24].

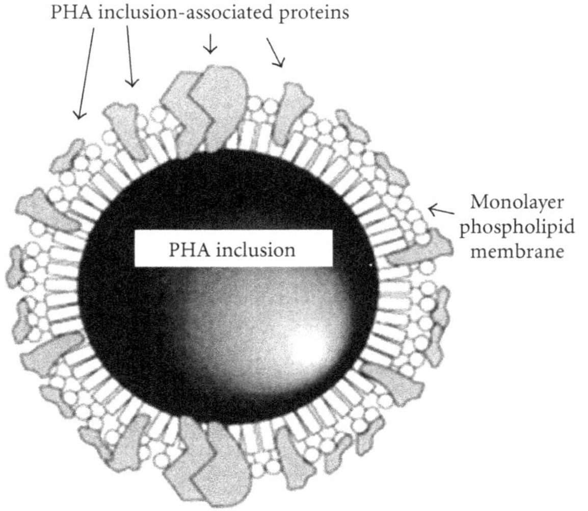

FIGURE 2: The structure of *in vivo* PHA inclusions and its association with specific proteins [7].

in the form of intracellular granules. The layer of phasins (granule associated proteins) stabilizes the granules and prevents coalescence of granules in the cytoplasm (Figures 1 and 2). Decades of PHA research were dedicated to understand the production, its material properties and potential applications [14–17]. The mechanical property of PHA is dependent on the side chain length of hydroxyalkanoates. Based on that, PHAs were classified as SCL, MCL, and LCL-PHA consisting of 3~5, 6~16 and >16 carbon atoms respectively [18] (Figure 3). Unlike SCL-PHAs, MCL-PHAs have low levels of crystallinity and are more elastic [19, 20]. Recently, reports on PHA consisting of both SCL and MCL 3-hydroxyalkanoate (3HA) monomers have demonstrated a broader spectrum of application properties [7]. Though many Pseudomonads belonging to rRNA-DNA homology group I produce PHA polymers containing Medium-Length alkyl side chains (MCL-PHA), only a few wild-type bacteria such as *Pseudomonas* sp.61-3 [21], *Pseudomonas oleovorans* strain B-778 [22], and *Pseudomonas stutzeri* [12] were found to produce a mixture of PHB and MCL-PHA. Although some of these monomers have been found in PHA produced by bacteria in their natural environment, a larger fraction of monomers have been incorporated into PHA following growth of bacteria under laboratory conditions in media containing exotic sources of carbon. *Cupriavidus necator, Rhodospirillum rubrum* and *Pseudomonas pseudoflava* are known to accumulate copolyesters composed of SCL monomer units only, while *Pseudomonas oleovorans, Pseudomonas putida* and other *Pseudomonas* strains biosynthesize copolyesters principally composed of MCL monomer units. To a considerable extent, the substrate specificity of the PHA synthases determines the composition of the accumulated PHA. Biosynthesis of PHA is possible due to PHA synthases exhibiting extraordinarily broad substrate ranges [23]. Table 1 shows bacterial strains capable of producing P (3HA) polymers, the types of polyhydroxyalkanoate (PHA) synthases associated with those strains and the type of P (3HA) polymers produced by those bacteria.

The size, core composition and surface functionality can be highly controlled and provide a platform technology for the production of functionalized, biocompatible and biodegradable particles, which can be applied for drug delivery, diagnostics, bioseparation, protein immobilization and so forth [25–27]. Thus, the value of the polymers can be increased by controlling the polymer's microstructure. The physical properties of PHA homopolymers as well as co- and heteropolymers have been the subject of study in various laboratories all over the world. By controlling the monomer composition of PHA, polymer scientists have shown that the polymer's physical properties can be regulated to a great extent. Furthermore, it is also clear that the rate of degradation of PHA in various environments can be controlled by judiciously altering its monomer composition.

4. Polymer-Based Drug Delivery Systems

In general, drug delivery systems can be classified into liposomal, electromechanical and polymeric delivery systems. Though these systems are promising, the recent focus is on degradable polymeric matrices as drug carriers. The drugs are incorporated in a polymer and applied to facilitate the targeted drug delivery. The release rate depends on various parameters like nature of the polymer matrix, matrix geometry, properties of the drug, initial drug load and polymer-drug interactions [6]. Currently biopolymers are attractive drug carriers as these polymers need not be removed after their function is over. Poly (3-hydroxybutyrate) was the first homopolymer of PHAs which are intensively used for various applications. Earlier studies have reported the use of PHAs in implant biomedical devices and controlled drug-release carriers [8, 28, 29]. In addition to this, there have been reports indicating that PHB and PHBV copolymer of 3-hydroxybutyrate (3HB) and 3-hydroxyvalerate (3HV) can be used as extracellular controlled drug release matrices [30–32]. Reports by Xiong et al. [33] demonstrated that lipid

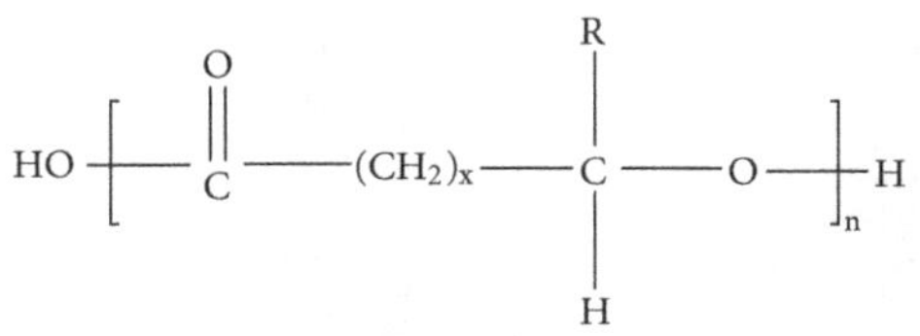

FIGURE 3: Chemical structure of PHA (a) and other classes (b) [10].

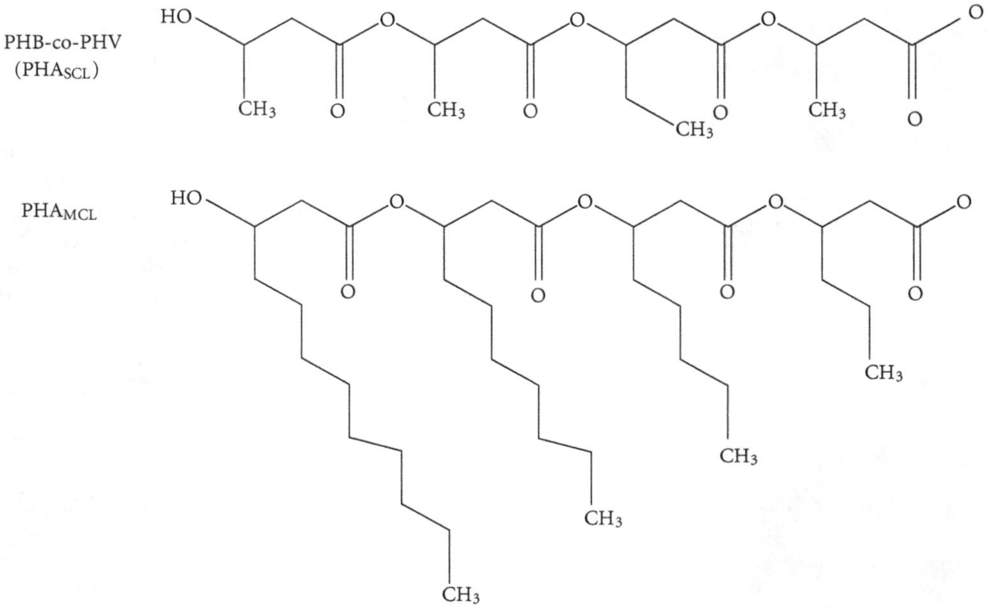

soluble colorant rhodamine poly (3-hydroxybutyrate-co-3-hydroxyalkanoate) (RBITC) encapsulated in PHB and PHB-HHX can be used for intracellular drug release. Above all, recent development in controlled drug delivery technology demands functionalized polymers for efficient performance. The following section will discuss the various strategies adopted to make functionalized polymers for drug delivery.

5. Improving the Biopolymer for Drug Delivery

Generally, drug delivery system to be designed in such a way is incapable of releasing its agent or agents (drugs) until it is placed in an appropriate biological environment, since biopolymers used in drug delivery are usually formed outside of the body and impregnated with drugs before placement the polymer plus drug complex in the body. In order to perform this, a wide range of cross-linking strategies can be used, including UV photopolymerization and various chemical cross-linking techniques. Such cross-linking methods are useful only if toxic reagents can be completely removed prior to implantation, which may be difficult to achieve without leaching loaded drug out of the polymer complex [34]. Though these approaches are advantageous, each method has its own advantages as well as limitations. In case of UV cross-linking due to their defined dimensionality and high elasticity excludes their extrusion through a needle. But these can be mitigated by making the complex into micro- or nanoparticles. Also, one has to consider the potential risks of exposure to UV and the cross-linking chemicals [34]. In general, the rate of drug release from a linear polymer matrix is inversely proportional to its viscosity.

6. Surface Functionalization of Biodegradable Polymers

Although many polymeric materials have been developed in the past decades to improve specific properties such as biocompatibility, degradability and drug delivery kinetics, there are also some limitations. In order to overcome this, copolymers with functional side groups which modify the surface with biologically active moieties may be useful [35–37]. Biomolecules such as fibronectin [38], collagen [39], insulin [40] and the epidermal growth factor [41] have often been introduced on polymer surfaces to enhance cell attachment or cell proliferation. The surface modification can produce controlled densities of hydroxyl groups on the surface and then these groups provide sites for the covalent attachment of specific biomaterials such as proteins or peptides [42]. Lee et al. [43] have performed modification of the biopolymers with various lengths of fluorocarbon (F-polyesters) end groups; this may improve the controllable biodegradability at initial stages by controlling the surface composition of fluorocarbon groups. Kang et al. [44, 45] emphasized the importance of plasma glow discharge technique in which partially ionized gas consisting of equal numbers of positive and negative charges and a different number of unionized neutral molecules is subjected to a DC or radio frequency (RF) potential. In order to introduce functional groups to polymeric surfaces (surface modification), glow discharge is widely used at reduced pressure. The characteristic glow of these plasmas is due to electronically excited species producing optical emission in the ultraviolet or visible regions of the spectrum and is characteristic of the composition of the glow discharge gas. Argon gives a bright blue colour and air or nitrogen gives a pink color that is due to excited nitrogen molecules.

7. Cross-Linking Strategies

In order to increase the stability, there is a growing trend towards the development of innovative biopolymeric material through the rational design of functional structures by cross-linking using various physical, chemical and enzymatic approaches depending on the specific characteristics of the biopolymers involved, rather than the use of the more traditional trial and error approach. These crosslinked polymers exhibit the required functional attributes, for example, optical properties, rheological properties, release characteristics, encapsulation properties and physicochemical stability [11].

8. Cross-Linking through Physical Methods

Polymer-polymer interaction without covalent bonding is known as physical cross-linking [46]. This includes ionic cross-linking, hydrogen bonding and hydrophobic bonding. Physical cross-linking can be achieved using a variety of environmental triggers such as pH, temperature, ionic strength and physiochemical interactions such as hydrophobic, charge condensation, hydrogen bonding and stereocomplexation [34]. These methods may be used for the induction of gelation processes among already formed discrete biopolymer particles.

9. Temperature

Temperature either strengthens or weakens the interactions of holding biopolymer molecules together, depending on the nature of the dominant forces involved. In general, hydrophobic forces are strengthened with increasing temperature, hydrogen bonding is weakened, and entropy effects are increased. Heating (heat-set-gelation) increases the hydrophobic driven association and cooling (cold-set-gelation) increases hydrogen bonding-driven association [47]. These temperature-associated reactions are either reversible or irreversible. Heat-set-gelation is used to cross-link globular proteins, such as milk, soy and egg as well as cross-link polysaccharides with some hydrophobic character. Cross-linking using heat-set-gelation of globular proteins tends to be irreversible. Once the aggregates are formed at higher temperatures, they remain intact when the system is cooled below the thermal denaturation temperature. In this reaction, the system is cooled below the thermal aggregation temperature as the molecules tend to dissociate [48]. Cold-set-gelation is stable at relatively low temperatures, but it tends to dissociate when heated above a critical temperature. This kind of gelation involves those biopolymers that are capable of forming helical regions that associate with each other through hydrogen bonding upon cooling. In case of mixed biopolymer systems that segregate, decreased temperatures can be utilized to favor separation, which in turn increases the biopolymer interaction within the dispersed phase [49]. Increased biopolymer interactions within these excluded volumes resulted into gel-like matrices; further heating may be utilized to further solidify these compact phases into particulates [50].

10. pH

In general, the response in different regions of the body is dependent on pH; the release of drugs in controlled fashion is possible by modifying the polymer cross-linking. Changes in pH and addition of mineral ions may be used to promote biopolymer association through alterations in electrostatic interactions. For example, calcium ions (Ca^{2+}) are frequently used to cross-link anionic polysaccharides such as pectins, alginates, or carrageenans [51]. Potassium ions (K^{+}) are used to cross-link anionic carrageenans through the formation of an "egg-box" structure [51]. Addition of mineral ion may be used to cross-link biopolymer particles in various ways [52]. An alternative method is to utilize slow-releasing salt devices to initiate cross-linking for controlled release of drug.

11. Chemical Cross-Linking

Introduction of a covalent linkage between polymer functional group is known as chemical cross-linking. The chemical agents may act as a bridge between similar or dissimilar amino acids [53]. They may also be used to directly bond 2 amino acids together. Functional groups capable of chemical bonds include amines, thiols, hydroxyl groups and phenyl rings. These chemical cross-linking strategies are widely used by researchers all over the world in comparison to physical cross-linking of functional groups in biopolymer.

12. Amphiphilic PHAs

PHAs are promising materials for biomedical applications in tissue engineering and drug delivery system because of their properties such as natural, renewable, biodegradable and biocompatible thermoplastics. The key to biocompatibility of biomedical implantable materials is to render their surface in a way that minimizes hydrophobic interaction with the surrounding tissue. Therefore, hydrophilic groups need to be introduced into the PHAs in order to obtain amphiphilic polymer. Amphiphilic polymers can be synthesized by introducing hydrophilic groups such as hydroxyl, carboxyl, amine, glycol and hydrophilic polymers such as PEG, poly(vinyl alcohol), polyacryl amide, poly acrylic acids, hydroxy ethyl methacrylate, poly vinyl pyridine and poly vinyl pyrrolidone to a hydrophobic moiety by means of functionalization and grafting [54]. Among the hydrophilic groups, PEG is a polyether known for its exceptional blood and tissue compatibility. It is used extensively as biomaterial in a variety of drug delivery vehicles and is also under investigation as surface coating for biomedical implants. PEG, when dissolved in water, has a low interfacial free energy and exhibits rapid chain motion and its large excluded volume leads to steric repulsion of approaching molecules [55].

13. Transesterification

Under certain conditions, esters and amides are capable of undergoing interchange reactions in aqueous solutions. Transesterification is carried out either in melt or in solution. Transesterification reactions in the melt between poly(ethylene glycol), mPEG, and PHB yield diblock amphiphilic copolymer with a dramatic decrease in molecular weight [56]. Genipin is a naturally occurring heterocyclic compound derived from *Genipa americana* that is able to form physical links between biopolymers containing amines. Mi et al. [57] reported that the primary amine groups located on biopolymers attack at either the α- or β-carbon of the genipin ester. During the attachment at the α-carbon that forms a simple amide linkage, attack at the β-carbon causes ring cleavage and further cross-linking capability. So far the genipin crosslinked biopolymers are chitosan, BSA, soy protein and gelatin [58]. Furthermore, the natural compound genipin is considered to be less cytotoxic than other cross-linking agents such as glutaraldehyde [57].

14. Maillard Reactions

Biopolymers are capable of loading both hydrophilic and hydrophobic drugs; in particular, modified polymers without synthetic chemical reagents are obviously desirable for biomedical applications. Several groups have studied the polymer fabrications for controlled release applications. One such fabrication is Maillard reaction; Maillard reaction is a natural, nontoxic reaction that occurs during the processing, cooking and storage of foods. As per, Oliver et al. [59], Maillard reaction is chemical linking of aldehydes and amines through a well-established oxidation-reduction pathway. High pH values favored these imido- and redox reactions. Maillarad conjugates have been used as emulsifiers and gelling agents. For example, the conjugation of hen egg lysozyme with dextran, galactomannan, or xyloglucan is effective in improving the emulsifying activity of the protein and it has been shown that the conjugated lysozyme has new antimicrobial characteristics [60, 61]. Rich and Foegeding (2000) [62] demonstrated that these Maillard reactions are useful to cross-link protein components with mono- and disaccharides. Recent, reports by Elzoghby et al. [63] have shown the significance of casein-based formulations as promising controlled release drug delivery systems. Casein, the major milk protein, is a good candidate for conventional and novel drug delivery systems for its property with high tensile strength. This Maillard reaction could also be used to cross-link proteins and polysaccharides within the biopolymer particles formed by spray-or freeze-drying. After drying, the system can then be subjected to dry-heating to promote the Maillard reaction.

15. Aldehyde Reactions

Glutaraldehyde and formaldehyde can be used to chemically cross-link protein particulate system. These may form compounds such as acetals, cyanohydrins, and oximes [64]. At the same time, studies by [65] showed that even 3 ppm of nonfood grade aldehydes exhibits cytotoxic effects to human fibroblasts. As a result, attempts were put forth to identify a number of food-grade alternatives to replace the potentially toxic compounds.

16. Quaternization and Sulfonation of the PHAs

Addition of the chlorine and bromine into the double bond is quantitative and halogenated PHAs can be easily obtained by this approach [66]. Chlorination can be done by either the addition to double bonds of the unsaturated PHA obtained from soybean oil (PHA-Sy) or substitution reactions with saturated hydrocarbon groups [67, 68]. Chlorination provides polyester with hard, brittle and crystalline physical properties depending on the chlorine content and also glass transition temperature has been shifted from -40°C to $+2^{\circ}$C [68]. For further functionalization, quaternization reactions of the chlorinated PHA with triethylamine (or triethanol amine) can be performed.

17. Enzymatic Methods

Apart from physical and chemical cross-linking, enzymes can also be utilized to catalyze specific cross-linking reactions between polymer. They are particularly useful in applications where alternative methods might cause damage to some encapsulated component. At the same time, sulfhydryl or phenolic residues can be oxidatively cross-linked using high levels of gaseous oxygen, but this would be deleterious to high-value lipids or phenolics [69]. In this case, usage of specific enzymes is a very good alternative. Recent reports on laccases (benzenediol: oxygen oxidoreductases; EC 1.10.3.2.) which are glycosylated polyphenol oxidases containing four copper ions per molecule, produced by white rot fungi in large amounts, have widespread applications such as effluent decolouration pulp bleaching and removal of phenolics from wines, organic synthesis, biosensors and synthesis of complex medical compounds, among others. Laccases are also able to cross-link biopolymers containing phenolic acids like ferulated arabinoxylans [70–72]. In particular, recently, enzymatic cross-linking and grafting of specific substances to the biopolymers can be exploited in food and nonfood applications allowing for generation of novel biomaterials [73]. De Jong and Koppelman [74] have reported the importance of transglutaminase which is commonly produced from bacterial sources that can be used as cross-linking agents for various types of proteins. This enzyme functions by a non-oxidative transamidation between glutamine and lysine, whether intra- or intermolecularly. Transglutaminase has been used to produce crosslinked protein films from gelatin [75], egg-white [76], gluten [77], soybean and whey proteins [78]. The extent of cross-linking can be controlled by changes in pH, enzyme inhibitors, or heating [74].

18. Grafting

Some of the naturally occurring polymers such as polyhydroxyalkanoate, alginate and chitosan find increasing application in biomedical research due to their biocompatible, biodegradable and nontoxic properties. However, to overcome the limitations posed by these polymers such as low moisture resistance, poor processability and incompatibility with some hydrophobic functional groups, the effective modification method grafting is adopted. This method is used to prepare multifunctional materials with improved chemical, physical and mechanical properties. Chitosan, sugar, PLA, gelatin, and PEG-mediated grafting are discussed in several studies [79–82]. Among these, glycopolymers are emerging as a novel class of neoglycoconjugates useful for biological studies and they are prepared by either copolymerization or grafting methods. One another, recently noted hydrophobic polymer with biodegradable ketal linkages in its backbone has an advantage over current biodegradable polymers for drug delivery is Polyketals. Polyketals do not release inflammatory byproducts compared to existing polymers [82].

19. Conclusion

Presently, novel polymeric materials have revolutionized the polymer applications in various fields including pharmaceutical, food and agricultural applications, pesticides, cosmetics, and household products. Particularly, in the pharmaceutical field, in addition to the importance of polymers, an understanding of the physiological barriers in the human body is also critical to develop appropriate controlled release systems. The skin, the gastrointestinal tract, the nose and the eye are of particular importance. In the immediate future, one of the dominant factors to be expected from human endeavor is environmental friendliness. Along this line, serious efforts are mounted to the developments of biopolymers with appropriate properties and processability, the so-called "green" polymers, that contrast to the conventional petrochemically originated polymers. In the future, custom-made prominent MCL-PHA synthases generated through *Pseudomonas* enzyme evolution will be utilized extensively to create high-performance P (3HA)s in various organisms from renewable carbon sources or through improved *in vitro* systems.

Acknowledgments

All the authors thank the University Grants Commission, Government of India, for sponsoring NRCBS in School of Biological Sciences, MKU. K. Sujatha thanks UGC for project Grant (No.F.No.39-206/2010 (SR)). J. Sridhar thanks the Department of Biotechnology, Government of India, for financial support. The authors thank Dr. Rajaiah Shenbagarathai of Lady Doak College forproviding the necessary input.

References

[1] O. Pillai, A. B. Dhanikula, and R. Panchagnula, "Polymers in drug delivery," *Current Opinion in Chemical Biology*, vol. 5, no. 4, pp. 439–446, 2001.

[2] N. Gupta Roop, R. Gupta, B. Pawan, and K. Rathore Garvendra, "Osmotically controlled oral drug delivery systems. A review," *International Journal of Pharmaceutical Sciences*, vol. 1, no. 2, pp. 269–275, 2009.

[3] M. Grassi and G. Grassi, "Mathematical modelling and controlled drug delivery: matrix systems," *Current Drug Delivery*, vol. 2, no. 1, pp. 97–116, 2005.

[4] K. Panduranga Rao, "New concepts in controlled drug delivery," *Pure and Applied Chemistry*, vol. 70, no. 6, pp. 1283–1287, 1998.

[5] L. L. Madison and G. W. Huisman, "Metabolic engineering of poly(3-hydroxyalkanoates): from DNA to plastic," *Microbiology and Molecular Biology Reviews*, vol. 63, no. 1, pp. 21–53, 1999.

[6] L. S. Nair and C. T. Laurencin, "Polymers as biomaterials for tissue engineering and controlled drug delivery," *Advances in Biochemical Engineering/Biotechnology*, vol. 102, pp. 47–90, 2006.

[7] K. Sudesh, H. Abe, and Y. Doi, "Synthesis, structure and properties of polyhydroxyalkanoates: biological polyesters," *Progress in Polymer Science*, vol. 25, no. 10, pp. 1503–1555, 2000.

[8] G. Q. Chen and Q. Wu, "The application of polyhydroxyalkanoates as tissue engineering materials," *Biomaterials*, vol. 26, no. 33, pp. 6565–6578, 2005.

[9] A. P. Bonartsev, G. A. Bonartseva, T. K. Makhina et al., "New poly-(3-hydroxybutyrate)-based systems for controlled release of dipyridamole and indomethacin," *Prikladnaia Biokhimiia i Mikrobiologiia*, vol. 42, no. 6, pp. 710–715, 2006.

[10] B. H. A. Rehm, "Biogenesis of microbial polyhydroxyalkanoate granules: a platform technology for the production of tailor-made bioparticles," *Current Issues in Molecular Biology*, vol. 9, no. 1, pp. 41–62, 2007.

[11] S. Gouin, "Microencapsulation: industrial appraisal of existing technologies and trends," *Trends in Food Science and Technology*, vol. 15, no. 7-8, pp. 330–347, 2004.

[12] L. Chen, G. E. Remondetto, and M. Subirade, "Food protein-based materials as nutraceutical delivery systems," *Trends in Food Science and Technology*, vol. 17, no. 5, pp. 272–283, 2006.

[13] C. P. Champagne and P. Fustier, "Microencapsulation for the improved delivery of bioactive compounds into foods," *Current Opinion in Biotechnology*, vol. 18, no. 2, pp. 184–190, 2007.

[14] H. Brandl, R. A. Gross, R. W. Lenz, and R. C. Fuller, "Plastics from bacteria and for bacteria: poly(beta-hydroxyalkanoates) as natural, biocompatible, and biodegradable polyesters," *Advances in Biochemical Engineering/Biotechnology*, vol. 41, pp. 77–93, 1990.

[15] S. Y. Lee and H. N. Chang, "Production of poly(hydroxyalkanoic acid)," *Advances in Biochemical Engineering/Biotechnology*, vol. 52, pp. 27–58, 1995.

[16] S. Y. Lee and J. I. Choi, "Production of microbial polyester by fermentation of recombinant microorganisms," *Advances in Biochemical Engineering/Biotechnology*, vol. 71, pp. 183–207, 2001.

[17] M. Zinn, B. Witholt, and T. Egli, "Occurrence, synthesis and medical application of bacterial polyhydroxyalkanoate," *Advanced Drug Delivery Reviews*, vol. 53, no. 1, pp. 5–21, 2001.

[18] G. Q. Chen, Q. Wu, J. Xi, H. P. Yu, and A. Chan, "Microbial the intermediates into other metabolic pathways, production of biopolyesters-polyhydroxyalkanoates," *Proceedings of the National Academy of Sciences of the United States of America*, vol. 10, pp. 843–850, 2000.

[19] R. A. Gross, C. DeMello, R. W. Lenz, H. Brandl, and R. C. Fuller, "Biosynthesis and characterization of poly(β-hydroxyalkanoates) produced by *Pseudomonas oleovorans*," *Macromolecules*, vol. 22, no. 3, pp. 1106–1115, 1989.

[20] H. Preusting, A. Nijenhuis, and B. Witholt, "Physical characteristics of poly(3-hydroxyalkanoates) and poly(3-hydroxyalkenoates) produced by *Pseudomonas oleovorans* grown on aliphatic hydrocarbons," *Macromolecules*, vol. 23, no. 19, pp. 4220–4224, 1990.

[21] H. Matsusaki, S. Manji, K. Taguchi, M. Kato, T. Fukui, and Y. Doi, "Cloning and molecular analysis of the poly(3-hydroxybutyrate) and poly(3-hydroxybutyrate-co-3-hydroxyalkanoate) biosynthesis genes in *Pseudomonas* sp. strain 61-3," *Journal of Bacteriology*, vol. 180, no. 24, pp. 6459–6467, 1998.

[22] R. Ashby, D. Solaiman, and T. Foglia, "The synthesis of short- and medium-chain-length poly(hydroxyalkanoate) mixtures from glucose- or alkanoic acid-grown *Pseudomonas oleovorans*," *Journal of Industrial Microbiology and Biotechnology*, vol. 28, no. 3, pp. 147–153, 2002.

[23] A. Steinbüchel and T. Lütke-Eversloh, "Metabolic engineering and pathway construction for biotechnological production of relevant polyhydroxyalkanoates in microorganisms," *Biochemical Engineering Journal*, vol. 16, no. 2, pp. 81–96, 2003.

[24] K. Sujatha, A. Mahalakshmi, and R. Shenbagarathai, "Molecular characterization of *Pseudomonas* sp. LDC-5 involved in accumulation of poly 3-hydroxybutyrate and medium-chain-length poly 3-hydroxyalkanoates," *Archives of Microbiology*, vol. 188, no. 5, pp. 451–462, 2007.

[25] J. A. Brockelbank, V. Peters, and B. H. A. Rehm, "Recombinant *Escherichia coli* strain produces a ZZ domain displaying biopolyester granules suitable for immunoglobulin G purification," *Applied and Environmental Microbiology*, vol. 72, no. 11, pp. 7394–7397, 2006.

[26] V. Peters and B. H. A. Rehm, "*In vivo* enzyme immobilization by use of engineered polyhydroxyalkanoate synthase," *Applied and Environmental Microbiology*, vol. 72, no. 3, pp. 1777–1783, 2006.

[27] B. H. A. Rehm, "Genetics and biochemistry of polyhydroxyalkanoate granule self-assembly: the key role of polyester synthases," *Biotechnology Letters*, vol. 28, no. 4, pp. 207–213, 2006.

[28] S. P. Valappil, S. K. Misra, A. Boccaccini, and I. Roy, "Biomedical applications of polyhydroxyalkanoates, an overview of animal testing and *in vivo* responses," *Expert Review of Medical Devices*, vol. 3, no. 6, pp. 853–868, 2006.

[29] J. Sun, Z. W. Dai, and G. Q. Chen, "Oligomers of polyhydroxyalkanoates stimulated calcium ion channels in mammalian cells," *Biomaterials*, vol. 28, pp. 3896–3903, 2007.

[30] F. Koosha, R. H. Muller, and S. S. Davis, "Polyhydroxybutyrate as a drug carrier," *Critical Reviews in Therapeutic Drug Carrier Systems*, vol. 6, no. 2, pp. 117–130, 1989.

[31] B. Saad, G. Ciardelli, S. Matter et al., "Characterization of the cell response of cultured macrophages and fibroblasts to particles of short-chain poly[(R)-3-hydroxybutyric acid]," *Journal of Biomedical Materials Research*, vol. 30, no. 4, pp. 429–439, 1996.

[32] I. Gürsel, F. Korkusuz, F. Türesin, N. Gürdal Alaeddinoğlu, and V. Hasirci, "*In vivo* application of biodegradable controlled antibiotic release systems for the treatment of implant-related osteomyelitis," *Biomaterials*, vol. 22, no. 1, pp. 73–80, 2001.

[33] Y. C. Xiong, Y. C. Yao, X. Y. Zhan, and G. Q. Chen, "Application of polyhydroxyalkanoates nanoparticles as intracellular sustained drug-release vectors," *Journal of Biomaterials Science, Polymer Edition*, vol. 21, no. 1, pp. 127–140, 2010.

[34] T. R. Hoare and D. S. Kohane, "Hydrogels in drug delivery: progress and challenges," *Polymer*, vol. 49, no. 8, pp. 1993–2007, 2008.

[35] P. J. A. In't Veld, P. J. Dijkstra, and J. Feijen, "Synthesis of biodegradable polyesteramides with pendant functional groups," *Makromolekulare Chemie*, vol. 193, pp. 2713–2730, 1992.

[36] P. J. A. In't Veld, P. J. Dijkstra, J. H. van Lochem, and J. Feijen, "Synthesis of alternating polydepsipeptides by ring-opening polymerization of morpholine-2,5-dione derivatives," *Makromolekulare Chemie*, vol. 191, no. 8, pp. 1813–1825, 1990.

[37] D. A. Barrera, E. Zylstra, P. T. Lansbury, and R. Langer, "Hydrogels and biodegradable polymers for bioapplications," *Macromolecules*, vol. 28, p. 425, 1995.

[38] Y. Ito, M. Inoue, S. Q. Liu, and Y. Imanishi, "Cell growth on immobilized cell growth factor. 6. Enhancement of fibroblast cell growth by immobilized insulin and/or fibronectin," *Journal of Biomedical Materials Research*, vol. 27, no. 7, pp. 901–907, 1993.

[39] H. W. Liu, F. A. Ofosu, and P. L. Chang, "Expression of human factor IX by microencapsulated recombinant fibroblasts," *Human Gene Therapy*, vol. 4, no. 3, pp. 291–301, 1993.

[40] Y. J. Kim, I. K. Kang, M. W. Huh, and S. C. Yoon, "Surface characterization and *in vitro* blood compatibility of poly(ethylene terephthalate) immobilized with insulin and/or heparin using plasma glow discharge," *Biomaterials*, vol. 21, no. 2, pp. 121–130, 2000.

[41] G. Chen, Y. Ito, and Y. Imanishi, "Regulation of growth and adhesion of cultured cells by insulin conjugated with thermo responsive polymers," *Biotechnology and Bioengineering*, vol. 53, pp. 339–344, 1997.

[42] H. V. Maulding, "Prolonged delivery of peptides by microcapsules," *Journal of Controlled Release*, vol. 6, pp. 167–176, 1987.

[43] W. K. Lee, I. Losito, J. A. Gardella, and W. L. Hicks, "Synthesis and surface properties of fluorocarbon end-capped biodegradable polyesters," *Macromolecules*, vol. 34, no. 9, pp. 3000–3006, 2001.

[44] I. K. Kang, B. K. Kwon, J. H. Lee, and H. B. Lee, "Immobilization of proteins on poly(methyl methacrylate) films," *Biomaterials*, vol. 14, no. 10, pp. 787–792, 1993.

[45] I. K. Kang, O. H. Kwon, Y. M. Lee, and Y. K. Sung, "Preparation and surface characterization of functional group-grafted and heparin-immobilized polyurethanes by plasma glow discharge," *Biomaterials*, vol. 17, no. 8, pp. 841–847, 1996.

[46] O. G. Jones and D. J. McClements, "Functional biopolymer particles: design, fabrication, and applications," *Comprehensive Reviews in Food Science and Food Safety*, vol. 9, no. 4, pp. 374–397, 2010.

[47] P. Burey, B. R. Bhandari, T. Howes, and M. J. Gidley, "Hydrocolloid gel particles: formation, characterization, and application," *Critical Reviews in Food Science and Nutrition*, vol. 48, no. 5, pp. 361–377, 2008.

[48] S. C. Joshi and Y. C. Lam, "Modeling heat and degree of gelation for methyl cellulose hydrogels with NaCl additives," *Journal of Applied Polymer Science*, vol. 101, no. 3, pp. 1620–1629, 2006.

[49] N. Lorén, A. M. Hermansson, M. A. K. Williams et al., "Phase separation induced by conformational ordering of gelatin in gelatin/maltodextrin mixtures," *Macromolecules*, vol. 34, no. 2, pp. 289–297, 2001.

[50] B. Wolf, R. Scirocco, W. J. Frith, and I. T. Norton, "Shear-induced anisotropic microstructure in phase-separated biopolymer mixtures," *Food Hydrocolloids*, vol. 14, no. 3, pp. 217–225, 2000.

[51] P. A. Williams, "Gelling agents," in *Handbook of Industrial Water Soluble Polymers*, P. A. Williams, Ed., pp. 73–97, Blackwell Publishing, Oxford, U.K., 2007.

[52] P. Burey, B. R. Bhandari, T. Howes, and M. J. Gidley, "Hydrocolloid gel particles: formation, characterization, and application," *Critical Reviews in Food Science and Nutrition*, vol. 48, no. 5, pp. 361–377, 2008.

[53] S. Andrea, "Chemical cross-linking and mass spectrometry for mapping three-dimensional structures of proteins and protein complexes," *Journal of Mass Spectrometry*, vol. 38, no. 12, pp. 1225–1237, 2003.

[54] S. Förster and M. Antonietti, "Amphiphilic block copolymers in structure-controlled nanomaterial hybrids," *Advanced Materials*, vol. 10, no. 3, pp. 195–217, 1998.

[55] K. J. Townsend, K. Busse, J. Kressler, and C. Scholz, "Contact angle, WAXS, and SAXS analysis of poly(β-hydroxybutyrate) and poly(ethylene glycol) block copolymers obtained via *Azotobacter vinelandii* UWD," *Biotechnology Progress*, vol. 21, no. 3, pp. 959–964, 2005.

[56] F. Ravenelle and R. H. Marchessault, "One-step synthesis of amphiphilic diblock copolymers from bacterial poly([R]-3-hydroxybutyric acid)," *Biomacromolecules*, vol. 3, no. 5, pp. 1057–1064, 2002.

[57] F. L. Mi, Y. C. Tan, H. C. Liang, R. N. Huang, and H. W. Sung, "*In vitro* evaluation of a chitosan membrane cross-linked with genipin," *Journal of Biomaterials Science, Polymer Edition*, vol. 12, no. 8, pp. 835–850, 2001.

[58] M. F. Butler, Y. F. Ng, and P. D. A. Pudney, "Mechanism and kinetics of the crosslinking reaction between biopolymers containing primary amine groups and genipin," *Journal of Polymer Science, Part A*, vol. 41, no. 24, pp. 3941–3953, 2003.

[59] C. M. Oliver, L. D. Melton, and R. A. Stanley, "Creating proteins with novel functionality via the maillard reaction: a review," *Critical Reviews in Food Science and Nutrition*, vol. 46, no. 4, pp. 337–350, 2006.

[60] S. Nakamura, A. Kato, and K. Kobayashi, "New antimicrobial characteristics of lysozyme-dextran conjugate," *Journal of Agricultural and Food Chemistry*, vol. 39, no. 4, pp. 647–650, 1991.

[61] M. Nakauma, T. Funami, S. Noda et al., "Comparison of sugar beet pectin, soybean soluble polysaccharide, and gum arabic as food emulsifiers. 1. Effect of concentration, pH, and salts on the emulsifying properties," *Food Hydrocolloids*, vol. 22, no. 7, pp. 1254–1267, 2008.

[62] L. M. Rich and E. A. Foegeding, "Effects of sugars on whey protein isolate gelation," *Journal of Agricultural and Food Chemistry*, vol. 48, no. 10, pp. 5046–5052, 2000.

[63] A. O. Elzoghby, W. S. Abo El-Fotoh, and N. A. Elgindy, "Casein-based formulations as promising controlled release drug delivery systems," *Journal of Controlled Release*, vol. 153, no. 3, pp. 206–216, 2011.

[64] D. McGregor, H. Bolt, V. Cogliano, and H. B. Richter-Reichhelm, "Formaldehyde and glutaraldehyde and nasal cytotoxicity: case study within the context of the 2006 IPCS human framework for the analysis of a cancer mode of action for humans," *Critical Reviews in Toxicology*, vol. 36, no. 10, pp. 821–835, 2006.

[65] D. P. Speer, M. Chvapil, C. D. Eskelson, and J. Ulreich, "Biological effects of residual glutaraldehyde in glutaraldehyde-tanned collagen biomaterials," *Journal of Biomedical Materials Research*, vol. 14, no. 6, pp. 753–764, 1980.

[66] D. J. Stigers and G. N. Tew, "Poly(3-hydroxyalkanoate)s functionalized with carboxylic acid groups in the side chain," *Biomacromolecules*, vol. 4, no. 2, pp. 193–195, 2003.

[67] Y. B. Kim, R. W. Lenz, and R. Clinton Fuller, "Poly(β-hydroxyalkanoate) copolymers containing brominated repeating units produced by *Pseudomonas oleovorans*," *Macromolecules*, vol. 25, no. 7, pp. 1852–1857, 1992.

[68] H. W. Kim, M. G. Chung, Y. B. Kim, and Y. H. Rhee, "Graft copolymerization of glycerol 1,3-diglycerolate diacrylate onto poly(3-hydroxyoctanoate) to improve physical properties and biocompatibility," *International Journal of Biological Macromolecules*, vol. 43, no. 3, pp. 307–313, 2008.

[69] G. Strauss and S. M. Gibson, "Plant phenolics as cross-linkers of gelatin gels and gelatin-based coacervates for use as food ingredients," *Food Hydrocolloids*, vol. 18, no. 1, pp. 81–89, 2004.

[70] S. Shleev, P. Persson, G. Shumakovich et al., "Interaction of fun gal laccases and laccase-mediator systems with lignin," *Enzyme and Microbial Technology*, vol. 39, no. 4, pp. 841–847, 2006.

[71] E. Rosales, S. Rodríguez Couto, and M. A. Sanromán, "Increased laccase production by Trametes hirsuta grown on ground orange peelings," *Enzyme and Microbial Technology*, vol. 40, no. 5, pp. 1286–1290, 2007.

[72] M. C. Figueroa-Espinoza and X. Rouau, "Oxidative cross-link ing of pentosans by a fungal laccase and horseradish peroxidase: mechanism of linkage between feruloylated arabinoxylans," *Cereal Chemistry*, vol. 75, no. 2, pp. 259–265, 1998.

[73] G. Freddi, A. Anghileri, S. Sampaio, J. Buchert, P. Monti, and P. Taddei, "Tyrosinase-catalyzed modification of *Bombyx mori* silk fibroin: grafting of chitosan under heterogeneous reaction conditions," *Journal of Biotechnology*, vol. 125, no. 2, pp. 281–294, 2006.

[74] G. A. H. De Jong and S. J. Koppelman, "Transglutaminase cata lyzed reactions: impact on food applications," *Journal of Food Science*, vol. 67, no. 8, pp. 2798–2806, 2002.

[75] L. T. Lim, Y. Mine, and M. A. Tung, "Barrier and tensile properties of transglutaminase cross-linked gelatin films as af fected by relative humidity, temperature, and glycerol content," *Journal of Food Science*, vol. 64, no. 4, pp. 616–622, 1999.

[76] L. T. Lim, Y. Mine, and M. A. Tung, "Transglutaminase cross-linked egg white protein films: tensile properties and oxygen permeability," *Journal of Agricultural and Food Chemistry*, vol. 46, no. 10, pp. 4022–4029, 1998.

[77] C. Larré, C. Desserme, J. Barbot, and J. Gueguen, "Properties of deamidated gluten films enzymatically cross-linked," *Journal of Agricultural and Food Chemistry*, vol. 48, no. 11, pp. 5444–5449, 2000.

[78] G. Su, H. Cai, C. Zhou, and Z. Wang, "Formation of edible soybean and soybean-complex protein films by a cross-linking treatment with a new *Streptomyces* transglutaminase," *Food Technology and Biotechnology*, vol. 45, no. 4, pp. 381–388, 2007.

[79] G. E. Yu, F. G. Morin, G. A. R. Nobes, and R. H. Marchessault, "Degree of acetylation of chitin and extent of grafting PHB on chitosan determined by solid state 15N NMR," *Macromolecules*, vol. 32, no. 2, pp. 518–520, 1999.

[80] M. Constantin, C. I. Simionescu, A. Carpov, E. Samain, and H. Driguez, "Chemical modification of poly(hydroxyalkanoates). Copolymers bearing pendant sugars," *Macromolecular Rapid Communications*, vol. 20, no. 2, pp. 91–94, 1999.

[81] Y. B. Kim, R. W. Lenz, and R. Clinton Fuller, "Poly(β-hydrox yalkanoate) copolymers containing brominated repeating units produced by *Pseudomonas oleovorans*," *Macromolecules*, vol. 25, no. 7, pp. 1852–1857, 1992.

[82] K. Avnesh, Y. Sudesh Kumar, and S. C. Yadav, "Biodegradable polymeric nanoparticles based drug delivery systems," *Colloids and Surfaces B*, vol. 75, no. 1, pp. 1–18, 2010.

14

Diversity of Cellulolytic Microbes and the Biodegradation of Municipal Solid Waste by a Potential Strain

S. P. Gautam,[1] P. S. Bundela,[2] A. K. Pandey,[3] Jamaluddin,[4] M. K. Awasthi,[2] and S. Sarsaiya[2,5]

[1] *Central Pollution Control Board, New Delhi, India*
[2] *Regional office, M. P. Pollution Control Board, Vijay Nagar, Jabalpur, India*
[3] *Mycological Research Laboratory, Department of Biological Sciences, Rani Durgavati University, Jabalpur, India*
[4] *Yeast and Mycorrhiza Biotechnology Laboratory, Department of Biological Sciences, Rani Durgavati University, Jabalpur, India*
[5] *International Institute of Waste Management (IIWM), Bhopal, India*

Correspondence should be addressed to S. Sarsaiya, surendra_sarsaiya@yahoo.co.in

Academic Editor: Thomas L. Kieft

Municipal solid waste contains high amounts of cellulose, which is an ideal organic waste for the growth of most of microorganism as well as composting by potential microbes. In the present study, Congo red test was performed for screening of microorganism, and, after selecting a potential strains, it was further used for biodegradation of organic municipal solid waste. Forty nine out of the 250 different microbes tested (165 belong to fungi and 85 to bacteria) produced cellulase enzyme and among these *Trichoderma viride* was found to be a potential strain in the secondary screening. During the biodegradation of organic waste, after 60 days, the average weight losses were 20.10% in the plates and 33.35% in the piles. There was an increase in pH until 20 days. pH however, stabilized after 30 days in the piles. Temperature also stabilized as the composting process progressed in the piles. The high temperature continued until 30 days of decomposition, after which the temperature dropped to 40°C and below during the maturation. Good quality compost was obtained in 60 days.

1. Introduction

In the present technoeconomic era, the energy and environmental crises developed due to huge amount of cellulosic materials are disposed as "waste." Municipal solid waste is composed of 40–50% cellulose, 9–12% hemicelluloses, and 10–15% lignin on a dry weight basis [1, 2]. Annually, Asia alone generates 4.4 billion tons of solid wastes, and municipal solid waste comprises 790 million tons of which about 48 million tons are generated in India. By the year 2047, municipal solid waste generation in India is expected to reach 300 million tons and land requirement for disposal of this waste would be 169.6 km^2. Unscientific disposal causes an adverse impact on all components of the environment and human health. Microorganism performs their metabolic processes rapidly and with remarkable specificity under ambient conditions, catalyzed by their diverse enzyme-mediated reactions. An enzyme alternative to harsh chemical technologies has led to intensive exploration of natural microbial biodiversity to discover enzymes. There is a wide spectrum of microorganisms which can produce the variety of enzymes like cellulase under appropriate conditions.

Cellulases are a consortium of free enzymes comprised of endoglucanases (β-1,4-D-glucan-4-glucanohydrolase, EC 3.2.1.4, carboxymethyl cellulase, EC), exoglucanases (β-1,4-D-glucan-4-glucohydrolase, EC 3.2.1.91, cellobiohydrolase, CBH), and cellobiases (β-D-glucoside glucohydrolase, EC 3.2.1.21, β-1,4-D-glucosidase) are found in many of the 57 glycosyl hydrolase families [3]. Several studies were carried out to produce cellulolytic enzymes in organic waste degradation process by several microorganism including fungi such as *Trichoderma* sp., *Penicillium* sp., and *Aspergillus* spp. respectively [4–6]. Many fungi capable of degrading cellulose synthesize large quantities of extracellular cellulases that are more efficient in depolymerising the cellulose substrate. Most commonly studied cellulolytic organisms

FIGURE 1: Collection of samples from different waste dump sites of Jabalpur.

include fungal species: *Trichoderma, Humicola, Penicillium,* and *Aspergillus* [7]. Among *Trichoderma* spp., *T. harzianum* [8–11] and *T. koningii* [12, 13] have been studied. Already an impressive collection of more than 14,000 fungi which were active against cellulose and other insoluble fibres were collected [14]. Many cellulases produced by bacteria appear to be bound to the cell wall and are unable to hydrolyze native lignocellulose preparations to any significant extent. A wide variety of Gram-positive and Gram-negative species are reported to produce cellulose, including *Clostridium thermocellum, Streptomyces* spp., *Ruminococcus* spp., *Pseudomonas* spp., *Cellulomonas* spp., *Bacillus* spp., *Serratia, Proteus, Staphylococcus* spp., and *Bacillus subtilis* [12, 15]. Various biological studies have been carried out to identify the major microbiological agents responsible for biodegradation. Today, environmental policies and regulation progress lead to the development of biodegradation processes to turn organic wastes into a valuable resource by potential microbes because only few strains are capable of secreting a complex of cellulase enzymes, which could have practical application in the enzymatic hydrolysis of cellulose as well as biodegradation of organic municipal solid waste.

Thus, the present work mainly focused on selecting a potential strain and utilization of cellulosic waste for value-added products.

2. Materials and Methods

2.1. Collection of Samples. Sampling sites were chosen, such that enough cellulose can be accessible naturally, whereby the resident microbial population could predominantly be cellulolytic by nature which could be isolated easily in large numbers. The samples were then brought to the laboratory for microbiological study. Municipal solid waste (MSW), compost, and soil samples collected from different areas of Jabalpur were screened for the isolation of cellulose-degrading microbes. All samples brought to laboratory for isolation were processed within 3–5 h of collection to minimize saprophytic developments (Figure 1).

2.2. Isolation, Identification, and Maintenance of Microbial Strains. Samples were collected from a depth of 2 cm and were carried to laboratory in sterilized polythene bags. Soil dilution plate method [16] was employed in the present work to isolate different microbes. The 100 μL portions (10^{-4} & 10^{-6}) of the suspensions were inoculated onto plates containing potato dextrose agar (PDA) and nutrient agar media (NAM). The plates were incubated at 30 ± 2°C for 4–8 days. All the isolates obtained from Jabalpur MSW, and composts and soils were identified according to morphological and biochemical basis [17–21]. The identified strains were maintained on PDA and NAM slants at low temperature (4 ± 1°C).

2.3. Primary Screening for Cellulolytic Activity. The isolated microbes were grown on basal salt media supplemented with 1% carboxymethylcellulose used for both bacteria and fungi [22]. The pure cultures were inoculated in the centre with almost equal amounts and incubated at 30 ± 2°C until substantial growth was recorded. The Petri plates were flooded with Congo red solution (0.1%), and after 5 min the Congo red solution was discarded, and the plates were washed with 1 M NaCl solution allowed to stand for 15–20 minutes. The clear zone was observed around the colony when the enzyme had utilized the cellulose.

2.4. Secondary Screening for Cellulolytic Activity. Potential microbes presenting large clearing zones in Congo red test were used for enzyme production on basal salt medium containing 1% cellulose as a sole carbon source [23]. Shake flask technique was used and 150 mL Erlenmeyer flask filled with 50 mL of the above medium. After autoclaving, each flask was inoculated with two discs (2 mm diameters) cut from the periphery of 6-day-old culture of fungi actively growing on PDA and a loopful of bacterial culture actively growing on NAM. The flasks were then incubated at 30°C in a shaker for 6 days. The flasks were filtered through Whatman number 1 filter paper to separate culture filtrates. The filtrate was analysed for enzyme activities.

2.5. Measurement of Enzyme Activity. Filter paper activity (FPase) for total cellulase activity in the culture filtrate was determined according to the standard method. Aliquots of appropriately diluted cultured filtrate as enzyme source were added to a Whatman number 1 filter paper strip (1 × 6 cm; 50 mg) immersed in one milliliter of 0.05 M sodium citrate buffer of pH 4.8. After incubation at 50 ± 2°C for 1 hrs, the reducing sugar released was estimated by dinitrosalicylic acid (DNS) method [24]. One unit of filter paper (FPU) activity was defined as the amount of enzyme-releasing 1 μmole of reducing sugar from filter paper per mL per min. Endoglucanase activity (CMCase) was measured using a reaction mixture containing 1 mL of 1% carboxymethyl cellulose (CMC) in 0.05 M sodium citrate buffer (pH 4.8) and aliquots of suitably diluted filtrate. The reaction mixture was incubated at 50 ± 2°C for 1 h, and the reducing sugar produced was determined by DNS method [24]. β-glucosidase activity was assayed by the method of pointing [25]. One unit (IU) of enzyme activity was defined as the amount of enzyme releasing 1 μmole of reducing sugar per min.

2.6. Biodegradation of Municipal Solid Waste

2.6.1. Inoculum Preparation. The fungus used in the study was *Trichoderma viride.* Two discs of fungal mycelium of *T. viride* were subcultured in PDA for mass cultivation and incubated at 30 ± 2°C for 6 days. After 144 hours of growth, 5% (v/v) of spore suspension of *T. viride* culture was mixed with municipal solid waste for bioconversion [26, 27].

2.6.2. Biodegradation of Municipal Solid Waste. Municipal solid wastes were collected from different waste dumping points of Jabalpur in polythene bags. Samples were cut into small pieces, and 5 g of each was aliquoted into petri plates, which were then wrapped by using polythene bags. The plates containing municipal solid waste were then autoclaved at 121°C for 15 min, and 25 kg of waste in polythene bags was also autoclaved for preparing piles. After sterilization, the inoculum was inoculated in a petri plate and piles in triplicate. Moisture content was maintained at 50–60% throughout the active biodegradation in the plates as well as piles. The pH and temperature were also measured periodically in the piles after 10 days intervals. Turning of the organic waste was provided once in every week to ensure aerobic condition both in the piles and plates. Changes in odour and weight loss of the decomposed organic solid waste were observed at 10-day intervals in the piles and at 30 days in the plates up to 60 days. For measurement of weight loss (%), the following formula was used:

$$\text{Weight loss (\%)} = W\frac{W_1}{W} \times 100, \quad (1)$$

where W is initial weight, and W_1 is final weight.

3. Results and Discussion

A total of 250 isolates were isolated; of these, 165 belonged to 37 fungal species, and 85 to 21 bacterial species. These were *Absidia* sp., *Alternaria alternata*, *Alternaria* sp., *Acremonium butyri, Aspergillus clavatus, Aspergillus flavus, Aspergillus fumigatus, Aspergillus nidulans, Aspergillus niger, Aspergillus candidus, Aspergillus luchuensis, Aspergillus terreus, Aspergillus* sp., *Chaetomium* sp., *Chrysosporium* sp., *Cladosporium* sp., *Colletotrichum* sp., *Curvularia lunata*, *Curvularia* sp., *Drechslera* sp., *Exserohilum* sp., *Fusarium oxysporum*, *Fusarium roseum*, *Gliocladium* sp., *Helminthosporium* sp., *Humicola* sp., *Mucor* sp., *Myrothecium* sp., *Penicillium digitatum*, *Penicillium* sp., *Rhizopus* sp., *Sclerotium rolfsii*, *Torula* sp., *Trichoderma viride*, *Trichoderma* sp., *Verticillium* sp., MRLB #38, MRLB #39, MRLB #40, MRLB #41, MRLB #42, MRLB #43, MRLB #44, MRLB #45, MRLB #46, MRLB #47, MRLB #48, MRLB #49, MRLB #50, MRLB #51, MRLB #52, MRLB #53, MRLB #54, MRLB #55, MRLB #56, MRLB #57, and MRLB #58 (Table 1). The most frequent fungi were *Aspergillus niger*, *Curvularia lunata*, *A. nidulans*, *A. fumigatus, Penicillium* sp., *Fusarium roseum*, and *Trichoderma viride*, and bacteria were MRLB #39, MRLB #42, and MRLB #44. Most of the above isolates have been reported as cellulase producers, but with variable capabilities by several workers [28–33]. Strom [34] reported that a large majority of the total number of bacterial isolates were members of the genus *Bacillus*. Proom and Knight [35] studied the bacteria required minimal nutritional requirement. Ezekiel et al. [36] isolated 22 different cellulolytic fungi from different sites. Duncan et al. [37] also isolated 72 fungi and screened for cellulase activity by using the carboxymethyl cellulose (CMC) Congo red plate technique.

3.1. Primary Screening for Cellulolytic Activity. The results of primary screening showed that degradation of cellulose by tested isolates differs from organism to organism. Out of two hundred fifty isolates tested, cellulolytic activity was detected in only 49 different isolates after 4 days of incubations, indicating them to be cellulose degraders. The diameter of the yellow halo varied from organism to organism. The data present in Table 1 revealed that *Alternaria alternata, Alternaria* sp., *Acremonium butyri, Aspergillus clavatus, Aspergillus flavus, Aspergillus candidus, Aspergillus luchuensis, Aspergillus fumigatus, Aspergillus nidulans, Aspergillus niger, Aspergillus terreus, Aspergillus* sp., *Chaetomium* sp., *Chrysosporium* sp., *Cladosporium* sp., *Curvularia lunata*, *Curvularia* sp., *Drechslera* sp., *Fusarium oxysporum*, *Fusarium roseum*,

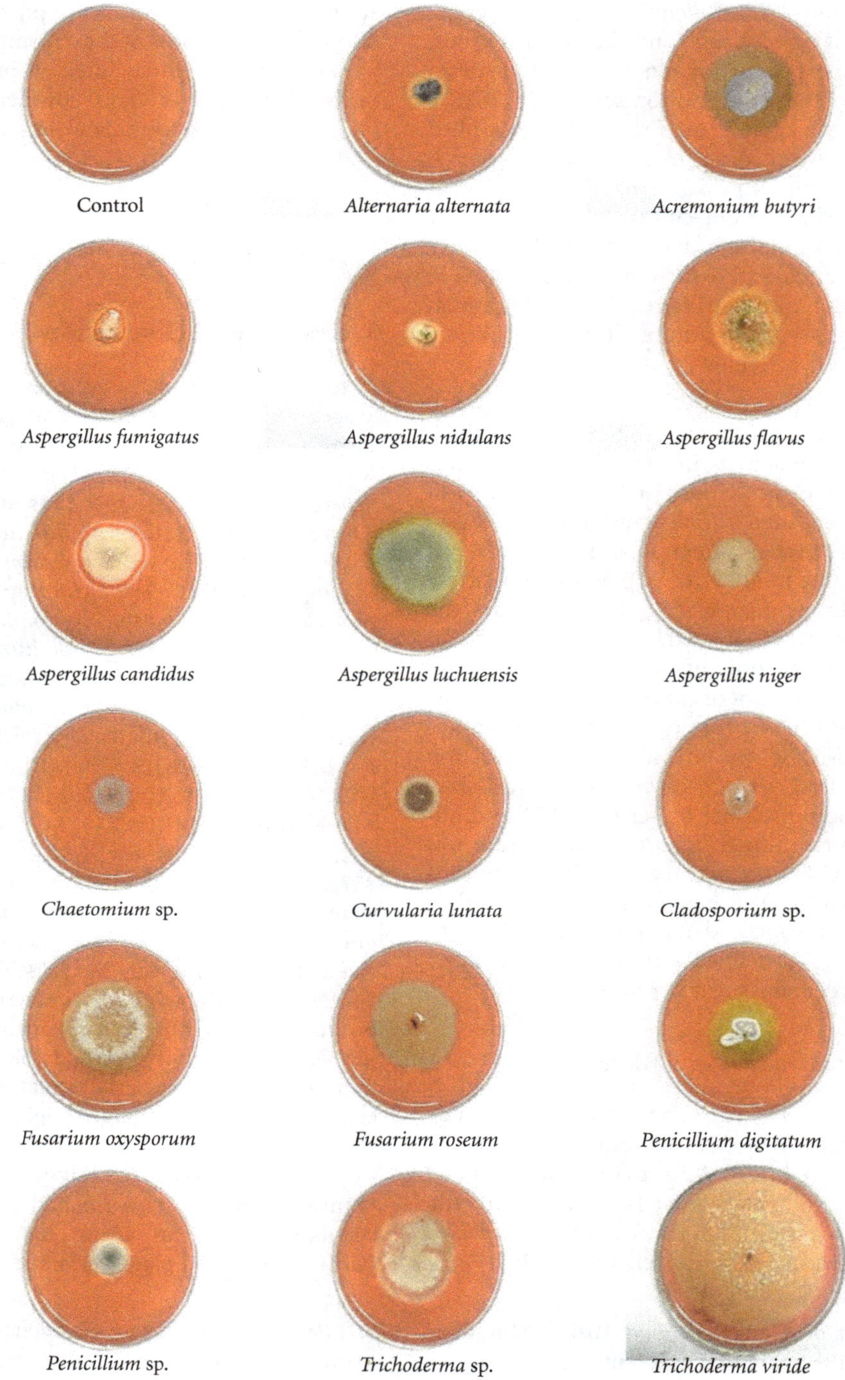

Figure 2: Primary screening of fungi isolated from different sources.

Gliocladium sp., *Humicola* sp., *Mucor* sp., *Myrothecium* sp., *Paecilomyces* sp., *Penicillium digitatum*, *Penicillium* sp., *Rhizopus* sp., *Sclerotium rolfsii*, *Trichoderma viride*, *Trichoderma* sp., *Verticillium* sp., MRLB #38, MRLB #39, MRLB #40, MRLB #42, MRLB #43, MRLB #44, MRLB #45, MRLB #46, MRLB #47, MRLB #49, MRLB #50, MRLB #51, MRLB #53, MRLB #54, MRLB #55, MRLB #56, and MRLB #57 were active cellulase producers (Figures 2 and 3). On the other hand, *Absidia* sp., *Colletotrichum* sp., *Exserohilum* sp., *Helminthosporium* sp., *Torula* sp., MRLB #41, MRLB #48, MRLB #52, and MRLB #58 were noncellulose producers. The above species were isolated with different numbers and frequencies from various sources in many places of the world by several workers [38–44]. Of the 250 isolates, 49 different

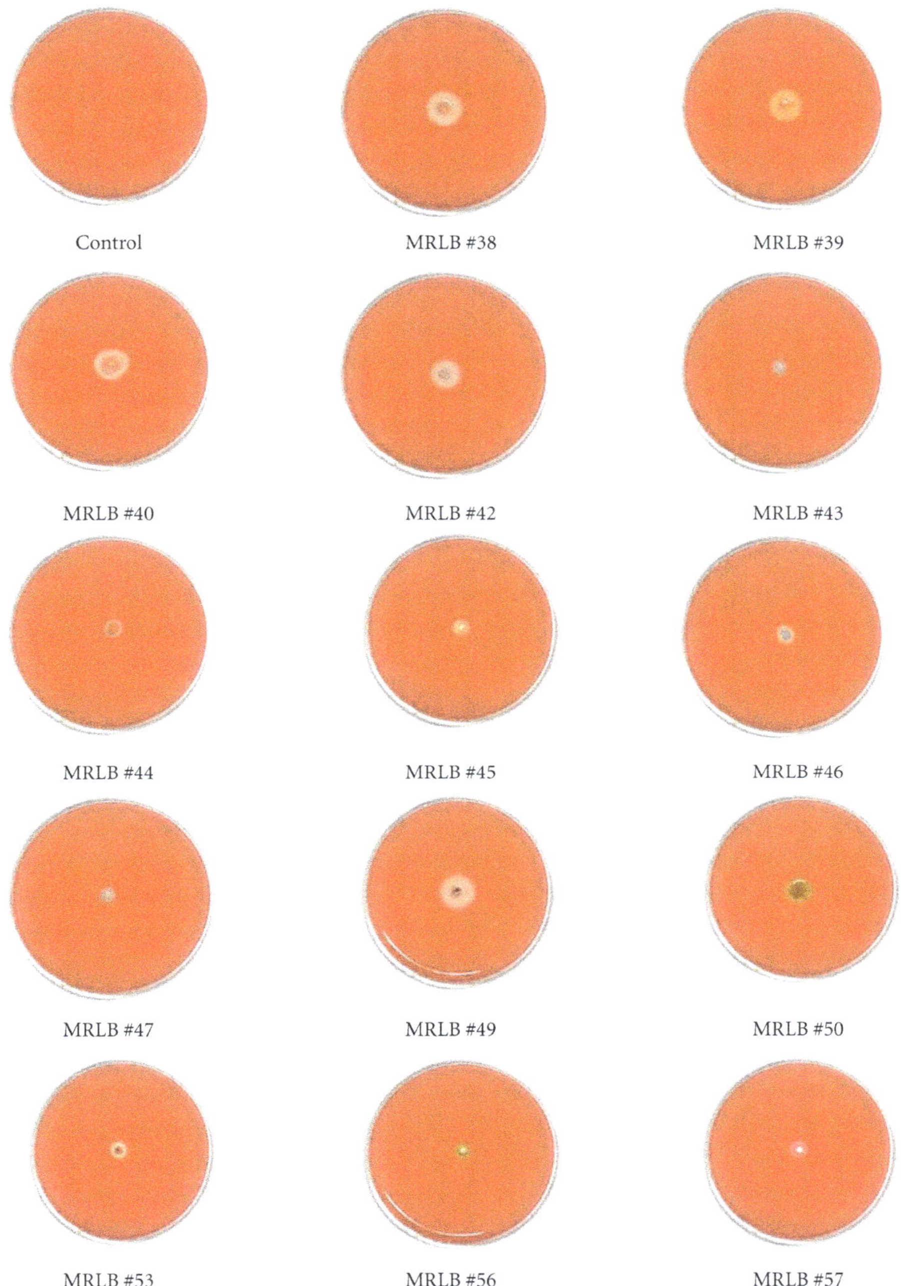

FIGURE 3: Primary screening of bacteria isolated from different sources.

isolates were tested for the production of cellulolytic enzyme (secondary screening).

3.2. Secondary Screening for Cellulolytic Activity. It is evident from the results that maximum cellulases activity was observed after the 6th day of incubation at 30°C in *Trichoderma viride*, *Penicillium digitatum*, *Aspergillus niger*, *Chaetomium* sp., and MRLB #38 and MRLB #40 followed by *Aspergillus fumigatus*, *Aspergillus nidulans*, *Alternaria alternata*, *Aspergillus* sp., *Fusarium oxysporum*, *Humicola* sp., MRLB #39, and MRLB #42. Out of the 49 isolates, exo-β-glucanase (C_1) activity in fourteen isolates (*Aspergillus flavus* (0.07 IU/mL), *Aspergillus terreus* (0.09 IU/mL), *Aspergillus* sp. (0.10 IU/mL), *Curvularia lunata* (0.11 IU/mL), *Drechslera* sp. (0.15 IU/mL), *Mucor* sp. (0.23 IU/mL), *Rhizopus* sp. (0.13 IU/mL), *Verticillium* sp. (0.20 IU/mL), MRLB #42 (0.08 IU/mL), MRLB #44 (0.16 IU/mL), MRLB #51 (0.08 IU/mL), MRLB #54 (0.14 IU/mL), MRLB #56 (0.09 IU/mL), and MRLB #57 (0.12 IU/mL)) were very low. Maximum exo-β-glucanase (C_1) activity was observed by *Trichoderma viride* (2.22 IU/mL), *Aspergillus niger* (2.05 IU/mL), *Fusarium oxysporum*

Table 1: Preliminary screening of cellulose-degrading microbes isolated from different sources.

Culture	Code no.	Source			Zone
		Soil	MSW	Compost	
Absidia sp.	MRLF #1	+	−	−	−
Alternaria alternata	MRLF #2	+	+	+	+
Alternaria sp.	MRLF #3	+	+	+	+
Acremonium butyri	MRLF #4	+	+	−	+
Aspergillus clavatus	MRLF #5	+	+	+	+
Aspergillus flavus	MRLF #6	+	+	−	+
Aspergillus candidus	MRLF #7	+	−	−	+
Aspergillus luchuensis	MRLF #8	+	+	−	+
Aspergillus fumigatus	MRLF #9	+	+	+	+
Aspergillus nidulans	MRLF #10	+	+	+	+
Aspergillus niger	MRLF #11	+	+	+	+
Aspergillus sp.	MRLF #12	+	−	−	+
Aspergillus terreus	MRLF #13	−	+	−	+
Chaetomium sp.	MRLF #14	−	+	−	+
Chrysosporium sp.	MRLF #15	+	−	−	+
Cladosporium sp.	MRLF #16	+	+	−	+
Colletotrichum sp.	MRLF #17	+	+	+	−
Curvularia lunata	MRLF #18	+	+	+	+
Curvularia sp.	MRLF #19	+	+	+	+
Drechslera sp.	MRLF #20	−	+	−	+
Exserohilum sp.	MRLF #21	−	+	−	−
Fusarium oxysporum	MRLF #22	+	+	+	+
Fusarium roseum	MRLF #23	−	+	+	+
Gliocladium sp.	MRLF #24	+	−	−	+
Helminthosporium sp.	MRLF #25	−	+	−	−
Humicola sp.	MRLF #26	+	+	−	+
Mucor sp.	MRLF #27	+	+	+	+
Myrothecium sp.	MRLF #28	−	+	−	+
Paecilomyces sp.	MRLF #29	+	+	−	+
Penicillium digitatum	MRLF #30	−	−	+	+
Penicillium sp.	MRLF #31	−	+	−	+
Rhizopus sp.	MRLF #32	+	+	+	+
Sclerotium rolfsii	MRLF #33	−	+	+	+
Torula sp.	MRLF #34	−	+	−	−
Trichoderma viride	MRLF #35	−	+	+	+
Trichoderma sp.	MRLF #36	+	+	−	+
Verticillium sp.	MRLB #37	−	+	+	+
Bacteria 1	MRLB #38	−	+	+	+
Bacteria 2	MRLB #39	+	+	+	+
Bacteria 3	MRLB #40	+	+	−	+
Bacteria 4	MRLB #41	+	−	−	−
Bacteria 5	MRLB #42	−	+	+	+
Bacteria 6	MRLB #43	+	+	+	+
Bacteria 7	MRLB #44	+	+	+	+
Bacteria 8	MRLB #45	−	−	+	+
Bacteria 9	MRLB #46	+	+	−	+
Bacteria 10	MRLB #47	+	+	−	+
Bacteria 11	MRLB #48	+	−	+	−
Bacteria 12	MRLB #49	−	−	+	+
Bacteria 13	MRLB #50	+	+	+	+

Table 1: Continued.

Culture	Code no.	Source			Zone
		Soil	MSW	Compost	
Bacteria 14	MRLB #51	−	+	+	+
Bacteria 15	MRLB #52	+	+	+	−
Bacteria 16	MRLB #53	+	−	−	+
Bacteria 17	MRLB #54	+	+	−	+
Bacteria 18	MRLB #55	+	+	+	+
Bacteria 19	MRLB #56	+	+	+	+
Bacteria 20	MRLB #57	+	−	+	+
Bacteria 21	MRLB #58	+	+	+	−

−: Absent; +: present; MSW: municipal solid waste.

(1.14 IU/mL), *Fusarium roseum* (1.67 IU/mL), *Penicillium digitatum* (1.38 IU/mL), MRLB #45 (1.66 IU/mL), MRLB #38 (1.65 IU/mL) followed by *Chaetomium* sp. (1.54 IU/mL), *Alternaria alternata* (0.30 IU/mL), *Aspergillus nidulans* (1.19 IU/mL), *Humicola* sp. (1.28 IU/mL), MRLB #39 (1.07 IU/mL), MRLB #47 (1.07 IU/mL), and MRLB #53 (1.34 IU/mL). Maximum endo-β-glucanase (C_x) activity was observed in *Trichoderma viride* (2.03 IU/mL), *Aspergillus niger* (1.76 IU/mL), *Aspergillus nidulans* (1.66 IU/mL) and MRLB #38 (1.81 IU/mL) followed by *Chaetomium* sp. (1.19 IU/mL), *Fusarium oxysporum* (1.62 IU/mL), *Fusarium roseum* (1.21 IU/mL), *Humicola* sp. (1.43 IU/mL), MRLB #38 (1.71 IU/mL), MRLB #45 (1.06 IU/mL), MRLB #47 (0.96 IU/mL), and MRLB #53 (1.29 IU/mL). On the other hand, *Alternaria* sp. (0.06 IU/mL), *Aspergillus flavus* (0.04 IU/mL), *Aspergillus terreus* (0.05 IU/mL), *Curvularia lunata* (0.09 IU/mL), *Rhizopus* sp. (0.09 IU/mL), MRLB #42 (0.05 IU/mL), MRLB #51 (0.11 IU/mL), MRLB #56 (0.11 IU/mL), and MRLB #57 (0.16 IU/mL) were showed very low endo-β-glucanase activity. β-glucosidase activity was detected in all the isolates, reaching to its maximum in *Trichoderma viride* (1.98 IU/mL), *Aspergillus niger* (1.96 IU/mL) and *Pencillium digitatum* (1.53 IU/mL) followed by *Aspergillus nidulans* (1.54 IU/mL), *Chaetomium* sp. (1.34 IU/mL), *Fusarium oxysporum* (1.86 IU/mL), *Fusarium roseum* (1.46 IU/mL), *Humicola* sp. (1.27 IU/mL), *Penicillium* sp. (1.08 IU/mL), MRLB #38 (1.59 IU/mL), MRLB #45 (1.12 IU/mL), and MRLB #53 (1.36 IU/mL). Reference [45] reported β-glucosidase production by 39 fungi and found *A. fumigatus* and *A. nidulans* as good β-glucosidase producers. The maximum β-glucosidase activity reported by [46] was in *Chaetomium cellulotricum*. In the present investigation (Table 2), *Trichoderma viride* was potential strain in the secondary screening and used further for utilization of municipal solid waste for production of value-added product (compost).

3.3. Bioconversion of Municipal Solid Waste. Fungi play an important role in the decomposition of organic waste and can be important contributors to optimal waste bioconversion. For decomposition of organic solid waste by using spore suspension treatment of *T. viride*, no bad smell was emitted after 60 days. It indicates the possible complete degradation of organic waste in plates that contained 5 g organic waste (Figure 4). In case of the control, the bad smell continued even after 60 days, and it indicates slow degradation. Data pre-sented in the Figure 6 showed that, after 60 days, the average weight loss in three trials (plates) was 12.51% and 20.10%. On the other hand, piles that contained 25 kg of waste in triplicates (Figures 5 and 7) showed that, after 60 days, the average weight loss in three piles was 33.35% and 11.24% in control.

During the bioconversion of organic waste, there was a shift in pH from the initial condition neutral (7.21 and 7.27) toward an alkaline condition in the piles. The occurrence of these conditions may be attributed to the bioconversion of the organic material into various intermediate types of organic acid and higher mineralization of the nitrogen and phosphorous into nitrites/nitrates and orthophosphate, respectively. This increase in pH during the biodegradation process could be due to the production of ammonium as a result of the ammonification process [47]. Our data show a similar trend (Figure 8) with a fast pH increase during the first ten days of bioconversion and then stabilization with pH fluctuations between 7.33 and 7.25 after 30 days. At the beginning of the experiment, the fluctuations can be explained by the industrial process used, involving periodic turnings and consequently the volatilization of NH_4^+. This acid production results from a lack of oxygen that can occur between two turnings. In such conditions, pH can reach values of about 6.0 [48]. Our results show that the pH of composts decreased to a final, mature pH of approximately 7.40 (control) and 7.26, which meets the compost regulations of pH 5.0–8.0 for the US, and pH 5.5–8.0 for the Council of European Communities (CEC) [49].

The initial average temperature of the turned piles was 29 and 31°C and rapidly rose to a peak of 58°C after 20 days of decomposing. The high temperature continued until 30 days of decomposition, after which the temperature dropped to 36°C by the 40th day of composting. Thereafter, the temperature varied within a narrow range (36–30°C) (Figure 9). The temperature levels in the compost piles tended to increase and reach 40–60°C due to the energy released from the biochemical reactions of the microorganisms in the compost piles, while the temperature levels in the compost piles tended to decrease after the thermophilic phase due to a loss of the substrate and a decrease in microbial activity [50]. It

TABLE 2: Secondary screening of cellulose-producing microorganism.

Culture	Code no.	Enzyme activity (IU/mL)		
		Exoglucanase	Endoglucanase	β-glucosidase
Alternaria alternata	MRLF #2	0.30 ± 0.02	0.58 ± 0.03	0.16 ± 0.01
Alternaria sp.	MRLF #3	0.19 ± 0.01	0.06 ± 0.007	0.12 ± 0.01
Acremonium butyri	MRLF #4	0.32 ± 0.01	0.45 ± 0.02	0.29 ± 0.01
Aspergillus clavatus	MRLF #5	0.46 ± 0.02	0.52 ± 0.03	0.40 ± 0.03
Aspergillus flavus	MRLF #6	0.07 ± 0.003	0.04 ± 0.005	0.10 ± 0.008
Aspergilluscandidus	MRLF #7	0.43 ± 0.02	0.36 ± 0.01	0.50 ± 0.02
Aspergillus luchuensis	MRLF #8	0.65 ± 0.02	0.71 ± 0.04	0.53 ± 0.04
Aspergillus fumigatus	MRLF #9	0.72 ± 0.03	0.58 ± 0.03	0.85 ± 0.05
Aspergillus nidulans	MRLF #10	1.19 ± 0.05	1.66 ± 0.06	1.54 ± 0.06
Aspergillus niger	MRLF #11	2.05 ± 0.06	1.76 ± 0.06	1.96 ± 0.06
Aspergillus terreus	MRLF #12	0.09 ± 0.005	0.05 ± 0.006	0.08 ± 0.005
Aspergillus sp.	MRLF #13	0.10 ± 0.007	0.14 ± 0.01	0.13 ± 0.01
Chaetomium sp.	MRLF #14	1.54 ± 0.04	1.19 ± 0.05	1.34 ± 0.05
Chrysosporium sp.	MRLF #15	0.29 ± 0.01	0.37 ± 0.02	0.26 ± 0.01
Cladosporium sp.	MRLF #16	0.82 ± 0.05	0.91 ± 0.05	0.65 ± 0.04
Curvularia lunata	MRLF #18	0.11 ± 0.01	0.09 ± 0.005	0.15 ± 0.01
Curvularia sp.	MRLF #19	0.32 ± 0.06	0.39 ± 0.04	0.27 ± 0.01
Drechslera sp.	MRLF #20	0.15 ± 0.04	0.17 ± 0.01	0.21 ± 0.02
Fusarium oxysporum	MRLF #22	1.14 ± 0.07	1.62 ± 0.06	1.86 ± 0.06
Fusarium roseum	MRLF #23	1.67 ± 0.06	1.21 ± 0.04	1.46 ± 0.05
Gliocladium sp.	MRLF #24	0.61 ± 0.03	0.78 ± 0.04	0.59 ± 0.03
Humicola sp.	MRLF #26	1.28 ± 0.04	1.43 ± 0.05	1.27 ± 0.06
Mucor sp.	MRLF #27	0.23 ± 0.01	0.41 ± 0.02	0.32 ± 0.02
Myrothecium sp.	MRLF #28	1.21 ± 0.05	0.99 ± 0.03	0.91 ± 0.04
Paecilomyces sp.	MRLF #29	0.51 ± 0.02	0.84 ± 0.03	0.64 ± 0.03
Penicillium digitatum	MRLF #30	1.38 ± 0.06	1.37 ± 0.05	1.53 ± 0.06
Penicillium sp.	MRLF #31	0.91 ± 0.03	0.72 ± 0.03	1.08 ± 0.04
Rhizopus sp.	MRLF #32	0.13 ± 0.009	0.09 ± 0.006	0.14 ± 0.007
Sclerotium rolfsii	MRLF #33	0.67 ± 0.03	0.71 ± 0.05	0.68 ± 0.04
Trichodermaviride	MRLF #35	2.22 ± 0.07	2.03 ± 0.06	1.98 ± 0.06
Trichoderma sp.	MRLF #36	0.76 ± 0.04	0.78 ± 0.04	0.69 ± 0.03
Verticillium sp.	MRLF #37	0.20 ± 0.02	0.17 ± 0.01	0.27 ± 0.01
Bacteria 1	MRLB #38	1.65 ± 0.06	1.71 ± 0.06	1.59 ± 0.05
Bacteria 2	MRLB #39	1.07 ± 0.05	0.96 ± 0.04	0.94 ± 0.04
Bacteria 3	MRLB #40	0.73 ± 0.04	0.67 ± 0.03	0.52 ± 0.03
Bacteria 5	MRLB #42	0.08 ± 0.006	0.05 ± 0.003	0.11 ± 0.006
Bacteria 6	MRLB #43	0.32 ± 0.02	0.24 ± 0.01	0.29 ± 0.01
Bacteria 7	MRLB #44	0.16 ± 0.01	0.19 ± 0.009	0.14 ± 0.008
Bacteria 8	MRLB #45	1.66 ± 0.04	1.06 ± 0.04	1.12 ± 0.06
Bacteria 9	MRLB #46	0.76 ± 0.04	0.61 ± 0.02	0.68 ± 0.04
Bacteria 10	MRLB #47	1.07 ± 0.06	0.96 ± 0.03	0.89 ± 0.05
Bacteria 12	MRLB #49	0.21 ± 0.02	0.30 ± 0.02	0.27 ± 0.01
Bacteria 13	MRLB #50	0.64 ± 0.03	0.72 ± 0.04	0.35 ± 0.02
Bacteria 14	MRLB #51	0.08 ± 0.007	0.11 ± 0.009	0.13 ± 0.008
Bacteria 16	MRLB #53	1.34 ± 0.06	1.29 ± 0.05	1.36 ± 0.04
Bacteria 17	MRLB #54	0.14 ± 0.01	0.20 ± 0.01	0.18 ± 0.009
Bacteria 18	MRLB #55	0.28 ± 0.02	0.16 ± 0.01	0.19 ± 0.007
Bacteria 19	MRLB #56	0.09 ± 0.007	0.11 ± 0.01	0.13 ± 0.01
Bacteria 20	MRLB #57	0.12 ± 0.009	0.16 ± 0.01	0.14 ± 0.004

Values are means ± SEm of the three observations.

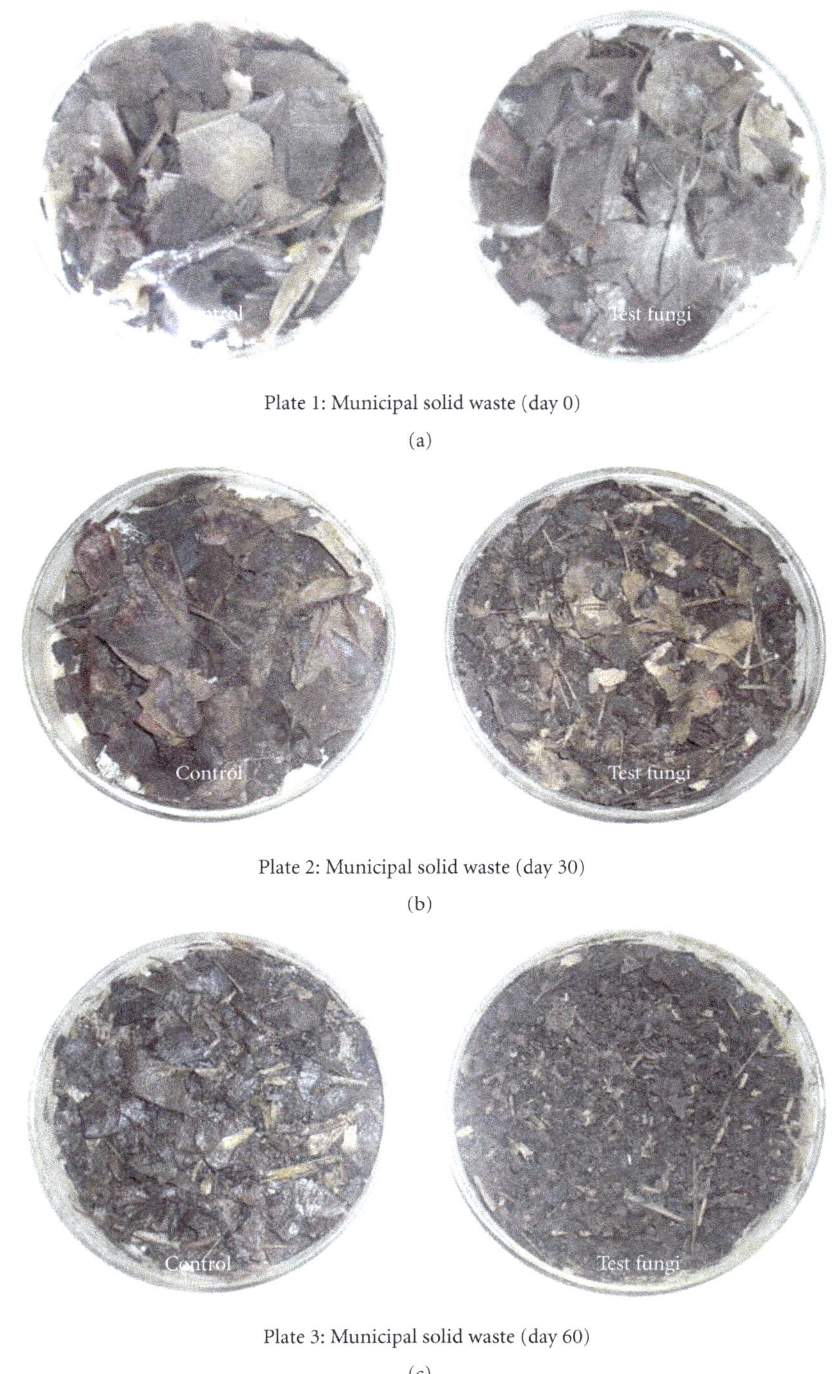

Plate 1: Municipal solid waste (day 0)

(a)

Plate 2: Municipal solid waste (day 30)

(b)

Plate 3: Municipal solid waste (day 60)

(c)

FIGURE 4: Bioconversion of municipal solid waste by using spore suspension of *T. viride* in petri plates.

was previously reported that the compost material can be considered mature when an ambient temperature of 28°C is reached [51]. Therefore, this parameter is considered as a good indicator for the end of the biooxidative phase in which the compost achieves some degree of maturity [52–54]. After 60 days, the mature compost was black in color, granular, and fibrous with a pleasant earthy smell. The appearance of black color indicated its maturity. In the case of the control (without culture), the biodegradation was very slow, and weight loss was also low. In order to assess the compost maturity, both the compost samples were placed separately in a sealed polythene bags for a week. After a week, the sealed bags were broken, and the odor was checked, which was found to have earthy smell in the compost produced by *T. viride*,

Figure 5: Bioconversion of municipal solid waste by *T. viride* using spore suspension in small piles.

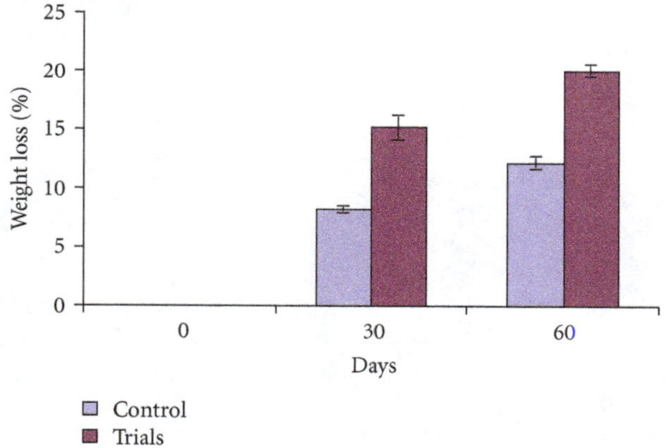

Figure 6: Percent weight loses of municipal solid waste by spore suspension of *T. viride* in plates (5 g).

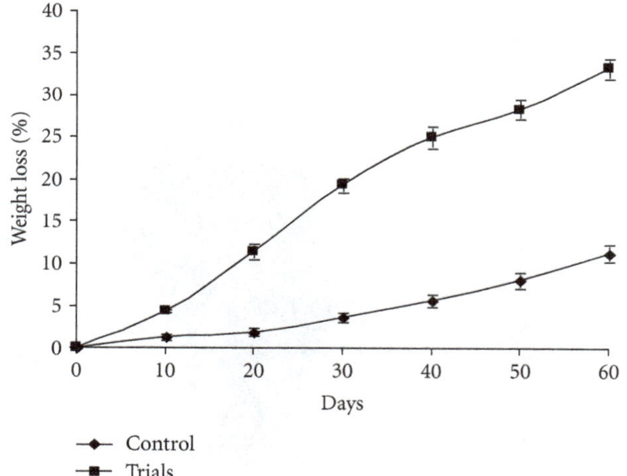

Figure 7: Percent weight loses of municipal solid waste by spore suspension of *T. viride* in open piles (25 Kg).

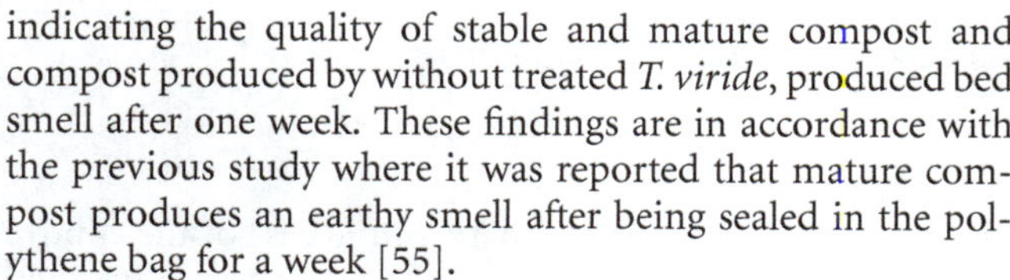

indicating the quality of stable and mature compost and compost produced by without treated *T. viride*, produced bed smell after one week. These findings are in accordance with the previous study where it was reported that mature compost produces an earthy smell after being sealed in the polythene bag for a week [55].

4. Conclusions

From the above findings, it can be concluded that a large number of microorganism were found in municipal solid waste, compost, and soil. Municipal solid waste is suitable for composting because of the presence of high percentages of organic matter. The *T. viride* had promising effects in the decomposition of organic municipal solid waste, resulting in a greater bioconversion of the original material than the control. Therefore, pH and temperature were considered as a good indicator for the end of the bioconversion of municipal solid waste in which the compost achieves some degree of maturity.

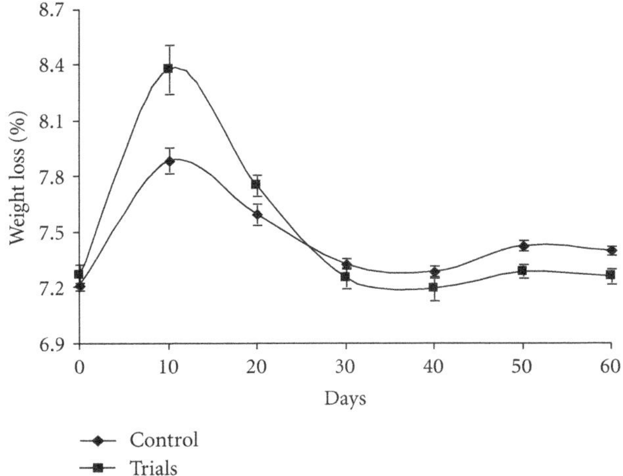

FIGURE 8: Variation of pH during the bioconversion of MSW.

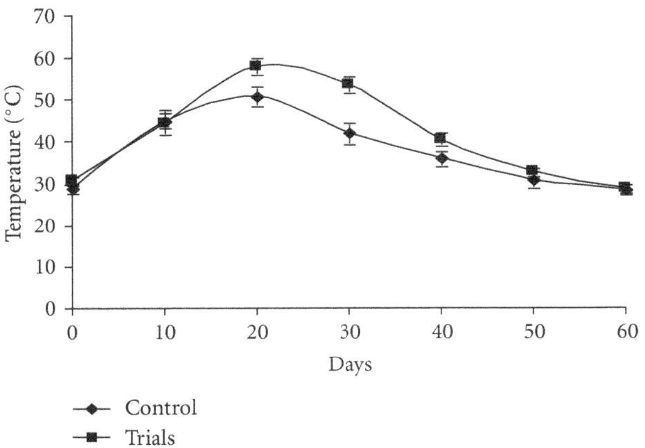

FIGURE 9: Variation of temperature during the bioconversion of MSW.

Acknowledgments

The authors are thankful to M. P. Pollution Control Board Bhopal and Head and Department of Biological Sciences, R.D. University, Jabalpur, for laboratory facilities. Ministry of Environment and Forest New Delhi is also thankfully acknowledged for financial support.

References

[1] D. S. Rani and K. Nand, "Production of thermostable cellulase-free xylanase by *Clostridium absonum* CFR-702," *Process Biochemistry*, vol. 36, no. 4, pp. 355–362, 2000.

[2] S. P. Gautam, P. S. Bundela, A. K. Pandey, M. K. Awasthi, and S. Sarsaiya, "Composting of municipal solid waste of Jabalpur City," *Global Journal of Environmental Research*, vol. 4, no. 1, pp. 43–46, 2010.

[3] K. S. Siddiqui, A. A. N. Saqio, M. H. Rashid, and M. I. Rajoka, "Carboxyl group modification significantly altered the kinetic properties of purified carboxymethyl cellulase from *Aspergillus niger*," *Enzyme Microbial Technology*, vol. 27, pp. 467–474, 2000.

[4] M. Mendals, "Microbial sources of cellulases," *Biotechnology Bioengineering*, no. 5, pp. 81–105, 1975.

[5] J. A. Brown, S. A. Collin, and T. M. Wood, "Development of a medium for high cellulase, xylanase and β-glucosidase production by a mutant strain (NTG III/6) of the cellulolytic fungus *Penicillium pinophilum*," *Enzyme and Microbial Technology*, vol. 9, no. 6, pp. 355–360, 1987.

[6] S. P. Gautam, P. S. Bundela, A. K. Pandey, M. K. Awasthi, and S. Sarsaiya, "Effect of different carbon sources on production of Cellulases by *Aspergillus niger*," *Journal of Applied Science Environmental Sanitation*, vol. 5, no. 3, pp. 277–281, 2010.

[7] S. P. Gautam, P. S. Bundela, A. K. Pandey, M. K. Awasthi, and S. Sarsaiya, "Prevalence of fungi in Municipal solid waste of Jabalpur city (M.P.)," *Journal of Basic & Applied Mycology*, vol. 8, no. 1-2, pp. 80–81, 2009.

[8] F. C. Deschamps, C. Giuliano, M. Asther, M. C. Huet, and S. Roussos, "Cellulase production by *Trichoderma harzianum* in static and mixed solid state fermentation reactors under non-aseptic conditions," *Biotechnology and Bioengineering*, vol. 27, no. 9, pp. 1385–1388, 1985.

[9] B. J. Macris, M. Paspaliari, and D. Kekos, "Production and cross-synergistic action of cellulolytic enzymes from certain fungal mutants grown on cotton and straw," *Biotechnology Letters*, vol. 7, no. 5, pp. 369–372, 1985.

[10] J. N. Saddler, M. K. H. Chan, M. Mes, Hartree, and C. Brevil, "Cellulase production and hydrolysis of pretreated lignocelluloses substrate in biomass conversion technology," in *Principles and Practice*, M. Moo Young, Ed., Pciqumon Press, 1987.

[11] D. B. Wilson, "Microbial diversity of cellulose hydrolysis," *Current Opinion in Microbiology*, vol. 14, pp. 259–263, 2011.

[12] T. M. Wood and K. M. Bhat, "Methods for measuring cellulase activities," in *Methods in Enzymology*, W. A. Wood and S. T. Kellogg, Eds., vol. 160, p. 87, Academic Press, New York, NY, USA, 1988.

[13] T. M. Wood and S. I. McCrae, "Purification and some properties of a $(1 \rightarrow 4)$-β-d-glucan glucohydrolase associated with the cellulase from the fungus *Penicillium funiculosum*," *Carbohydrate Research*, vol. 110, no. 2, pp. 291–303, 1982.

[14] M. Mandels and D. Sternberg, "Recent advances in cellulases technology," *Journal of Fermentation Technology*, vol. 54, no. 4, pp. 267–286, 1976.

[15] S. P. Gautam, P. S. Bundela, A. K. Pandey, Jamaluddin, M. K. Awasthi, and S. Sarsaiya, "Cellulase production by *Pseudomonas* sp. isolated from municipal solid waste compost," *International Journal of Academic Research*, vol. 2, no. 6, pp. 330–333, 2010.

[16] S. A. Waksman, *Principles of Soil Microbiology*, Williams & Wilkins, Baltimore, Md, USA, 1927.

[17] K. B. Raper and C. Thom, *A Manual of the Penicillia*, Williams & Wilkins, Baltimore, Md, USA, 1949.

[18] J. L. Barnett and B. B. Hunter, *Illustrated Genera of Imperfect Fungi*, Burgess Publishing Company, Minneapolis, Minn, USA, 1972.

[19] M. B. Ellis, *Demataceous Hyphomycetes*, Commonwealth Mycological Institute, Kew, UK, 1971.

[20] M. B. Ellis, *Demataceous Hyphomycetes*, Commonwealth Mycological Institute, Kew, UK, 1976.

[21] M. Goodfellow and S. T. Williams, *Bergey's Manual of Systematic Bacteriology*, vol. 4, 1989.

[22] L. Hankin and L. Anagnostaksis, "The use of solid media for detection of enzyme production by fungi," *Mycologia*, vol. 47, pp. 597–607, 1975.

[23] A. A. Sherief, A. B. El-Tanash, and N. Atia, "Cellulase production by *Aspergillus fumigatus* grown on mixed substrate of

rice straw and wheat bran," *Research Journal of Microbiology*, vol. 5, no. 3, pp. 199–211, 2010.

[24] G. L. Miller, "Use of dinitrosalicylic acid reagent for determination of reducing sugar," *Analytical Chemistry*, vol. 31, no. 3, pp. 426–428, 1959.

[25] S. B. Pointing, "Qualitative methods for the determination of lignocellulolytic enzyme production by tropical fungi," *Fungal Diversity*, vol. 2, pp. 17–33, 1999.

[26] R. Elango and J. Divakaran, "Microbial consortium for effective composting of coffee pulp waste by enzymatic activities," *Global Journal of Environmental Research*, vol. 3, no. 2, pp. 92–95, 2009.

[27] A. Rahman, M. F. Begum, M. Rahman, and M. A. Bari, "Isolation and identification of *Trichoderma* species from different habitats and their use for bioconversion of solid waste," *Turk Journal of Biology*, vol. 34, pp. 1–12, 2010.

[28] B. Abrha and B. A. Gashe, "Cellulase production and activity in a species of *Cladosporium*," *World Journal of Microbiology & Biotechnology*, vol. 8, no. 2, pp. 164–166, 1992.

[29] S. I. I. Abdel-Hafez, A. H. M. El-Said, and Y. A. M. H. Gherabawy, "Mycoflora of leaf surface, stem, bagasse and juice of adult sugarcane (*Saccharum officinarum* L.) plant and cellulolytic ability in Egypt," *A Bulletin of the Faculty of Science*, vol. 24, pp. 113–130, 1995.

[30] A. I. I. Abdel-Hafez, S. I. I. Abdel-Hafez, S. M. Mohawed, and A. H. M. El-Said, "Composition, occurrence and cellulolytic activities of fungi in habiting soils along Idfu-Marsa Alam road at Eastern desert, Egypt," *A Bulletin of the Faculty of Science*, vol. 20, pp. 21–48, 1991.

[31] A. M. Moharram, S. I. I. Abdel-Hafez, and M. A. Abdel-Star, "Cellulolytic activity of fungi isolated from different substrates from the New Valley Governorate, Egypt," *Pure Science Engineering*, vol. 1, pp. 101–114, 1995.

[32] A. H. M. El-Said, "Phyllosphere and phylloplane fungi of banana cultivated in Upper Egypt and their cellulolytic ability," *Mycobiology*, vol. 29, pp. 210–217, 2001.

[33] A. Berlin, N. Gilkes, D. Kilburn et al., "Evaluation of novel fungal cellulase preparations for ability to hydrolyze softwood substrates-Evidence for the role of accessory enzymes," *Enzyme and Microbial Technology*, vol. 37, no. 2, pp. 175–184, 2005.

[34] P. F. Strom, "Identification of thermophilic bacteria in solid-waste composting," *Applied and Environmental Microbiology*, vol. 50, no. 4, pp. 906–913, 1985.

[35] H. Proom and B. C. J. G. Knight, "The minimal nutritional requirements of some species in the genus *Bacillus*," *Journal of General Microbiology*, vol. 13, no. 3, pp. 474–480, 1955.

[36] C. N. Ezekiel, A. C. Odebode, R. O. Omenka, and F. A. Adesioye, "Growth response and comparative cellulase induction in soil fungi grown on different cellulose media," *Acta SATECH*, vol. 3, no. 2, pp. 52–59, 2010.

[37] S. M. Duncan, R. L. Farrell, J. M. Thwaites et al., "Endoglucanase-producing fungi isolated from Cape Evans historic expedition hut on Ross Island, Antartica," *Environonmental Microbialogy*, vol. 8, no. 7, pp. 1212–1219, 2006.

[38] S. P. Gautam, P. S. Bundela, A. K. Pandey, M. K. Awasthi, and S. Sarsaiya, "Screening of cellulolytic fungi for management of municipal solid waste," *Journal of Applied Science for Environmental Sanitation*, vol. 5, no. 4, pp. 367–371, 2010.

[39] A. I. I. Abdel-Hafez, S. I. I. Abdel-Hafez, S. M. Mohawed, and T. A. M. Ahmed, "Seasonal fluctuations of soil and air borne fungi at Qena, Upper Egypt," *A Bulletin of the Faculty of Science*, vol. 19, pp. 47–63, 1990.

[40] S. P. Gautam, P. S. Bundela, A. K. Pandey, Jamaluddin, M. K. Awasthi, and S. Sarsaiya, "*Fusarium* sp. from municipal solid waste compost: ability to produce extracellular cellulases enzyme," *International Journal of Biotechnology and Bioengineering Research*, vol. 2, no. 2, pp. 239–245, 2010.

[41] A. H. Moubasher and M. B. Mazen, "Assay of cellulolytic activity of cellulose-decomposing fungi isolated from Egyptian soils," *Journal of Basic Microbiology*, vol. 31, pp. 59–66, 1991.

[42] S. I. I. Abdel-Hafez, "Studies on soil mycoflora of desert soils in Saudi Arabia," *Mycopathologia*, vol. 80, pp. 3–8, 1994.

[43] R. Karl and M. Y. Iain, "Interactions between soil structure and fungi," *Mycologist*, vol. 18, no. 2, pp. 52–59, 2004.

[44] J. S. Lalley and H. A. Viles, "Terricolous lichens in the northern Namib desert of Namibia: distribution and community composition," *Lichenologist*, vol. 37, no. 1, pp. 77–91, 2005.

[45] A. P. S. Sandhu, G. S. Randhawa, and K. S. Dhugga, "Plant cell wall matrix polysaccharide biosynthesis," *Molecular Plant*, vol. 2, no. 5, pp. 840–850, 2009.

[46] D. K. Sandhu and M. K. Kalra, "Effect of cultural conditions on production of cellulases in *Trichoderma longibrachiatum*," *Transactions of the British Mycological Society*, vol. 84, pp. 251–258, 1985.

[47] G. F. Huang, J. W. C. Wong, Q. T. Wu, and B. B. Nagar, "Effect of C/N on composting of pig manure with sawdust," *Waste Management Research*, vol. 24, no. 8, pp. 805–813, 2004.

[48] C. Sundberg, S. Smars, and H. Jonsson, "Low pH as an inhibiting factor in the transition from mesophilic to thermophilic phase in composting," *Bioresource Technology*, vol. 95, no. 2, pp. 145–150, 2004.

[49] B. Felicita, H. Nina, V. Marija, and G. Zoran, "Aerobic composting of tobacco industry solid waste-simulation of the process," *Clean Technology Environmental Policy*, vol. 5, pp. 295–301, 2003.

[50] M. Bertoldi, G. Vallini, and A. Pera, "The biology of composting: a review," *Waste Management and Research*, vol. 1, no. 2, pp. 157–176, 1983.

[51] S. M. Tiquia, N. F. Y. Tam, and I. J. Hodgkiss, "Effects of bacterial inoculum and moisture adjustment on composting of pig manure," *Environmental Pollution*, vol. 96, no. 2, pp. 161–171, 1997.

[52] S. P. Gautam, P. S. Bundela, A. K. Pandey, M. K. Awasthi, and S. Sarsaiya, "Isolation, identification and cultural optimization of indigenous fungal isolates as a potential bioconversion agent of municipal solid waste," *Annals of Environmental Science*, vol. 5, 2010.

[53] S. P. Gautam, P. S. Bundela, A. K. Pandey, Jamaluddin, M. K. Awasthi, and S. Sarsaiya, "Optimization of the medium for the production of cellulase by the Trichoderma viride using submerged fermentation," *International Journal of Environmental Science*, vol. 1, no. 4, pp. 656–665, 2010.

[54] E. I. Jimenez and V. P. Garcia, "Evaluation of city refuse compost maturity: a review," *Biological Wastes*, vol. 27, no. 2, pp. 115–142, 1989.

[55] S. Jilani, "Municipal solid waste composting and its assessment for reuse in plant production," *Pakistan Journal of Botany*, vol. 39, no. 1, pp. 271–277, 2007.

15

Advances in Bacteriophage-Mediated Control of Plant Pathogens

Rebekah A. Frampton,[1] Andrew R. Pitman,[2] and Peter C. Fineran[1]

[1] *Department of Microbiology & Immunology, University of Otago, P.O. Box 56, Dunedin 9054, New Zealand*
[2] *New Zealand Institute for Plant & Food Research, Private Bag 4704, Christchurch 8140, New Zealand*

Correspondence should be addressed to Peter C. Fineran, peter.fineran@otago.ac.nz

Academic Editor: Beatriz Martinez

There is continuing pressure to maximise food production given a growing global human population. Bacterial pathogens that infect important agricultural plants (phytopathogens) can reduce plant growth and the subsequent crop yield. Currently, phytopathogens are controlled through management programmes, which can include the application of antibiotics and copper sprays. However, the emergence of resistant bacteria and the desire to reduce usage of toxic products that accumulate in the environment mean there is a need to develop alternative control agents. An attractive option is the use of specific bacteriophages (phages), viruses that specifically kill bacteria, providing a more targeted approach. Typically, phages that target the phytopathogen are isolated and characterised to determine that they have features required for biocontrol. In addition, suitable formulation and delivery to affected plants are necessary to ensure the phages survive in the environment and do not have a deleterious effect on the plant or target beneficial bacteria. Phages have been isolated for different phytopathogens and have been used successfully in a number of trials and commercially. In this paper, we address recent progress in phage-mediated control of plant pathogens and overcoming the challenges, including those posed by CRISPR/Cas and abortive infection resistance systems.

1. Introduction

In October 2011 the United Nations announced that the global human population had reached 7 billion. The world is facing not only this increase in population, but also a decrease in land availability for agriculture and a changing climate [1]. It is apparent that there is a requirement to increase food production to feed the growing population and a need to achieve this with diminished land and water resources [1]. A major threat to food production is plant diseases, which are influenced by changing agricultural practices and more global trade [2]. Recent topical examples include citrus greening of oranges caused by psyllids that transmit bacteria belonging to the genus *Candidatus* Liberibacter [3] and canker of kiwifruit caused by the bacterium *Pseudomonas syringae* pv. *actinidiae* [4]. Citrus greening has doubled the cost of orange production for growers in Florida, where the disease was first identified in 2005 [3]. In New Zealand, where *Pseudomonas syringae* pv. *actinidiae* was discovered in late 2010, 40% of orchards have been infected resulting in a significant economic cost to the industry [5].

A variety of approaches are required to minimise the impact of bacterial plant diseases on the quantity, quality, and economy of food production. Conventional control measures involve the implementation of operating practices to prevent further infections, removal of infected plant tissue, and appropriate disposal to stop the transmission of the pathogen from one site to another. Other methods to control phytopathogens include chemicals such as pesticides, to control insect vectors, antibiotics (e.g., tetracycline and streptomycin), and copper. Copper has been used for over 100 years and antibiotics such as streptomycin have been used since the 1950s [6, 7]. Streptomycin has been used for many years for the control of pathogens, including *Pseudomonas syringae* pathovars, and resistance has been regularly reported following use [6–8]. Another concern with antibiotics is the spread of resistance genes to other bacteria, including other pathogens or nonpathogenic bacteria present in the environment [7]. Copper resistance has also been documented for plant pathogens, including *Xanthomonas* and *Pseudomonas* species [9–11]. Continual use of copper sprays can lead to toxic levels in the environment [12, 13].

Therefore, it is favourable to replace or integrate chemical control methods with less toxic biological methods.

There is mounting interest in using bacteriophages (phages) as biocontrol agents (BCAs) to target phytopathogens. Phages are viruses that specifically infect bacteria, yet have no direct negative effects on animals or plants. Infection of a bacterium by a virulent phage typically results in rapid viral replication, followed by the lysis of the bacterium and the release of numerous progeny phages. These phages can then proceed to infect neighbouring bacteria. Therefore, the numbers of phage will expand when target pathogens are encountered and the therapy will essentially be amplified in response to the bacterial infection. This is a distinct advantage over other treatments, such as antibiotics. Since the 1920s, within a decade of the first discovery of phages, their potential as therapeutic agents for use in agriculture was under investigation and provided some promising results (see [14] for review of the early literature). Recent years have seen a resurgence of interest in phage therapy for the control of phytopathogens (see [15–17] for recent reviews). In part, this renewed interest is due to the nontoxic nature of phages and their ability to infect antibiotic or heavy metal resistant bacteria. Successful phage therapy is being applied commercially to processed and packaged foods by Intralytix and Micreos Food Safety (formerly EBI Food Safety) and to agricultural crops by Omnilytics. There is also interest in the use of phage in the detection of phytopathogens. Indeed, many of the first phages against plant pathogens were isolated for diagnosis and strain typing [18] and recently genetic advances are yielding effective phage-based reporter systems [19, 20]. In this paper we will examine the use of phage as BCAs for bacterial plant diseases. First, we will address the initial isolation and laboratory characterisation of phages. Next, we will discuss the transition from *in vitro* analyses to bioassays and field/greenhouse trials to commercialisation and application. Finally, we will provide an analysis of phage resistance in phytopathogens and address how this can be avoided or minimised when developing phage BCAs.

2. Initial Phage Characterisation

2.1. Isolation and Host Range of Phages. The initial stage in developing a phage-based BCA involves the isolation of phages. Phages can be isolated relatively simply from soil, water, and plant material collected from multiple locations using the soft agar overlay technique and a range of host pathogens. Isolation methods are well established and covered in several reviews [21, 22], and many successful studies have resulted from this approach (e.g., [23–25]). To isolate diverse phages it is important to include a range of host bacterial strains that represent the diversity of the pathogens involved in the disease [17, 26]. Phages that produce clear plaques should be chosen preferentially to reduce the isolation of temperate phages as certain temperate phage can cause lysogenic conversion, a process whereby virulence genes carried by the prophage contribute to pathogenicity of the bacterial lysogen [27]. This approach increases the isolation of phages from the order Caudovirales (Myoviridae, Siphoviridae, and Podoviridae). Filamentous phage from the Inoviridaefamily produce a chronic infection that results in the continuous release of phage from growing bacterial cells. Filamentous phage could also be a BCA option and this is discussed in more detail in Section 3.5 [28].

Determining the host range of each phage enables the design of a cocktail capable of infecting all known pathogenic strains involved in the disease. Although the host range of most phages is usually narrow, bacteria isolated from the plant environment should be tested for lysis by the putative BCA phages to ensure a minimal impact of the phages on the wider microbial community and potential commensal strains.

2.2. Basic Characterisation of Phages. Once isolated, the phages need to be characterised to ensure they are appropriate BCAs. This information will allow for rational design of a phage cocktail and enable the tracking of phages during bioassays and field trials. Transmission electron microscopy and molecular methods, such as restriction pattern analysis of phage DNA, enable assessment of phage diversity. Identification of the phage receptors can assist in the rational selection of phages that target through different mechanisms to reduce the frequency of resistance (discussed in Section 5).

The development of next-generation sequencing has dramatically reduced the cost of determining the complete DNA sequence of a phage genome. The sequence also enables the design of quantitative PCR strategies for phage tracking in field trials [26] and if present, genes required for integration/lysogeny or that encode known toxins, antibiotic resistance, or virulence factors can be identified. Phage-mediated horizontal transfer of bacterial genes by specialised and generalised transduction should be avoided [27]. The possibility of specialised transduction is eliminated by removing temperate phages but assays are necessary to examine generalised transduction. These assays cannot rule out transduction but enable the identification and elimination of phages with a high transducing frequency [27]. It has been proposed that if a phage cocktail is used then any bacteria that receive additional DNA through transduction can still be killed by another phage in the mixture [17]. An understanding of the basic growth parameters of the phage will aid in the design of a phage BCA. Therefore, investigating the length of infection and burst size with one-step growth curves at temperatures and conditions likely to be encountered in the field iss important. It is also important to identify conditions under which the phage can be successfully stored [15]. Following phage characterisation the next step is to perform bioassays and/or field trials which are discussed in Section 3.

2.3. Alternative Biocontrol Phage Technologies. Through phage genomics, genes encoding other potential biocontrol options have been identified. One example is phage endolysins, which have been investigated for use against antibiotic resistant bacteria that colonize human mucosal membranes [29]. Endolysins are phage-encoded peptidoglycan hydrolases that work in concert with holins to ensure cell lysis occurs following phage maturation [30]. Generally, the C-terminus binds the bacterial cell wall and positions the enzymatically active N-terminus close to its target. Due to their high level of specificity, phages CMP1 and CN77 were originally used for detection and identification of plant pathogenic *Clavibacter michiganensis* strains [31, 32]. *Clavibacter michiganensis* subsp. *michiganensis* is a recalcitrant tomato pathogen because there are no resistant cultivars and chemical control agents are ineffective. The endolysins of CMP1 and CN77 responsible for degrading the bacterial cell wall have been pursued as potential antimicrobials [33]. Sequence analysis revealed that the similarity of the catalytic domains of the CMP1 and CN77 endolysins was low, suggesting they target different covalent bonds within the peptidoglycan. However, both endolysins are specific for *Clavibacter michiganensis* subsp. *michiganensis*, as observed for the phages [34]. The use of endolysins is still highly specific but avoids concerns of using a replicating BCA. Compared with the application of whole phage preparations, the development of lysine-based therapies is more technically challenging. For example, the generation of lysins requires greater molecular insight than phage therapy because sequencing, identification, cloning, characterization, and purification of the lysins must be performed.

3. Trials of Phage Biocontrol

Once a selection of phytopathogen-specific phages is obtained and the initial characterisation has indicated their potential for biocontrol, the next step is to test their efficacy in relation to plant disease. First it is necessary to scale-up phage preparations, which requires knowledge about the characteristics and lifestyle of the phages [35]. To test phytopathogen control, a range of approaches have been taken, from laboratory-based bioassays through greenhouse and field trials. Table 1 summarises some results of phage trials that have been performed on a range of phytopathogens including *A. tumefaciens*, *Dickeya solani*, *Pectobacterium carotovorum*, *Erwinia amylovora*, *Pseudomonas syringae*, *Ralstonia solanacearum*, *Streptomyces scabies*, and *Xanthomonas* species. Many of these studies have given promising results. However, a number of factors can contribute to phage biocontrol trials that have failed. Firstly, to understand the reason for success or failure it is important to measure the dynamics of phage and host, for example, via quantitative PCR or traditional enumeration assays [26]. Secondly, field trials are biologically complex and the presence of other microbes, including other pathogens, can influence the effectiveness of the phages. In some cases pathogens of a different genus or species will cause very similar disease symptoms but are not killed by the phages [23]. Therefore, it is important to check which organism(s) caused the disease symptoms and if it is the targeted bacterial host, whether phage-resistant mutants account for the lack of phage killing or not. It is important to confirm that phage preparations are free of any pathogenic bacteria used in the production process, which highlights the requirement for phage-only controls in trials. The absence of a "gold standard" positive treatment, where available, in some studies makes evaluating the efficacy less informative. Other factors including the type of water and the presence of certain components in some fertilizers can influence phage viability and the trial's success [36]. Despite these challenges, many of the studies in Table 1 demonstrated a positive effect of phage treatment.

Each phage-phytopathogen-plant system has unique features and requires characterisation and optimisation of the phage biocontrol. External and environmental factors can play a role and cause variable results. For example, greenhouse-grown crops are in a more stable environment, whereas plants grown outside are exposed to more variable weather conditions and these factors vary between geographic locations. Phage survival and their persistence at the required site of action are affected by conditions such as pH, temperature, desiccation, rain, and UV. The most damaging factor appears to be UV irradiation in sunlight [52]. Interestingly, some plant extracts negatively affect the growth or viability of phage *in vitro* [53, 54] but it is unclear whether this is relevant during phage therapy on plants. Various approaches to address these challenges in phage biocontrol are available and will be discussed in the following sections.

3.1. Protective Formulations and Application Timing. In order to minimise UV damage, some researchers have investigated protective formulations. Few published studies have addressed stabilising agents, but one group found that certain combinations of sucrose, Casecrete NH-400, pregelatinised corn flour or skim milk, increased the persistence of phages active against *Xanthomonas campestris* pv. *vesicatoria* in greenhouse and field trials and improved treatment efficacy [51]. However, in a trial with phages against *Xanthomonas axonopodis* pv. *citri* and *citrumelo*, skim milk inhibited the action of the phages even though the phages persisted on the leaf surface for longer [25]. Balogh et al. also investigated the effect of the time of day of phage application and demonstrated that evening applications increased phage survival [51]. The lower UV intensity at dusk was attributed to this improved viral longevity. The frequency of phage application also appears to be specific to the particular phage-phytopathogen-plant system and the best results vary from daily to weekly application [25, 44, 48, 50, 51, 55]. Therefore, these conditions must be optimised for each newly developed phage BCA.

3.2. Coapplication with Other Control Strategies. Researchers have begun examining the effects of combining phage with existing or new control measures. Hypersensitive response and systemic acquired resistance plant activators have been tested against *Xanthomonas* spp. in two disease models in the presence or absence of phages. The combined approach provided disease control equal to, or greater than, either treatment alone [44, 48, 49] (Table 1) and comparable results were obtained with phage and copper hydroxide with mancozeb [48]. Recently, it was demonstrated that lipid-containing phages were the most susceptible to copper, whereas most dsDNA phages were unaffected [56]. This indicates that dsDNA phages are the best candidates for plant disease therapies in combination with copper. Phages can also

Table 1: Phage therapy trials of plant pathogens.

Pathogen	Host	Disease		References
Agrobacterium tumefaciens	Tomato	Crown gall	Bioassay with infected tissue—bacteriophage had no effect	[37]
Dickeya solani	Potato	Soft rot	Small effect seen when seed tubers were treated with phage prior to planting	[23]
Erwinia amylovora	Pome fruits	Fire blight	Phages isolated and characterised *in vitro*. Some testing *in planta*, promising results in combination with non-pathogenic carrier *Pantoea agglomerans*	[24, 38, 39]
Pectobacterium carotovorum subsp. *caratovorum*	Calla lily	Bacterial soft rot	Bacterial load reduction by phages but inhibition of killing by fertiliser solutions	[36]
Ralstonia solanacearum	Tobacco	Bacterial wilt	In a greenhouse trial pretreatment of plant roots with an avirulent strain and application of phage to the plants protected plants against bacterial wilt. No comparison was made to conventional chemical control methods	[40]
Ralstonia solanacearum	Tomato	Bacterial wilt	In a greenhouse trial pretreatment of tomato seedlings with φRSL1 prevented bacterial wilt in all plants, untreated plants all wilted. φRSL1 inhibited growth of bacteria but did not completely kill bacteria.	[41]
Streptomyces scabies	Potato	Potato scab	Phage treatment of seed tubers prior to planting reduced scab lesion coverage	[42]
Xanthomonas arboricola pv. *pruni*	Stonefruits	Bacterial spot	Application of phage to peach leaves prior to infection resulted in a 42% disease reduction compared to a nontreated control. Application of phage after infection had no effect	[43]
Xanthomonas axonopodis pv. *allii*	Onion	Xanthomonas leaf blight	Field and greenhouse trials of phage and plant activator provided equivalent protection to copper	[44]
Xanthomonas axonopodis pv. *vignaeradiatae*	Mungbean	Bacterial leaf spot	Synergistic effect of phage and streptomycin on mungbean seeds reduced seedling infection	[45]
Xanthomonas campestris pv. *juglandis*	Walnut	Walnut blight	Phage did not survive on walnut leaves in a greenhouse trial; pathogen was not included on leaves	[46]
Xanthomonas campestris pv. *pruni*	Peach	Leaf and fruit spot	A significant reduction in disease was seen in one out of three orchards with weekly application of a single phage	[47]
Xanthomonas campestris pv. *vesicatoria*	Tomato and Pepper	Bacterial spot	These studies have led to the successful development of a phage BCA (http://www.omnilytics.com/).	[48–51]
Xanthomonas citri subsp. *citri*	Citrus	Citrus canker	Mixed results in greenhouse and nursery trials when compared to copper bacteriocides	[25]
Xanthomonas fuscans subsp. *citrumelonis*	Citrus	Citrus bacterial spot	Mixed results in greenhouse and nursery trials when compared to copper bacteriocides	[25]

be used in combination with streptomycin as they are not a direct target of antibiotics [45]. Cotherapy with copper or streptomycin could enable a "belt and braces" approach to minimise streptomycin/copper or phage resistance. However, this approach is not possible in some parts of the world. For example, regulations that disallow the use of streptomycin for the control of plant diseases have been introduced in the EU [57]. Phage compatibility with other agrichemicals has not been thoroughly examined. Phages used would require testing for viability and persistence in the presence of any agrichemical treatment used simultaneously on the affected plants.

A range of bacterial and fungal BCAs that include pathogen antagonists have been developed and are commercially available. These include BlightBan A506 (*Pseudomonas fluorescens* A506; Nufarm Americas Inc), Blossom Bless (*Pantoea agglomerans* P10c; Gro-Chem NZ Ltd), and Superzyme (*Bacillus subtilis*, *Trichoderma*, and *Pseudomonas putida*; J H Biotech Inc). When used on their own, the protection provided by some of these products often varies and does not compare favourably with streptomycin, which is considered the benchmark [58]. These BCAs can complement conventional control strategies. For example, use of BCAs in the control of fire blight [59, 60] in apple trees reduced the number of streptomycin sprays required to provide the same amount of protection against *E. amylovora* infection in apple blossoms [58]. Growth of these BCAs is thought to be required for their action as competitors. Variable environmental conditions influence growth and the regulation of the production of secondary metabolites, such as antibiotics, which might account for some of the inconsistent responses to BCAs. One approach to improve plant protection is to incorporate phages into existing BCA products as part of the pest management plan.

3.3. Use of Carrier Bacteria. The abundance of host bacteria, in addition to the environmental factors mentioned above, causes fluctuations in phage numbers [61]. To improve phage persistence and efficacy the idea of coapplication of phages with nonpathogenic host bacteria to support phage replication has been raised [26, 62]. It is necessary to identify a suitable nonpathogenic carrier bacterium that does not affect the plant and to isolate phages that infect both the target phytopathogen and the carrier strain. Therefore, isolation of broad host range phages can also be beneficial for the purposes of biological control. Care is needed to use phages that do not infect plant beneficial bacteria in the phyllosphere or rhizosphere.

The use of competitive antagonists has been investigated in the control of fire blight for many years and can prevent or reduce infection [24, 26, 63]. An extensive panel of *Erwinia amylovora* phages have been isolated and characterised [24, 38, 39, 62, 64] and their use with carrier bacteria investigated [17, 26, 62]. In an early study a temperate phage that infected both a saprophyte and *Erwinia amylovora* provided increased protection in a pear slice bioassay [62]. However, current opinion is that temperate phages should be avoided for phage therapy. Recently, a *Pantoea agglomerans* carrier system was developed which amplified the phages and reduced disease in blossom bioassays [17, 26]. The application of lytic phages against *Erwinia amylovora* in conjunction with the *Pantoea agglomerans* carrier strain to potted apple trees reduced the incidence of fire blight to the same level as streptomycin [24]. Application of the carrier strain itself also helped prevent infection. In orchard trials, the coapplication of phages with carrier bacteria resulted in a significant reduction in disease incidence compared with the no phage control or the *P. agglomerans* carrier alone, and the protective effect was similar to that observed with streptomycin [17, 26].

Due to difficulties in isolating broad host range phages that infect suitable non-pathogenic strains, other strategies should be considered. For example, non-pathogenic mutant derivatives of the phytopathogen can be used as hosts for phage replication in the environment [40]. There are some risks if the pathogen is not fully attenuated or if reversion is possible. An alternative use for non-pathogenic or attenuated pathogenic hosts is for production of phages. Non-pathogenic hosts might enable safer and more economical phage amplification due to reduced purification requirements for removal of pathogenic bacteria.

3.4. Different Plant Disease Systems. The disease symptoms and location of infection within or on the plant can pose challenges for phage biocontrol. For example, *Pseudomonas syringae* pv. *actinidiae* and *Erwinia amylovora* spend some of the disease cycle inside the host plant following infection through plant openings or wounds [4, 65]. High numbers of bacteria can accumulate within cankers in the plant and are protected from any control agent applied to the outside of the plant that cannot penetrate to deeper tissues. One study has shown a preventative and beneficial effect on disease progression with phage treatment of citrus canker [25]. However, the ability of phage to act as a curative agent of cankers was not directly assessed. High-pressure injection of phage is a possible strategy that is under consideration. In cases where treatment is problematic, phage prophylaxis would be the preferred application at a time when the chance of infection is greatest [17].

Phages survive in soil for at least one month [66] but the type of soil, pH, moisture content, and nutrients all influence persistence and can affect phage bioavailability [22, 67]. There is a desire to use phages in soil against bacterial pathogens that infect plant tubers, such as *Pectobacterium* spp. and *Dickeya* spp. In a recent study, treatment of potato tubers with phages prevented soft rot disease caused by *Dickeya solani* in a controlled environment [23]. Only a small protective effect was seen when the tubers were planted and only when the tubers were allowed to dry completely after treatment. A mixture of phages against *Dickeya solani* and *Pectobacterium atrosepticum* would be useful as Adriaenssens et al. observed soft rot caused by both species during their trials [23].

3.5. Use of Filamentous Phages. The filamentous phages φRSS1 and φRSM3 have almost opposite effects on the virulence of *Ralstonia solanacearum*. Infection of *Ralstonia solanacearum* with *φ*RSS1 increased twitching motility, EPS production, the expression of the virulence regulator *phcA*, and the rate of tomato plant wilting compared with a noninfected control [68]. In contrast, infection with *φ*RSM3 decreased twitching, EPS and *phcA* expression, growth, and movement in tomato plant stems and caused less wilting [69]. This difference was due to a repressor in the genome of *φ*RSM3 that is absent in *φ*RSS1 [69]. Tomato plants exposed to *φ*RSM3-infected bacterial cells increased the expression of genes to help the plant resist infection [28]. Indeed, pretreatment of tomato plants with *φ*RSM3-infected bacteria prevented infection by noninfected cells and prevented bacterial wilt. The authors suggest that a mixture of *φ*RSM3 and a lytic phage, such as *φ*RSL1 [41], would be a good BCA [28].

4. Application

Safety, efficacy, intellectual property, and a market are important factors for the commercial success of phage control of phytopathogens. As phages are ubiquitous and are consumed with food and water without any negative effects, their safety as BCAs is not an issue. Phages for control of *Listeria monocytogenes* have FDA approval for use on food [70] and have been granted GRAS (generally recognised as safe) status. Interestingly, in the EU [57] and the USA LISTEX (Micreos Food Safety) [71] is considered organic, suggesting that organic growers could be a valuable market for phage biocontrol products. Furthermore, phages are used to treat bacterial infections in humans in Russia and Georgia [72, 73]. Clinical trials have tested the effectiveness of phages to treat ear infections [74] and have assessed the safety of phages for treating burn victims [75]. These strictly controlled medical trials demonstrate that phage therapy is safe and effective, indicating that phage control of phytopathogens will not cause adverse health problems. Phages are being developed as BCAs for control of human pathogens in animals and on food products, which has been reviewed elsewhere [76, 77] and in other articles in this special issue. Patent protection and intellectual property are important factors in commercialisation of phage BCAs. Despite the concept of phage therapy having existed for over 90 years, multiple companies have acquired patents and established commercial platforms. This has been thoroughly reviewed recently by Gill et al. [78]. In the agricultural sector Omnilytics has developed AgriPhage, a range of phage products for the control of *Xanthomonas campestris* pv. *vesicatoria*, for the treatment of bacterial spot of tomatoes and peppers, and *Pseudomonas syringae* pv. *tomato*, which is the causative agent of bacterial speck on tomatoes. We expect further growth in the area of phage control of plant pathogens, which is the least examined application of phage control and has the advantage of less regulatory and safety hurdles.

5. Phage Resistance

Bacterial mechanisms of phage resistance are well understood and should be considered when designing a BCA to reduce resistance and/or to help develop alternative BCAs. Most stages during phage infection can be affected by resistance development (Figure 1) [19, 79]. Briefly, these mechanisms include prevention of phage adsorption, blocking DNA entry, abortive infection, CRISPR/Cas systems, and restriction modification systems.

5.1. Receptors and Inhibition of Phage Adsorption. Cell surface receptors are essential for phage attachment. Receptor identification is important because receptor mutation is a common cause of phage resistance [80]. However, development of phage resistance can be beneficial. For example, mutants of *Pectobacterium atrosepticum* resistant to *φ*S32 had mutations in LPS, which reduced their virulence in a potato tuber rot assay [81]. Likewise, *Pectobacterium atrosepticum* mutants resistant to *φ*AT1 contained flagella mutations and were attenuated for motility and virulence [82]. The double-stranded RNA phages, *φ*6 and *φ*2954, that infect *Pseudomonas syringae* use Type IV pili for attachment [83–85]. Mutations in Type IV pili cause reduced survival in the phyllosphere and pathogenicity [86]. Phage *φ*6 also uses host cell phospholipids as secondary receptors during infection [87]. Because phages often recognize these important components (e.g., LPS and flagella), the resistant mutants are frequently less competitive or pathogenic. Therefore, careful choice of the phage cocktail can ensure that if resistant bacteria arise they will be attenuated. This approach has been utilised in phage therapy of *E. coli* infections in animal trials [88].

It is generally accepted that a phage mixture that uses different receptors is better for biocontrol because resistance to a carefully chosen range of phages cannot usually be acquired with a single point mutation [89, 90]. It is possible to use mutant bacterial hosts to enrich for the isolation of phages with alternative receptors. For example, a phage that targeted the flagellum was isolated using an LPS mutant host [82]. However, in most studies where multiple phages are used the exact receptors are unknown, limiting the potential benefit of such an approach. Cocktails are not always the most effective treatment. One study by Fujiwara et al. [41] characterised resistant strains generated by infection with three phages. Resistant strains were isolated for two phages, whereas no phage-resistant mutants were observed for a third phage. Use of this phage (*φ*RSL1) was more effective in tomato plant assays and greenhouse experiments compared with either of the other phages used singularly or as a mixture. Discovering the receptors used by these phages might shed some light on these results. To overcome resistance, Omnilytics has developed a management plan, which involves monitoring the pathogen and updating the phage mixture if, or when, bacterial resistance emerges [91]. This involves the selection of host range (h-) mutants, which can be evolved to avoid resistance [55].

5.2. Intracellular Factors and Abortive Infection/Toxin-Antitoxin Systems. Cytosolic factors are important for phage infection and their mutations can lead to resistance. *Pseudomonas syringae* phage *φ*2954 requires host glutaredoxin 3

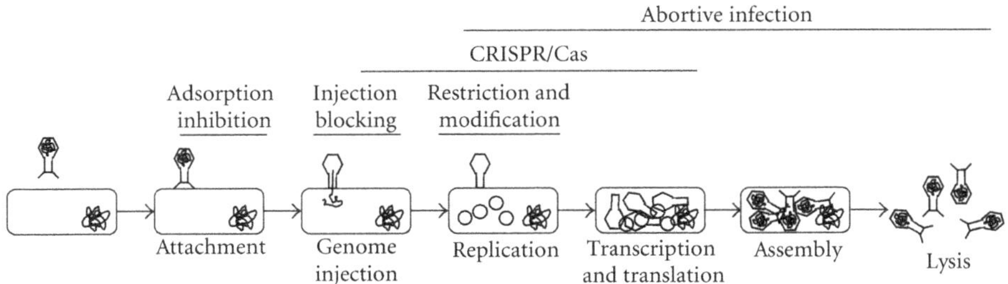

Figure 1: Bacteria can acquire phage resistance against most stages of the phage lifecycle. Fortunately, we can use our knowledge of these systems and the ability to evolve or isolate phage that infect resistant strains to minimise or avoid resistance (see text for details).

(GrxC) for transcription of the third L segment of RNA and deletion of *grxC* led to resistance [92]. Phages with gene 1 mutations were easily selected that had overcome the loss of GrxC [92]. This indicates that phage can be isolated to overcome resistance caused by mutations of intracellular factors. Therefore, in theory, if φ2954 was to be used as a BCA, these escape mutants could be selected as part of the cocktail to avoid the impact of *grxC* mutants.

Other mechanisms interfere with phage reproduction such as the abortive infection (Abi) systems [19, 93]. Abi systems cause the "suicide" of infected cells and the inhibition of phage reproduction. Most Abi systems were identified in dairy bacteria but, recently, an Abi system, termed ToxIN, was isolated in *Pectobacterium atrosepticum* [94]. ToxIN inhibited infection by multiple phages and worked in different genera [94]. Theoretically, this broad efficacy and the presence of some of these systems on plasmids [94, 95] might pose a threat for phage as biocontrol agents. Fortunately, the only Abi system shown to function in a phytopathogen is ToxIN. ToxIN acts as a novel Type III protein-RNA toxin-antitoxin (TA) system [94, 96, 97]. TA systems consist of a toxic protein and an antitoxin and are found in most bacterial genomes [98]. Despite debated roles, TAs have been shown to provide resistance against phages [99]. Whether TA systems influence use of phage as biocontrol agents is at present unclear. Reassuringly, phage mutants can be isolated that avoid Abi/TA systems [97], demonstrating that resistance can be overcome if encountered.

5.3. CRISPR/Cas Resistance. Recently, roughly 40% of sequenced bacteria have been shown to possess an adaptable phage resistance system [100]. These systems contain a clustered regularly interspaced short palindromic repeat (CRISPR) array and CRISPR associated (Cas) proteins. CRISPR arrays acquire short stretches of nucleic acids (termed spacers) from invading phages. The arrays are transcribed and processed into small RNAs, which, with the assistance of Cas proteins, target and degrade spacer-complementary viral nucleic acids. In short, CRISPR/Cas provides a heritable memory of past invaders and elicits a sequence-specific immunity. The experimental analysis of CRISPR/Cas systems in phytopathogens is limited to *Pectobacterium atrosepticum* [101], *Erwinia amylovora* [102], and *Xanthomonas oryzae* [103]. *Pectobacterium atrosepticum* has three CRISPR arrays and the *cas* operon is expressed *in planta* and *in vitro*, indicating that phage resistance could be active during plant infection [101]. In *Erwinia amylovora* the CRISPR/Cas was used to study the evolutionary history of these strains and some spacers matched viral sequences but not to any sequenced *Erwinia* phages [102]. In *Xanthomonas oryzae*, sequence analyses suggested that the system had previously provided resistance against phage Xop411 but the phage had acquired a mutation to avoid the resistance system [103]. The role of CRISPR/Cas systems in plant pathogens is not well characterised but phage can be selected to avoid CRISPR resistance if, and when, it arises [104]. In summary, despite the presence of multiple resistance mechanisms, our understanding of these systems and the ability to easily select phage escape mutants and create intelligent cocktails can minimise any possible impact of resistance development in an effective therapy.

6. Conclusions and Future Directions

As plant diseases continue to have a serious impact on food production worldwide, new approaches for control are sought. This has seen a resurgence of studies into the use of phage for prophylaxis and treatment of phytopathogens. As highlighted in this paper, multiple phage-phytopathogen-plant systems have been studied and promising results are beginning to emerge. However, although available, commercial application of phages to treat plant disease is still uncommon. Alternative strategies for phage-based control of plant pathogens are being developed. For example, one idea is to insert phage genes into plant genomes, especially for the control of systemic pathogens. To avoid the requirement of producing large amounts of phages or purified enzymes, transgenic tomato plants that express the CMP1 and CN77 endolysins in the xylem are being developed to kill invading *C. michiganensis* [33, 34]. Despite the possible advantages of this approach, the regulatory approval needed because of the use of transgenic tomato plants may present a challenge in certain countries and to the consumers. In conclusion, studies into phage BCAs will not only aid in tackling the problems of plant diseases but will also continue to shed light on the basic biology of phages and their pathogenic bacterial hosts.

Acknowledgments

The authors thank members of the Fineran laboratory for useful discussions. Research in our laboratories on phage control of plant pathogens is supported by Zespri International Ltd and phage resistance in the Fineran laboratory is funded by the Marsden Fund and a Rutherford Discovery Fellowship (P. C. Fineran) from the Royal Society of New Zealand.

References

[1] P. Ronald, "Plant genetics, sustainable agriculture and global food security," *Genetics*, vol. 188, no. 1, pp. 11–20, 2011.

[2] C. A. Gilligan, "Sustainable agriculture and plant diseases: an epidemiological perspective," *Philosophical Transactions of the Royal Society B*, vol. 363, no. 1492, pp. 741–759, 2008.

[3] E. Stokstad, "Agriculture. Dread citrus disease turns up in California, Texas," *Science*, vol. 336, no. 6079, pp. 283–284, 2012.

[4] M. Scortichini, S. Marcelletti, P. Ferrante Petriccione, and G. Firrao, "*Pseudomonas syringae* pv. *actinidiae*: a re-emerging, multi-faceted, pandemic pathogen," *Molecular Plant Pathology*, vol. 13, no. 7, pp. 631–640, 2012.

[5] "Kiwifruit Vine Health," 2012, http://www.kvh.org.nz/.

[6] D. A. Cooksey, "Genetics of bactericide resistance in plant pathogenic bacteria," *Annual Review of Phytopathology*, vol. 28, pp. 201–219, 1990.

[7] P. S. McManus, V. O. Stockwell, G. W. Sundin, and A. L. Jones, "Antibiotic use in plant agriculture," *Annual Review of Phytopathology*, vol. 40, pp. 443–465, 2002.

[8] D. W. Dye, "Control of *Pseudomonas syringae* with streptomycin," *Nature*, vol. 172, no. 4380, pp. 683–684, 1953.

[9] F. Behlau, B. I. Canteros, G. V. Minsavage, J. B. Jones, and J. H. Graham, "Molecular characterization of copper resistance genes from *Xanthomonas citri* subsp. citri and *Xanthomonas alfalfae* subsp. *citrumelonis*," *Applied and Environmental Microbiology*, vol. 77, no. 12, pp. 4089–4096, 2011.

[10] M. S. H. Hwang, R. L. Morgan, S. F. Sarkar, P. W. Wang, and D. S. Guttman, "Phylogenetic characterization of virulence and resistance phenotypes of *Pseudomonas syringae*," *Applied and Environmental Microbiology*, vol. 71, no. 9, pp. 5182–5191, 2005.

[11] D. A. Cooksey, "Molecular mechanisms of copper resistance and accumulation in bacteria," *FEMS Microbiology Reviews*, vol. 14, no. 4, pp. 381–386, 1994.

[12] J. M. Hirst, H. H. Riche, and C. L. Bascomb, "Copper accumulation in the soils of apple orchards near Wisbech," *Plant Pathology*, vol. 10, no. 3, pp. 105–108, 1961.

[13] U. Pietrzak and D. C. McPhail, "Copper accumulation, distribution and fractionation in vineyard soils of Victoria, Australia," *Geoderma*, vol. 122, no. 2–4, pp. 151–166, 2004.

[14] N. Okabe and M. Goto, "Bacteriophages of Plant Pathogens," *Annual Review of Phytopathology*, vol. 1, pp. 397–418, 1963.

[15] B. Balogh, J. B. Jones, F. B. Iriarte, and M. T. Momol, "Phage therapy for plant disease control," *Current Pharmaceutical Biotechnology*, vol. 11, no. 1, pp. 48–57, 2010.

[16] J. B. Jones, L. E. Jackson, B. Balogh, A. Obradovic, F. B. Iriarte, and M. T. Momol, "Bacteriophages for plant disease control," *Annual Review of Phytopathology*, vol. 45, pp. 245–262, 2007.

[17] A. M. Svircev, A.J. Castle, and S. M. Lehman, "Bacteriophages for control of phytopathogens in food production systems," in *Bacteriophages in the Control of Food- and Waterborne Pathogens*, P. M. Sabour and M.W. Griffiths, Eds., pp. 79–102, ASM Press, Washington, DC, USA, 2010.

[18] Z. Klement, "Some new specific bacteriophages for plant pathogenic *Xanthomonas* spp," *Nature*, vol. 184, no. 4694, pp. 1248–1249, 1959.

[19] N. K. Petty, T. J. Evans, P. C. Fineran, and G. P. C. Salmond, "Biotechnological exploitation of bacteriophage research," *Trends in Biotechnology*, vol. 25, no. 1, pp. 7–15, 2007.

[20] D. A. Schofield, C. T. Bull, I. Rubio, W. P. Wechter, C. Westwater, and I. J. Molineux, "Development of an engineered "bioluminescent" reporter phage for the detection of bacterial blight of crucifers," *Applied and Environmental Microbiology*, vol. 78, no. 10, pp. 3592–3598, 2012.

[21] M. R. J. Clokie and A. M. Kropinski, Eds., *Bacteriophages Methods and Protocols, Volume 1: Isolation, Characterization, and Interactions*, Humana Press, New York, NY, USA, 2009.

[22] A. M. Svircev, S.M. Lehman, P. Sholberg, D. Roach, and A. J. Castle, "Phage biopesticides and soil bacteria: multilayered and complex interactions," in *Biocommunication in Soil Microorganisms*, G. Witzany, Ed., pp. 215–235, Springer, Berlin, Germany, 2011.

[23] E. M. Adriaenssens, J. Van Vaerenbergh, D. Vandenheuvel et al., "T4-related bacteriophage LIMEstone isolates for the control of soft rot on potato caused by "Dickeya solani"," *PLoS ONE*, vol. 7, no. 3, Article ID e33227, 2012.

[24] J. Boulé, P. L. Sholberg, S. M. Lehman, D. T. O'gorman, and A. M. Svircev, "Isolation and characterization of eight bacteriophages infecting *Erwinia amylovora* and their potential as biological control agents in British Columbia, Canada," *Canadian Journal of Plant Pathology*, vol. 33, no. 3, pp. 308–317, 2011.

[25] B. Balogh, B. I. Canteros, R. E. Stall, and J. B. Jones, "Control of citrus canker and citrus bacterial spot with bacteriophages," *Plant Disease*, vol. 92, no. 7, pp. 1048–1052, 2008.

[26] S. M. Lehman, *Development of a bacteriophage-based biopesticide for fire blight [Ph.D. thesis]*, Brock University, St Catharines, Canada, 2007.

[27] P. C. Fineran, N. K. Petty, and G. P. C. Salmond, "Transduction: host DNA transfer by bacteriophages," in *The Encyclopedia of Microbiology*, M. Schaechter, Ed., Elsevier, 2009.

[28] H. S. Addy, A. Askora, T. Kawasaki, M. Fujie, and T. Yamada, "Utilization of filamentous phage ϕRSM3 to control bacterial wilt caused by *Ralstonia solanacearum*," *Plant Disease*, vol. 96, no. 8, pp. 1204–1208, 2012.

[29] J. Borysowski, B. Weber-Dabrowska, and A. Górski, "Bacteriophage endolysins as a novel class of antibacterial agents," *Experimental Biology and Medicine*, vol. 231, no. 4, pp. 366–377, 2006.

[30] M. J. Loessner, "Bacteriophage endolysins—current state of research and applications," *Current Opinion in Microbiology*, vol. 8, no. 4, pp. 480–487, 2005.

[31] E. Echandi and M. Sun, "Isolation and characterization of a bacteriophage for the identification of *Corynebacterium michiganense*," *Phytopathology*, vol. 63, pp. 1398–1401, 1973.

[32] F. D. Cook and H. Katznelson, "Isolation of bacteriophages for the detection of *Corynebacterium insidiosum*, agent of bacterial wilt of alfalfa," *Canadian Journal of Microbiology*, vol. 6, pp. 121–125, 1960.

[33] J. Wittmann, R. Eichenlaub, and B. Dreiseikelmann, "The endolysins of bacteriophages CMP1 and CN77 are specific for the lysis of *Clavibacter michiganensis* strains," *Microbiology*, vol. 156, no. 8, pp. 2366–2373, 2010.

[34] J. Wittmann, K.-H. Gartemann, R. Eichenlaub, and B. Dreiseikelmann, "Genomic and molecular analysis of phage CMP1 from *Clavibacter michiganensis* subspecies michiganensis," *Bacteriophage*, vol. 1, no. 1, pp. 6–14, 2011.

[35] H. W. Ackermann, D. Tremblay, and S. Moineau, "Long-term bacteriophage preservation," *World Federation For Culture Collections Newsletter*, vol. 38, pp. 35–40, 2004.

[36] M. Ravensdale, T. J. Blom, J. A. Gracia-Garza, A. M. Svircev, and R. J. Smith, "Bacteriophages and the control of *Erwinia carotovora* subsp. *carotovora*," *Canadian Journal of Plant Pathology*, vol. 29, no. 2, pp. 121–130, 2007.

[37] T. Stonier, J. McSharry, and T. Speitel, "*Agrobacterium tumefaciens* Conn. IV. Bacteriophage PB21 and its inhibitory effect on tumor induction," *Journal of Virology*, vol. 1, no. 2, pp. 268–273, 1967.

[38] J. J. Gill, A. M. Svircev, R. Smith, and A. J. Castle, "Bacteriophages of *Erwinia amylovora*," *Applied and Environmental Microbiology*, vol. 69, no. 4, pp. 2133–2138, 2003.

[39] E. L. Schnabel and A. L. Jones, "Isolation and characterization of five *Erwinia amylovora* bacteriophages and assessment of phage resistance in strains of *Erwinia amylovora*," *Applied and Environmental Microbiology*, vol. 67, no. 1, pp. 59–64, 2001.

[40] H. Tanaka Negishi and H. Maeda, "Control of tobacco bacterial wilt by an avirulent strain of *Pseudomonas solanacearum* M4S and its bacteriophage," *Annals of the Phytopathological Society of Japan*, vol. 56, no. 2, pp. 243–246, 1990.

[41] A. Fujiwara, M. Fujisawa, R. Hamasaki, T. Kawasaki, M. Fujie, and T. Yamada, "Biocontrol of *Ralstonia solanacearum* by treatment with lytic bacteriophages," *Applied and Environmental Microbiology*, vol. 77, no. 12, pp. 4155–4162, 2011.

[42] F. McKenna, K. A. El-Tarabily, G. E. S. T. J. Hardy, and B. Dell, "Novel in vivo use of a polyvalent *Streptomyces* phage to disinfest *Streptomyces scabies*-infected seed potatoes," *Plant Pathology*, vol. 50, no. 6, pp. 666–675, 2001.

[43] E. L. Civerolo and H. L. Kiel, "Inhibition of bacterial spot of peach foliage by *Xanthomonas pruni* bacteriophage," *Phytopathology*, vol. 59, pp. 1966–1967, 1969.

[44] J. M. Lang, D. H. Gent, and H. F. Schwartz, "Management of *Xanthomonas* leaf blight of onion with bacteriophages and a plant activator," *Plant Disease*, vol. 91, no. 7, pp. 871–878, 2007.

[45] P. K. Borah, J. K. Jindal, and J. P. Verma, "Integrated management of bacterial leaf spot of mungbean with bacteriophages of Xav and chemicals," *Journal of Mycology and Plant Pathology*, vol. 30, no. 1, pp. 19–21, 2000.

[46] D. L. McNeil, S. Romero, J. Kandula, C. Stark, A. Stewart, and S. Larsen, "Bacteriophages: a potential biocontrol agent against walnut blight (*Xanthomonas campestris* pv. *juglandis*)," *New Zealand Plant Protection*, vol. 54, pp. 220–224, 2001.

[47] A. Saccardi, E. Gambin, M. Zaccardelli, G. Barone, and U. Mazzucchi, "*Xanthomonas campestris* pv. pruni control trials with phage treatments on peaches in the orchard," *Phytopathologia Mediterranea*, vol. 32, pp. 206–210, 1993.

[48] A. Obradovic, J. B. Jones, M. T. Momol, B. Balogh, and S. M. Olson, "Management of tomato bacterial spot in the field by foliar applications of bacteriophages and SAR inducers," *Plant Disease*, vol. 88, no. 7, pp. 736–740, 2004.

[49] A. Obradovic, J. B. Jones, M. T. Momol et al., "Integration of biological control agents and systemic acquired resistance inducers against bacterial spot on tomato," *Plant Disease*, vol. 89, no. 7, pp. 712–716, 2005.

[50] J. E. Flaherty, J. B. Jones, B. K. Harbaugh, G. C. Somodi, and L. E. Jackson, "Control of bacterial spot on tomato in the greenhouse and field with H-mutant bacteriophages," *HortScience*, vol. 35, no. 5, pp. 882–884, 2000.

[51] B. Balogh, J. B. Jones, M. T. Momol et al., "Improved efficacy of newly formulated bacteriophages for management of bacterial spot on tomato," *Plant Disease*, vol. 87, no. 8, pp. 949–954, 2003.

[52] F. B. Iriarte, B. Balogh, M. T. Momol, L. M. Smith, M. Wilson, and J. B. Jones, "Factors affecting survival of bacteriophage on tomato leaf surfaces," *Applied and Environmental Microbiology*, vol. 73, no. 6, pp. 1704–1711, 2007.

[53] B. H. Chantrill, C. E. Coulthard, L. Dickinson, G. W. Inkley, W. Morris, and A. H. Pyle, "The action of plant extracts on a bacteriophage of *Pseudomonas pyocyanea* and on influenza A virus," *Journal of General Microbiology*, vol. 6, no. 1-2, pp. 74–84, 1952.

[54] A. Delitheos, E. Tiligada, A. Yannitsaros, and I. Bazos, "Antiphage activity in extracts of plants growing in Greece," *Phytomedicine*, vol. 4, no. 2, pp. 117–124, 1997.

[55] J. E. Flaherty, B. K. Harbaugh, J. B. Jones, G. C. Somodi, and L. E. Jackson, "H-mutant bacteriophages as a potential biocontrol of bacterial blight of geranium," *HortScience*, vol. 36, no. 1, pp. 98–100, 2001.

[56] J. Li Dennehy, "Differential bacteriophage mortality on exposure to copper," *Applied and Environmental Microbiology*, vol. 77, no. 19, pp. 6878–6883, 2011.

[57] "Commission decision of 30 January 2004 concerning the non-inclusion of certain active substances in Annex I to Council Directive 91/414/EEC and the withdrawl of authorisations for plant protection products containing these substances," Official Journal of the European Union, 2004, http://eur-lex.europa.eu/LexUriServ/LexUriServ.do?uri=OJ:L:2004:037:0027:0031:EN:PDF.

[58] G. W. Sundin, N. A. Werner, K. S. Yoder, and H. S. Aldwinckle, "Field evaluation of biological control of fire blight in the Eastern United States," *Plant Disease*, vol. 93, no. 4, pp. 386–394, 2009.

[59] E. Billing, "Fire blight. Why do views on host invasion by *Erwinia amylovora* differ?" *Plant Pathology*, vol. 60, no. 2, pp. 178–189, 2011.

[60] A. Palacio-Bielsa, M. Roselló, P. Llop, and M. M. López, "*Erwinia* spp. from pome fruit trees: similarities and differences among pathogenic and non-pathogenic species," *Trees*, vol. 26, no. 1, pp. 13–29, 2012.

[61] P. Gómez and A. Buckling, "Bacteria-phage antagonistic coevolution in soil," *Science*, vol. 332, no. 6025, pp. 106–109, 2011.

[62] J. M. Erskine, "Characteristics of *Erwinia amylovora* bacteriophage and its possible role in the epidemiology of fire blight," *Canadian Journal of Microbiology*, vol. 19, no. 7, pp. 837–845, 1973.

[63] M. J. Hattingh, S.V. Beer, and E. W. Lawson, "Scanning electron microscopy of apple blossoms colonized by *Erwinia amylovora* and *E. herbicola*," *Phytopathology*, vol. 76, pp. 900–904, 1986.

[64] S. M. Lehman, A. M. Kropinski, A. J. Castle, and A. M. Svircev, "Complete genome of the broad-host-range *Erwinia amylovora* phage φEa21-4 and its relationship to *Salmonella*

phage f elix O1," *Applied and Environmental Microbiology*, vol. 75, no. 7, pp. 2139–2147, 2009.

[65] J. L. Norelli, A. L. Jones, and H. S. Aldwinckle, "Fire blight management in the twenty-first century: using new technologies that enhance host resistance in apple," *Plant Disease*, vol. 87, no. 7, pp. 756–765, 2003.

[66] N. W. Assadian, G. D. Di Giovanni, J. Enciso, J. Iglesias, and W. Lindemann, "The transport of waterborne solutes and bacteriophage in soil subirrigated with a wastewater blend," *Agriculture, Ecosystems and Environment*, vol. 111, no. 1–4, pp. 279–291, 2005.

[67] M. Kimura, Z. J. Jia, N. Nakayama, and S. Asakawa, "Ecology of viruses in soils: past, present and future perspectives," *Soil Science and Plant Nutrition*, vol. 54, no. 1, pp. 1–32, 2008.

[68] H. S. Addy, A. Askora, T. Kawasaki, M. Fujie, and T. Yamada, "The filamentous phage ϕRSS1 enhances virulence of phytopathogenic *Ralstonia solanacearum* on tomato," *Phytopathology*, vol. 102, no. 3, pp. 244–251, 2012.

[69] H. S. Addy, A. Askora, T. Kawasaki, M. Fujie, and T. Yamada, "Loss of virulence of the phytopathogen *Ralstonia solanacearum* through infection by phiRSM filamentous phages," *Phytopathology*, vol. 102, no. 5, pp. 469–477, 2012.

[70] L. Bren, "Bacteria-eating virus approved as food additive," *FDA Consumer*, vol. 41, no. 1, 2007.

[71] "Regulatory position Listex^{TM} ," 2012, http://www.micreos-foodsafety.com/en/listex-regulatory.aspx.

[72] M. Kutateladze and R. Adamia, "Bacteriophages as potential new therapeutics to replace or supplement antibiotics," *Trends in Biotechnology*, vol. 28, no. 12, pp. 591–595, 2010.

[73] C. Loc-Carrillo and S. T. Abedon, "Pros and cons of phage therapy," *Bacteriophage*, vol. 1, no. 2, pp. 111–114, 2011.

[74] A. Wright, C. H. Hawkins, E. E. Änggård, and D. R. Harper, "A controlled clinical trial of a therapeutic bacteriophage preparation in chronic otitis due to antibiotic-resistant *Pseudomonas aeruginosa*; A preliminary report of efficacy," *Clinical Otolaryngology*, vol. 34, no. 4, pp. 349–357, 2009.

[75] M. Merabishvili, J. P. Pirnay, G. Verbeken et al., "Quality-controlled small-scale production of a well-defined bacteriophage cocktail for use in human clinical trials," *PLoS ONE*, vol. 4, no. 3, Article ID e4944, 2009.

[76] R. J. Atterbury, "Bacteriophage biocontrol in animals and meat products," *Microbial Biotechnology*, vol. 2, no. 6, pp. 601–612, 2009.

[77] J. Mahony, O. McAuliffe, R. P. Ross, and D. van Sinderen, "Bacteriophages as biocontrol agents of food pathogens," *Current Opinion in Biotechnology*, vol. 22, no. 2, pp. 157–163, 2011.

[78] J. J. Gill, T. Hollyer, and P. M. Sabour, "Bacteriophages and phage-derived products as antibacterial therapeutics," *Expert Opinion on Therapeutic Patents*, vol. 17, no. 11, pp. 1341–1350, 2007.

[79] S. J. Labrie, J. E. Samson, and S. Moineau, "Bacteriophage resistance mechanisms," *Nature Reviews Microbiology*, vol. 8, no. 5, pp. 317–327, 2010.

[80] T. R. Blower, T.J. Evans, P.C. Fineran, I.K. Toth, I.J. Foulds, and G. P. C. Salmond, "Phage-receptor interactions and phage abortive infection: potential biocontrol factors in a bacterial plant pathogen," in *Biology of Molecular Plant-Microbe Interactions*, H. Antoun et al., Ed., pp. 1–7, International Society for Molecular Plant-Microbe Interactions, St Paul, Minn, USA, 2010.

[81] T. J. Evans, A. Ind, E. Komitopoulou, and G. P. C. Salmond, "Phage-selected lipopolysaccharide mutants of *Pectobacterium atrosepticum* exhibit different impacts on virulence," *Journal of Applied Microbiology*, vol. 109, no. 2, pp. 505–514, 2010.

[82] T. J. Evans, A. Trauner, E. Komitopoulou, and G. P. C. Salmond, "Exploitation of a new flagellatropic phage of *Erwinia* for positive selection of bacterial mutants attenuated in plant virulence: towards phage therapy," *Journal of Applied Microbiology*, vol. 108, no. 2, pp. 676–685, 2010.

[83] A. K. Vidaver, R. K. Koski, and J. L. Van Etten, "Bacteriophage phi6: a lipid containing virus of Pseudomonas phaseolicola," *Journal of Virology*, vol. 11, no. 5, pp. 799–805, 1973.

[84] J. S. Semancik, A. K. Vidaver, and J. L. Van Etten, "Characterization of a segmented double helical RNA from bacteriophage ϕ6," *Journal of Molecular Biology*, vol. 78, no. 4, pp. 617–625, 1973.

[85] X. Qiao, Y. Sun, J. Qiao, F. Di Sanzo, and L. Mindich, "Characterization of Φ2954, a newly isolated bacteriophage containing three dsRNA genomic segments," *BMC Microbiology*, vol. 10, article 55, 2010.

[86] S. S. Hirano and C. D. Upper, "Bacteria in the leaf ecosystem with emphasis on *Pseudomonas syringae*—a pathogen, ice nucleus, and epiphyte," *Microbiology and Molecular Biology Reviews*, vol. 64, no. 3, pp. 624–653, 2000.

[87] V. Cvirkaite-Krupovič, M. M. Poranen, and D. H. Bamford, "Phospholipids act as secondary receptor during the entry of the enveloped, double-stranded RNA bacteriophage φ6," *Journal of General Virology*, vol. 91, no. 8, pp. 2116–2120, 2010.

[88] H. W. Smith and M. B. Huggins, "Successful treatment of experimental *Escherichia coli* infections in mice using phage: its general superiority over antibiotics," *Journal of General Microbiology*, vol. 128, no. 2, pp. 307–318, 1982.

[89] Y. Tanji, T. Shimada, M. Yoichi, K. Miyanaga, K. Hori, and H. Unno, "Toward rational control of *Escherichia coli* O157:H7 by a phage cocktail," *Applied Microbiology and Biotechnology*, vol. 64, no. 2, pp. 270–274, 2004.

[90] Y. Tanji, T. Shimada, H. Fukudomi, K. Miyanaga, Y. Nakai, and H. Unno, "Therapeutic use of phage cocktail for controlling *Escherichia coli* O157:H7 in gastrointestinal tract of mice," *Journal of Bioscience and Bioengineering*, vol. 100, no. 3, pp. 280–287, 2005.

[91] Agriphage, 2004, http://www.omnilytics.com/products/agriphage/agriphage_info/agriphage_overview.html.

[92] J. Qiao, X. Qiao, Y. Sun, and L. Mindich, "Role of host protein glutaredoxin 3 in the control of transcription during bacteriophage ϕ2954 infection," *Proceedings of the National Academy of Sciences of the United States of America*, vol. 107, no. 13, pp. 6000–6004, 2010.

[93] M. C. Chopin, A. Chopin, and E. Bidnenko, "Phage abortive infection in lactococci: variations on a theme," *Current Opinion in Microbiology*, vol. 8, no. 4, pp. 473–479, 2005.

[94] P. C. Fineran, T. R. Blower, I. J. Foulds, D. P. Humphreys, K. S. Lilley, and G. P. C. Salmond, "The phage abortive infection system, ToxIN, functions as a protein-RNA toxin-antitoxin pair," *Proceedings of the National Academy of Sciences of the United States of America*, vol. 106, no. 3, pp. 894–899, 2009.

[95] T. R. Blower, F. L. Short, F. Rao et al., "Identification and classification of bacterial Type III toxin-antitoxin systems encoded in chromosomal and plasmid genomes," *Nucleic Acids Research*, vol. 40, no. 13, pp. 6158–6173, 2012.

[96] T. R. Blower, X. Y. Pei, F. L. Short et al., "A processed noncoding RNA regulates an altruistic bacterial antiviral

system," *Nature Structural and Molecular Biology*, vol. 18, no. 2, pp. 185–190, 2011.

[97] T. R. Blower, P. C. Fineran, M. J. Johnson, I. K. Toth, D. P. Humphreys, and G. P. C. Salmond, "Mutagenesis and functional characterization of the RNA and protein components of the *toxIN* abortive infection and toxin-antitoxin locus of *Erwinia*," *Journal of Bacteriology*, vol. 191, no. 19, pp. 6029–6039, 2009.

[98] Y. Yamaguchi, J. H. Park, and M. Inouye, "Toxin-antitoxin systems in bacteria and archaea," *Annual Review of Genetics*, vol. 45, pp. 61–79, 2011.

[99] F. L. Short, T. R. Blower, and G. P. Salmond, "A promiscuous antitoxin of bacteriophage T4 ensures successful viral replication," *Molecular Microbiology*, vol. 83, no. 4, pp. 665–668, 2012.

[100] S. Al-Attar, E. R. Westra, J. Van Der Oost, and S. J. J. Brouns, "Clustered regularly interspaced short palindromic repeats (CRISPRs): the hallmark of an ingenious antiviral defense mechanism in prokaryotes," *Biological Chemistry*, vol. 392, no. 4, pp. 277–289, 2011.

[101] R. Przybilski, C. Richter, T. Gristwood, J. S. Clulow, R. B. Vercoe, and P. C. Fineran, "Csy4 is responsible for CRISPR RNA processing in *Pectobacterium atrosepticum*," *RNA Biology*, vol. 8, no. 3, pp. 517–528, 2011.

[102] F. Rezzonico, T. H. M. Smits, and B. Duffy, "Diversity, evolution, and functionality of clustered regularly interspaced short palindromic repeat (CRISPR) regions in the fire blight pathogen *Erwinia amylovora*," *Applied and Environmental Microbiology*, vol. 77, no. 11, pp. 3819–3829, 2011.

[103] E. Semenova, M. Nagornykh, M. Pyatnitskiy, I. I. Artamonova, and K. Severinov, "Analysis of CRISPR system function in plant pathogen *Xanthomonas oryzae*," *FEMS Microbiology Letters*, vol. 296, no. 1, pp. 110–116, 2009.

[104] H. Deveau, R. Barrangou, J. E. Garneau et al., "Phage response to CRISPR-encoded resistance in *Streptococcus thermophilus*," *Journal of Bacteriology*, vol. 190, no. 4, pp. 1390–1400, 2008.

16

Invasive Mold Infections: Virulence and Pathogenesis of *Mucorales*

Giulia Morace and Elisa Borghi

Department of Public Health, Microbiology, and Virology, Università degli Studi di Milano, Via C. Pascal 36, 20133 Milan, Italy

Correspondence should be addressed to Giulia Morace, giulia.morace@unimi.it

Academic Editor: Arianna Tavanti

Mucorales have been increasingly reported as cause of invasive fungal infections in immunocompromised subjects, particularly in patients with haematological malignancies or uncontrolled diabetes mellitus and in those under deferoxamine treatment or undergoing dialysis. The disease often leads to a fatal outcome, but the pathogenesis of the infection is still poorly understood as well as the role of specific virulence determinants and the interaction with the host immune system. Members of the order *Mucorales* are responsible of almost all cases of invasive mucormycoses, the majority of the etiological agents belonging to the *Mucoraceae* family. *Mucorales* are able to produce various proteins and metabolic products toxic to animals and humans, but the pathogenic role of these potential virulence factors is unknown. The availability of free iron in plasma and tissues is believed to be crucial for the pathogenesis of these mycoses. Vascular invasion and neurotropism are considered common pathogenic features of invasive mucormycoses.

1. Introduction

The *Mucorales*, which is the core group of the traditional *Zygomycota* [1–3], have been recently reclassified into the subphylum *Mucoromycotina* of the *Glomeromycota* phylum of the kingdom Fungi [4]. This new classification does not include *Zygomycota,* because the authors consider the phylum polyphyletic, indeed the name zygomycosis, which encompassed infections caused by members of *Mucorales* and *Entomophthorales*, has become obsolete [4]. The *Mucorales* are characterized by aseptate (coenocytic) hyaline hyphae, sexual reproduction with the formation of zygospores, and asexual reproduction with nonmotile sporangiospores. They are ubiquitous in nature, being found in food, vegetation, and soil [1–3]. The majority of the invasive diseases are caused by genera of the *Mucoraceae* family, and the resulted disease is called mucormycosis [1–3, 5–7]. Transmission occurs by inhalation of aerosolized spores, ingestion of contaminated foodstuffs, or through cutaneous exposure, the latter being the most important mode of acquisition of mucormycosis in immunocompetent hosts [6, 8]. Risk factors for invasive diseases include uncontrolled diabetes mellitus, haematological malignancies, bone marrow and solid organ transplantation, deferoxamine therapy, corticosteroid therapy, or other underlying conditions impairing the immune system [9]. Limited activity of some principal classes of antifungal drugs (i.e., echinocandins and azole derivatives) as well as vascular invasion and neurotropic activity could explain the high mortality seen in mucormycosis [9, 10]. This paper, together with others published in this special issue, reviews the clinical spectrum of and risk factors for mucormycosis with particular emphasis on the role of fungal traits interacting with human host defences.

2. Epidemiology

A few members of the *Mucorales* (Table 1) are able to grow in human tissues causing a wide spectrum of clinical diseases. The entity and severity of the disease depends on the interaction between the fungus and the host immune

Table 1: Agents[a] of mucormycosis belonging to *Mucorales* order of the *Glomeromycota* phylum.

Order	Family	Genus	Species	Maximum growth temp (°C)
Mucorales	*Mucoraceae*	*Rhizopus*	*oryzae*	>37°C
			microsporus	>37°C
			azygosporus	>37°C
			schipperae	>37°C
		Mucor	*circinelloides*	>37°C
			indicus	>37°C
			racemosus	32°C
			ramosissimus	36°C
		Rhizomucor	*pusillus*	>37°C
		Lichteimia (*Absidia*)	*corymbifera*	>37°C
		Apophysomyces	*elegans*	>37°C
	Cunninghamellaceae	*Cunninghamella*	*bertholletiae*	>37°C
	Saksenaeaceae	*Saksenaea*	*vasiformis*	>37°C
	Syncephalastraceae	*Syncephalastrum*	*racemosum*	>37°C

defences [5, 7, 11]. In their exhaustive review, Roden and coworkers analysed 929 cases of documented infections caused by members of the former *Zygomycota* since 1885 [9]. They found that 19% of patients did not have any underlying disease at time of infection, while diabetes (36% of cases) was the main risk factor for developing the infection among patients with underlying conditions [9]. More recently, 230 cases of infections were collected in 13 European countries between 2005 and 2007 [12]. The majority of patients (53%) had haematological malignancies (44%) and haematopoietic stem-cell transplantation (9%) as underlying conditions, while only 9% of patients presented diabetes mellitus as the main risk factor [12]. *Rhizopus* spp. are the most common causative agents of invasive mucormycosis, *Mucor* spp. and *Lichteimia* (formerly *Absidia*) spp. rank as second and/or third cause [6–9]. Although mucormycosis remains a highly fatal disease, its burden is still low, as well documented by Pagano and coworkers [13]. They were able to demonstrate that mucormycosis affected about 0.1% of 11,802 patients with hematologic malignancies. Among the 346 cases of proven and probable mold infections, only 14 (4%) were caused by members of *Mucorales* [13]. In immune-competent subjects, mucormycosis generally develops as a consequence of traumatic injuries, and the disease commonly involves skin even if possible dissemination from skin to contiguous organs can occur [9, 11].

3. The Infection

Mucormycosis can be classified in rhinocerebral, pulmonary and disseminated, abdominal-pelvic and gastric, and cutaneous or chronic subcutaneous diseases. Common features of rhinocerebral, pulmonary, and disseminated diseases include blood vessel invasion, hemorrhagic necrosis, thrombosis, and a rapid fatal outcome.

Rhinocerebral mucormycosis is more often associated with uncontrolled diabetes mellitus and ketoacidosis than malignancies or deferoxamine therapy. Inhaled spores colonize at first the upper turbinates and paranasal sinuses and cause sinusitis. Depending on the underlying disease, the fungus can rapidly invade the central nervous system, causing symptoms like an altered mental state, progression to coma, and death within a few days [1–3, 5–11].

Pulmonary mucormycosis is commonly seen in patients with leukemia, lymphoma, solid organ or bone marrow transplantation, and diabetes but is occasionally reported also in apparently healthy subjects. Disease manifestations vary from a localized nodular lesion to cavitary lesions and dissemination; in the latter case, massive hemoptysis generally occurs. Crude mortality is lower (60%) in cases of isolated lesions than in severe pulmonary (87%) and disseminated (95%) diseases [9].

Gastrointestinal disease is a rare manifestation of mucormycosis, and it is mainly associated with malnutrition in presence of predisposing factors, especially in children with amoebic colitis, typhoid, and pellagra [11]. In the most severe cases, the disease can be characterized by ulceration of the mucosa and invasion of blood vessels with subsequent production of necrotic ulcers, this form of the disease is fatal [3, 11].

Cutaneous mucormycosis may be a primary disease following skin barrier break or may occur as a consequence of hematogenous dissemination from other sites, and the outcome of the disease is strictly dependent on the patients' conditions. Primary cutaneous mucormycosis can involve the subcutaneous tissue as well as the fat, muscle and fascial layers [3].

4. Treatment

Treatment of mucormycosis combines surgical intervention and antifungal therapy. Liposomal amphotericin B is the drug of choice for the therapy of mucormycosis. The *in vitro* susceptibility testing for amphotericin B gives a broad range of values according to the genus and the species. With the exception of posaconazole, the azole derivatives show a limited *in vitro* activity against *Mucorales*, and the echinocandins have a limited activity against these fungi [14]. Studies of *in vitro* combination of posaconazole with amphotericin B showed synergistic effects against hyphae of some species [15]. In addition, combination therapy with liposomal amphotericin B plus caspofungin or posaconazole and posaconazole with colony-stimulating factor has been successfully used in experimental infections [10, 16–18]. In humans, combination therapy (liposomal amphotericin B plus echinocandins or posaconazole with or without iron chelation) has been used as aggressive antifungal treatment following surgical resection of the damaged tissue [19–23]. Patients treated with combination of antifungal drugs had a better survival outcome than those treated with amphotericin B alone [20, 21]. A promising therapeutic approach consists of the use of iron chelation. Although deferoxamine therapy is associated with a high risk to develop mucormycosis [2, 3, 5–7, 9–11, 24], newer iron chelators (deferiprone and deferasirox) have not been associated with increased risk of mucormycosis and have been used as therapeutic agents in cases of experimental mucormycosis [24].

5. Virulence Traits and Pathogenesis

According to Casadevall and Pirofski [25]: "*Quantitative and qualitative measures of virulence vary as a function of host factors, microbial factors, environmental factors, social factors and interactions amongst them*". This concept is especially true if we consider opportunistic microorganisms such as fungi. Macrophages and neutrophils play the major role in immune defence against agents of mucormycosis. Prolonged neutropenia is thus the main risk factor for developing the disease. Moreover, therapeutic interventions (i.e., corticosteroid therapy), that cause functional defects in macrophages and neutrophils, represent additional risk factors for mucormycosis. Diabetes itself can impair the function of neutrophils contributing to the severity of the mucormycosis in patients with ketoacidosis [26]. An important protective factor against mucormycosis is the low concentration of free iron in plasma and tissues. Many of the underlying diseases listed above as predisposing factors for developing mucormycosis share an iron overload as a consequence of iron tissue burden, elevated serum transferring, or increased nontransferrin-bound iron [24]. Iron is essential for *Mucorales* either enhancing their growth and hyphal development *in vitro* or increasing their pathogenicity *in vivo* [27]. Hemodialysis patients under treatment with deferoxamine (DFO), an iron chelator, are particularly at risk for mucormycosis, and Boelaert and coworkers [28] reported a high mortality (89%) in 46 patients who developed severe mucormycosis during DFO treatment. The same group [27, 29] was able to demonstrate that *Mucorales* use DFO as a xeno-siderophore, being capable to detach iron from DFO in a very efficient manner. More recently, other investigators confirmed the importance of iron in the pathogenicity of *Mucorales* by studying the expression of the FTR1 (high-affinity iron permease of *R. oryzae*) gene and its product [30]. The authors were able to demonstrate the effect of gene disruption and gene silencing on *R. oryzae*, which was unable to acquire iron *in vitro* and showed a reduced virulence in mice. Consistently, anti-Ftr1p antibodies protected mice from *R. oryzae* infection [30]. Angioinvasion with subsequent infarction of the surrounding tissue is uniformly present in all cases of severe disseminated mucormycosis [31]. Specific adhesion to endothelial cells and internalization of the fungus by the endothelial cells are important for the pathogenic strategy of *Mucorales* [32]. More recently, Liu and coworkers [33] demonstrated that a novel host receptor (the glucose-regulated protein 78 [GPR78]) facilitates the invasion of human endothelial cells by *Rhizopus oryzae*. This study demonstrated that in the presence of high iron and glucose concentrations, such as in diabetic subjects, there is a direct relationship between an increased expression of GPR78 and an increased damage to endothelial cells in diabetic mice [33]. Involvement of the CNS is common in invasive mucormycosis, *Mucorales* are capable to gain access to the central nervous system (CNS) by local vessels invasion or direct extension from paranasal sinuses [1–3, 5–11]. Another possible mechanism, involving a retrograde extension of the fungi into CNS by means of the nerves, was hypothesized by Frater and coworkers [31]. By evaluating the histologic features of 20 patients with invasive disease, they found a high percentage of perineural invasion. A further fascinating hypothesis concerning the virulence of *Mucorales*, in particular of *Rhizopus* species—the most common etiological agents of disseminated mucormycosis—is a possible involvement of endosymbiotic bacteria in the pathogenesis of the disease [34]. The authors formulate their hypothesis on the basis of the ability of *Rhizopus* species to live with endosymbiotic toxin-producing bacteria [35] and of the existing link between emergence of mucormycosis and the increased drug resistance of Gram-negative bacteria seen in the recent decades. Later on, both the groups of researchers demonstrated that endosymbiotic toxin-producing bacteria were not essential for the pathogenesis of mucormycosis [36, 37]. Other potential virulence factors of *Mucorales* could be proteolytic, lipolytic, and glycosidic enzymes as well as metabolites like alkaloids or mycotoxins as agroclavine. However, their direct involvement in human cases of mucormycosis has been still to be documented [3].

6. Diagnosis

Histology and culture are still the most important diagnostic approaches for mucormycosis because of the lacking of molecular diagnosis methods standardized or commercially available. Moreover the β-1–3 glucan detection is not useful in this kind of infection due to the extremely low content of this molecule in the *Mucorales* [38, 39]. Timely diagnosis of invasive mucormycosis is essential due to the rapid progression of the disease, and because signs and symptoms of the infection could mimic other invasive fungal infections. Tissue biopsies are the clinical specimens of choice and

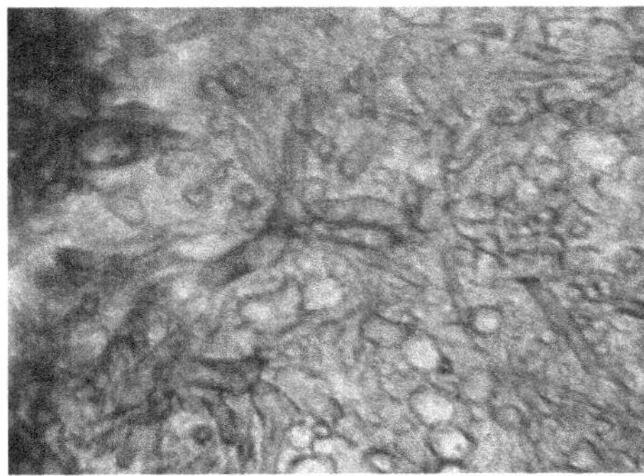

Figure 1: Aseptate hyphae with wide branching angles and large diameter from a lung fungus ball suggestive of mucormycosis (GMS stain 400×).

should be submitted to histopathological and microbiological examination. When cultures are performed, it should be remembered that slicing rather than grinding of the samples should be adopted, because grinding could result in the loss of viability due to the coenocytic characteristics of the mycelium. Microscopic detection of aseptate or pauciseptate hyphae with a large diameter and wide branching angles is suggestive of mucormycosis (Figure 1). Histopathological examination of the infected tissues reveals inflammatory response, often entirely filled with neutrophils, invasion of arterial and venous walls (angioinvasion) with subsequent infarction, and perineural invasion [31].

7. Conclusion

Invasive mucormycosis is an important cause of morbidity and mortality in patients with impaired immune defence and severe underlying diseases. In immunocompromised or debilitated patients, the disease is rapidly progressive, refractory to antifungal therapy, and often cause of death. Several characteristics of *Mucorales* have been involved in the pathogenesis of the infection as potential virulence factors, but a trait that can be considered a specific determinant of virulence has not been defined yet. Angioinvasion, neurotropism, and iron uptake are common characteristics of *Mucorales* that trigger diseases in humans. Many open issues remain to be clarified on the interaction between members of the *Mucorales* order and the host immune response. Different therapeutic approaches, especially the combination therapy, seem to have a promising impact on the clinical outcome of this infection. However, the development of the most severe forms of mucormycosis and the subsequent outcome is strictly dependent on the efficiency of the host immune system.

References

[1] G. S. De Hoog, J. Guarro, J. Genè, and M. J. Figueras, "Zygomycota," in *Atlas of Clinical Fungi*, pp. 58–124, Centraalbureau voor Schimmelcultures, Universitat Rovira I Virgili, Utrecht, The Netherlands, 2000.

[2] C. A. Kauffman, "Zygomycosis: reemergence of an old pathogen," *Clinical Infectious Diseases*, vol. 39, no. 4, pp. 588–590, 2004.

[3] J. A. Ribes, C. L. Vanover-Sams, and D. J. Baker, "Zygomycetes in human disease," *Clinical Microbiology Reviews*, vol. 13, no. 2, pp. 236–301, 2000.

[4] D. S. Hibbett, M. Binder, J. F. Bischoff et al., "A higher-level phylogenetic classification of the Fungi," *Mycological Research*, vol. 111, no. 5, pp. 509–547, 2007.

[5] E. Mantadakis and G. Samonis, "Clinical presentation of zygomycosis," *Clinical Microbiology and Infection*, vol. 15, no. 5, supplement, pp. 15–20, 2009.

[6] M. Chayakulkeeree, M. A. Ghannoum, and J. R. Perfect, "Zygomycosis: the re-emerging fungal infection," *European Journal of Clinical Microbiology and Infectious Diseases*, vol. 25, no. 4, pp. 215–229, 2006.

[7] R. R. Klont, J. F. G. M. Meis, and P. E. Verweij, "Uncommon opportunistic fungi: new nosocomial threats," *Clinical Microbiology and Infection*, vol. 7, supplement 2, pp. 8–24, 2001.

[8] L. Pagano, C. G. Valentini, B. Posteraro et al., "Zygomycosis in Italy: a survey of FIMUA-ECMM (Federazione Italiana di Micopatologia Umana ed Animale and European Confederation of Medical Mycology)," *Journal of Chemotherapy*, vol. 21, no. 3, pp. 322–329, 2009.

[9] M. M. Roden, T. E. Zaoutis, W. L. Buchanan et al., "Epidemiology and outcome of zygomycosis: a review of 929 reported cases," *Clinical Infectious Diseases*, vol. 41, no. 5, pp. 634–653, 2005.

[10] B. Spellberg, J. Edwards Jr., and A. Ibrahim, "Novel perspectives on mucormycosis: pathophysiology, presentation, and management," *Clinical Microbiology Reviews*, vol. 18, no. 3, pp. 556–569, 2005.

[11] J. W. Rippon, "Zygomycosis," in *Medical Mycology: The Pathogenic Fungi and the Pathogenic Actinomycetes*, WB Saunders, Philadelphia, Pa, USA, 3rd edition, 1988.

[12] A. Skiada, L. Pagano, A. Groll et al., "Zygomycosis in Europe: analysis of 230 cases accrued by the registry of the European Confederation of Medical Mycology (ECMM) Working Group on Zygomycosis between 2005 and 2007," *Clinical Microbiology and Infection*. In press.

[13] L. Pagano, M. Caira, A. Candoni et al., "The epidemiology of fungal infections in patients with hematologic malignancies: the SEIFEM-2004 study," *Haematologica*, vol. 91, no. 8, pp. 1068–1075, 2006.

[14] A. Alastruey-Izquierdo, M. V. Castelli, I. Cuesta et al., "In vitro activity of antifungals against zygomycetes," *Clinical Microbiology and Infection*, vol. 15, no. 5, pp. 71–76, 2009.

[15] S. Perkhofer, M. Locher, M. Cuenca-Estrella et al., "Posaconazole enhances the activity of amphotericin B against hyphae of zygomycetes in vitro," *Antimicrobial Agents and Chemotherapy*, vol. 52, no. 7, pp. 2636–2638, 2008.

[16] B. Spellberg, Y. Fu, J. E. Edwards Jr., and A. S. Ibrahim, "Combination therapy with amphotericin B lipid complex and caspofungin acetate of disseminated zygomycosis in diabetic ketoacidotic mice," *Antimicrobial Agents and Chemotherapy*, vol. 49, no. 2, pp. 830–832, 2005.

[17] M. M. Rodríguez, C. Serena, M. Mariné, F. J. Pastor, and J. Guarro, "Posaconazole combined with amphotericin B, an effective therapy for a murine disseminated infection caused by Rhizopus oryzae," *Antimicrobial Agents and Chemotherapy*, vol. 52, no. 10, pp. 3786–3788, 2008.

[18] S. Saoulidis, M. Simitsopoulou, M. Dalakiouridou et al., "Antifungalactivity of posaconazole and granulocytecolony-stimulatingfactor in the treatment of disseminatedzygomycosis

(mucormycosis) in a neutropaenic murine model," *Mycoses*, vol. 54, pp. e486–e492, 2011.

[19] E. M. Uy, T. Rustagi, and S. Khera, "Cerebralmucormycosis in a diabetic man," *Connecticut Medicine*, vol. 75, pp. 273–279, 2011.

[20] N. Van Sickels, J. Hoffman, L. Stuke, and K. Kempe, "Survival of a patient with trauma-induced mucormycosis using an aggressive surgical and medical approach," *Journal of Trauma*, vol. 70, no. 2, pp. 507–509, 2011.

[21] J. Y. Ting, S. Y. Chan, D. C. Lung et al., "Intra-abdominal rhizopus microsporus infection successfully treated by combined aggressive surgical, antifungal, and iron chelating therapy," *Journal of Pediatric Hematology/Oncology*, vol. 32, no. 6, pp. e238–e240, 2010.

[22] O. Lebeau, C. Van Delden, J. Garbino et al., "Disseminated Rhizopus microsporus infection cured by salvage allogeneic hematopoietic stem cell transplantation, antifungal combination therapy, and surgical resection," *Transplant Infectious Disease*, vol. 12, no. 3, pp. 269–272, 2010.

[23] B. G. L. Roux, F. Méchinaud, F. Gay-Andrieu et al., "Successful triple combination therapy of disseminated Absidia corymbifera infection in an adolescent with osteosarcoma," *Journal of Pediatric Hematology/Oncology*, vol. 32, no. 2, pp. 131–133, 2010.

[24] A. S. Symeonidis, "The role of iron and iron chelators in zygomycosis," *Clinical Microbiology and Infection*, vol. 15, no. 5, supplement, pp. 26–32, 2009.

[25] A. Casadevall and L. A. Pirofski, "On virulence," *Virulence*, vol. 1, no. 1, pp. 1–2, 2010.

[26] R. Y. W. Chinn and R. D. Diamond, "Generation of chemotactic factors by Rhizopus oryzae in the presence and absence of serum: relationship to hyphal damage mediated by human neutrophils and effects of hyperglycemia and ketoacidosis," *Infection and Immunity*, vol. 38, no. 3, pp. 1123–1129, 1982.

[27] J. R. Boelaert, M. de Locht, J. Van Cutsem et al., "Mucormycosis during deferoxamine therapy is a siderophore-mediated infection: in vitro and in vivo animal studies," *Journal of Clinical Investigation*, vol. 91, no. 5, pp. 1979–1986, 1993.

[28] J. R. Boelaert, A. Z. Fenves, and J. W. Coburn, "Deferoxamine therapy and mucormycosis in dialysis patients: report of an international registry," *American Journal of Kidney Diseases*, vol. 18, no. 6, pp. 660–667, 1991.

[29] J. R. Boelaert, J. Van Cutsem, M. de Locht, Y. J. Schneider, and R. R. Crichton, "Deferoxamine augments growth and pathogenicity of Rhizopus, while hydroxypyridinone chelators have no effect," *Kidney International*, vol. 45, no. 3, pp. 667–671, 1994.

[30] A. S. Ibrahim, T. Gebremariam, L. Lin et al., "The high affinity iron permease is a key virulence factor required for Rhizopus oryzae pathogenesis," *Molecular Microbiology*, vol. 77, no. 3, pp. 587–604, 2010.

[31] J. L. Frater, G. S. Hall, and G. W. Procop, "Histologic features of zygomycosis: emphasis on perineural invasion and fungal morphology," *Archives of Pathology and Laboratory Medicine*, vol. 125, no. 3, pp. 375–378, 2001.

[32] A. S. Ibrahim, B. Spellberg, V. Avanessian, Y. Fu, and J. E. Edwards, "Rhizopus oryzae adheres to, is phagocytosed by, and damages endothelial cells in vitro," *Infection and Immunity*, vol. 73, no. 2, pp. 778–783, 2005.

[33] M. Liu, B. Spellberg, Q. T. Phan et al., "The endothelial cell receptor GRP78 is required for mucormycosis pathogenesis in diabetic mice," *Journal of Clinical Investigation*, vol. 120, no. 6, pp. 1914–1924, 2010.

[34] G. Chamilos, R. E. Lewis, and D. P. Kontoyiannis, "Multidrug-resistant endosymbiotic bacteria account for the emergence of zygomycosis: a hypothesis," *Fungal Genetics and Biology*, vol. 44, no. 2, pp. 88–92, 2007.

[35] L. P. Partida-Martinez and C. Hertweck, "Pathogenic fungus harbours endosymbiotic bacteria for toxin production," *Nature*, vol. 437, no. 7060, pp. 884–888, 2005.

[36] A. S. Ibrahim, T. Gebremariam, M. Liu et al., "Bacterial endosymbiosis is widely present among zygomycetes but does not contribute to the pathogenesis of mucormycosis," *Journal of Infectious Diseases*, vol. 198, no. 7, pp. 1083–1090, 2008.

[37] L. P. Partida-Martinez, S. Bandemer, R. Rüchel, E. Dannaoui, and C. Hertweck, "Lack of evidence of endosymbiotic toxin-producing bacteria in clinical *Rhizopus* isolates," *Mycoses*, vol. 51, no. 3, pp. 266–269, 2008.

[38] Z. Odabasi, V. L. Paetznick, J. R. Rodriguez, E. Chen, M. R. McGinnis, and L. Ostrosky-Zeichner, "Differences in beta-glucan levels in culture supernatants of a variety of fungi," *Medical Mycology*, vol. 44, no. 3, pp. 267–272, 2006.

[39] M. D. Richardson and P. Koukila-Kahkola, "Rhizopus, Rhizomucor, Absidia and other agents of systemic and subcutaneous zygomycoses," in *Manual of Clinical Microbiology*, P. A. Murray, E. J. Baron, J. H. Jorgensen, M. L. Landry, and M. A. Pfaller, Eds., pp. 1839–1856, ASM Press, Washington, DC, USA, 2007.

17

Communication in Fungi

Fabien Cottier[1,2] and Fritz A. Mühlschlegel[1,3]

[1] *School of Biosciences, University of Kent, Canterbury, Kent CT2 7NJ, UK*
[2] *Singapore Immunology Network, Agency for Science, Technology, and Research, Singapore 138648*
[3] *Clinical Microbiology Service, William Harvey Hospital, East Kent Hospitals University NHS Foundation Trust, Ashford, Kent TN24 0LZ, UK*

Correspondence should be addressed to Fritz A. Mühlschlegel, f.a.muhlschlegel@kent.ac.uk

Academic Editor: Julian R. Naglik

We will discuss fungal communication in the context of fundamental biological functions including mating, growth, morphogenesis, and the regulation of fungal virulence determinants. We will address intraspecies but also interkingdom signaling by systematically discussing the sender of the message, the molecular message, and receiver. Analyzing communication shows the close coevolution of fungi with organisms present in their environment giving insights into multispecies communication. A better understanding of the molecular mechanisms underlying microbial communication will promote our understanding of the "fungal communicome."

1. Introduction

Any form of communication requires the existence of three obligatory components: a sender, a message, and a receiver. The process starts with the release of a message by a sender and ends with the understanding of the message by a receiver. This type of cycle has been developed with different degrees of complexity from prokaryote to higher eukaryotes optimizing fitness and adaptation for individual members and populations. The nature and mode of action of communication is as diverse as the response to the information it carries. Inter- and intraspecies communication has been widely studied analyzing the exchange of information between fungi and bacteria or fungi and plant cells [1, 2]. This review will focus predominantly on intraspecies fungal communication addressing key biological functions including mating, growth, morphological switching, or the regulation of virulence factor expression (Figure 1). We will show that in the fungal kingdom most of these mechanisms are controlled by a variety of messengers including small peptides, alcohols, lipids, and volatile compounds.

2. Peptides: Pheromones

Pheromones have been known to act as an informative molecule since 1959 [3] and were reported to be involved in the sexual cycle of fungi in 1974 [4]. In the fungal kingdom, they are involved in the reconnaissance of compatible sexual partner to promote plasmogamy and karyogamy between two opposite mating types followed by meiosis. Taking the example of the extensively described sexual cycle of *Saccharomyces cerevisiae*, pheromones are diffusible peptides called a-factor (12 aa) when produced by a cells, and α-factor (13 aa) when produced by α cells. Each mating type responds to the opposite factor, and is able to produce only one of the two peptide pheromones depending on the alleles present at the *MAT* locus. Indeed, *MATa* or *MAT*α controls the expression of a and α specific genes, respectively, such as genes encoding the prepro-factor and the pheromone receptor (for a comprehensive description of the *MAT* locus, see review [5]). In the example of α cells, *MF*α*1* encodes the pheromone precursor, prepro-α-factor, which undergoes several proteolytic reactions in the classical secretory pathway before releasing the mature pheromone.

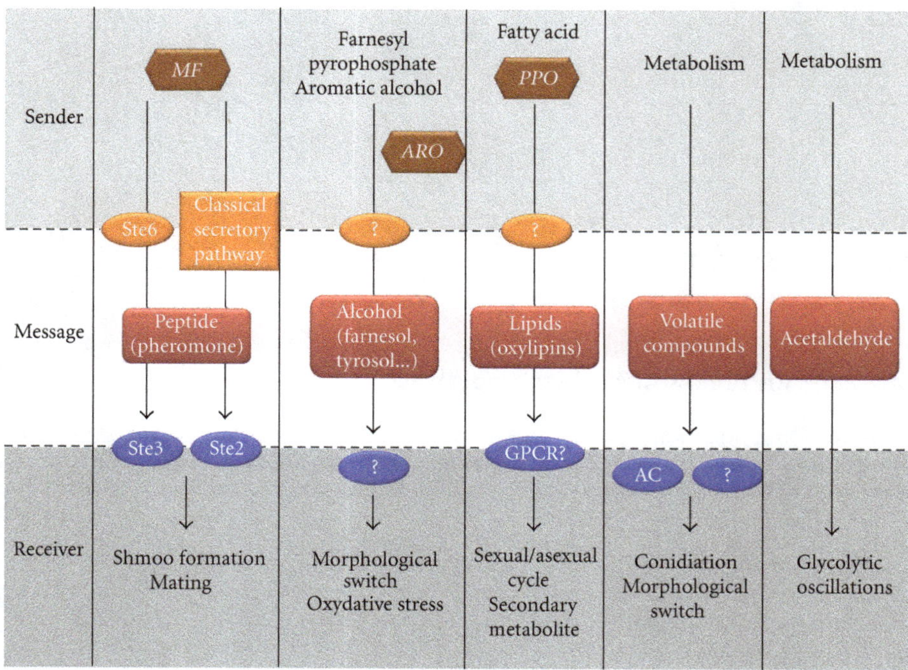

FIGURE 1: Schematic representation of fungal intra and interspecies communication. The "sender" is an organism from the fungal kingdom and the "receiver" can be from any kingdom. Genes involved in messenger synthesis are represented as brown hexagons. Proteins involved in secretion or receiving the message are in orange and blue.

Contrary to the α-factor, the ABC transporter Ste6p is required to secrete a-factor [6]. This difference could be due to the fact that a-factors are farnesylated [7]. Once released, pheromones freely diffuse in the environment and create a concentration gradient. These peptides are subsequently recognized by a 7 transmembrane receptor present on the surface of cells: Ste2p on a cells binds the α-factor and Ste3p on α cells binds the a-factor. Ste2p and Ste3p are G-protein coupled receptors (GPCR) and the binding of pheromone induces the separation of the associated heterotrimeric G-protein into a monomeric α subunit GTPase (Gpa1p) and a $\beta\gamma$ dimer (Ste4p–Ste18p). This mechanism results in the recruitment of Ste5p by Ste4p to the membrane, which activates a protein kinase cascade ultimately resulting in the phosphorylation of the MAP kinases Fus3p and Kss1p [8]. Once phosphorylated Fus3p migrates to the nucleus were it activates the transcriptional factor Ste12p leading to the expression of pheromone responsive genes. Phenotypically, the morphological response of cells to opposite mating pheromone is the development of a shmoo, that is, directional cell growth in response to the pheromone gradient. As each opposite cell develops a shmoo, the plasmogamy between the two cells occurs when the shmoos establish contact, starting the first step of the sexual cycle.

Contrary to *S. cerevisiae*, where karyogamy is followed by meiosis to end the sexual cycle, *Candida albicans* has not yet been described to undergo meiosis. This yeast displays only an imperfect sexual cycle, where karyogamy results in the formation of tetraploid cells that restore natural diploidy via random loss of chromosomes [9]. This process occurs only after *C. albicans* has undergone a so-called white-to-opaque switch (see review [10]), where opaque cells are the sole mating competent form of this yeast. The opaque cells morphologically respond to pheromone by producing a shmoo, like in *S. cerevisiae*, but the mating incompetent form, white cells, is also sensitive to pheromone [11]. Indeed, *C. albicans* α-factor but also a-factor promotes the formation of biofilm by white cells via enhancing their adhesiveness. A process which uses the same receptor (Ste2p or Ste3p) and transduction pathway as the response of opaque cells to pheromone [12]. The formation of fungal biomass by white cells facilitates the establishment of a pheromone gradient in a population of individual cells and assists the mating process of opaque cells. This process involves another molecule, farnesol, as the production of this molecule under aerobic conditions induces the death of the mating competent opaque cells. Anaerobic conditions that prevent production of farnesol facilitate mating between *C. albicans* opaque cells. These observations suggest that the gastrointestinal tract of humans could promote *C. albicans* mating [13]. The mechanism of pheromone communication has a broad significance in diverse fungi including ascomycetes like *Histoplasma capsulatum* [14] or *Aspergillus fumigatus* [15, 16], to basidiomycetes such as *Cryptococcus neoformans* [17] and *Ustilago maydis*, which possess a tetrapolar mating system [18]. Pheromone communication appears to be a critical mechanism for fungi as it supports the exchange of genetic material between cells and by extension the ability of the organism to evolve in response to their environment.

3. Alcohols: Quorum Sensing

Quorum sensing is a mechanism of communication based on the accumulation of a messenger molecule in the medium of culture [19]. As the production of messenger molecules increases with cell number, this system reflects population size. Initially discovered in bacteria, quorum sensing in fungi became relevant for the control of virulence factor expression in *C. albicans*. In 1979, Hazen and Cutler showed that the supernatant from a 48 h culture of *C. albicans* prevents the yeast to hyphae switch of a fresh culture [20]. The quorum sensing molecule (QSM) responsible for this effect has since been identified as an acyclic sesquiterpene alcohol called farnesol [21].

C. albicans produces farnesol at a rate of 0.12–0.133 mg/g of cells dry weight [22] from an intermediate of the mevalonate pathway (sterol biosynthesis), farnesyl pyrophosphate [23]. At concentrations of 10–250 μM farnesol inhibits the formation of hyphae when induced with proline, N-acetylglucosamine, and serum, but does not suppress further elongation of preexisting hyphae [24]. Farnesol-dependent quorum sensing involves the histidine kinase Chk1p [25] and the Ras1-Cyr1 pathway [26] but the receptor for farnesol remains to be identified. Farnesol regulates the expression of several genes and induces *TUP1*, a transcriptional cofactor repressing filamentation [27], while repressing *CPH1* and *HST7* expression, which are both activators of the morphological switch [28]. The oxidized form of farnesol, farnesoic acid, has also been reported to inhibit hyphal growth by acting via *PHO81* [29]. However, morphological inhibition is stronger with farnesol, although farnesoic acid is less toxic at high concentration [30], it displays only 3% of farnesols QSM activity [31]. While the function of farnesol as a cell density regulator remains to be established, farnesol has been described to inhibit *C. albicans* biofilm formation due to its repressing function on the morphological switch [32]. Additionally it has been shown to increase resistance to oxidative stress by suppressing the Ras1-cAMP pathway [33].

Notably, farnesol also acts as an interspecies QSM that impacts on growth of other *Candida* species including *Candida tropicalis* or *Candida parapsilosis* [34] as well as *S. cerevisiae* or the mould *Aspergillus nidulans* and *A. fumigatus* [35–38]. In the case of *A. fumigatus*, farnesol has been described to alter the localization of AfRho1p and AfRho3p, proteins involved in the cell wall integrity (CWI) pathway and cytoskeleton regulation [35]. This phenotype is explained by the fact that farnesyl derivatives interfere with prenylated proteins such as the two Rho GTPases [39, 40]. The CWI pathway implies the activation of AfPkcA by AfRho1p, which leads on to the MAP kinase cascade and subsequent AfMpkA phosphorylation. Dichtl et al. showed that in the presence of only 40 μM farnesol, phosphorylation of AfMpkA in response to Congo red was completely inhibited [35]. In *S. cerevisiae*, farnesol prevents growth via a different mechanism, which involves an increase of mitochondrial reactive oxygen species (ROS) [37]. The latter observation was also reported for *A. nidulans* where ROS augmentation induced cellular apoptosis but had no role on hyphal morphogenesis [38]. Two proteins have been identified in this response; the Gα subunit FadA of a heterotrimeric G protein, where hyperactivation leads to a strong increase in farnesol sensitivity [38], and the kinase PkcA. Mutation of PkcA increases resistance to farnesol while overexpression results in a higher rate of cell death in response to the QSM [41]. Finally, farnesol has also been described to induce apoptosis of cancerous cells *in vivo* (see review [42]), as well as increasing antibiotic sensitivity of *Staphylococcus aureus* [43]. Thus, farnesol appears to function as both an intraspecies and inter-species communication molecule.

Farnesol is not the only continuously released messenger molecule by *C. albicans*. Tyrosol, an aromatic alcohol, is produced from aromatic amino acids undergoing the processes of transamination (*ARO8, ARO9*), decarboxylation (*ARO10*), and reduction by alcohol dehydrogenase (*ADH*) [44]. This synthesis pathway is strongly dependant on growth conditions including environmental pH, availability of aromatic amino acids, oxygen levels, or presence of ammonium salts [44]. Similar to farnesol, tyrosol's sensor has not yet been identified. Fungal responses to tyrosol include the induction of germ tubes in planktonically growing cells and during the early stages of biofilm formation, as well as a reduction in the lag phase of *C. albicans* growth following dilution of a highly concentrated culture to fresh minimal medium [45, 46]. The latter phenotype occurs predominantly at low concentrations of cells (5×10^3 cell/mL) by promoting the expression of genes involved in DNA replication, chromosome segregation, and cell cycle processes [45].

Aromatic alcohol synthesis is not exclusive to *C. albicans* but can also be found in *S. cerevisiae*, which produces phenylethanol and tryptophol via a similar pathway involving AROs genes [47]. Both molecules stimulate diploid pseudohyphal growth at concentrations above 20 μM on low-ammonium agar (SLAD) by inducing the PKA pathway resulting in *FLO11* induction [48]. Recently, response to phenylethanol and tryptophol has been proposed to involve two main transcriptional regulators: Cat8p and Mig1p [49]. Interestingly, *C. albicans* is insensitive to phenylethanol and tryptophol [48]. *H. capsulatum* and *Ceratocystis ulmi* are two fungi also displaying quorum sensing phenotypes. However, the messenger molecule is not yet characterized [50, 51]. At low density, *H. capsulatum* cells have low amounts of α-(1,3)-glucan in their cell walls and addition of supernatant from a stationary phase culture induces α-(1,3)-glucan incorporation into the cell wall [50]. Similarly addition of *C. ulmi* spent medium to a fresh culture promotes a switch from hyphae to yeast growth [51].

4. Lipids: Oxylipins

Oxylipins are oxygenated fatty acids used as cell messengers and have been intensely studied in plants and mammalian cells (see review [52]). They also appear to be widely synthesized and secreted by fungi. *A. nidulans* was reported to produce one of the first oxylipins called psi factor (precocious sexual inducer), which is composed of a series of different oxylipin derivates from oleic acid (C18:1), linoleic acid (C18:2), and linolenic acid (C18:3) called psiA, psiB, and psiC, respectively [53]. The genes involved in the production of psi factor are called *Ppos* (for psi-producing oxygenases) [54]. In the case of *A. nidulans*, *PpoA* is involved in psiBα synthesis and *PpoB* and *PpoC* contribute to psiBβ biogenesis [54, 55]. Inactivation of these genes results in perturbations not only of psi factor production but also mycotoxins production, as well as in the ratio between the development of sexual and asexual ascospores [54, 55]. The latter phenotype is due to the fact that oxylipins control the expression of *NsdD* and *BrlA*, transcription factors required for meiotic and mitotic sporulation, respectively [54, 56]. Overexpression or addition of psiBα or psiCα to the culture medium stimulates sexual sporulation and represses asexual spore development while an opposite effect is observed for psiAα and psiBβ [55]. Secondary metabolite mycotoxin sterigmatocystin (ST) and antibiotic penicillin (PN) production are also dependent on oxylipin [57]. Indeed, inactivation of the *Ppos* genes results in the inability to secrete ST as a result of downregulation of ST biosynthesis genes including *aflR* and *stcU*, and the overproduction of PN through induction of the gene involved in its biosynthesis: *ipnA* [57]. Interestingly, the exact opposite observations are found when FadA, an α subunit of a GPCR, is constitutively activated due to a G42R mutation, a reaction which is mediated by the PkaA enzyme [58]. This result, in addition to the fact that oxylipins in mammalian cells are sensed via GPCR complexes [59] led to the hypothesis that fungi use the same system to detect oxylipin, ultimately activating the cAMP/PKA pathway [58]. The *ppo* and GPCR encoding genes have been identified in the genomes of several filamentous fungi predicting a broader role of oxylipin in fungal biology [60]. In fact, inactivation of the *ppo* genes in *Aspergillus flavus* and *Fusarium sporotrichioides* has already been shown to perturb mycotoxin and spore production [61, 62]. Finally, using confocal laser scanning microscopy, oxylipins have been described to accumulate in the capsule of *C. neoformans* before being released into the external medium under the form of hydrophobic droplets that are transported via tubular protuberances [63].

Recently, Nigam et al. have described a 3(*R*)-Hydroxy-tetradecanoic acid, a derivate of linoleic acid as a novel QSM of *C. albicans* [64]. Previously known to be produced during the sexual phase of *Dipodascopsis uninucleata* [65], this oxylipin increases filamentation in *C. albicans* in response to N-acetylglucosamine at a concentration of 1 μM. Although the receptor of 3(*R*)-Hydroxy-tetradecanoic is not known, this QSM induces *HWP1* and *CAP1* mRNA transcripts [64]. Interestingly, 3(*R*)-Hydroxy-tetradecanoic is metabolized inside cells to generate two new compounds that could also act as messenger molecules [64].

Another family of oxylipin are the eicosanoids, which are molecules containing a 20 carbon backbone [66]. PGE_2 is produced by *C. albicans* from exogenous arachidonic acid via enzymes not yet characterized [67]. PGE_2 is also produced by humans, similar to other prostaglandins (PG), and it appears that fungal PGE_2 can enter in competition with human PG impacting on the host's immune response [67]. Indeed, PGE_2 is known to balance Th1/Th2 differentiation as this molecule decreases the expression of IL-12R and inactivates Th1 differentiation while activating the Th2-related immune responses [68]. PGE_2 also enhances the production of IgE in stimulated B cells [69].

5. Volatile Compounds and Gas

In addition to releasing mediators into solution or onto solid growth media, organisms also exchange information via the liberation of messenger molecules into air. For example, insects have been thoroughly studied for their secretion of pheromones into air to attract mating partners [70]. In the fungal kingdom, as early as in the 1970s, volatile compounds from fungi and others organism have been described to impact on fungal growth (review [71, 72]). More recently, Palkova et al. observed that *S. cerevisiae* colonies grown on complex agar form a turbid path in the vicinity of another colony. Subsequently, they discovered that this reaction is induced by the small volatile messenger molecule, later described as ammonia [73], which also required amino-acid uptake for its production. Indeed, inactivation of *SHR3*, which is responsible for the correct localization of several amino-acid permeases, disrupts the turbid path between colonies [73].

Trichoderma species have been described to produce the volatile molecule 6-Pentyl-α-pyrone, a secondary metabolite with antifungal activity [74]. However, more recently the induction of conidiation in *Trichoderma* species, which is known to be regulated by a circadian cycle, has also been shown to be controlled via a volatile agent. Solid-phase microextraction linked with gas chromatography and mass spectrometry has allowed the identification of the chemical profiles of volatile molecules produced from nonconidiated colonies grown in darkness and conidiating colonies grown in light [75]. Comparison of the two profiles identified production of the 8-carbon compounds molecules 1-octen-3-ol, 3-octanol and 3-octanone specifically during conidiation [75]. Each of these three compounds induces conidiation in colonies placed in the dark. This regulation could involve a calcium-dependant signaling pathway as it has been shown that high concentration of calcium can induce conidiation of *Penicillium isariaeform* in darkness [76]. 1-octen-3-ol is the most efficient molecule being active at concentrations of only 0.1 μM. Interestingly, concentrations above 500 μM of any of the three compounds suppress conidiation and growth of *Trichoderma* species. These observations are consistent with a previously described putative fungistatic and fungicidal role of the molecules [77, 78]. Notably, the same compounds have previously been shown to function as insect attractants improving fungal spore dispersal [77, 78], and inter-species communication has already been described between *Epichloë* species and the female *Botanophila* flies [79].

Fungi are not only responsive to volatile compounds that they produce but also, in at least one example, to a gas liberated during respiration: carbon dioxide (CO_2). As early as 1961, Vakil et al. demonstrated that the optimum CO_2 concentration for the germination of *Aspergillus niger* conidiospores is reached not under normal atmospheric concentrations of CO_2 (0.033%) but at 0.5% [80]. Since then several additional phenotypes in fungi have been attributed to changes in the concentration of environmental CO_2 including the sporulation of *Alternaria crassa* and *Alternaria cassiae* [81], conidiation of *Neurospora crassa* [82], or capsule formation and mating in *C. neoformans* [83, 84].

Recently, significant advances have been made in the understanding of CO_2 sensing in fungi. It was already known that the yeast to hyphae morphological switch in *C. albicans* is triggered by elevated environmental CO_2 [85]. Furthermore, the frequency of white-to-opaque switching can be increased 16-fold in hypercapnic conditions as opposed to atmospheric CO_2 [86]. Two different studies show that both phenotypes involve the *C. albicans* adenylyl cyclase Cyr1, first fungal CO_2 sensor. This enzyme generates the secondary messenger cAMP, which in the context of the cAMP/PKA signaling pathway has a fundamental impact on *C. albicans* morphogenesis [86, 87]. CO_2 activation of Cyr1p depends on the concentration of bicarbonate, the hydrated form of CO_2 [87]. CO_2 hydration occurs naturally at a very low rate, but is enhanced by the enzyme carbonic anhydrase [88]. Inactivation of *CYR1* results in a loss of filamentation and white to opaque switching frequency in response to hypercapnia [86, 87]. Hall et al. have now demonstrated that Lysine 1373 of the Cyr1 catalytic domain is essential for CO_2 sensing in *C. albicans* as mutation of this amino acid leads to a loss of filamentation in response to CO_2 but not to serum, another morphological cue [89]. These data show that in fungi environmental CO_2 is sensed via the adenylyl cyclase, which transduces the message via the regulation of the cAMP/PKA pathway. Hall et al. also showed that hypercapnia is not a condition solely encountered inside the host but can also establish itself as a population event, such as the center of a colony grown under normal atmospheric conditions [89]. Another study demonstrated that *C. albicans* produces CO_2 via the conversion of arginine to urea. Urea is ultimately degraded to generate CO_2 by the enzyme urea amidolyase (Dur1,2). Inactivation of the latter interferes with *C. albicans* filamentation in response to arginine and urea compared to the control strain but not to elevated CO_2 [90].

Control of the white to opaque switch-frequency in *C. albicans* by environmental CO_2 also involves the GTPAse Ras1 and the transcriptional factor Wor1. Indeed, Ras1, Cyr1, and Wor1 are critical for increasing the white to opaque switch in response to concentrations of CO_2 at 1%, but Ras1 and Cyr1 become optional for the induction at higher concentrations (20%). However, Wor1 remains essential for the switch even at high CO_2 [86]. These results imply that an alternative CO_2 sensing pathway is involved in the regulation of Wor1 at high CO_2 in *C. albicans*. However, it is important to note that under this condition a significant increase of the internal pH may occur which could also be a component of this alternative CO_2 sensing pathway.

6. Small Molecule: Acetaldehyde

Acetaldehyde, an organic compound involved in several cellular pathways, has been shown to impact on cell-density-dependent glycolytic oscillations of *S. cerevisiae* [91]. In 1964, Chance et al. described that the level of NADH in yeast oscillated when starved cells endure a pulse of glucose after a switch to anaerobic conditions [92]. Since then other metabolites have been described to oscillate in yeast including glucose-6-phosphate, fructose-6-phosphate, fructose-1,6-biphosphate, AMP, ADP, and ATP (for a comprehensive review see [93]). Interestingly, at a population level these oscillations are not chaotic but appear to be subject to synchronization. The most striking observation was achieved when mixing two populations with a 180° out-of-phase oscillation showing that within minutes the oscillation of the new population were synchronized [91]. Acetaldehyde was identified as the active molecule in the synchronization of these oscillations, as the use of acetaldehyde traps induced the oscillation to be damped and addition of acetaldehyde to the medium produced a phase shift in the oscillation [91]. Acetaldehyde is a small molecule that can passively diffuse through the cell membrane. No specific target for acetaldehyde is known; however, this compound has an important impact on the NAD^+/NADH balance [94].

Acetaldehyde is also a volatile molecule, a property used to study inter and intraspecies communication in a synthetic ecosystem [95]. By engineering sender cells that liberate volatile acetaldehyde and receiver cells that contain a construct under an acetaldehyde-inducible promoter, it was possible to study volatile cell communication in a controlled environment. Using mammalian (CHO-K1), bacterial (*Escherichia coli*), yeast (*S. cerevisiae*), or plant (*Lepidium sativum*) cells, all combination of sender/receiver for inter and intraspecies resulted in a positive communication between cells [95]. These results show that virtually all cells can communicate with themselves or different species. Clearly such models could bring new insight in the understanding of communication in complex living systems.

7. Concluding Remarks and Outlooks

We are currently at an interesting stage in the understanding of fungal communication. Many essential compounds of the communication process have been identified, the sender (in our case fungi), the message (protein, alcohol, lipid, gas), and the receiver (bacteria, fungi, plant, mammalian). However, in most cases the actual molecular mechanism of such communication remains for most parts unknown. The determination of these pathways is of substantial significance as molecular messengers control the expression of fungal virulence determinants including the yeast-to-hyphae switch and biofilm formation in *C. albicans*, capsule formation in *C. neoformans*, or mycotoxin synthesis in *A. nidulans*, but also the propagation of these organisms via the regulation of their sexual and asexual cycle. A better knowledge of fungal communication is now required to permit the development of innovative strategies aiming to control disease or toxin production of these organisms.

Fungi have already taken advantage of the different communication processes and particularly inter-species communication to gain competitive advantages over other species. Good examples are the production of pollinators attracting insects to give phytopathogenic fungi a better chance for dispersal of their spores [79]. Additionally, synthesis of PGE_2 by the human pathogens *C. albicans* and *C. neoformans* modify the host immune response and may enhance fungal survival [67]. Such mechanisms reveal the close coevolution of fungi with their environmental partner and give insights into multispecies communication. The remarkable versatility of communication in the fungal kingdom also raises the question how these organisms integrate intra- and inter-species messages that can have opposing effects. As the molecular mechanisms of fungal communication unravel further, they will promote our understanding of the highly attractive but challenging topic of the fungal "communicome."

References

[1] M. T. Tarkka, A. Sarniguet, and P. Frey-Klett, "Inter-kingdom encounters: recent advances in molecular bacterium-fungus interactions," *Current Genetics*, vol. 55, no. 3, pp. 233–243, 2009.

[2] S. A. Christensen and M. V. Kolomiets, "The lipid language of plant-fungal interactions," *Fungal Genetics and Biology*, vol. 48, no. 1, pp. 4–14, 2011.

[3] P. Karlson and M. Lüscher, "'Pheromones': a new term for a class of biologically active substances," *Nature*, vol. 183, no. 4653, pp. 55–56, 1959.

[4] L. M. Hereford and L. H. Hartwell, "Sequential gene function in the initiation of Saccharomyces cerevisiae DNA synthesis," *Journal of Molecular Biology*, vol. 84, no. 3, pp. 445–461, 1974.

[5] J. E. Haber, "Mating-type gene switching in Saccharomyces cerevisiae," *Annual Review of Genetics*, vol. 32, pp. 561–599, 1998.

[6] J. P. McGrath and A. Varshavsky, "The yeast STE6 gene encodes a homologue of the mammalian multidrug resistance P-glycoprotein," *Nature*, vol. 340, no. 6232, pp. 400–404, 1989.

[7] R. J. Anderegg, J. M. Becker, Y. Jiang et al., "Structure of Saccharomyces cerevisiae mating hormone a-factor. Identification of S-farnesyl cysteine as a structural component," *Journal of Biological Chemistry*, vol. 263, no. 34, pp. 18236–18240, 1988.

[8] A. Gartner, K. Nasmyth, and G. Ammerer, "Signal transduction in Saccharomyces cerevisiae requires tyrosine and threonine phosphorylation of FUS3 and KSS1," *Genes and Development*, vol. 6, no. 7, pp. 1280–1292, 1992.

[9] R. J. Bennett and A. D. Johnson, "Completion of a parasexual cycle in Candida albicans by induced chromosome loss in tetraploid strains," *EMBO Journal*, vol. 22, no. 10, pp. 2505–2515, 2003.

[10] D. R. Soll, "Why does Candida albicans switch?" *FEMS Yeast Research*, vol. 9, no. 7, pp. 973–989, 2009.

[11] K. J. Daniels, T. Srikantha, S. R. Lockhart, C. Pujol, and D. R. Soll, "Opaque cells signal white cells to form biofilms in Candida albicans," *EMBO Journal*, vol. 25, no. 10, pp. 2240–2252, 2006.

[12] S. Yi, N. Sahni, K. J. Daniels et al., "Self-induction of a/a or alpha/alpha biofilms in candida albicans is a pheromone-based paracrine system requiring switching," *Eukaryot Cell*, vol. 10, no. 6, pp. 753–760, 2011.

[13] R. Dumitru, D. H. M. L. P. Navarathna, C. P. Semighini et al., "In vivo and in vitro anaerobic mating in Candida albicans," *Eukaryotic Cell*, vol. 6, no. 3, pp. 465–472, 2007.

[14] M. Bubnick and A. G. Smulian, "The MAT1 locus of Histoplasma capsulatum is responsive in a mating type-specific manner," *Eukaryotic Cell*, vol. 6, no. 4, pp. 616–621, 2007.

[15] S. Poggeler, "Genomic evidence for mating abilities in the asexual pathogen Aspergillus fumigates," *Current Genetics*, vol. 42, no. 3, pp. 153–160, 2002.

[16] M. Paoletti, C. Rydholm, E. U. Schwier et al., "Evidence for sexuality in the opportunistic fungal pathogen Aspergillus fumigates," *Current Biology*, vol. 15, no. 13, pp. 1242–1248, 2005.

[17] J. Heitman, B. Allen, J. A. Alspaugh, and K. J. Kwon-Chung, "On the origins of congenic MATα and MATa strains of the pathogenic yeast Cryptococcus neoformans," *Fungal Genetics and Biology*, vol. 28, no. 1, pp. 1–5, 1999.

[18] G. Bakkeren, J. Kämper, and J. Schirawski, "Sex in smut fungi: structure, function and evolution of mating-type complexes," *Fungal Genetics and Biology*, vol. 45, supplement 1, pp. S15–S21, 2008.

[19] C. M. Waters and B. L. Bassler, "Quorum sensing: cell-to-cell communication in bacteria," *Annual Review of Cell and Developmental Biology*, vol. 21, pp. 319–346, 2005.

[20] K. C. Hazen and J. E. Cutler, "Autoregulation of germ tube formation by Candida albicans," *Infection and Immunity*, vol. 24, no. 3, pp. 661–666, 1979.

[21] J. M. Hornby, E. C. Jensen, A. D. Lisec et al., "Quorum sensing in the dimorphic fungus Candida albicans is mediated by farnesol," *Applied and Environmental Microbiology*, vol. 67, no. 7, pp. 298229–298292, 2001.

[22] J. M. Hornby and K. W. Nickerson, "Enhanced production of farnesol by Candida albicans treated with four azoles," *Antimicrobial Agents and Chemotherapy*, vol. 48, no. 6, pp. 2305–2307, 2004.

[23] J. M. Hornby, B. W. Kebaara, and K. W. Nickerson, "Farnesol biosynthesis in Candida albicans: cellular response to sterol inhibition by zaragozic acid B," *Antimicrobial Agents and Chemotherapy*, vol. 47, no. 7, pp. 2366–2369, 2003.

[24] D. D. Mosel, R. Dumitru, J. M. Hornby, A. L. Atkin, and K. W. Nickerson, "Farnesol concentrations required to block germ tube formation in Candida albicans in the presence and absence of serum," *Applied and Environmental Microbiology*, vol. 71, no. 8, pp. 4938–4940, 2005.

[25] M. Kruppa, B. P. Krom, N. Chauhan, A. V. Bambach, R. L. Cihlar, and R. A. Calderone, "The two-component signal transduction protein Chk1p regulates quorum sensing in Candida albicans," *Eukaryotic Cell*, vol. 3, no. 4, pp. 1062–1065, 2004.

[26] A. Davis-Hanna, A. E. Piispanen, L. I. Stateva, and D. A. Hogan, "Farnesol and dodecanol effects on the Candida albicans Ras1-cAMP signalling pathway and the regulation of morphogenesis," *Molecular Microbiology*, vol. 67, no. 1, pp. 47–62, 2008.

[27] B. W. Kebaara, M. L. Langford, D. H. M. L. P. Navarathna, R. Dumitru, K. W. Nickerson, and A. L. Atkin, "Candida albicans Tup1 is involved in farnesol-mediated inhibition of filamentous-growth induction," *Eukaryotic Cell*, vol. 7, no. 6, pp. 980–987, 2008.

[28] T. Sato, T. Watanabe, T. Mikami, and T. Matsumoto, "Farnesol, a morphogenetic autoregulatory substance in the dimorphic fungus Candida albicans, inhibits hyphae growth through suppression of a mitogen-activated protein kinase cascade," *Biological and Pharmaceutical Bulletin*, vol. 27, no. 5, pp. 751–752, 2004.

[29] S. C. Chung, T. I. Kim, C. H. Ahn, J. Shin, and K. B. Oh, "Candida albicans PHO81 is required for the inhibition of

hyphal development by farnesoic acid," *FEBS Letters*, vol. 584, no. 22, pp. 4639–4645, 2010.

[30] S. Kim, E. Kim, D. S. Shin, H. Kang, and K. B. Oh, "Evaluation of morphogenic regulatory activity of farnesoic acid and its derivatives against Candida albicans dimorphism," *Bioorganic and Medicinal Chemistry Letters*, vol. 12, no. 6, pp. 895–898, 2002.

[31] R. Shchepin, J. M. Hornby, E. Burger, T. Niessen, P. Dussault, and K. W. Nickerson, "Quorum sensing in Candida albicans: probing farnesol's mode of action with 40 natural and synthetic farnesol analogs," *Chemistry and Biology*, vol. 10, no. 8, pp. 743–750, 2003.

[32] G. Ramage, S. P. Saville, B. L. Wickes, and J. L. López-Ribot, "Inhibition of Candida albicans biofilm formation by farnesol, a quorum-sensing molecule," *Applied and Environmental Microbiology*, vol. 68, no. 11, pp. 5459–5463, 2002.

[33] A. Deveau, A. E. Piispanen, A. A. Jackson, and D. A. Hogan, "Farnesol induces hydrogen peroxide resistance in Candida albicans yeast by inhibiting the Ras-cyclic AMP signaling pathway," *Eukaryotic Cell*, vol. 9, no. 4, pp. 569–577, 2010.

[34] K. Weber, B. Schulz, and M. Ruhnke, "The quorum-sensing molecule E, E-farnesol—its variable secretion and its impact on the growth and metabolism of Candida species," *Yeast*, vol. 27, no. 9, pp. 727–739, 2010.

[35] K. Dichtl, F. Ebel, F. Dirr, F. H. Routier, J. Heesemann, and J. Wagener, "Farnesol misplaces tip-localized Rho proteins and inhibits cell wall integrity signalling in Aspergillus fumigates," *Molecular Microbiology*, vol. 76, no. 5, pp. 1191–1204, 2010.

[36] K. Machida, T. Tanaka, Y. Yano, S. Otani, and M. Taniguchi, "Farnesol-induced growth inhibition in Saccharomyces cerevisiae by a cell cycle mechanism," *Microbiology*, vol. 145, no. 2, pp. 293–299, 1999.

[37] K. Machida, T. Tanaka, K. I. Fujita, and M. Taniguchi, "Farnesol-induced generation of reactive oxygen species via indirect inhibition of the mitochondrial electron transport chain in the yeast Saccharomyces cerevisiae," *Journal of Bacteriology*, vol. 180, no. 17, pp. 4460–4465, 1998.

[38] C. P. Semighini, J. M. Hornby, R. Dumitru, K. W. Nickerson, and S. D. Harris, "Farnesol-induced apoptosis in Aspergillus nidulans reveals a possible mechanism for antagonistic interactions between fungi," *Molecular Microbiology*, vol. 59, no. 3, pp. 753–764, 2006.

[39] C. Volker, R. A. Miller, W. R. McCleary et al., "Effects of farnesylcysteine analogs on protein carboxyl methylation and signal transduction," *Journal of Biological Chemistry*, vol. 266, no. 32, pp. 21515–21522, 1991.

[40] A. Dietrich, A. Scheer, D. Illenberger, Y. Kloog, Y. I. Henis, and P. Gierschik, "Studies on G-protein $\alpha \cdot \beta\gamma$ heterotrimer formation reveal a putative S-prenyl-binding site in the α subunit," *Biochemical Journal*, vol. 376, no. 2, pp. 449–456, 2003.

[41] A. C. Colabardini, P. A. De Castro, P. F. De Gouvêa et al., "Involvement of the Aspergillus nidulans protein kinase C with farnesol tolerance is related to the unfolded protein response," *Molecular Microbiology*, vol. 78, no. 5, pp. 1259–1279, 2010.

[42] J. H. Joo and A. M. Jetten, "Molecular mechanisms involved in farnesol-induced apoptosis," *Cancer Letters*, vol. 287, no. 2, pp. 123–135, 2010.

[43] M. A. Jabra-Rizk, T. F. Meiller, C. E. James, and M. E. Shirtliff, "Effect of farnesol on Staphylococcus aureus biofilm formation and antimicrobial susceptibility," *Antimicrobial Agents and Chemotherapy*, vol. 50, no. 4, pp. 1463–1469, 2006.

[44] S. Ghosh, B. W. Kebaara, A. L. Atkin, and K. W. Nickerson, "Regulation of aromatic alcohol production in Candida albicans," *Applied and Environmental Microbiology*, vol. 74, no. 23, pp. 7211–7218, 2008.

[45] H. Chen, M. Fujita, Q. Feng, J. Clardy, and G. R. Fink, "Tyrosol is a quorum-sensing molecule in Candida albicans," *Proceedings of the National Academy of Sciences of the United States of America*, vol. 101, no. 14, pp. 5048–5052, 2004.

[46] M. A. Alem, M. D. Oteef, T. H. Flowers, and L. J. Douglas, "Production of tyrosol by Candida albicans biofilms and its role in quorum sensing and biofilm development," *Eukaryotic Cell*, vol. 5, no. 10, pp. 1770–1779, 2006.

[47] L. A. Hazelwood, J. M. Daran, A. J. A. Van Maris, J. T. Pronk, and J. R. Dickinson, "The Ehrlich pathway for fusel alcohol production: a century of research on Saccharomyces cerevisiae metabolism," *Applied and Environmental Microbiology*, vol. 74, no. 8, pp. 2259–2266, 2008.

[48] H. Chen and G. R. Fink, "Feedback control of morphogenesis in fungi by aromatic alcohols," *Genes and Development*, vol. 20, no. 9, pp. 1150–1161, 2006.

[49] A. Wuster and M. M. Babu, "Transcriptional control of the quorum sensing response in yeast," *Molecular BioSystems*, vol. 6, no. 1, pp. 134–141, 2009.

[50] S. Kügler, T. S. Sebghati, L. G. Eissenberg, and W. E. Goldman, "Phenotypic variation and intracellular parasitism by Histoplasma capsulatum," *Proceedings of the National Academy of Sciences of the United States of America*, vol. 97, no. 16, pp. 8794–8798, 2000.

[51] J. M. Hornby, S. M. Jacobitz-Kizzier, D. J. McNeel, E. C. Jensen, D. S. Treves, and K. W. Nickerson, "Inoculum size effect in dimorphic fungi: extracellular control of yeast-mycelium dimorphism in Ceratocystis ulmi," *Applied and Environmental Microbiology*, vol. 70, no. 3, pp. 1356–1359, 2004.

[52] A. Mosblech, I. Feussner, and I. Heilmann, "Oxylipins: structurally diverse metabolites from fatty acid oxidation," *Plant Physiology and Biochemistry*, vol. 47, no. 6, pp. 511–517, 2009.

[53] A. M. Calvo, L. L. Hinze, H. W. Gardner, and N. P. Keller, "Sporogenic effect of polyunsaturated fatty acids on development of Aspergillus spp," *Applied and Environmental Microbiology*, vol. 65, no. 8, pp. 3668–3673, 1999.

[54] D. I. Tsitsigiannis, T. M. Kowieski, R. Zarnowski, and N. P. Keller, "Three putative oxylipin biosynthetic genes integrate sexual and asexual development in Aspergillus nidulans," *Microbiology*, vol. 151, no. 6, pp. 1809–1821, 2005.

[55] D. I. Tsitsigiannis, R. Zarnowski, and N. P. Keller, "The lipid body protein, PpoA, coordinates sexual and asexual sporulation in Aspergillus nidulans," *Journal of Biological Chemistry*, vol. 279, no. 12, pp. 11344–11353, 2004.

[56] D. I. Tsitsigiannis, T. M. Kowieski, R. Zarnowski, and N. P. Keller, "Endogenous lipogenic regulators of spore balance in Aspergillus nidulans," *Eukaryotic Cell*, vol. 3, no. 6, pp. 1398–1411, 2004.

[57] D. I. Tsitsigiannis and N. P. Keller, "Oxylipins act as determinants of natural product biosynthesis and seed colonization in Aspergillus nidulans," *Molecular Microbiology*, vol. 59, no. 3, pp. 882–892, 2006.

[58] A. Tag, J. Hicks, G. Garifullina et al., "G-protein signalling mediates differential production of toxic secondary metabolites," *Molecular Microbiology*, vol. 38, no. 3, pp. 658–665, 2000.

[59] T. Shimizu, "Lipid mediators in health and disease: enzymes and receptors as therapeutic targets for the regulation of immunity and inflammation," *Annual Review of Pharmacology and Toxicology*, vol. 49, pp. 123–150, 2009.

[60] J. H. Yu and N. Keller, "Regulation of secondary metabolism in filamentous fungi," *Annual Review of Phytopathology*, vol. 43, pp. 437–458, 2005.

[61] S. H. Brown, J. B. Scott, J. Bhaheetharan et al., "Oxygenase coordination is required for morphological transition and the host-fungus interaction of Aspergillus flavus," *Molecular Plant-Microbe Interactions*, vol. 22, no. 7, pp. 882–894, 2009.

[62] T. E. A. McDonald, "Signaling events connecting mycotoxin biosynthesis and sporulation in Aspergillus and Fusarium spp. Takamatsu," in *Proceedings of the International Symposium of Mycotoxicology*, Co. Bookish, Ed., New Horizon of Mycotoxicology for Assuring Food Safety, 2004.

[63] O. M. Sebolai, C. H. Pohl, P. J. Botes et al., "3-hydroxy fatty acids found in capsules of Cryptococcus neoformans," *Canadian Journal of Microbiology*, vol. 53, no. 6, pp. 809–812, 2007.

[64] S. Nigam, R. Ciccoli, I. Ivanov, M. Sczepanski, and R. Deva, "On mechanism of quorum sensing in Candida albicans by 3(R)-hydroxy-tetradecaenoic acid," *Current Microbiology*, vol. 62, no. 1, pp. 55–63, 2011.

[65] P. Venter, J. L. F. Kock, G. Sravan Kumar et al., "Production of 3R-hydroxy-polyenoic fatty acids by the yeast Dipodascopsis uninucleata," *Lipids*, vol. 32, no. 12, pp. 1277–1283, 1997.

[66] G. Pohnert, "Phospholipase A2 activity triggers the wound-activated chemical defense in the diatom Thalassiosira rotula," *Plant Physiology*, vol. 129, no. 1, pp. 103–111, 2002.

[67] J. R. Erb-Downward and M. C. Noverr, "Characterization of prostaglandin E2 production by Candida albicans," *Infection and Immunity*, vol. 75, no. 7, pp. 3498–3505, 2007.

[68] Y. Shibata, R. A. Henriksen, I. Honda, R. M. Nakamura, and Q. N. Myrvik, "Splenic PGE2-releasing macrophages regulate Th1 and Th2 immune responses in mice treated with heat-killed BCG," *Journal of Leukocyte Biology*, vol. 78, no. 6, pp. 1281–1290, 2005.

[69] E. R. Fedyk and R. P. Phipps, "Prostaglandin E2 receptors of the EP2 and EP4 subtypes regulate activation and differentiation of mouse B lymphocytes to IgE-secreting cells," *Proceedings of the National Academy of Sciences of the United States of America*, vol. 93, no. 20, pp. 10978–10983, 1996.

[70] M. Ayasse, R. J. Paxton, and J. Tengö, "Mating behavior and chemical communication in the order Hymenoptera," *Annual Review of Entomology*, vol. 46, pp. 31–78, 2001.

[71] N. Fries, "Effects of volatile organic compounds on the growth and development of fungi," *Transactions of the British Mycological Society*, vol. 60, pp. 1–21, 1973.

[72] S. A. Hutchinson, "Biological activities of volatile fungal metabolites," *Annual Reviews of Phytopathology*, vol. 11, pp. 223–246, 1973.

[73] Z. Palkova, B. Janderova, J. Gabriel, B. Zikanova, M. Pospisek, and J. Forstova, "Ammonia mediates communication between yeast colonies," *Nature*, vol. 390, no. 6659, pp. 532–536, 1997.

[74] J. M. Cooney and D. R. Lauren, "Trichoderma/pathogen interactions: measurement of antagonistic chemicals produced at the antagonist/pathogen interface using a tubular bioassay," *Letters in Applied Microbiology*, vol. 27, no. 5, pp. 283–286, 1998.

[75] M. Nemcovic, L. Jakubikova, I. Viden, and V. Farkas, "Induction of conidiation by endogenous volatile compounds in Trichoderma spp," *FEMS Microbiology Letters*, vol. 284, no. 2, pp. 231–236, 2008.

[76] G. Muthukumar, E. C. Jensen, A. W. Nickerson, M. K. Eckles, and K. W. Nickerson, "Photomorphogenesis in Penicillium isariaeforme: exogenous calcium substitytes for light," *Photochemistry and Photobiology*, vol. 53, no. 2, pp. 297–291, 1991.

[77] G. S. Chitarra, T. Abee, F. M. Rombouts, M. A. Posthumus, and J. Dijksterhuis, "Germination of penicillium paneum Conidia is regulated by 1-octen-3-ol, a volatile self-inhibitor," *Applied and Environmental Microbiology*, vol. 70, no. 5, pp. 2823–2829, 2004.

[78] G. S. Chitarra, T. Abee, F. M. Rombouts, and J. Dijksterhuis, "1-Octen-3-ol inhibits conidia germination of Penicillium paneum despite of mild effects on membrane permeability, respiration, intracellular pH, and changes the protein composition," *FEMS Microbiology Ecology*, vol. 54, no. 1, pp. 67–75, 2005.

[79] F. P. Schiestl, F. Steinebrunner, C. Schulz et al., "Evolution of "pollinator"-attracting signals in fungi," *Biology Letters*, vol. 2, no. 3, pp. 401–404, 2006.

[80] J. R. Vakil, M. R. Raghavendra Rao, and P. K. Bhattacharyya, "Effect of CO2 on the germination of conidiospores of Aspergillus niger," *Archiv für Mikrobiologie*, vol. 39, no. 1, pp. 53–57, 1961.

[81] M. G. Smart, K. M. Howard, and R. J. Bothast, "Effect of carbon dioxide on sporulation of Alternaria crassa and Alternaria cassia," *Mycopathologia*, vol. 118, no. 3, pp. 167–171, 1992.

[82] S. Park and K. Lee, "Inverted race tube assay for circadian clock studies of the neurospora accessions," *Fungal Genetics Newsletter*, vol. 51, pp. 12–14, 2004.

[83] D. L. Granger, J. R. Perfect, and D. T. Durack, "Virulence of Cryptococcus neoformans. Regulation of capsule synthesis by carbon dioxide," *Journal of Clinical Investigation*, vol. 76, no. 2, pp. 508–516, 1985.

[84] Y. S. Bahn, G. M. Cox, J. R. Perfect, and J. Heitman, "Carbonic anhydrase and CO2 sensing during Cryptococcus neoformans growth, differentiation, and virulence," *Current Biology*, vol. 15, no. 22, pp. 2013–2020, 2005.

[85] W. Sims, "Effect of carbon dioxide on the growth and form of Candida albicans," *Journal of Medical Microbiology*, vol. 22, no. 3, pp. 203–208, 1986.

[86] G. Huang, T. Srikantha, N. Sahni, S. Yi, and D. R. Soll, "CO(2) regulates white-to-opaque switching in Candida albicans," *Current Biology*, vol. 19, no. 4, pp. 330–334, 2009.

[87] T. Klengel, W. J. Liang, J. Chaloupka et al., "Fungal adenylyl cyclase integrates CO2 sensing with cAMP signaling and virulence," *Current Biology*, vol. 15, no. 22, pp. 2021–2026, 2005.

[88] S. Lindskog, "Structure and mechanism of Carbonic Anhydrase," *Pharmacology and Therapeutics*, vol. 74, no. 1, pp. 1–20, 1997.

[89] R. A. Hall, L. De Sordi, D. M. Maccallum et al., "CO(2) acts as a signalling molecule in populations of the fungal pathogen Candida albicans," *PLoS Pathogens*, vol. 6, no. 11, Article ID e1001193, 2010.

[90] S. Ghosh, D. H. M. L. P. Navarathna, D. D. Roberts et al., "Arginine-induced germ tube formation in Candida albicans is essential for escape from murine macrophage line RAW 264.7," *Infection and Immunity*, vol. 77, no. 4, pp. 1596–1605, 2009.

[91] P. Richard, B. M. Bakker, B. Teusink, K. Van Dam, and H. V. Westerhoff, "Acetaldehyde mediates the synchronization of sustained glycolytic oscillations in populations of yeast cells," *European Journal of Biochemistry*, vol. 235, no. 1-2, pp. 238–241, 1996.

[92] B. Chance, R. W. Estabrook, and A. Ghosh, "Damped sinusoidal oscillations of cytoplasmic reduced pyridine nucleotide in yeast cells," *Proceedings of the National Academy of Sciences of the United States of America*, vol. 51, pp. 1244–1251, 1964.

[93] P. Richard, "The rhythm of yeast," *FEMS Microbiology Reviews*, vol. 27, no. 4, pp. 547–557, 2003.

[94] A. Betz and J. U. Becker, "Phase dependent phase shifts induced by pyruvate and acetaldehyde in oscillating NADH of yeast cells," *Journal of Interdisciplinary Cycle Research*, vol. 6, no. 2, pp. 167–173, 1975.

[95] W. Weber, M. Daoud-El Baba, and M. Fussenegger, "Synthetic ecosystems based on airborne inter-and intrakingdom communication," *Proceedings of the National Academy of Sciences of the United States of America*, vol. 104, no. 25, pp. 10435–10440, 2007.

18

Dermatophyte Virulence Factors: Identifying and Analyzing Genes That May Contribute to Chronic or Acute Skin Infections

Rebecca Rashid Achterman[1] and Theodore C. White[1,2]

[1] *Seattle Biomedical Research Institute, 307 Westlake Avenue North, Suite 500, Seattle, WA 98109-5219, USA*
[2] *School of Biological Sciences, University of Missouri-Kansas City, 5007 Rockhill Road, BSB 404, Kansas City, MO 64110, USA*

Correspondence should be addressed to Theodore C. White, whitetc@umkc.edu

Academic Editor: Julian R. Naglik

Dermatophytes are prevalent causes of cutaneous mycoses and, unlike many other fungal pathogens, are able to cause disease in immunocompetent individuals. They infect keratinized tissue such as skin, hair, and nails, resulting in tinea infections, including ringworm. Little is known about the molecular mechanisms that underlie the ability of these organisms to establish and maintain infection. The recent availability of genome sequence information and improved genetic manipulation have enabled researchers to begin to identify and study the role of virulence factors of dermatophytes. This paper will summarize our current understanding of dermatophyte virulence factors and discuss future directions for identifying and testing virulence factors.

1. Introduction

Dermatophytes are the most common cause of fungal infections worldwide, resulting in treatment costs of close to half a billion dollars annually in the USA [1, 2]. The World Health Organization estimates global prevalence of dermatomycoses to be approaching 20% [3]. Despite this, researchers lack a sophisticated understanding of dermatophyte pathogenesis [4].

Dermatophytes are the group of filamentous fungi that are the most common cause of cutaneous mycoses. The diseases caused by these organisms are generally named after the part of the body that is infected rather than the infecting organism. For example, tinea pedis refers to athlete's foot and tinea unguium refers to a nail infection. Dermatophyte infections are generally superficial, but immunocompromised patients can experience severe, disseminated disease [5]. Although dermatophyte infections are treatable, there is a high rate of reinfection; it remains to be determined whether this is due to relapse (the fungus not being completely eradicated during treatment) or a new infection [6].

The dermatophytes include three genera of molds in the class Euascomycetes: *Trichophyton*, *Microsporum*, and *Epidermophyton*. Dermatophytes are grouped according to their habitat as being either anthropophilic (human associated), zoophilic (animal associated), or geophilic (soil dwelling). Anthropophilic species are responsible for the majority of human infections; however, species from all three groups of dermatophytes have been associated with clinical disease. Human infections caused by anthropophiles tend to be chronic, with little inflammation, whereas infections caused by geophiles and zoophiles are often associated with acute inflammation and are self-healing [7].

The recent sequencing and annotation of several dermatophyte genomes, as well as advances in techniques for genetic manipulation of dermatophytes, provide resources that will aid in elucidating the mechanisms of virulence of these ubiquitous organisms. An understanding of the specific virulence factors that contribute to dermatophyte pathogenesis would aid in the design of effective therapeutics. This paper will summarize the current state of understanding of dermatophyte virulence factors and comment on future directions for their studies.

2. Virulence Factor Identification

2.1. Virulence Factor Identification Using Bioinformatics Approaches. The sequencing of seven dermatophyte genomes has recently been completed, and the sequences have been made publically available via the Broad Institute database [8]. The Broad Institute sequenced and annotated the genomes of five dermatophyte species. *Trichophyton rubrum* (*Tr*) is anthropophilic and the most common cause of dermatophyte infections in humans worldwide. *Trichophyton tonsurans* (*Tt*) is also anthropophilic and is a major cause of tinea capitis (scalp ringworm). *Tt* was recently found to be present in more than 30% of students in some grade levels at US schools [9]. *Trichophyton equinum* (*Te*) is closely related to *Tt* but is zoophilic and primarily associated with disease in horses. *Microsporum canis* (*Mc*) is also zoophilic and is the most common cause of tinea capitis in Europe [10]. *Microsporum gypseum* (*Mg*) is a geophile that is associated with gardener's ringworm. The strains selected for sequencing are all clinically relevant (associated with human disease) [4] and have been characterized with respect to growth rate, conidiation, and drug susceptibility [11]. Genome sequences of the remaining two species, the phylogenetically related zoophiles *Arthroderma benhamiae* (*Ab*, a teleomorph of *Trichophyton mentagrophytes*) and *Trichophyton verrucosum* (*Tv*), were recently completed by the Hans Knoell Institute (Jena, Germany) and published [12]. These organisms cause a highly inflammatory infection in humans. As expected, comparison of the seven dermatophyte genomes indicates that these species are closely related phylogenetically and each is more closely related to the others than to *Coccidioides immitis* or the dimorphic fungi (unpublished data). This is in agreement with a comparative study of five dermatophyte mitochondrial genomes, which suggested a recent divergence of the dermatophyte clade [13].

All seven genomes were found to encode high numbers of protease-encoding genes compared to related, nondermatophytic fungi ([12] and unpublished data). In particular, dermatophytes appear to have expanded sets of endopeptidases, exopeptidases, and secreted proteases. In contrast, there is little difference in abundance of carbohydrate enzymes of the CAZy family designation [14, 15] between dermatophytes and dimorphic fungi (unpublished data). This highlights the important role of protein degradation in the lifestyle of dermatophytes.

Secretome analysis of *Ab* during growth on keratin confirmed that proteases made up the largest group of identified secreted proteins [12]. The *Ab* and *Tv* genome sequences also revealed a relatively high number of secondary metabolite gene clusters, and expression of some of these genes were confirmed to be up- or downregulated during keratinocyte infection by *Ab* [12].

As described above, disease caused by human-adapted organisms *Tr* and *Tt* tends to be chronic with low inflammation, whereas zoophiles (*Te*, *Mc*, *Ab*) and geophiles (*Mg*) generally cause an inflammatory infection. The availability of sequence information now allows researchers to use a bioinformatics approach to make predictions about which genes are involved in virulence, as well as differences between species that have adapted to different ecological niches. Preliminary unpublished analysis indicates that among the five genomes sequenced by the Broad Institute, most genes are conserved in all five species, the majority of which are annotated. This is not surprising and confirms the genetic relatedness of the dermatophytes. Of the annotated genes that are unique to a particular species, there does not appear to be any trend. However, there are a number of hypothetical genes unique to each species that potentially play a role in niche adaptation and pathogenicity (unpublished data).

In many pathogenic eukaryotes, including protozoan parasites and fungi, there is a trend for clinical isolates to have cryptic, modified sexual cycles. Although originally thought to be asexual, *Aspergillus fumigatus*, *Candida albicans*, and *Cryptococcus neoformans* are all examples of fungi that cause systemic disease and were shown to have sexual cycles [16, 17]. It is therefore possible that other pathogenic fungi, including anthropophilic dermatophytes, have sexual cycles that have not yet been identified under laboratory conditions. Bioinformatic analysis identified the mating locus in each of the sequenced dermatophyte species, and PCR analysis with additional strains of the species has identified the other mating type, except for *Tr*, where the second mating type has not yet been identified [18]. Recent work has shown normal mating with progeny in the geophilic species *Mg* [19] and the ability of *Tr* to initiate but not complete a mating cycle with a related species of the opposite mating type [20]. Whether *Tr* is able to mate during growth on human skin remains to be determined, and the potential contributions of mating to virulence represent an area of active research.

Knowledge of the mechanisms of pathogenesis of other fungi also leads to predictions of virulence factors in dermatophytes. For example, the dipeptidyl peptidase DppIV was identified in *Mc* based on sequence similarity [21]. Additionally, dermatophytes have recently been shown to produce melanin or melanin-like compounds, which are predicted to play a role in virulence based on the known role of melanins in other pathogenic fungi [22]. Similarly, *Tr* has been shown to produce xanthomegnin, a toxin previously known to be produced by *Aspergillus*, in culture and during human infection [23]. The complete annotated sequences of dermatophyte genomes will aid in identifying additional putative virulence factors based on sequence similarity, which will then need to be tested experimentally to confirm expression and role during infection.

2.2. Virulence Factor Identification Using High-Throughput Screens. The *T. rubrum* Expression Database [24, 25] provided an important starting point for transcriptional analysis by collating expressed sequence tags (ESTs) and transcriptional profiles from microarrays, resulting in the identification of numerous genes involved in growth during a variety of conditions. Complete genome sequencing will enhance our ability to identify open reading frames (ORFs) in future studies.

Screens have historically been used to identify gene products likely to play a role in virulence. Dermatophytes are known to infect keratinized structures such as skin, hair, and nails, and therefore the ability of dermatophytes to degrade keratin is considered a major virulence attribute. In support of this, a correlation between keratinase activity and pathogenesis has been observed for *Mc* [26] and dermatophytes have been shown to secrete more than 20 proteases *in vitro* when grown in medium containing protein as the sole nitrogen source (reviewed in [27]). As discussed above, genome analysis confirmed expansion of protease genes in the seven dermatophyte genomes ([12] and unpublished data).

Given the importance of keratin to the pathogenic lifestyle of dermatophytes, studies that aimed to identify virulence factors have often examined the response of dermatophytes to growth on keratin. For example, subtractive suppression hybridization (SSH) approaches have been used to compare *Tr* during growth on keratin as compared to glucose [28] or minimal medium [29]. Select genes identified in this manner were confirmed to be upregulated during interaction with keratinocytes [29]. These included a homeobox transcription factor and a zinc-finger protein, which are candidates for acting as transcriptional regulators during infection.

Kaufman et al. found that thioredoxin, cellobiohydrolase, and the protease-encoding gene *Tri m 4* had increased transcription during growth of *T. mentagrophytes* (*Tm*) with keratin as compared to glucose alone [30]. Zaugg et al. constructed a cDNA microarray for *Tr* and examined gene expression during growth on soy and soy + keratin as compared to rich medium (Sabouraud) to find factors induced by one or both proteins [31]. They found that growth in soy or soy + keratin activated a large set of genes encoding secreted endo- and exoproteases, as well as other proteins potentially implicated in protein degradation, some of which appeared to be keratin specific. Interestingly, the authors noted that upregulation of enzymes in the glyoxylate cycle was also observed during growth on soy or soy + keratin, as compared to Sabouraud. The glyoxylate cycle has been implicated in virulence of other microorganisms [32], and its upregulation during dermatophyte growth on keratin was confirmed for *Ab* by Staib et al., who examined gene expression of *Ab* under the same conditions [33]. They found induction of similar sets of orthologous protease-encoding genes as compared to the *Tr* data.

However, a growing body of evidence suggests that not all keratin-induced proteases play a role during infection. For example, although *Tri m 4* was induced by the presence of keratin, it was not significantly upregulated when homogenized skin was provided as the sole nutrient source [30]. Furthermore, some of the prominent keratin-induced genes of *Ab* were not found to be strongly upregulated *in vivo* during guinea pig infection and the protease-encoding gene *SUB6* was strongly upregulated *in vivo* but not detectably activated during growth on keratin [33]. Instead, Staib et al. found just one protease-encoding gene, *MCPA*, which was strongly induced during both infection and growth on keratin [33]. Their study identified nonprotease genes, such as those encoding a putative opsin-related protein and enzymes of the glyoxylate cycle, which were upregulated *in vivo*. Likewise, Burmester et al. found that only some of the keratin-induced proteases were strongly expressed during fungus-keratinocyte interaction [12]. Interestingly, they found that secondary metabolites were induced during interaction with keratinocytes. Previous work has shown that antibiotic substances are produced by dermatophytes that may help the fungi compete against bacteria also present on the skin [34, 35]. It is possible that differentially regulated secondary metabolites, perhaps including antibiotics or pigment production, play a role in dermatophyte infection by providing the fungi with an advantage over other microorganisms present on the skin.

Expression of specific secreted subtilisins and metalloproteases was monitored by RT-PCR during *Tr* growth *in vitro* in the presence of keratin, elastin, collagen, or human skin sections [36]. *SUB3*, *SUB4*, and *MEP4* had increased expression under all four conditions (compared to growth in glucose medium). The increased expression of *SUB3* is consistent with *in vivo* findings in *Ab*, although *SUB3* of *Ab* was upregulated only at a low level [12]. Furthermore, *SUB4* was not found to be upregulated during guinea pig infection by *Ab*.

Together, these results indicate that secreted proteases are not the only virulence factors of dermatophytes and indeed that not all proteases play an overlapping role during infection. Some proteases may be used only during specific stages of infection or might have a more general role in growth that is not specific to virulence. Furthermore, it is possible that each of the dermatophyte species will have a unique program of expression for proteases and other putative virulence factors during infection.

In order to identify additional factors that play a role during infection, investigators have examined the transcriptional response of dermatophytes exposed to environmental factors such as growth on lipids [37, 38], changes in pH [38, 39], and the presence of antifungal drugs [38, 40–45]. They have also assessed transcript abundance during different stages of growth [46–49]. The relationship of these environmental factors to the pathogenesis of dermatophytes is a continuing area of research.

3. Testing the Role of Putative Virulence Factors

3.1. Models for Testing Virulence. Although dermatophytes were initially studied in experimental human infections [50], the current most common animal model for studying virulence factors of dermatophytes is the guinea pig [51]. This model has been useful for zoophiles [33, 52–54] but does not provide an accurate infection model for most anthropophilic species [4]. A murine model has been useful for studying the immune response to dermatophytes [55, 56], but again the mice only develop disease in response to infection by zoophiles. An alternative to mammalian models has been to rely on growth of the dermatophyte on keratinized surfaces such as sterilized nail fragments as an indication of pathogenicity [57, 58]. Despite its relative ease, this is a nonquantitative model based solely on the observed (qualitative) ability of the dermatophyte to grow. Its continued use in virulence studies highlights the need for development of a more appropriate model of anthropophilic dermatophyte infection.

Galleria mellonella (wax moth) larvae are an established virulence model for several fungal pathogens, including *Candida*, *Cryptococcus*, and the mold *Aspergillus* [59–65].

The *Galleria* model has several advantages as a virulence model, including an immune system with some similarities to the human innate immune system [61, 66]. However, *Galleria* does not appear to be a useful model to study the pathogenic mechanisms of dermatophytes [11].

Recently, researchers have used skin explants as a model for dermatophyte adherence and invasion [67–71]. *Tm* and *Tr* have been tested in this model. Conidial suspensions or pieces of mycelium are applied to a skin explant, and germination and hyphal invasion are monitored microscopically. Dermatophytes can adhere to and invade *ex vivo* skin explants [67–69], and dermatophyte growth is inhibited by antifungal drugs [70]. A reconstructed feline epidermal model [72] and a feline *ex vivo* epidermal model [52] have also been reported for the study of *Mc*, a zoophile whose natural host is cats.

Human epidermal tissues are commercially available, which abrogates the need for researchers to have access to clinical samples and provides a greater degree of standardization between labs that wish to use skin explants as a virulence model. Dermatophyte microconidia are able to germinate and cause damage to these tissues (our own unpublished data). Skin explants therefore represent a possible virulence model to study the initial stages of dermatophyte infection.

3.2. Virulence Factor Genes That Have Been Tested. There are a few cases in which a gene hypothesized to play a role in virulence has been deleted or knocked down. Due to the historical difficulties of genetic manipulation of dermatophytes, gene deletions are often not complemented. However, recent genetic advances have provided a foundation for genetic manipulation of dermatophytes [53, 73–77]. Ideally, future studies of dermatophytes should include both a deletion and a complementation of the mutation to definitively prove the role of a gene product in virulence.

A study on the gene encoding malate synthase (AcuE, a key enzyme of the glyoxylate cycle) provided an excellent step towards this goal. *ACUE* was identified as being upregulated during infection of guinea pigs by *Ab* [33]. Grumbt et al. constructed a deletion, Δ*acuE*, and compared its growth to the parental strain on different carbon sources as well as during guinea pig infection [53]. Although they did not see a difference in pathogenicity between the mutant and the wild type, they did see differences in growth when provided with 0.5% olive oil as the sole carbon source, with Δ*acuE* being unable to grow. Complementation with the wild-type allele restored growth.

TruMDR2 is a gene identified in *Tr* that is predicted by sequence similarity to encode an ATP-binding cassette (ABC) transporter protein [78]. Deletion of this gene results in increased susceptibility to some antifungal compounds [78]. *TruMDR2* was found to be upregulated during growth in the presence of keratin as compared to glucose [28, 58], suggesting a role for the transporter protein during infection. To confirm this, the wild-type and deletion mutants were compared for growth on nail fragments. As predicted, the deletion mutant showed a reduced ability to grow [58].

Tr also encodes a protein with sequence similarity to *pacC*, a pH-regulated transcription factor in *Aspergillus nidulans* [57]. Disruption of *pacC* results in reduction of keratinolytic activity and a reduction in the ability to grow on nail fragments [57].

As discussed above, *SUB3* has been identified in screens as a putative virulence factor in *Tr* and *Ab*. *SUB3* expression and activity has also been monitored in *Mc*, where *SUB3* was identified as a 31.5 kDa secreted protein with *in vitro* keratinolytic activity [79]. It was found to be expressed during natural *Mc* infection of cats, although presence of *SUB3* did not correlate to disease state as *SUB3* was found in both symptomatic and asymptomatic infections [79, 80]. Expression of *SUB3* was experimentally reduced using RNA-mediated silencing [21], and the resulting strain was tested in a feline *ex vivo* adherence model [52]. Arthroconidia from the *SUB3* RNA-silenced *M. canis* strain had reduced adherence to feline epidermis compared to the control strain. Although the control-strain arthroconidia did not adhere equally well to epidermis from each of three different cats, for each cat the *SUB3* RNA-silenced strain had a reduction in adherence that was statistically significant. The strains were also tested for their ability to cause lesions in the guinea pig model, but no difference in virulence was observed [52]. The authors conclude that *SUB3* is required for adherence to, but not invasion of, the epidermis. It is also possible that the function of other secreted subtilisins masked the loss of *SUB3* or that the guinea pig model does not completely mimic feline infection.

In *Mc*, the *dnr1* gene, which has sequence similarity to nitrogen regulatory genes of other filamentous fungi, is able to complement a loss-of-function nitrogen regulatory mutation (*areA*) in *Aspergillus nidulans*. Disruption of *dnr1* in *Mc* caused a reduction in the ability of the fungus to grow on medium containing keratin as the sole nitrogen source [81]. A similar result was seen for *Tm* (teleomorph: *A. vanbreuseghemii*) when *tnr1*, a gene with sequence similarity to *areA* and *dnr1*, was disrupted [77]. Neither of these mutants have been studied in a virulence model. Virulence studies of these and other factors identified, for example in screens or through bioinformatics approaches, will be essential to determining the contribution of each factor to disease.

4. The Role of the Immune System

Fungal virulence is the result of interplay between the infecting organism and the host. During dermatophyte infection, cell-mediated immunity is widely considered to be responsible for modulating dermatophyte disease [7, 82–87] and fungal antigens activate T-suppressor and T-helper cells [56]. Differences specific to the host are thought to be important in determining the relative susceptibility of individuals, with factors such as age, gender, and genetics all likely to play a role [85, 87]. These clinical factors will not be reviewed here.

A recent review of host-dermatophyte interactions is available [83]. We will briefly describe these as they relate to dermatophyte virulence factors. The most numerous cells in the epidermis are keratinocytes, indicating that dermatophytes must primarily interact with these cells. Interestingly, keratinocytes seem to exhibit a differential response following exposure to different dermatophyte species.

Tani et al. measured cytokine production by epidermal keratinocytes following coculture with *Tm*, *Tt*, and *Tr* [88]. Of these, *Tm* causes an acute inflammatory response, whereas *Tt* and *Tr* are anthropophiles that cause minimal inflammation. Although all three species induced Interleukin- (IL-) 8 secretion, coculture of keratinocytes with *Tm* resulted in higher levels of IL-8 production than coculture with *Tt* or *Tr*. Additionally, *Tm* but not *Tt* or *Tr* was able to induce secretion of TNFα. Similar results were found when cytokine secretion profiles of human keratinocytes were compared during dermatophyte infection with *Ab*, a zoophile and causes a severe inflammatory response, and *Tt* [89]. They found that both species caused an increase in expression of IL-8, which was confirmed by reverse transcriptase-polymerase chain reaction (RT-PCR). However, *Ab* also induced secretion of a broad spectrum of cytokines, whereas *Tt* did not. These studies support the hypothesis that different dermatophyte species induce different immune responses in the host, contributing to the relative severity of the infection.

Keratinocytes are not the only cells that will interact with the dermatophytes. Phagocytic cells are attracted to the site of infection (reviewed in [87]), and an ability of dermatophytes to survive those interactions would contribute to pathogenesis. Campos et al. determined that *Tr* conidia could germinate in a macrophage, resulting in macrophage death [90].

Certainly, it would be advantageous for anthropophiles to downregulate the immune response so as to facilitate chronic infection, and fungal factors that modulate the host immune response represent potential virulence factors. Indeed, mannan extracted from the cell wall of *Tr* inhibited lymphoproliferation of human mononuclear leukocytes *in vitro* [91]. Addition of filtrate solution from *Tm* or *Tr* was shown to induce secretion of IL-1α and basic fibroblast growth factor in keratinocytes [85], with more IL-1α being secreted in cells exposed to the *Tm* filtrate. This suggests that at least some of the putative virulence factors involved in modulating the immune response might be secreted by the fungi. It is likely that cell-associated as well as secreted factors contribute to the dermatophyte's ability to exaggerate or suppress an inflammatory response.

Temporal expression of proteases has also been postulated to contribute to the relative intensity of the inflammatory response [84]. Recent data comparing secreted proteolytic and lipolytic enzymes in *Tt* and *Te* support this idea [92]. *Tt* and *Te* are adapted to different host species (humans and horses, resp.), and *Tt* causes a mild chronic disease whereas *Te* causes an inflammatory disease in humans. Of the 31 genes studied, each had ≥99.5% sequence identity between the two species; however, transcriptional analysis identified differences in expression during growth on keratin [92]. For example, of the subtilisin-like proteases examined, *Sub6* and *Sub7* had significantly higher expression in *Tt* compared to *Te*, whereas *Sub1* and *Sub5* had significantly higher expression in *Te* compared to *Tt* [92].

Arthroconidia are produced during some infections and might aid survival in the nail and transmission of the infection to a new host. Arthroconidia can be formed by a majority of *Tr* clinical isolates during growth on nail powder under specific laboratory conditions [93] and have decreased susceptibility to some antifungals compared to microconidia [94]. The precise contribution of arthroconidia to infection and the mechanisms by which arthroconidia interact with host cells remains an area of investigation.

Few studies have examined dermatophyte gene expression in response to human cells [12, 30], and those that have serve to highlight the fact that growth in the presence of keratin does not necessarily reflect conditions during infection. The precise mechanisms by which dermatophyte species interact with host cells at the molecular level remain unknown. There is a clear need to identify the dermatophyte factors involved in pathogenesis, and a logical starting point is to identify dermatophyte genes and proteins that are upregulated during interactions with epithelial cells. To this end, transcriptional and proteomic profiling of dermatophytes during infection of human epidermal tissue, in addition to a bioinformatic approach, may identify additional potential virulence factors. These studies must go further, though. It is imperative that we test the expression and role that these factors play during infection. Only then can we expand our list of true virulence factors of dermatophytes and use the knowledge to inform directions for therapy and preventative measures.

Acknowledgments

The autors thank Drs. C. Cuomo, D. Martinez, M. Henn, and the rest of the staff at the Broad Institute who have contributed to the sequencing and annotation of the dermatophyte genomes. They thank Drs. J. Heitman and W. Li (Duke U.) for sharing their unpublished data about *Tr* mating types. They thank Drs. B. Oliver and D. Vinh (Seattle BioMed) for reading the paper. The authors were supported in part by R21-AI081235 awarded to T. C. White.

References

[1] E. S. Smith, S. R. Feldman, A. B. Fleischer, B. Leshin, and A. McMichael, "Characteristics of office-based visits for skin cancer: dermatologists have more experience than other physicians in managing malignant and premalignant skin conditions," *Dermatologic Surgery*, vol. 24, no. 9, pp. 981–985, 1998.

[2] L. A. Drake, S. M. Dinehart, E. R. Farmer et al., "Guidelines of care for superficial mycotic infections of the skin: tinea corporis, tinea cruris, tinea faciei, tinea manuum, and tinea pedis," *Journal of the American Academy of Dermatology*, vol. 34, no. 2 I, pp. 282–286, 1996.

[3] S. A. Marques, A. M. Robles, A. M. Tortorano, M. A. Tuculet, R. Negroni, and R. P. Mendes, "Mycoses associated with AIDS in the third world," *Medical Mycology*, vol. 38, no. 1, pp. 269–279, 2000.

[4] T. C. White, B. G. Oliver, Y. Graser, and M. R. Henn, "Generating and testing molecular hypotheses in the dermatophytes," *Eukaryotic Cell*, vol. 7, no. 8, pp. 1238–1245, 2008.

[5] G. E. J. Rodwell, C. L. Bayles, L. Towersey, and R. Aly, "The prevalence of dermatophyte infection in patients infected with human immunodeficiency virus," *International Journal of Dermatology*, vol. 47, no. 4, pp. 339–343, 2008.

[6] A. K. Gupta and E. A. Cooper, "Update in antifungal therapy of dermatophytosis," *Mycopathologia*, vol. 166, no. 5-6, pp. 353–367, 2008.

[7] I. Weitzman and R. C. Summerbell, "The dermatophytes," *Clinical Microbiology Reviews*, vol. 8, no. 2, pp. 240–259, 1995.

[8] The Broad Institute, "Dermatophyte comparative database," 2011, http://www.broadinstitute.org/annotation/genome/dermatophyte_comparative/MultiHome.html.

[9] S. M. Abdel-Rahman, N. Farrand, E. Schuenemann et al., "The prevalence of infections with Trichophyton tonsurans in schoolchildren: the CAPITIS study," *Pediatrics*, vol. 125, no. 5, pp. 966–973, 2010.

[10] R. Sharma, S. De Hoog, W. Presber, and Y. Gräser, "A virulent genotype of Microsporum canis is responsible for the majority of human infections," *Journal of Medical Microbiology*, vol. 56, no. 10, pp. 1377–1385, 2007.

[11] R. R. Achterman, A. R. Smith, B. G. Oliver, and T. C. White, "Sequenced dermatophyte strains: growth rate, conidiation, drug susceptibilities, and virulence in an invertebrate model," *Fungal Genetics and Biology*, vol. 48, no. 3, pp. 335–341, 2011.

[12] A. Burmester, E. Shelest, G. Glockner et al., "Comparative and functional genomics provide insights into the pathogenicity of dermatophytic fungi," *Genome Biology*, vol. 12, no. 1, article R7, 2011.

[13] Y. Wu, J. Yang, F. Yang et al., "Recent dermatophyte divergence revealed by comparative and phylogenetic analysis of mitochondrial genomes," *BMC Genomics*, vol. 10, article no. 238, 2009.

[14] B. I. Cantarel, P. M. Coutinho, C. Rancurel, T. Bernard, V. Lombard, and B. Henrissat, "The carbohydrate-active enZymes database (CAZy): an expert resource for glycogenomics," *Nucleic Acids Research*, vol. 37, no. 1, pp. D233–D238, 2009.

[15] "Carbohydrate active enzymes database," 2011, http://www.cazy.org.

[16] J. Heitman, "Evolution of eukaryotic microbial pathogens via covert sexual reproduction," *Cell host & Microbe*, vol. 8, no. 1, pp. 86–99, 2010.

[17] J. Heitman, "Microbial pathogens in the fungal kingdom," *Fungal Biology Reviews*, vol. 25, no. 1, pp. 48–60, 2011.

[18] J. Heitman and W. Li, Duke University, personal communication, 2011.

[19] W. Li, B. Metin, T. C. White, and J. Heitman, "Organization and evolutionary trajectory of the mating type (MAT) locus in dermatophyte and dimorphic fungal pathogens," *Eukaryotic Cell*, vol. 9, no. 1, pp. 46–58, 2010.

[20] K. Anzawa, M. Kawasaki, T. Mochizuki, and H. Ishizaki, "Successful mating of Trichophyton rubrum with Arthroderma simii," *Medical Mycology*, vol. 48, no. 4, pp. 629–634, 2010.

[21] S. Vermout, J. Tabart, A. Baldo, M. Monod, B. Losson, and B. Mignon, "RNA silencing in the dermatophyte Microsporum canis," *FEMS Microbiology Letters*, vol. 275, no. 1, pp. 38–45, 2007.

[22] S. Youngchim, S. Pornsuwan, J. D. Nosanchuk, W. Dankai, and N. Vanittanakom, "Melanogenesis in dermatophyte species in vitro and during infection," *Microbiology*, vol. 157, pp. 2348–2356, 2011.

[23] A. K. Gupta, I. Ahmad, I. Borst, and R. C. Summerbell, "Detection of xanthomegnin in epidermal materials infected with Trichophyton rubrum," *Journal of Investigative Dermatology*, vol. 115, no. 5, pp. 901–905, 2000.

[24] TrEd, "T. rubrum expression database (TrED)," 2011.

[25] J. Yang, L. Chen, L. Wang, W. Zhang, T. Liu, and Q. Jin, "TrED: the Trichophyton rubrum expression database," *BMC Genomics*, vol. 8, article no. 250, 2007.

[26] F. C. Viani, J. I. Dos Santos, C. R. Paula, C. E. Larson, and W. Gambale, "Production of extracellular enzymes by Microsporum canis and their role in its virulence," *Medical Mycology*, vol. 39, no. 5, pp. 463–468, 2001.

[27] M. Monod, "Secreted proteases from dermatophytes," *Mycopathologia*, vol. 166, no. 5-6, pp. 285–294, 2008.

[28] F. C.A. Maranhão, F. G. Paião, and N. M. Martinez-Rossi, "Isolation of transcripts over-expressed in human pathogen Trichophyton rubrum during growth in keratin," *Microbial Pathogenesis*, vol. 43, no. 4, pp. 166–172, 2007.

[29] L. C. Baeza, A. M. Bailão, C. L. Borges, M. Pereira, C. M. D. A. Soares, and M. J. S. Mendes Giannini, "cDNA representational difference analysis used in the identification of genes expressed by Trichophyton rubrum during contact with keratin," *Microbes and Infection*, vol. 9, no. 12-13, pp. 1415–1421, 2007.

[30] G. Kaufman, I. Berdicevsky, J. A. Woodfolk, and B. A. Horwitz, "Markers for host-induced gene expression in Trichophyton dermatophytosis," *Infection and Immunity*, vol. 73, no. 10, pp. 6584–6590, 2005.

[31] C. Zaugg, M. Monod, J. Weber et al., "Gene expression profiling in the human pathogenic dermatophyte Trichophyton rubrum during growth on proteins," *Eukaryotic Cell*, vol. 8, no. 2, pp. 241–250, 2009.

[32] M. C. Lorenz and G. R. Fink, "Life and death in a macrophage: role of the glyoxylate cycle in virulence," *Eukaryotic Cell*, vol. 1, no. 5, pp. 657–662, 2002.

[33] P. Staib, C. Zaugg, B. Mignon et al., "Differential gene expression in the pathogenic dermatophyte Arthroderma benhamiae in vitro versus during infection," *Microbiology*, vol. 156, no. 3, pp. 884–895, 2010.

[34] N. Youssef, C. H. E. Wyborn, and G. Holt, "Antibiotic production by dermatophyte fungi," *Journal of General Microbiology*, vol. 105, no. 1, pp. 105–111, 1978.

[35] H. M. Lappin-Scott, M. E. Rogers, and M. W. Adlard, "High-performance liquid chromatographic identification of beta-lactam antibiotics produced by dermatophytes," *Journal of Applied Bacteriology*, vol. 59, no. 5, pp. 437–441, 1985.

[36] W. Leng, T. Liu, J. Wang, R. Li, and Q. Jin, "Expression dynamics of secreted protease genes in Trichophyton rubrum induced by key host's proteinaceous components," *Medical Mycology*, vol. 47, no. 7, pp. 759–765, 2009.

[37] F. C. Maranhao, H. C. Silveira, A. Rossi, and N. M. Martinez-Rossi, "Isolation of transcripts overexpressed in the human pathogen Trichophyton rubrum grown in lipid as carbon source," *Canadian Journal of Microbiology*, vol. 57, no. 4, pp. 333–338, 2011.

[38] N. T. Peres, P. R. Sanches, J. P. Falco et al., "Transcriptional profiling reveals the expression of novel genes in response to various stimuli in the human dermatophyte Trichophyton rubrum," *BMC Microbiology*, vol. 10, article no. 39, 2010.

[39] H. C. S. Silveira, D. E. Gras, R. A. Cazzaniga, P. R. Sanches, A. Rossi, and N. M. Martinez-Rossi, "Transcriptional profiling reveals genes in the human pathogen Trichophyton rubrum that are expressed in response to pH signaling," *Microbial Pathogenesis*, vol. 48, no. 2, pp. 91–96, 2010.

[40] Y. Diao, R. Zhao, X. Deng, W. Leng, J. Peng, and Q. Jin, "Transcriptional profiles of Trichophyton rubrum in response to itraconazole," *Medical Mycology*, vol. 47, no. 3, pp. 237–247, 2009.

[41] W. Zhang, L. Yu, W. Leng et al., "cDNA microarray analysis of the expression profiles of Trichophyton rubrum in response to novel synthetic fatty acid synthase inhibitor PHS11A," *Fungal Genetics and Biology*, vol. 44, no. 12, pp. 1252–1261, 2007.

[42] W. Zhang, L. Yu, J. Yang, L. Wang, J. Peng, and Q. Jin, "Transcriptional profiles of response to terbinafine in Trichophyton rubrum," *Applied Microbiology and Biotechnology*, vol. 82, no. 6, pp. 1123–1130, 2009.

[43] L. Yu, W. Zhang, L. Wang et al., "Transcriptional profiles of the response to ketoconazole and amphotericin B in Trichophyton rubrum," *Antimicrobial Agents and Chemotherapy*, vol. 51, no. 1, pp. 144–153, 2007.

[44] F. G. Paião, F. Segato, J. R. Cursino-Santos, N. T. A. Peres, and N. M. Martinez-Rossi, "Analysis of Trichophyton rubrum gene expression in response to cytotoxic drugs," *FEMS Microbiology Letters*, vol. 271, no. 2, pp. 180–186, 2007.

[45] F. Segato, S. R. Nozawa, A. Rossi, and N. M. Martinez-Rossi, "Over-expression of genes coding for proline oxidase, riboflavin kinase, cytochrome c oxidase and an MFS transporter induced by acriflavin in Trichophyton rubrum," *Medical Mycology*, vol. 46, no. 2, pp. 135–139, 2008.

[46] X. Xu, T. Liu, W. Leng et al., "Global gene expression profiles for the growth phases of Trichophyton rubrum," *Science China Life Sciences*, vol. 54, no. 7, pp. 675–682, 2011.

[47] W. Leng, T. Liu, R. Li et al., "Proteomic profile of dormant Trichophyton Rubrum conidia," *BMC Genomics*, vol. 9, article 303, 2008.

[48] L. Yang, L. Wang, J. Peng et al., "Comparison between gene expression of conidia and germinating phase in Trichophyton rubrum," *Science in China. Series C*, vol. 50, no. 3, pp. 377–384, 2007.

[49] T. Liu, Q. Zhang, L. Wang et al., "The use of global transcriptional analysis to reveal the biological and cellular events involved in distinct development phases of Trichophyton rubrum conidial germination," *BMC Genomics*, vol. 8, article 100, 2007.

[50] J. H. Reinhardt, A. M. Allen, D. Gunnison, and W. A. Akers, "Experimental human Trichophyton mentagrophytes infections," *Journal of Investigative Dermatology*, vol. 63, no. 5, pp. 419–422, 1974.

[51] J. H. Greenberg, R. D. King, S. Krebs, and R. Field, "A quantitative dermatophyte infection model in the guinea pig: a parallel to the quantitated human infection model," *Journal of Investigative Dermatology*, vol. 67, no. 6, pp. 704–708, 1976.

[52] A. Baldo, A. Mathy, J. Tabart et al., "Secreted subtilisin Sub3 from Microsporum canis is required for adherence to but not for invasion of the epidermis," *British Journal of Dermatology*, vol. 162, no. 5, pp. 990–997, 2010.

[53] M. Grumbt, V. Defaweux, B. Mignon et al., "Targeted gene deletion and in vivo analysis of putative virulence gene function in the pathogenic dermatophyte Arthroderma benhamiae," *Eukaryotic Cell*, vol. 10, no. 6, pp. 842–853, 2011.

[54] B. R. Mignon, T. Leclipteux, C. Focant, A. J. Nikkels, G. E. Piérard, and B. J. Losson, "Humoral and cellular immune response to a crude exo-antigen and purified keratinase of Microsporum canis in experimentally infected guinea pigs," *Medical Mycology*, vol. 37, no. 2, pp. 123–129, 1999.

[55] R. J. Hay, R. A. Calderon, and M. J. Collins, "Experimental dermatophytosis: the clinical and histopathologic features of a mouse model using Trichophyton quinckeanum (mouse favus)," *Journal of Investigative Dermatology*, vol. 81, no. 3, pp. 270–274, 1983.

[56] R. A. Calderon, "Immunoregulation of dermatophytosis," *Critical Reviews in Microbiology*, vol. 16, no. 5, pp. 339–368, 1989.

[57] M. S. Ferreira-Nozawa, H. C. S. Silveira, C. J. Ono, A. L. Fachin, A. Rossi, and N. M. Martinez-Rossi, "The pH signaling transcription factor PacC mediates the growth of Trichophyton rubrum on human nail in vitro," *Medical Mycology*, vol. 44, no. 7, pp. 641–645, 2006.

[58] F. C.A. Maranhão, F. G. Paião, A. L. Fachin, and N. M. Martinez-Rossi, "Membrane transporter proteins are involved in Trichophyton rubrum pathogenesis," *Journal of Medical Microbiology*, vol. 58, no. 2, pp. 163–168, 2009.

[59] G. Cotter, S. Doyle, and K. Kavanagh, "Development of an insect model for the in vivo pathogenicity testing of yeasts," *FEMS Immunology and Medical Microbiology*, vol. 27, no. 2, pp. 163–169, 2000.

[60] J. C. Jackson, L. A. Higgins, and X. Lin, "Conidiation color mutants of Aspergillus fumigatus are highly pathogenic to the heterologous insect host Galleria mellonella," *PLoS ONE*, vol. 4, no. 1, Article ID e4224, 2009.

[61] E. Mylonakis, "Galleria mellonella and the study of fungal pathogenesis: making the case for another genetically tractable model host," *Mycopathologia*, vol. 165, no. 1, pp. 1–3, 2008.

[62] E. Mylonakis, R. Moreno, J. B. El Khoury et al., "Galleria mellonella as a model system to study Cryptococcus neoformans pathogenesis," *Infection and Immunity*, vol. 73, no. 7, pp. 3842–3850, 2005.

[63] E. P. Reeves, C. G. M. Messina, S. Doyle, and K. Kavanagh, "Correlation between gliotoxin production and virulence of Aspergillus fumigatus in Galleria mellonella," *Mycopathologia*, vol. 158, no. 1, pp. 73–79, 2004.

[64] J. Renwick, P. Daly, E. P. Reeves, and K. Kavanagh, "Susceptibility of larvae of Galleria mellonella to infection by Aspergillus fumigatus is dependent upon stage of conidial germination," *Mycopathologia*, vol. 161, no. 6, pp. 377–384, 2006.

[65] R. J. S. Leger, S. E. Screen, and B. Shams-Pirzadeh, "Lack of host specialization in Aspergillus flavus," *Applied and Environmental Microbiology*, vol. 66, no. 1, pp. 320–324, 2000.

[66] K. Kavanagh and E. P. Reeves, "Exploiting the potential of insects for in vivo pathogenicity testing of microbial pathogens," *FEMS Microbiology Reviews*, vol. 28, no. 1, pp. 101–112, 2004.

[67] L. Duek, G. Kaufman, Y. Ulman, and I. Berdicevsky, "The pathogenesis of dermatophyte infections in human skin sections," *Journal of Infection*, vol. 48, no. 2, pp. 175–180, 2004.

[68] G. Kaufman, B. Horwitz, L. Duek, Y. Ullman, and I. Berdicevsky, "Infection stages of the dermatophyte pathogen Trichophyton: microscopic characterization and proteolytic enzymes," *Medical Mycology*, vol. 45, no. 2, pp. 149–155, 2007.

[69] G. Kaufman, B. A. Horwitz, R. Hadar, Y. Ullmann, and I. Berdicevsky, "Green fluorescent protein (GFP) as a vital marker for pathogenic development of the dermatophyte Trichophyton mentagrophytes," *Microbiology*, vol. 150, no. 8, pp. 2785–2790, 2004.

[70] C. Onyewu, E. Eads, W. A. Schell et al., "Targeting the calcineurin pathway enhances ergosterol biosynthesis inhibitors against Trichophyton mentagrophytes in vitro and in a human skin infection model," *Antimicrobial Agents and Chemotherapy*, vol. 51, no. 10, pp. 3743–3746, 2007.

[71] T. G. M. Smijs, J. A. Bouwstra, H. J. Schuitmaker, M. Talebi, and S. Pavel, "A novel ex vivo skin model to study the susceptibility of the dermatophyte Trichophyton rubrum to photodynamic treatment in different growth phases," *Journal of Antimicrobial Chemotherapy*, vol. 59, no. 3, pp. 433–440, 2007.

[72] J. Tabart, A. Baldo, S. Vermout et al., "Reconstructed interfollicular feline epidermis as a model for Microsporum canis dermatophytosis," *Journal of Medical Microbiology*, vol. 56, no. 7, pp. 971–975, 2007.

[73] M. M. Alshahni, T. Yamada, K. Takatori, T. Sawada, and K. Makimura, "Insights into a nonhomologous integration pathway in the dermatophyte Trichophyton mentagrophytes: efficient targeted gene disruption by use of mutants lacking ligase IV," *Microbiology and Immunology*, vol. 55, no. 1, pp. 34–43, 2011.

[74] M. Grumbt, M. Monod, and P. Staib, "Genetic advances in dermatophytes," *FEMS Microbiology Letters*, vol. 320, no. 2, pp. 79–86, 2011.

[75] T. Yamada, "Development of efficient tools for genetic manipulation of dermatophytes," *Nihon Ishinkin Gakkai Zasshi*, vol. 51, no. 2, pp. 87–92, 2010.

[76] T. Yamada, K. Makimura, T. Hisajima, Y. Ishihara, Y. Umeda, and S. Abe, "Enhanced gene replacements in Ku80 disruption mutants of the dermatophyte, Trichophyton mentagrophytes," *FEMS Microbiology Letters*, vol. 298, no. 2, pp. 208–217, 2009.

[77] T. Yamada, K. Makimura, K. Satoh, Y. Umeda, Y. Ishihara, and S. Abe, "Agrobacterium tumefaciens-mediated transformation of the dermatophyte, Trichophyton mentagrophytes: an efficient tool for gene transfer," *Medical Mycology*, vol. 47, no. 5, pp. 485–495, 2009.

[78] A. L. Fachin, M. S. Ferreira-Nozawa, W. Maccheroni, and N. M. Martinez-Rossi, "Role of the ABC transporter TruMDR2 in terbinafine, 4-nitroquinoline N-oxide and ethidium bromide susceptibility in Trichophyton rubrum," *Journal of Medical Microbiology*, vol. 55, no. 8, pp. 1093–1099, 2006.

[79] B. Mignon, M. Swinnen, J. P. Bouchara et al., "Purification and characterization of a 315 kDa keratinolytic subtilisin-like serine protease from Microsporum canis and evidence of its secretion in naturally infected cats," *Medical Mycology*, vol. 36, no. 6, pp. 395–404, 1998.

[80] B. R. Mignon, A. F. Nikkels, G. E. Piérard, and B. J. Losson, "The in vitro and in vivo production of a 31.5-kD keratinolytic subtilase from Microsporum canis and the clinical status in naturally infected cats," *Dermatology*, vol. 196, no. 4, pp. 438–441, 1998.

[81] T. Yamada, K. Makimura, and S. Abe, "Isolation, characterization, and disruption of dnr1, the areA/nit-2 -like nitrogen regulatory gene of the zoophilic dermatophyte, Microsporum canis," *Medical Mycology*, vol. 44, no. 3, pp. 243–252, 2006.

[82] S. R. Almeida, "Immunology of dermatophytosis," *Mycopathologia*, vol. 166, no. 5-6, pp. 277–283, 2008.

[83] N. T. De Aguiar Peres, F. C. A. Maranhão, A. Rossi, and N. M. Martinez-Rossi, "Dermatophytes: host-pathogen interaction and antifungal resistance," *Anais Brasileiros de Dermatologia*, vol. 85, no. 5, pp. 657–667, 2010.

[84] S. Vermout, J. Tabart, A. Baldo, A. Mathy, B. Losson, and B. Mignon, "Pathogenesis of dermatophytosis," *Mycopathologia*, vol. 166, no. 5-6, pp. 267–275, 2008.

[85] H. Ogawa, R. C. Summerbell, K. V. Clemons et al., "Dermatophytes and host defence in cutaneous mycoses," *Medical Mycology*, vol. 36, no. 1, Supplement 1, pp. 166–173, 1998.

[86] M. V. Dahl, "Suppression of immunity and inflammation by products produced by dermatophytes," *Journal of the American Academy of Dermatology*, vol. 28, no. 5 I, pp. S19–S23, 1993.

[87] J. Brasch, "Current knowledge of host response in human tinea," *Mycoses*, vol. 52, no. 4, pp. 304–312, 2009.

[88] K. Tani, M. Adachi, Y. Nakamura et al., "The effect of dermatophytes on cytokine production by human keratinocytes," *Archives of Dermatological Research*, vol. 299, no. 8, pp. 381–387, 2007.

[89] Y. Shiraki, Y. Ishibashi, M. Hiruma, A. Nishikawa, and S. Ikeda, "Cytokine secretion profiles of human keratinocytes during Trichophyton tonsurans and Arthroderma benhamiae infections," *Journal of Medical Microbiology*, vol. 55, no. 9, pp. 1175–1185, 2006.

[90] M. R. M. Campos, M. Russo, E. Gomes, and S. R. Almeida, "Stimulation, inhibition and death of macrophages infected with Trichophyton rubrum," *Microbes and Infection*, vol. 8, no. 2, pp. 372–379, 2006.

[91] J. S. Blake, M. V. Dahl, M. J. Herron, and R. D. Nelson, "An immunoinhibitory cell wall glycoprotein (mannan) from Trichophyton rubrum," *Journal of Investigative Dermatology*, vol. 96, no. 5, pp. 657–661, 1991.

[92] B. L. Preuett, E. Schuenemann, J. T. Brown, M. E. Kovac, S. K. Krishnan, and S. M. Abdel-Rahman, "Comparative analysis of secreted enzymes between the anthropophilic-zoophilic sister species Trichophyton tonsurans and Trichophyton equinum," *Fungal Biology*, vol. 114, no. 5-6, pp. 429–437, 2010.

[93] S. A. Yazdanparast and R. C. Barton, "Arthroconidia production in Trichophyton rubrum and a new ex vivo model of onychomycosis," *Journal of Medical Microbiology*, vol. 55, no. 11, pp. 1577–1581, 2006.

[94] L. M. Coelho, R. Aquino-Ferreira, C. M. L. Maffei, and N. M. Martinez-Rossi, "In vitro antifungal drug susceptibilities of dermatophytes microconidia and arthroconidia," *Journal of Antimicrobial Chemotherapy*, vol. 62, no. 4, pp. 758–761, 2008.

Hosting Infection: Experimental Models to Assay *Candida* Virulence

Donna M. MacCallum

Aberdeen Fungal Group, School of Medical Sciences, Institute of Medical Sciences, University of Aberdeen, Foresterhill, Aberdeen, AB25 2ZD, UK

Correspondence should be addressed to Donna M. MacCallum, d.m.maccallum@abdn.ac.uk

Academic Editor: Arianna Tavanti

Although normally commensals in humans, *Candida albicans*, *Candida tropicalis*, *Candida parapsilosis*, *Candida glabrata*, and *Candida krusei* are capable of causing opportunistic infections in individuals with altered physiological and/or immunological responses. These fungal species are linked with a variety of infections, including oral, vaginal, gastrointestinal, and systemic infections, with *C. albicans* the major cause of infection. To assess the ability of different *Candida* species and strains to cause infection and disease requires the use of experimental infection models. This paper discusses the mucosal and systemic models of infection available to assay *Candida* virulence and gives examples of some of the knowledge that has been gained to date from these models.

1. *Candida* and Man

1.1. Carriage of Candida *Species.* In healthy individuals *Candida* species are harmless members of the normal gastrointestinal (GI), oral, and vaginal microbial flora. It is assumed that everyone carries *Candida* in their GI tract (reviewed in [1]), with *C. albicans* the species most frequently identified in faecal sampling, representing 40–70% of isolates [2–4]. Other isolates are usually identified as *C. parapsilosis*, *C. glabrata*, *C. tropicalis*, or *C. krusei* [2–4].

In comparison to GI carriage, oral carriage is observed in only ~40% of healthy individuals, with considerable variation found between studies (reviewed in [1]). Higher carriage levels are generally associated with diabetes, cancer, HIV, or denture use (reviewed in [1]). Again, the majority of isolates (~80%) are identified as *C. albicans*, with *C. glabrata* or *C. parapsilosis* making up the remainder [5–9].

Vaginal carriage occurs in an even smaller proportion of the healthy population, with only ~20% of healthy women found to have vaginal *Candida* carriage [10–13]. *C. albicans* is again the most commonly identified species, with *C. glabrata* the only other species usually found [10, 12, 14–17].

Therefore, *C. albicans* is the major species found as a commensal in healthy individuals, with four other species, *C. tropicalis*, *C. parapsilosis*, *C. glabrata*, and *C. krusei*, also found.

1.2. Candida *and Disease.* *Candida* species, however, have an alternative lifestyle, causing opportunistic infection in hosts with altered physiological or immune response. The infections caused by *Candida* species range from self-limiting, superficial mucosal lesions (commonly referred to as thrush), chronic and/or recurrent mucosal, skin, and nail infections, through to life-threatening invasive or disseminated infection [1, 18–21].

In humans, the most common infections caused by *Candida* species are superficial infections of the mucosa, skin, and nails [20–24]. Pseudomembranous oral thrush is common in babies and in the elderly, but is also found in HIV-positive individuals and cancer patients (reviewed in [1, 25]). Denture stomatitis is also a significant infection, occurring in approximately 60% of denture wearers [26, 27]. In oral candidiasis most infections are caused by *C. albicans* (58%), with the remainder caused by *C. parapsilosis, C. tropicalis, C. glabrata*, and *C. krusei* [28, 29].

Vaginal candidiasis, or thrush, another form of superficial infection, affects approximately 75% of women of child-bearing age [30, 31]. *C. albicans* is most commonly isolated, with *C. glabrata* also found, but at a lower frequency [17, 30, 32–35], reflecting the species normally carried in the vulvo-vaginal area.

An additional form of candidiasis involving the mucous membranes, as well as the skin and nails, is chronic mucocutaneous candidiasis. Unlike other forms of candidiasis, there is evidence that this condition can be inherited or is associated with thymoma, with almost every infection caused by *C. albicans* [20–24, 36].

The most serious infections caused by *Candida* species, however, are invasive or disseminated infections. *Candida* species cause ~11% of all bloodstream infections and 20% of those occurring in the ICU population [37–39]. However, in comparison to bacterial infections occurring in the same patient population, these infections are much more serious as mortality rates remain high (~45%) [1, 40]. This is due, in part, to diagnostic difficulties and limited antifungal therapies. Invasive infections occur in those patients who are already seriously ill, with major risk factors including admission to ICU, surgery (especially abdominal surgery), and neutropenia (reviewed in [1]). The five *Candida* species commonly isolated from the human GI tract are also responsible for 90% of invasive *Candida* infections [1, 41]. Geographical variations in the epidemiology of these infections do occur, with *C. tropicalis* the most common cause of invasive *Candida* infection in both India and Singapore [42–44]. In addition, in patients with haematological malignancies and in young children and babies, there is increased incidence of *C. tropicalis* and *C. parapsilosis* [45–49].

Patients with invasive *Candida* infection usually present with clinical symptoms similar to those associated with invasive bacterial infection and can eventually develop sepsis [50]. From autopsy reports, it is evident that the lungs and the kidneys are the organs most commonly affected, with fungal lesions also found in the heart, liver, and spleen [51–55]. Infection most likely originates from the GI tract, as the majority of invasive infections show GI involvement (oesophagus, stomach, and intestines) [51, 53] and *Candida* isolates from the bloodstream are identical, or closely related, to isolates from nonsterile sites of the same patient [56].

Increasing numbers of patients suffering immunosuppression and undergoing invasive treatments, for example, for cancers and organ transplants, mean that there is an ever-increasing population at risk of invasive fungal infection. With a medical need for the development of new and more efficient diagnostics and therapies for fungal infection, we need a better understanding of *Candida* pathogenesis, that is, how do the major *Candida* species cause opportunistic infections?

2. Experimental Models of *Candida* Infection

Experimental infection models allow disease development to be followed from the moment that fungal cells are introduced into the host. To be a good model, a model should be reproducible, relatively easy to set up, and should reproduce the major clinical symptoms seen in the human disease. It is also an added advantage if the model is cost effective. Models which satisfy these conditions allow further in-depth investigation of *Candida* virulence to be carried out and, subsequently, allow inferences about *Candida* virulence in human disease to be made.

Although a great deal of preliminary research on virulence can be carried out by laboratory experiment, infection modelling requires the involvement of a host organism. It is only in a whole organism that the complex host-fungus interactions that determine whether or not disease will occur can be investigated. Although larger animals have been used to study *Candida* infections, for example, macaques [57, 58], piglets [59], rabbits [60–62], and guinea pigs [63, 64], the majority of *Candida* virulence studies use rodent infection models. This is due to economic factors, ease of handling, and the availability of genetically modified mouse strains, which allow human genetic conditions to be mimicked.

In this paper, experimental animal models that have been developed for *Candida* virulence assays are discussed. It should be noted that the majority of models focus on *C. albicans* as this is the major species associated with human *Candida* infections.

2.1. Mucosal Infection Models. To model *Candida* oral and vaginal infections, mucosal models have been developed mainly in rats and mice. The procedures used in rats and mice are generally similar. However, the larger animal has the added advantage that denture-associated fungal biofilms formation can also be studied in a host [65]. Establishment of infection at mucosal sites generally requires treatment with immunosuppressive agents, oestrogen, or antibiotics prior to infection, or the use of germ-free animals [66–68]. However, the nude (*Foxn1^{nu}*) mouse model of oral infection allows infection to be established without any immunosuppression or other pretreatment [69]. Greater detail can be found in more extensive reviews of these infection models [67, 68, 70, 71].

In order to assess virulence in mice using the oral infection model, mice are routinely pretreated with corticosteroids and *Candida* cells are administered into the oral cavity of anaesthetised animals either by applying a *Candida*-soaked cotton bud under the tongue or by applying the inoculum directly onto the teeth, gums, and oral cavity [67, 70, 72]. Virulence in this model is usually determined by fungal organ burden and histopathology.

Both rat and mouse models have been used to compare the virulence of *C. albicans* mutant strains and also clinical isolates [73–77]. Using these models, *C. albicans* mutant strains which are unable to switch between the yeast and hyphal growth forms were found to be unable to cause oral infection, demonstrating a requirement for yeast-hypha switching in oral infection [75]. In addition, protein kinase Ck2 was also shown to be required for oropharyngeal *C. albicans* infections [77].

Mouse and rat models have also been developed to assay *Candida* virulence in vaginal infection. In these models the rodents are maintained in oestrus in order to maintain colonisation and infection, which probably mimics pregnancy-associated candidiasis [78–81]. In rats, this generally involves surgery to remove the ovaries, with subsequent administration of oestrogen [81]. Recently, however, a new rat model has been developed, similar to the mouse model, where oestrus is maintained merely through administration of oestrogen [82], which will increase the ease of setting up the infection model. Immunosuppression of the host can also prolong colonisation by *Candida* species [83]. These models allow us to examine single vaginitis episodes; however, a satisfactory model of recurrent, chronic vaginitis is not yet available.

The virulence of *C. albicans* clinical isolates has been compared in rodent vaginitis models, demonstrating that isolates have varying capacities to cause disease [84, 85]. This model has also been used to assess virulence of genetically modified *C. albicans* mutants [85–87].

In addition to assessing *C. albicans* virulence, this model can be used to examine virulence of other *Candida* species. As *C. glabrata* is also associated with human vaginal infection, researchers have used the rat vaginitis model to evaluate the virulence of a *C. glabrata* petite mutant, discovering than the mutant was more virulent that the parental strain [88]. In addition, *C. parapsilosis* isolates have also been assessed for their ability to cause vaginal infection in the rat model [80]. In this study only a single isolate, recently obtained from a woman with active vaginal infection, was capable of initiating infection [80].

A major development in *Candida* virulence testing at mucosal surfaces occurred recently with the development of a concurrent oral and vaginal infection model by Rahman et al. [72]. This mouse model allows both oral and vaginal infections to be initiated in the same host, greatly reducing the numbers of animals required for these virulence assays. A comparison of the virulence of three different *C. albicans* isolates in this model clearly demonstrated that *C. albicans* isolates were not equally virulent, with obvious differences in their ability to initiate mucosal infections [72].

2.2. Invasive Infection Models. Mouse models of invasive fungal infection have been the most popular methods to assess *Candida* virulence up until the present day, although assays have also been carried out in rabbits, guinea pigs, and rats also used in some studies. There are two major models of *Candida* invasive infection, the intravenous (IV) challenge model and the gastrointestinal (GI) colonisation with subsequent dissemination model. These models were recently reviewed [89].

2.2.1. Intravenous Challenge Model. The mouse IV challenge model has been used to study *Candida* virulence since the 1960s and is both well characterised and reproducible [90–92]. *Candida* cells are injected directly into the lateral tail vein, bypassing any requirement of the fungus to cross epithelial and endothelial barriers to gain entry into the bloodstream. In this mouse model, which is similar to human invasive infection occurring with catheter involvement, fungal cells are found in all organs, but disease progresses only in the kidneys and brain, which depends upon inoculum level and mouse strain [91–93]. Sepsis develops as invasive disease progresses, which eventually leads to the death of the mouse [92, 94, 95].

In these models of *Candida* invasive infection, virulence is determined by monitoring survival of infected mice and/or by quantifying fungal organ burdens at predetermined times after infection. Drug treatments can also be administered to the host to allow host conditions to be mimicked, for example, immunosuppression [88, 96–110] or diabetes [99], with greater *Candida* virulence in both of these treatments.

Using immunocompetent mice, the IV challenge model has been used to compare the virulence of different *Candida* species [97–99, 107, 111–114]. *C. albicans* is clearly the most virulent species [97, 98, 111, 112, 114], followed closely by *C. tropicalis* [97, 98, 111, 112, 114]. In contrast, *C. krusei* and *C. parapsilosis* were unable to kill the infected animals, even at high inoculum levels, and fungi were eventually cleared from the host [98, 111, 114].

In immunosuppressed mice, *C. tropicalis* showed greater virulence, with disease progressing in the kidneys, rather than infection being controlled which occurs in immunocompetent mice [96, 98, 99, 107, 115]. *C. parapsilosis* and *C. krusei* remained unable to initiate progressive infections, even with addition of immunosuppressive treatments [98, 107], although administration of a very high inoculum potentially allows some *C. parapsilosis* isolates to initiate disease [108, 110].

Within each *Candida* species, clinical isolates were found to show considerable virulence differences in the IV challenge model. This was true for *C. albicans* [97, 107, 116, 117], *C. tropicalis* [97, 99, 112, 115, 118], and *C. parapsilosis* [108, 119], with some isolates unable to initiate invasive infections. This raises questions as to whether virulence results found for a single strain or isolate are representative of the entire species. This could be of particular importance for *C. albicans* studies where the vast majority of gene disruption studies have been carried out in a single strain, SC5314, background.

Numerous studies have evaluated *C. tropicalis* clinical isolate virulence differences; however, there are very few studies published on the virulence of genetically modified *C. tropicalis* strains. One study which has been published was able to demonstrate that a secreted acid protease was required for full virulence of *C. tropicalis* in immunocompetent mice [120]. In contrast to *C. tropicalis,* vast numbers of studies have been published on the virulence of *C. albicans* mutants, with over 200 genes identified as contributing to the *C. albicans* virulence in this model (reviewed in [89]).

C. glabrata behaves very differently from the other *Candida* species in the mouse model of invasive infection. Although *C. glabrata* is maintained, or tolerated, at high levels in the kidneys of immunocompetent mice, the mice did not die and there was little inflammation associated with the fungal cells [113, 114]. Immunosuppression appears to increase virulence of *C. glabrata* in terms of higher fungal organ burdens, but mouse survival is only increased in some *C. glabrata* infections [100, 103–106]. However, because immunosuppression may allow invasive disease to develop in *C. glabrata*-infected mice, these treatments have been added to an infection model used in some studies to compare the virulence of genetically modified *C. glabrata*, with fungal burdens used as the virulence estimate [88, 101, 102, 105]. The immunosuppressed mouse infection model has demonstrated the importance of hypertonic stress responses, the cell wall integrity pathway, and nitrogen starvation responses in *C. glabrata* virulence [103, 104, 106]. In addition, this model has identified a petite mutant, strains expressing hyperactive alleles of the transcription factor gene *PDR1* and the *ace2* null mutant as being more virulent than their parent strains [88, 105, 121]. However, it should be noted that the hypervirulent phenotype of the *C. glabrata ace2* null was completely lost in immunocompetent mice [122]. In other virulence experiments in immunocompetent mice, where virulence was determined from fungal organ burdens at day 7 after infection, researchers were able to demonstrate that the cell wall integrity pathway [123, 124] and oxidative stress response [125], as well as the transcription factor Pdr1p and some of the genes that it regulates [101, 121], contribute to *C. glabrata* virulence.

2.2.2. Gastrointestinal Colonisation and Dissemination Model. Gastrointestinal models can either be set up in neonatal or adult mice. Intragastric infection of neonatal mice leads to persistent colonisation, without any requirement for pretreatment of the mice. However, to obtain colonisation of adult mice, the natural mouse gastrointestinal flora must first be removed by treatment with broad spectrum antibiotics. Adult mice can either be infected by gavage (intragastrically) or orally via their chow or drinking water. Subsequent treatment of *Candida* colonised mice with immunosuppressants and/or drugs which damage the gut wall allow fungal dissemination to occur (reviewed in [70, 126]).

In the gastrointestinal models fungal colonisation is highest in the stomach, caecum, and small intestine [107, 127–129], reflecting some of the clinical findings seen in human invasive infection. During the model, persistent colonisation is routinely monitored by noninvasive faecal fungal counts, and after dissemination *Candida* cells can be cultured from the liver, kidneys, and spleen [128–130]. However, differences may be seen between mouse strains [131].

This murine model is believed to be a more accurate reflection of the events occurring in the human patient, with broad spectrum antibiotics allowing fungal overgrowth and later invasive therapies causing mucosal damage. Mucosal damage then allows *Candida* to enter the bloodstream and disseminate to the internal organs. In the mouse, similar to human patients, there is increased animal-to-animal variation compared to the intravenous challenge model, requiring higher numbers of animals per group to obtain statistically significant results [128–130].

Comparison of *Candida* species virulence in this model demonstrated that *C. parapsilosis* had lower virulence compared to *C. albicans* and *C. tropicalis,* as there was little evidence of dissemination from the gut [107, 132]. However, *C. parapsilosis* was successful in establishing persistent colonisation of the GI tract [107]. In separate studies, *C. tropicalis* appeared to be more virulent than *C. albicans* in the gastrointestinal model, with greater dissemination to the internal organs [133, 134] and higher mortality rates [97, 134]. However, given the levels of variation observed in other models for the virulence of strains of different *Candida* species, further isolates will require to be assayed before a definitive conclusion on the relative virulence of the two species can be made.

To date, only a limited number of *C. albicans* mutant strains have been tested in the gastrointestinal colonisation and dissemination infection model, with only 6 mutants identified so far as contributing to virulence [89, 135]. However, this model has demonstrated that a constitutively filamentous *C. albicans* mutant was unable to disseminate, suggesting that the ability to switch between morphological forms may be more important for dissemination [136].

C. glabrata also behaved differently from the other four major *Candida* species in this model, being unable to colonise the oesophageal tissue in the neonatal mouse gastrointestinal colonisation and dissemination model [137]. Again, there was little host inflammatory response to *C. glabrata* [137], suggesting that *C. glabrata* virulence mechanisms may be quite different from those of the other species studied.

3. Beyond the Genome: Challenges of *Candida* Virulence Testing in the Postgenomic Era

The genome sequences of *C. albicans*, *C. glabrata*, *C. tropicalis,* and *C. parapsilosis* are now available [138, 139], encouraging the creation of large-scale mutant libraries. The challenge comes, however, when these large libraries are to be screened for genes involved in fungal virulence, with logistical, financial, and ethical issues to be considered.

In library screening programmes carried out to date different virulence testing strategies have been taken. Noble et al. [140] used signature-tagged mutagenesis to allow pools of mutants to be assayed in small numbers of animals, significantly reducing the animal numbers required for testing. By contrast, in order to screen a library of 177 *C. albicans* strains for altered virulence, Becker et al. [141] assayed each strain in 15 mice. From these two examples it is clear that traditional testing methods can lead to large numbers of mice being required to assay virulence. However, researchers have recently begun to address the issues of virulence testing large numbers of *Candida* strains by developing a range of minihosts, which are mainly based on invertebrate hosts.

Minihosts may not initially appear relevant to the human disease, but these hosts do possess an innate immune system and this is known to be critical in the development of *Candida* infections [142]. However, many of the minihosts do not possess an adaptive immune system, which may limit their usefulness. In addition, the majority of invertebrate models have the disadvantage that they must be kept at temperatures below normal human body temperature, with the exception of *Galleria* which can be incubated at 37°C. Potentially, incubation at lower temperatures may induce physiological changes in the fungus, affecting host-fungus interactions during disease development.

3.1. Wax Moth and Silk Worm Larval Models. The first minihost model developed for *Candida* virulence testing was the *Galleria mellonella* (wax moth) larval model [143]. In this model fungi are injected into larvae, via a proleg, and survival is monitored over a short time period. The model is relatively cheap and has the added advantage that large numbers of larvae can be infected with each mutant strain, increasing the statistical power of the assay. The *Galleria* model has been successfully used to model *C. albicans* virulence, with results roughly similar to those found in mouse infection models [143–146]. A similar model has also been developed using the silk worm (*Bombyx mori*) [147, 148]. Both *C. albicans* and *C. tropicalis* are capable of killing silk worm larvae within two days [148], and *C. albicans* virulence differences were shown to correlate with results previously found in a mouse model [147].

3.2. Drosophila melanogaster. The fruit fly, *Drosophila melanogaster*, has also been used to assay *Candida* virulence [149–152]. The susceptibility of wild-type *D. melanogaster* continues to be debated; however, both Toll- and Spätzle-deficient fruit flies are susceptible to infection by *Candida* species when fungi are injected directly into the thorax [149–151]. Again, *D. melanogaster* models also have the advantage that large numbers of flies (>30 flies) can be infected with each *Candida* strain, increasing the statistical power of the assay.

In fruit flies, *C. albicans* was shown to be more virulent than other *Candida* species, confirming the results found in mammalian models (see above; [149]). In addition, virulence results for *C. albicans* clinical isolates and mutants were broadly similar to those found in the mouse systemic model [149–151]. However, differences do occur. In the fruit fly, CO_2 sensing is important for virulence, but this was not the case in the mouse IV challenge model [153]. This model has already been successfully used to screen a *C. albicans* mutant library, identifying Cas5, a transcription factor involved in cell wall integrity, as being required for full virulence [154].

In addition to the systemic *D. melanogaster* infection model, a new gastrointestinal infection model has also been developed recently, which should provide new options for virulence screening in a gastrointestinal model [152].

3.3. Caenorhabditis elegans. In addition to fly and larval models, the nematode *Caenorhabditis elegans* has also been evaluated as an infection model for assaying *Candida* virulence [155]. This model is particularly suited to high-throughput screening, as the *Candida* cells are fed to the nematodes in their food and assays are carried out in multi-well plates. This model has also been used successfully to screen a *C. albicans* transcription factor mutant library, allowing identification of transcription factor genes involved in hypha formation [155].

3.4. A Vertebrate Minihost: Zebrafish (Danio rerio). Zebrafish are the first vertebrate minihost model developed for virulence testing of *Candida*. This organism has the added advantage of having both innate and adaptive immune systems [156], and methods are also available to allow fish gene expression to be manipulated to mimic human genetic conditions [157].

The first virulence assay developed in zebrafish involved intraperitoneal injection of *C. albicans* into 7-month-old zebrafish [158]. In this model, similar to mouse models, progressive infection depends upon dose and is associated with increased proinflammatory gene expression. This model also allows increased group sizes, with group sizes of 20 fish being used to date. Using this model, researchers demonstrated that a clinical isolate with reduced virulence in a mouse model also showed reduced virulence in this model [158]. In addition, a *C. albicans* mutant (*efg1/cph1*) known to have attenuated virulence due to filamentation defects also had reduced virulence in this model [159, 160]. Of greater interest was the finding that, although these mutants were unable to form filaments *in vitro*, they clearly formed filaments when growing within fish. This model also allows interactions between zebrafish immune cells and *Candida* cells to be imaged, which will be made even easier in the future with the development of the new transparent adult (*casper*) zebrafish [161].

A second zebrafish infection model has also been described, where each fish larva (36 h after fertilization) is infected directly into the hindbrain ventricle with approximately 10 fungal cells [162]. In this model the *C. albicans efg1/cph1* mutant again demonstrated attenuated virulence, similar to results found in the mouse IV challenge model [162].

There are, however, disadvantages to the zebrafish infection models. One of the major drawbacks of this model, in common with the majority of other minihosts, is that the fish need to be kept at 28-29°C, which does not allow accurate mimicking of human infection.

4. Assaying Virulence in Experimental Models: Final Considerations

There are some important points to remember when evaluating *Candida* virulence in experimental infections. The first concerns the *Candida* species of interest. Although *C. albicans, C. tropicalis, C. glabrata,* and *C. krusei* are all associated with human carriage and infection, they are not natural mouse commensals or pathogens [163]. As such, there may be different interactions occurring between the fungus and the two different host species. This is of particular relevance when considering *C. glabrata* and its inability to initiate disseminated infection in the IV challenge models, especially when we know that *C. glabrata* can cause lethal infections in severely ill humans [164].

The second point to consider is that, although the immune systems of mice and men are similar, there are differences that could affect how the host and fungus interact [165–168]. Of particular relevance to *Candida* infections are differences in proportions of neutrophils and lymphocytes in the blood, complement receptor expression, and T-cell differentiation, to name but a few (reviewed in [168]). In addition, different mouse strains show differing susceptibility to infection, which could potentially alter virulence results [93, 169–172].

The third point to consider is which model should be used to evaluate *Candida* virulence. Some *C. albicans* isolates exhibit virulence differences depending upon the infection model being used [72, 134, 173]. A good example is the *C. albicans* genome sequenced strain SC5314. In the IV challenge model, SC5314 is one of the most virulent *C. albicans* isolates, causing lethal infection in a relatively short time [92, 116]; however, in a vaginal infection model, SC5314 is a very poor coloniser of the vaginal mucosa [72]. In addition, a nongerminative *C. albicans* strain [173] and a *ura3* minus *C. albicans* strain [174], both of which were attenuated in systemic infection models [173, 175–177], successfully established mucosal infections [173, 174].

Only careful consideration of the above points will allow the *Candida* researcher to select the appropriate experimental *Candida* infection model to answer a particular research question. These models remain essential for increasing our understanding of fungal pathogenesis since both fungal attributes and host responses are known to contribute to the development of clinical disease.

Acknowledgment

The author would like to apologise to those whose work could not be included in this review due to lack of space.

References

[1] D. M. MacCallum, "*Candida* infections and modelling disease," in *Pathogenic Yeasts, The Yeast Handbook*, H. R. Ashbee and E. Bignell, Eds., pp. 41–67, Springer, 2010.

[2] M. E. Bougnoux, D. Diogo, N. François et al., "Multilocus sequence typing reveals intrafamilial transmission and microevolutions of *Candida albicans* isolates from the human digestive tract," *Journal of Clinical Microbiology*, vol. 44, no. 5, pp. 1810–1820, 2006.

[3] S. Kusne, D. Tobin, A. W. Pasculle, D. H. Van Thiel, M. Ho, and T. E. Starzl, "*Candida* carriage in the alimentary tract of liver transplant candidates," *Transplantation*, vol. 57, no. 3, pp. 398–402, 1994.

[4] P. D. Scanlan and J. R. Marchesi, "Micro-eukaryotic diversity of the human distal gut microbiota: qualitative assessment using culture-dependent and -independent analysis of faeces," *ISME Journal*, vol. 2, no. 12, pp. 1183–1193, 2008.

[5] M. Belazi, A. Velegraki, A. Fleva et al., "Candidal overgrowth in diabetic patients: potential predisposing factors," *Mycoses*, vol. 48, no. 3, pp. 192–196, 2005.

[6] H. Ben-Aryeh, E. Blumfield, R. Szargel, D. Laufer, and I. Berdicevsky, "Oral *Candida* carriage and blood group antigen secretor status," *Mycoses*, vol. 38, no. 9-10, pp. 355–358, 1995.

[7] G. Campisi, G. Pizzo, M. E. Milici, S. Mancuso, and V. Margiotta, "Candidal carriage in the oral cavity of human immunodeficiency virus-infected subjects," *Oral Surgery, Oral Medicine, Oral Pathology, Oral Radiology, and Endodontics*, vol. 93, no. 3, pp. 281–286, 2002.

[8] S. Thaweboon, B. Thaweboon, T. Srithavaj, and S. Choonharuangdej, "Oral colonization of *Candida* species in patients receiving radiotherapy in the head and neck area," *Quintessence International*, vol. 39, no. 2, pp. e52–57, 2008.

[9] J. Wang, T. Ohshima, U. Yasunari et al., "The carriage of *Candida* species on the dorsal surface of the tongue: the correlation with the dental, periodontal and prosthetic status in elderly subjects," *Gerodontology*, vol. 23, no. 3, pp. 157–163, 2006.

[10] M. Dan, R. Segal, V. Marder, and A. Leibovitz, "*Candida* colonization of the vagina in elderly residents of a long-term-care hospital," *European Journal of Clinical Microbiology and Infectious Diseases*, vol. 25, no. 6, pp. 394–396, 2006.

[11] I. W. Fong, "The rectal carriage of yeast in patients with vaginal candidiasis," *Clinical and Investigative Medicine*, vol. 17, no. 5, pp. 426–431, 1994.

[12] M. V. Pirotta and S. M. Garland, "Genital *Candida* species detected in samples from women in Melbourne, Australia, before and after treatment with antibiotics," *Journal of Clinical Microbiology*, vol. 44, no. 9, pp. 3213–3217, 2006.

[13] E. Rylander, A. L. Berglund, C. Krassny, and B. Petrini, "Vulvovaginal *Candida* in a young sexually active population: prevalence and association with oro-genital sex and frequent pain at intercourse," *Sexually Transmitted Infections*, vol. 80, no. 1, pp. 54–57, 2004.

[14] E. M. de Leon, S. J. Jacober, J. D. Sobel, and B. Foxman, "Prevalence and risk factors for vaginal *Candida* colonization in women with type 1 and type 2 diabetes," *BMC Infectious Diseases*, vol. 2, article 1, 2002.

[15] A. Beltrame, A. Matteelli, A. C. C. Carvalho et al., "Vaginal colonization with *Candida* spp. in human immunodeficiency virus - Infected women: a cohort study," *International Journal of STD and AIDS*, vol. 17, no. 4, pp. 260–266, 2006.

[16] O. Grigoriou, S. Baka, E. Makrakis, D. Hassiakos, G. Kapparos, and E. Kouskouni, "Prevalence of clinical vaginal candidiasis in a university hospital and possible risk factors," *European Journal of Obstetrics Gynecology and Reproductive Biology*, vol. 126, no. 1, pp. 121–125, 2006.

[17] A. Paulitsch, W. Weger, G. Ginter-Hanselmayer, E. Marth, and W. Buzina, "A 5-year (2000–2004) epidemiological survey of *Candida* and non-*Candida* yeast species causing vulvovaginal candidiasis in Graz, Austria," *Mycoses*, vol. 49, no. 6, pp. 471–475, 2006.

[18] F. C. Odds, *Candida and Candidosis*, Bailliere Tindall, London, UK, 1988.

[19] B. Havlickova, V. A. Czaika, and M. Friedrich, "Epidemiological trends in skin mycoses worldwide," *Mycoses*, vol. 51, supplement 4, pp. 2–15, 2008.

[20] C. H. Kirkpatrick and H. R. Hill, "Chronic mucocutaneous candidiasis," *Pediatric Infectious Disease Journal*, vol. 20, no. 2, pp. 197–206, 2001.

[21] C. H. Kirkpatrick, "Chronic mucocutaneous candidiasis," *European Journal of Clinical Microbiology and Infectious Diseases*, vol. 8, no. 5, pp. 448–456, 1989.

[22] A. Puel, C. Picard, S. Cypowyj, D. Lilic, L. Abel, and J. L. Casanova, "Inborn errors of mucocutaneous immunity to *Candida albicans* in humans: a role for IL-17 cytokines?" *Current Opinion in Immunology*, vol. 22, no. 4, pp. 467–474, 2010.

[23] K. Kisand, D. Lilic, J. L. Casanova, P. Peterson, A. Meager, and N. Willcox, "Mucocutaneous candidiasis and autoimmunity against cytokines in APECED and thymoma patients: clinical and pathogenetic implications," *European Journal of Immunology*, vol. 41, no. 6, pp. 1517–1527, 2011.

[24] A. Puel, S. Cypowyj, J. Bustamante et al., "Chronic mucocutaneous candidiasis in humans with inborn errors of interleukin-17 immunity," *Science*, vol. 332, no. 6025, pp. 65–68, 2011.

[25] M. D. Richardson and D. W. Warnock, "Superficial candidosis," in *Fungal Infection: Diagnosis and Management*, pp. 78–93, Blackwell Science, London, UK, 1997.

[26] T. Daniluk, G. Tokajuk, W. Stokowska et al., "Occurrence rate of oral *Candida albicans* in denture wearer patients," *Advances in Medical Sciences*, vol. 51, pp. 77–80, 2006.

[27] M. H. Figueiral, A. Azul, E. Pinto, P. A. Fonseca, F. M. Branco, and C. Scully, "Denture-related stomatitis: identification of aetiological and predisposing factors—a large cohort," *Journal of Oral Rehabilitation*, vol. 34, no. 6, pp. 448–455, 2007.

[28] J. P. Lyon, S. C. da Costa, V. M. G. Totti, M. F. V. Munhoz, and M. A. De Resende, "Predisposing conditions for *Candida* spp. carriage in the oral cavity of denture wearers and individuals with natural teeth," *Canadian Journal of Microbiology*, vol. 52, no. 5, pp. 462–467, 2006.

[29] R. H. Pires-Gonçalves, E. T. Miranda, L. C. Baeza, M. T. Matsumoto, J. E. Zaia, and M. J. S. Mendes-Giannini, "Genetic relatedness of commensal strains of *Candida albicans* carried in the oral cavity of patients' dental prosthesis users in Brazil," *Mycopathologia*, vol. 164, no. 6, pp. 255–263, 2007.

[30] J. D. Sobel, "Vulvovaginal candidosis," *Lancet*, vol. 369, no. 9577, pp. 1961–1971, 2007.

[31] J. D. Sobel, S. Faro, R. W. Force et al., "Vulvovaginal candidiasis: epidemiologic, diagnostic, and therapeutic considerations," *American Journal of Obstetrics and Gynecology*, vol. 178, no. 2, pp. 203–211, 1998.

[32] A. B. Guzel, M. Ilkit, T. Akar, R. Burgut, and S. C. Demir, "Evaluation of risk factors in patients with vulvovaginal candidiasis and the value of chromID *Candida* agar versus CHROMagar *Candida* for recovery and presumptive identification of vaginal yeast species," *Medical Mycology*, vol. 49, no. 1, pp. 16–25, 2010.

[33] M. A. Kennedy and J. D. Sobel, "Vulvovaginal Candidiasis caused by non-*albicans Candida* species: new insights," *Current Infectious Disease Reports*, vol. 12, no. 6, pp. 465–470, 2010.

[34] S. Asticcioli, L. Sacco, R. Daturi et al., "Trends in frequency and in vitro antifungal susceptibility patterns of *Candida* isolates from women attending the STD outpatients clinic of a tertiary care hospital in Northern Italy during the years 2002—2007," *New Microbiologica*, vol. 32, no. 2, pp. 199–204, 2009.

[35] S. Corsello, A. Spinillo, G. Osnengo et al., "An epidemiological survey of vulvovaginal candidiasis in Italy," *European Journal of Obstetrics Gynecology and Reproductive Biology*, vol. 110, no. 1, pp. 66–72, 2003.

[36] F. L. van de Veerdonk, T. S. Plantinga, A. Hoischen et al., "STAT1 mutations in autosomal dominant chronic mucocutaneous candidiasis," *New England Journal of Medicine*, vol. 365, no. 1, pp. 54–61, 2011.

[37] H. Markogiannakis, N. Pachylaki, E. Samara et al., "Infections in a surgical intensive care unit of a university hospital in Greece," *International Journal of Infectious Diseases*, vol. 13, no. 2, pp. 145–153, 200.

[38] G. B. Orsi, L. Scorzolini, C. Franchi, V. Mondillo, G. Rosa, and M. Venditti, "Hospital-acquired infection surveillance in a neurosurgical intensive care unit," *Journal of Hospital Infection*, vol. 64, no. 1, pp. 23–29, 2006.

[39] E. Sarvikivi, O. Lyytikäinen, M. Vaara, and H. Saxén, "Nosocomial bloodstream infections in children: an 8-year experience at a tertiary-care hospital in Finland," *Clinical Microbiology and Infection*, vol. 14, no. 11, pp. 1072–1075, 2008.

[40] M. Morrell, V. J. Fraser, and M. H. Kollef, "Delaying the empiric treatment of *Candida* bloodstream infection until positive blood culture results are obtained: a potential risk factor for hospital mortality," *Antimicrobial Agents and Chemotherapy*, vol. 49, no. 9, pp. 3640–3645, 2005.

[41] H. Wisplinghoff, H. Seifert, R. P. Wenzel, and M. B. Edmond, "Inflammatory response and clinical course of adult patients with nosocomial bloodstream infections caused by *Candida* spp," *Clinical Microbiology and Infection*, vol. 12, no. 2, pp. 170–177, 2006.

[42] Y. A. L. Chai, Y. Wang, A. L. Khoo et al., "Predominance of *Candida tropicalis* bloodstream infections in a Singapore teaching hospital," *Medical Mycology*, vol. 45, no. 5, pp. 435–439, 2007.

[43] S. Shivaprakasha, K. Radhakrishnan, and P. Karim, "*Candida* spp. other than *Candida albicans*: a major cause of fungaemia in a tertiary care centre," *Indian Journal of Medical Microbiology*, vol. 25, no. 4, pp. 405–407, 2007.

[44] I. Xess, N. Jain, F. Hasan, P. Mandal, and U. Banerjee, "Epidemiology of candidemia in a tertiary care centre of North India: 5-Year study," *Infection*, vol. 35, no. 4, pp. 256–259, 2007.

[45] A. C. Pasqualotto, D. D. Rosa, L. R. Medeiros, and L. C. Severo, "Candidaemia and cancer: patients are not all the same," *BMC Infectious Diseases*, pp. 50–56, 2006.

[46] R. Hachem, H. Hanna, D. Kontoyiannis, Y. Jiang, and I. Raad, "The changing epidemiology of invasive candidiasis: *Candida glabrata* and *Candida krusei* as the leading causes of candidemia in hematologic malignancy," *Cancer*, vol. 112, no. 11, pp. 2493–2499, 2008.

[47] E. Presterl, F. Daxböck, W. Graninger, and B. Willinger, "Changing pattern of candidaemia 2001-2006 and use of antifungal therapy at the University Hospital of Vienna, Austria," *Clinical Microbiology and Infection*, vol. 13, no. 11, pp. 1072–1076, 2007.

[48] S. Vigouroux, O. Morin, P. Moreau, J. L. Harousseau, and N. Milpied, "Candidemia in patients with hematologic malignancies: analysis of 7 years' experience in a single center," *Haematologica*, vol. 91, no. 5, pp. 717–718, 2006.

[49] R. Saha, S. Das Das, A. Kumar, and I. R. Kaur, "Pattern of *Candida* isolates in hospitalized children," *Indian Journal of Pediatrics*, vol. 75, no. 8, pp. 858–860, 2008.

[50] B. Spellberg and J. E. Edwards, "The pathophysiology and treatment of *Candida* Sepsis," *Current Infectious Disease Reports*, vol. 4, no. 5, pp. 387–399, 2002.

[51] J. Berenguer, M. Buck, F. Witebsky, F. Stock, P. A. Pizzo, and T. J. Walsh, "Lysis-centrifugation blood cultures in the detection of tissue-proven invasive candidiasis: disseminated versus single-organ infection," *Diagnostic Microbiology and Infectious Disease*, vol. 17, no. 2, pp. 103–109, 1993.

[52] T. Lehrnbecher, C. Frank, K. Engels, S. Kriener, A. H. Groll, and D. Schwabe, "Trends in the postmortem epidemiology of invasive fungal infections at a university hospital," *Journal of Infection*, vol. 61, no. 3, pp. 259–265, 2010.

[53] K. Donhuijsen, P. Petersen, and K. W. Schmid, "Trend reversal in the frequency of mycoses in hematological neoplasias: autopsy results from 1976 to 2005," *Deutsches Arzteblatt*, vol. 105, no. 28-29, pp. 501–506, 2008.

[54] S. Antinori, M. Nebuloni, C. Magni et al., "Trends in the postmortem diagnosis of opportunistic invasive fungal infections in patients with AIDS: a retrospective study of 1,630 autopsies performed between 1984 and 2002," *American Journal of Clinical Pathology*, vol. 132, no. 2, pp. 221–227, 2009.

[55] G. Schwesinger, D. Junghans, G. Schröder, H. Bernhardt, and M. Knoke, "Candidosis and aspergillosis as autopsy findings

from 1994 to 2003," *Mycoses*, vol. 48, no. 3, pp. 176–180, 2005.

[56] F. C. Odds, A. D. Davidson, M. D. Jacobsen et al., "*Candida albicans* strain maintenance, replacement, and microvariation demonstrated by multilocus sequence typing," *Journal of Clinical Microbiology*, vol. 44, no. 10, pp. 3647–3658, 2006.

[57] E. Budtz-Jorgensen, "Denture stomatitis. IV. An experimental model in monkeys," *Acta Odontologica Scandinavica*, vol. 29, no. 5, pp. 513–526, 1971.

[58] C. Steele, M. Ratterree, and P. L. Fidel, "Differential susceptibility of two species of macaques to experimental vaginal candidiasis," *Journal of Infectious Diseases*, vol. 180, no. 3, pp. 802–810, 1999.

[59] K. A. Andrutis, P. J. Riggle, C. A. Kumamoto, and S. Tzipori, "Intestinal lesions associated with disseminated candidiasis in an experimental animal model," *Journal of Clinical Microbiology*, vol. 38, no. 6, pp. 2317–2323, 2000.

[60] S. G. Filler, M. A. Crislip, C. L. Mayer, and J. E. Edwards, "Comparison of fluconazole and amphotericin B for treatment of disseminated candidiasis and endophthalmitis in rabbits," *Antimicrobial Agents and Chemotherapy*, vol. 35, no. 2, pp. 288–292, 1991.

[61] C. A. Lyman, C. Gonzalez, M. Schneider, J. Lee, and T. J. Walsh, "Effects of the hematoregulatory peptide SKandF 107647 alone and in combination with amphotericin B against disseminated candidiasis in persistently neutropenic rabbits," *Antimicrobial Agents and Chemotherapy*, vol. 43, no. 9, pp. 2165–2169, 1999.

[62] A. Polanco, E. Mellado, C. Castilla, and J. L. Rodriguez-Tudela, "Detection of *Candida albicans* in blood by PCR in a rabbit animal model of disseminated candidiasis," *Diagnostic Microbiology and Infectious Disease*, vol. 34, no. 3, pp. 177–183, 1999.

[63] J. Fransen, J. van Cutsem, R. Vandesteene, and P. A. J. Janssen, "Histopathology of experimental systemic candidosis in guinea-pigs," *Sabouraudia*, vol. 22, no. 6, pp. 455–469, 1984.

[64] J. van Cutsem and D. Thienpont, "Experimental cutaneous *Candida albicans* infection in guinea-pigs," *Sabouraudia*, vol. 9, no. 1, pp. 17–20, 1971.

[65] J. E. Nett, K. Marchillo, C. A. Spiegel, and D. R. Andes, "Development and validation of an *in vivo Candida albicans* biofilm denture model," *Infection and Immunity*, vol. 78, no. 9, pp. 3650–3659, 2010.

[66] Y. Kamai, M. Kubota, Y. Kamai, T. Hosokawa, T. Fukuoka, and S. G. Filler, "New model of oropharyngeal candidiasis in mice," *Antimicrobial Agents and Chemotherapy*, vol. 45, no. 11, pp. 3195–3197, 2001.

[67] Y. H. Samaranayake and L. P. Samaranayake, "Experimental oral candidiasis in animal models," *Clinical Microbiology Reviews*, vol. 14, no. 2, pp. 398–429, 2001.

[68] P. L. Fidel Jr. and J. D. Sobel, "Murine models of *Candida* vaginal infections," in *Handbook of Animal Models of Infection: Experimental Models in Antimicrobial Chemotherapy*, O. Zak and M. A. Sande, Eds., pp. 741–748, Academic Press, New York, NY, USA, 1999.

[69] C. S. Farah, S. Elahi, K. Drysdale et al., "Primary role for CD4+ T lymphocytes in recovery from oropharyngeal candidiasis," *Infection and Immunity*, vol. 70, no. 2, pp. 724–731, 2002.

[70] J. R. Naglik, P. L. Fidel, and F. C. Odds, "Animal models of mucosal *Candida* infection," *FEMS Microbiology Letters*, vol. 283, no. 2, pp. 129–139, 2008.

[71] F. de Bernardis, R. Lorenzini, and A. Cassone, "Rat model of *Candida* vaginal infection," in *Handbook of Animal Models of Infection: Experimental Models in Antimicrobial Chemotherapy*, O. Zak and M. A. Sande, Eds., pp. 735–740, Academic Press, New York, NY, USA, 1999.

[72] D. Rahman, M. Mistry, S. Thavaraj, S. J. Challacombe, and J. R. Naglik, "Murine model of concurrent oral and vaginal *Candida albicans* colonization to study epithelial host-pathogen interactions," *Microbes and Infection*, vol. 9, no. 5, pp. 615–622, 2007.

[73] H. Badrane, M. H. Nguyen, S. Cheng et al., "The *Candida albicans* phosphatase Inp51 p interacts with the EH domain protein Irs4p, regulates phosphatidylinositol-4,5-bisphosphate levels and influences hyphal formation, the cell integrity pathway and virulence," *Microbiology*, vol. 154, no. 11, pp. 3296–3308, 2008.

[74] W. P. Holbrook, J. A. Sofaer, and J. C. Southam, "Experimental oral infection of mice with a pathogenic and a non-pathogenic strain of the yeast *Candida albicans*," *Archives of Oral Biology*, vol. 28, no. 12, pp. 1089–1091, 1983.

[75] C. J. Nobile, N. Solis, C. L. Myers et al., "*Candida albicans* transcription factor Rim101 mediates pathogenic interactions through cell wall functions," *Cellular Microbiology*, vol. 10, no. 11, pp. 2180–2196, 2008.

[76] H. Park, C. L. Myers, D. C. Sheppard et al., "Role of the fungal ras-protein kinase a pathway in governing epithelial cell interactions during oropharyngeal candidiasis," *Cellular Microbiology*, vol. 7, no. 4, pp. 499–510, 2005.

[77] L. Y. Chiang, D. C. Sheppard, V. M. Bruno, A. P. Mitchell, J. E. Edwards, and S. G. Filler, "*Candida albicans* protein kinase CK2 governs virulence during oropharyngeal candidiasis," *Cellular Microbiology*, vol. 9, no. 1, pp. 233–245, 2007.

[78] Z. Chen and X. Kong, "Study of *Candida albicans* vaginitis model in Kunming mice," *Journal of Huazhong University of Science and Technology, Medical Science*, vol. 27, no. 3, pp. 307–310, 2007.

[79] K. V. Clemons, J. L. Spearow, R. Parmar, M. Espiritu, and D. A. Stevens, "Genetic susceptibility of mice to *Candida albicans* vaginitis correlates with host estrogen sensitivity," *Infection and Immunity*, vol. 72, no. 8, pp. 4878–4880, 2004.

[80] F. de Bernardis, R. Lorenzini, L. Morelli, and A. Cassone, "Experimental rat vaginal infection with *Candida parapsilosis*," *FEMS Microbiology Letters*, vol. 53, no. 1-2, pp. 137–141, 1989.

[81] J. D. Sobel, G. Muller, and J. F. McCormick, "Experimental chronic vaginal candidosis in rats," *Sabouraudia*, vol. 23, no. 3, pp. 199–206, 1985.

[82] M. A. Carrara, L. Donatti, E. Damke, T. I. E. Svidizinski, M. E. L. Consolaro, and M. R. Batista, "A new model of vaginal infection by *Candida albicans* in rats," *Mycopathologia*, vol. 170, no. 5, pp. 331–338, 2010.

[83] M. Foldvari, J. Radhi, G. Yang, Z. He, R. Rennie, and L. Wearley, "Acute vaginal candidosis model in the immunocompromized rat to evaluate delivery systems for antimycotics," *Mycoses*, vol. 43, no. 11-12, pp. 393–401, 2000.

[84] A. Tavanti, D. Campa, S. Arancia, L. A. M. Hensgens, F. de Bernardis, and S. Senesi, "Outcome of experimental rat vaginitis by *Candida albicans* isolates with different karyotypes," *Microbial Pathogenesis*, vol. 49, no. 1-2, pp. 47–50, 2010.

[85] B. N. Taylor, C. Fichtenbaum, M. Saavedra et al., "*In vivo* virulence of *Candida albicans* isolates causing mucosal infections in people infected with the human immunodeficiency virus," *Journal of Infectious Diseases*, vol. 182, no. 3, pp. 955–959, 2000.

[86] T. Bader, K. Schröppel, S. Bentink, N. Agabian, G. Köhler, and J. Morschhäuser, "Role of calcineurin in stress resistance, morphogenesis, and virulence of a *Candida albicans* wild-type strain," *Infection and Immunity*, vol. 74, no. 7, pp. 4366–4369, 2006.

[87] Y. Fu, G. Luo, B. J. Spellberg, J. E. Edwards, and A. S. Ibrahim, "Gene overexpression/suppression analysis of candidate virulence factors of *Candida albicans*," *Eukaryotic Cell*, vol. 7, no. 3, pp. 483–492, 2008.

[88] S. Ferrari, M. Sanguinetti, F. De Bernardis et al., "Loss of mitochondrial functions associated with azole resistance in *Candida glabrata* results in enhanced virulence in mice," *Antimicrobial Agents and Chemotherapy*, vol. 55, no. 5, pp. 1852–1860, 2011.

[89] E. K. Szabo and D. M. MacCallum, "The contribution of mouse models to our understanding of systemic candidiasis," *FEMS Microbiology Letters*, vol. 320, no. 1, pp. 1–8, 2011.

[90] D. B. Louria, R. G. Brayton, and G. Finkel, "Studies on the pathogenesis of experimental *Candida albicans* infections in mice," *Sabouraudia*, vol. 2, pp. 271–283, 1963.

[91] J. M. Papadimitriou and R. B. Ashman, "The pathogenesis of acute systemic candidiasis in a susceptible inbred mouse strain," *Journal of Pathology*, vol. 150, no. 4, pp. 257–265, 1986.

[92] D. M. MacCallum and F. C. Odds, "Temporal events in the intravenous challenge model for experimental *Candida albicans* infections in female mice," *Mycoses*, vol. 48, no. 3, pp. 151–161, 2005.

[93] R. B. Ashman, A. Fulurija, and J. M. Papadimitriou, "Strain-dependent differences in host response to *Candida albicans* infection in mice are related to organ susceptibility infectious load," *Infection and Immunity*, vol. 64, no. 5, pp. 1866–1869, 1996.

[94] B. Spellberg, A. S. Ibrahim, J. E. Edwards, and S. G. Filler, "Mice with disseminated candidiasis die of progressive sepsis," *Journal of Infectious Diseases*, vol. 192, no. 2, pp. 336–343, 2005.

[95] D. M. MacCallum, L. Castillo, A. J. P. Brown, N. A. R. Gow, and F. C. Odds, "Early-expressed chemokines predict kidney immunopathology in experimental disseminated *Candida albicans* infections," *PLoS ONE*, vol. 4, no. 7, Article ID e6420, 2009.

[96] J. R. Graybill, L. K. Najvar, J. D. Holmberg, and M. F. Luther, "Fluconazole, D0870, and flucytosine treatment of disseminated *Candida tropicalis* infections in mice," *Antimicrobial Agents and Chemotherapy*, vol. 39, no. 4, pp. 924–929, 1995.

[97] L. De Repentigny, M. Phaneuf, and L. G. Mathieu, "Gastrointestinal colonization and systemic dissemination by *Candida albicans* and *Candida tropicalis* in intact and immunocompromised mice," *Infection and Immunity*, vol. 60, no. 11, pp. 4907–4914, 1992.

[98] F. Bistoni, A. Vecchiarelli, and E. Cenci, "A comparison of experimental pathogenicity of *Candida* species in cyclophosphamide-immunodepressed mice," *Sabouraudia*, vol. 22, no. 5, pp. 409–418, 1984.

[99] D. B. Louria, M. Busé, R. G. Brayton, and G. Finkel, "The pathogenesis of *Candida tropicalis* infections in mice," *Sabouraudia*, vol. 5, no. 1, pp. 14–25, 1967.

[100] I. D. Jacobsen, S. Brunke, K. Seider et al., "*Candida glabrata* persistence in mice does not depend on host immunosuppression and is unaffected by fungal amino acid auxotrophy," *Infection and Immunity*, vol. 78, no. 3, pp. 1066–1077, 2010.

[101] S. Ferrari, M. Sanguinetti, R. Torelli, B. Posteraro, and D. Sanglard, "Contribution of *CgPDR1*-regulated genes in enhanced virulence of azole-resistant *Candida glabrata*," *PLoS ONE*, vol. 6, no. 3, 2011.

[102] H. Nakayama, K. Ueno, J. Uno et al., "Growth defects resulting from inhibiting *ERG20* and *RAM2* in *Candida glabrata*," *FEMS Microbiology Letters*, vol. 317, no. 1, pp. 27–33, 2011.

[103] A. M. Calcagno, E. Bignell, T. R. Rogers, M. D. Jones, F. A. Mühlschlegel, and K. Haynes, "*Candida glabrata* Ste11 is involved in adaptation to hypertonic stress, maintenance of wild-type levels of filamentation and plays a role in virulence," *Medical Mycology*, vol. 43, no. 4, pp. 355–364, 2005.

[104] A. M. Calcagno, E. Bignell, T. R. Rogers, M. Canedo, F. A. Mühlschleger, and K. Haynes, "*Candida glabrata* Ste20 is involved in maintaining cell wall integrity and adaptation to hypertonic stress, and is required for wild-type levels of virulence," *Yeast*, vol. 21, no. 7, pp. 557–568, 2004.

[105] M. Kamran, A. M. Calcagno, H. Findon et al., "Inactivation of transcription factor gene *ACE2* in the fungal pathogen *Candida glabrata* results in hypervirulence," *Eukaryotic Cell*, vol. 3, no. 2, pp. 546–552, 2004.

[106] A. M. Calcagno, E. Bignell, P. Warn et al., "*Candida glabrata STE12* is required for wild-type levels of virulence and nitrogen starvation induced filamentation," *Molecular Microbiology*, vol. 50, no. 4, pp. 1309–1318, 2003.

[107] E. Mellado, M. Cuenca-Estrella, J. Regadera, M. González, T. M. Díaz-Guerra, and J. L. Rodríguez-Tudela, "Sustained gastrointestinal colonization and systemic dissemination by *Candida albicans*, *Candida tropicalis* and *Candida parapsilosis* in adult mice," *Diagnostic Microbiology and Infectious Disease*, vol. 38, no. 1, pp. 21–28, 2000.

[108] F. De Bernardis, L. Morelli, T. Ceddia, R. Lorenzini, and A. Cassone, "Experimental pathogenicity and acid proteinase secretion of vaginal isolates of *Candida parapsilosis*," *Journal of Medical and Veterinary Mycology*, vol. 28, no. 2, pp. 125–137, 1990.

[109] C. Girmenia, P. Martine, F. De Bernardis et al., "Rising incidence of *Candida parapsilosis* fungemia in patients with hematologic malignancies: clinical aspects, predisposing factors, and differential pathogenicity of the causative strains," *Clinical Infectious Diseases*, vol. 23, no. 3, pp. 506–514, 1996.

[110] E. Anaissie, R. Hachem, U. C. K-Tin, L. C. Stephens, and G. P. Bodey, "Experimental hematogenous candidiasis caused by *candida krusei* and *Candida albicans*: species differences in pathogenicity," *Infection and Immunity*, vol. 61, no. 4, pp. 1268–1271, 1993.

[111] C. Y. Koga-Ito, E. Y. Komiyama, C. A. de Paiva Martins et al., "Experimental systemic virulence of oral *Candida dubliniensis* isolates in comparison with *Candida albicans*, *Candida tropicalis* and *Candida krusei*," *Mycoses*, vol. 54, no. 5, pp. e278–e285, 2010.

[112] H.F. Hasenclever and W. O. Mitchell, "Pathogenicity of *C. albicans* and *C. tropicalis*," *Sabouraudia*, vol. 1, pp. 16–21, 1961.

[113] J. Brieland, D. Essig, C. Jackson et al., "Comparison of pathogenesis and host immune responses to *Candida glabrata* and *Candida albicans* in systemically infected immunocompetent mice," *Infection and Immunity*, vol. 69, no. 8, pp. 5046–5055, 2001.

[114] M. Arendrup, T. Horn, and N. Frimodt-Møller, "*In vivo* pathogenicity of eight medically relevant *Candida* species in an animal model," *Infection*, vol. 30, no. 5, pp. 286–291, 2002.

[115] R. A. Fromtling, G. K. Abruzzo, and D. M. Giltinan, "*Candida tropicalis* infection in normal, diabetic and neutropenic mice," *Journal of Clinical Microbiology*, vol. 25, no. 8, pp. 1416–1420, 1987.

[116] D. M. MacCallum, L. Castillo, K. Nather et al., "Property differences among the four major *Candida albicans* strain clades," *Eukaryotic Cell*, vol. 8, no. 3, pp. 373–387, 2009.

[117] P. Sampaio, M. Santos, A. Correia et al., "Virulence attenuation of *Candida albicans* genetic variants isolated from a patient with a recurrent bloodstream infection," *PLoS ONE*, vol. 5, no. 4, Article ID e10155, 2010.

[118] Y. Okawa, M. Miyauchi, and H. Kobayashi, "Comparison of pathogenicity of various *Candida tropicalis* strains," *Biological and Pharmaceutical Bulletin*, vol. 31, no. 8, pp. 1507–1510, 2008.

[119] A. Cassone, F. De Bernardis, E. Pontieri et al., "Biotype diversity of *Candida parapsilosis* and its relationship to the clinical source and experimental pathogenicity," *Journal of Infectious Diseases*, vol. 171, no. 4, pp. 967–975, 1995.

[120] G. Togni, D. Sanglard, and M. Monod, "Acid proteinase secreted by *Candida tropicalis*: virulence in mice of a proteinase negative mutant," *Journal of Medical and Veterinary Mycology*, vol. 32, no. 4, pp. 257–265, 1994.

[121] S. Ferrari, F. Ischer, D. Calabrese et al., "Gain of function mutations in *CgPDR1* of *Candida glabrata* not only mediate antifungal resistance but also enhance virulence," *PLoS Pathogens*, vol. 5, no. 1, Article ID e1000268, 2009.

[122] D. M. MacCallum, H. Findon, C. C. Kenny, G. Butler, K. Haynes, and F. C. Odds, "Different consequences of *ACE2* and *SWI5* gene disruptions for virulence of pathogenic and nonpathogenic yeasts," *Infection and Immunity*, vol. 74, no. 9, pp. 5244–5248, 2006.

[123] T. Miyazaki, T. Inamine, S. Yamauchi et al., "Role of the Slt2 mitogen-activated protein kinase pathway in cell wall integrity and virulence in *Candida glabrata*," *FEMS Yeast Research*, vol. 10, no. 3, pp. 343–352, 2010.

[124] T. Miyazaki, S. Yamauchi, T. Inamine et al., "Roles of calcineurin and Crz1 in antifungal susceptibility and virulence of *Candida glabrata*," *Antimicrobial Agents and Chemotherapy*, vol. 54, no. 4, pp. 1639–1643, 2010.

[125] T. Saijo, T. Miyazaki, K. Izumikawa et al., "Skn7p is involved in oxidative stress response and virulence of *Candida glabrata*," *Mycopathologia*, vol. 169, no. 2, pp. 81–90, 2010.

[126] G. T. Cole, A. A. Halawa, and E. J. Anaissie, "The role of the gastrointestinal tract in hematogenous candidiasis: from the laboratory to the bedside," *Clinical Infectious Diseases*, vol. 22, supplement 2, pp. S73–S88, 1996.

[127] S. M. Wiesner, R. P. Jechorek, R. M. Garni, C. M. Bendel, and C. L. Wells, "Gastrointestinal colonization by *Candida albicans* mutant strains in antibiotic-treated mice," *Clinical and Diagnostic Laboratory Immunology*, vol. 8, no. 1, pp. 192–195, 2001.

[128] K. V. Clemons, G. M. Gonzalez, G. Singh et al., "Development of an orogastrointestinal mucosal model of candidiasis with dissemination to visceral organs," *Antimicrobial Agents and Chemotherapy*, vol. 50, no. 8, pp. 2650–2657, 2006.

[129] H. Sandovsky-Losica, L. Barr-Nea, and E. Segal, "Fatal systemic candidiasis of gastrointestinal origin: an experimental model in mice compromised by anti-cancer treatment," *Journal of Medical and Veterinary Mycology*, vol. 30, no. 3, pp. 219–231, 1992.

[130] A. Y. Koh, J. R. Köhler, K. T. Coggshall, N. Van Rooijen, and G. B. Pier, "Mucosal damage and neutropenia are required for *Candida albicans* dissemination," *PLoS Pathogens*, vol. 4, no. 2, p. e35, 2008.

[131] M. T. Cantorna and E. Balish, "Mucosal and systemic candidiasis in congenitally immunodeficient mice," *Infection and Immunity*, vol. 58, no. 4, pp. 1093–1100, 1990.

[132] M. J. Kennedy and P. A. Volz, "Dissemination of yeasts after gastrointestinal inoculation in antibiotic-treated mice," *Sabouraudia*, vol. 21, no. 1, pp. 27–33, 1983.

[133] J. R. Wingard, J. D. Dick, and W. G. Merz, "Pathogenicity of *Candida tropicalis* and *Candida albicans* after gastrointestinal inoculation in mice," *Infection and Immunity*, vol. 29, no. 2, pp. 808–813, 1980.

[134] J. R. Wingard, J. D. Dick, and W. G. Merz, "Differences in virulence of clinical isolates of *Candida tropicalis* and *Candida albicans* in mice," *Infection and Immunity*, vol. 37, no. 2, pp. 833–836, 1982.

[135] M. S. Skrzypek, M. B. Arnaud, M. C. Costanzo et al., "New tools at the *Candida* genome database: biochemical pathways and full-text literature search," *Nucleic Acids Research*, vol. 38, pp. D428–D432, 2010.

[136] C. M. Bendel, D. J. Hess, R. M. Garni, M. Henry-Stanley, and C. L. Wells, "Comparative virulence of *Candida albicans* yeast and filamentous forms in orally and intravenously inoculated mice," *Critical Care Medicine*, vol. 31, no. 2, pp. 501–507, 2003.

[137] C. Westwater, D. A. Schofield, P. J. Nicholas, E. E. Paulling, and E. Balish, "*Candida glabrata* and *Candida albicans*; dissimilar tissue tropism and infectivity in a gnotobiotic model of mucosal candidiasis," *FEMS Immunology and Medical Microbiology*, vol. 51, no. 1, pp. 134–139, 2007.

[138] G. Butler, M. D. Rasmussen, M. F. Lin et al., "Evolution of pathogenicity and sexual reproduction in eight *Candida* genomes," *Nature*, vol. 459, no. 7247, pp. 657–662, 2009.

[139] B. Dujon, D. Sherman, G. Fischer et al., "Genome evolution in yeasts," *Nature*, vol. 430, no. 6995, pp. 35–44, 2004.

[140] S. M. Noble, S. French, L. A. Kohn, V. Chen, and A. D. Johnson, "Systematic screens of a *Candida albicans* homozygous deletion library decouple morphogenetic switching and pathogenicity," *Nature Genetics*, vol. 42, no. 7, pp. 590–598, 2010.

[141] J. M. Becker, S. J. Kauffman, M. Hauser et al., "Pathway analysis of *Candida albicans* survival and virulence determinants in a murine infection model," *Proceedings of the National Academy of Sciences of the United States of America*, vol. 107, no. 51, pp. 22044–22049, 2010.

[142] K. Kavanagh and E. P. Reeves, "Exploiting the potential of insects for *in vivo* pathogenicity testing of microbial pathogens," *FEMS Microbiology Reviews*, vol. 28, no. 1, pp. 101–112, 2004.

[143] G. Cotter, S. Doyle, and K. Kavanagh, "Development of an insect model for the *in vivo* pathogenicity testing of yeasts," *FEMS Immunology and Medical Microbiology*, vol. 27, no. 2, pp. 163–169, 2000.

[144] M. Brennan, D. Y. Thomas, M. Whiteway, and K. Kavanagh, "Correlation between virulence of *Candida albicans* mutants in mice and *Galleria mellonella* larvae," *FEMS Immunology and Medical Microbiology*, vol. 34, no. 2, pp. 153–157, 2002.

[145] B. B. Fuchs, J. Eby, C. J. Nobile, J. B. El Khoury, A. P. Mitchell, and E. Mylonakis, "Role of filamentation in *Galleria mellonella* killing by *Candida albicans*," *Microbes and Infection*, vol. 12, no. 6, pp. 488–496, 2010.

[146] G. B. Dunphy, U. Oberholzer, M. Whiteway, R. J. Zakarian, and I. Boomer, "Virulence of *Candida albicans* mutants

toward larval *Galleria mellonella* (Insecta, Lepidoptera, Galleridae)," *Canadian Journal of Microbiology*, vol. 49, no. 8, pp. 514–524, 2003.

[147] N. Hanaoka, Y. Takano, K. Shibuya, H. Fugo, Y. Uehara, and M. Niimi, "Identification of the putative protein phosphatase gene *PTC1* as a virulence-related gene using a silkworm model of *Candida albicans* infection," *Eukaryotic Cell*, vol. 7, no. 10, pp. 1640–1648, 2008.

[148] H. Hamamoto, K. Kurokawa, C. Kaito et al., "Quantitative evaluation of the therapeutic effects of antibiotics using silkworms infected with human pathogenic microorganisms," *Antimicrobial Agents and Chemotherapy*, vol. 48, no. 3, pp. 774–779, 2004.

[149] G. Chamilos, M. S. Lionakis, R. E. Lewis et al., "*Drosophila melanogaster* as a facile model for large-scale studies of virulence mechanisms and antifungal drug efficacy in *Candida* species," *Journal of Infectious Diseases*, vol. 193, no. 7, pp. 1014–1022, 2006.

[150] A. M. Alarco, A. Marcil, J. Chen, B. Suter, D. Thomas, and M. Whiteway, "Immune-Deficient *Drosophila melanogaster*: a model for the innate immune response to human fungal pathogens," *Journal of Immunology*, vol. 172, no. 9, pp. 5622–5628, 2004.

[151] M. T. Glittenberg, S. Silas, D. M. MacCallum, N. A.R. Gow, and P. Ligoxygakis, "Wild-type *Drosophila melanogaster* as an alternative model system for investigating the pathogenicity of *Candida albicans*," *DMM Disease Models and Mechanisms*, vol. 4, no. 4, pp. 504–514, 2011.

[152] M. T. Glittenberg, I. Kounatidis, D. Christensen et al., "Pathogen and host factors are needed to provoke a systemic host response to gastrointestinal infection of *Drosophila larvae* by *Candida albicans*," *DMM Disease Models and Mechanisms*, vol. 4, no. 4, pp. 515–525, 2011.

[153] R. A. Hall, L. de Sordi, D. M. MacCallum et al., "CO_2 acts as a signalling molecule in populations of the fungal pathogen *Candida albicans*," *PLoS Pathogens*, vol. 6, no. 11, 2010.

[154] G. Chamilos, C. J. Nobile, V. M. Bruno, R. E. Lewis, A. P. Mitchell, and D. P. Kontoyiannis, "*Candida albicans* Cas5, a regulator of cell wall integrity, is required for virulence in murine and toll mutant fly models," *Journal of Infectious Diseases*, vol. 200, no. 1, pp. 152–157, 2009.

[155] R. Pukkila-Worley, A. Y. Peleg, E. Tampakakis, and E. Mylonakis, "*Candida albicans* hyphal formation and virulence assessed using a *Caenorhabditis elegans* infection model," *Eukaryotic Cell*, vol. 8, no. 11, pp. 1750–1758, 2009.

[156] N. D. Meeker and N. S. Trede, "Immunology and zebrafish: spawning new models of human disease," *Developmental and Comparative Immunology*, vol. 32, no. 7, pp. 745–757, 2008.

[157] J. P. Levraud, E. Colucci-Guyon, M. J. Redd, G. Lutfalla, and P. Herbomel, "*In vivo* analysis of zebrafish innate immunity," *Methods in Molecular Biology*, vol. 415, pp. 337–363, 2008.

[158] C. C. Chao, P. C. Hsu, C. F. Jen et al., "Zebrafish as a model host for *Candida albicans* infection," *Infection and Immunity*, vol. 78, no. 6, pp. 2512–2521, 2010.

[159] H. J. Lo, J. R. Köhler, B. Didomenico, D. Loebenberg, A. Cacciapuoti, and G. R. Fink, "Nonfilamentous *C. albicans* mutants are avirulent," *Cell*, vol. 90, no. 5, pp. 939–949, 1997.

[160] G. G. Chen, Y. L. Yang, H. H. Cheng et al., "Non-lethal *Candida albicans cph1/cph1 efg1/efg1* transcription factor mutant establishing restricted zone of infection in a mouse model of systemic infection," *International Journal of Immunopathology and Pharmacology*, vol. 19, no. 3, pp. 561–565, 2006.

[161] R. M. White, A. Sessa, C. Burke et al., "Transparent adult zebrafish as a tool for *in vivo* transplantation analysis," *Cell Stem Cell*, vol. 2, no. 2, pp. 183–189, 2008.

[162] K. M. Brothers, Z. R. Newman, and R. T. Wheeler, "Live imaging of disseminated candidiasis in zebrafish reveals role of phagocyte oxidase in limiting filamentous growth," *Eukaryotic Cell*, vol. 10, no. 7, pp. 932–944, 2011.

[163] D. C. Savage and R. J. Dubos, "Localization of indigenous yeast in the murine stomach," *The Journal of Bacteriology*, vol. 94, no. 6, pp. 1811–1816, 1967.

[164] N. V. Sipsas, R. E. Lewis, J. Tarrand et al., "Candidemia in patients with hematologic malignancies in the era of new antifungal agents (2001–2007): stable incidence but changing epidemiology of a still frequently lethal infection," *Cancer*, vol. 115, no. 20, pp. 4745–4752, 2009.

[165] M. Rehli, "Of mice and men: species variations of Toll-like receptor expression," *Trends in Immunology*, vol. 23, no. 8, pp. 375–378, 2002.

[166] X. Jiang, C. Shen, H. Yu, K. P. Karunakaran, and R. C. Brunham, "Differences in innate immune responses correlate with differences in murine susceptibility to *Chlamydia muridarum* pulmonary infection," *Immunology*, vol. 129, no. 4, pp. 556–566, 2010.

[167] D. L. Gibbons and J. Spencer, "Mouse and human intestinal immunity: same ballpark, different players; different rules, same score," *Mucosal Immunology*, vol. 4, no. 2, pp. 148–157, 2011.

[168] J. Mestas and C. C. W. Hughes, "Of mice and not men: differences between mouse and human immunology," *Journal of Immunology*, vol. 172, no. 5, pp. 2731–2738, 2004.

[169] G. Marquis, S. Montplaisir, M. Pelletier, P. Auger, and W. S. Lapp, "Genetics of resistance to infection with *Candida albicans* in mice," *British Journal of Experimental Pathology*, vol. 69, no. 5, pp. 651–660, 1988.

[170] G. Marquis, S. Montplaisir, and M. Pelletier, "Strain-dependent differences in susceptibility of mice to experimental candidosis," *Journal of Infectious Diseases*, vol. 154, no. 5, pp. 906–909, 1986.

[171] R. B. Ashman, E. M. Bolitho, and J. M. Papadimitriou, "Patterns of resistance to *Candida albicans* in inbred mouse strains," *Immunology and Cell Biology*, vol. 71, no. 3, pp. 221–225, 1993.

[172] I. Radovanovic, A. Mullick, and P. Gros, "Genetic control of susceptibility to infection with *Candida albicans* in mice," *PLoS ONE*, vol. 6, no. 4, 2011.

[173] F. De Bernardis, D. Adriani, R. Lorenzini, E. Pontieri, G. Carruba, and A. Cassone, "Filamentous growth and elevated vaginopathic potential of a nongerminative variant of *Candida albicans* expressing low virulence in systemic infection," *Infection and Immunity*, vol. 61, no. 4, pp. 1500–1508, 1993.

[174] E. Balish, "A *URA3* null mutant of *Candida albicans* (CAI-4) causes oro-oesophageal and gastric candidiasis and is lethal for gnotobiotic, transgenic mice (Tgε26) that are deficient in both natural killer and T cells," *Journal of Medical Microbiology*, vol. 58, no. 3, pp. 290–295, 2009.

[175] A. Brand, D. M. MacCallum, A. J. P. Brown, N. A. R. Gow, and F. C. Odds, "Ectopic expression of *URA3* can influence the virulence phenotypes and proteome of *Candida albicans* but can be overcome by targeted reintegration of *URA3* at the *RPS10* locus," *Eukaryotic Cell*, vol. 3, no. 4, pp. 900–909, 2004.

[176] J. Lay, L. K. Henry, J. Clifford, Y. Koltin, C. E. Bulawa, and J. M. Becker, "Altered expression of selectable marker *URA3*

in gene-disrupted *Candida albicans* strains complicates interpretation of virulence studies," *Infection and Immunity*, vol. 66, no. 11, pp. 5301–5306, 1998.

[177] P. Sundstrom, J. E. Cutler, and J. F. Staab, "Reevaluation of the role of *HWP1* in systemic candidiasis by use of *Candida albicans* strains with selectable marker *URA3* targeted to the *ENO1* locus," *Infection and Immunity*, vol. 70, no. 6, pp. 3281–3283, 2002.

Point Mutations in the *folP* Gene Partly Explain Sulfonamide Resistance of *Streptococcus mutans*

W. Buwembo,[1] S. Aery,[2] C. M. Rwenyonyi,[3] G. Swedberg,[2] and F. Kironde[4]

[1] *Department of Anatomy, Makerere University, P.O. Box 7072, Kampala, Uganda*
[2] *Department of Medical Biochemistry and Microbiology, Uppsala University, Husargaten 3, Building D7 Level 3, P.O. Box 582, SE-75123 Uppsala, Sweden*
[3] *Department of Dentistry, Makerere University, P.O. Box 7072, Kampala, Uganda*
[4] *Department of Biochemistry, Makerere University, P.O. Box 7072, Kampala, Uganda*

Correspondence should be addressed to F. Kironde; kironde@starcom.co.ug

Academic Editor: Marco Gobbetti

Cotrimoxazole inhibits dhfr and dhps and reportedly selects for drug resistance in pathogens. Here, *Streptococcus mutans* isolates were obtained from saliva of HIV/AIDS patients taking cotrimoxazole prophylaxis in Uganda. The isolates were tested for resistance to cotrimoxazole and their *folP* DNA (which encodes sulfonamide-targeted enzyme dhps) cloned in pUC19. A set of recombinant plasmids carrying different point mutations in cloned folP were separately transformed into *folP*-deficient *Escherichia coli*. Using sulfonamide-containing media, we assessed the growth of *folP*-deficient bacteria harbouring plasmids with differing *folP* point mutations. Interestingly, cloned *folP* with three mutations (A37V, N172D, R193Q) derived from *Streptococcus mutans* 8 conferred substantial resistance against sulfonamide to *folP*-deficient bacteria. Indeed, change of any of the three residues (A37V, N172D, and R193Q) in plasmid-encoded *folP* diminished the bacterial resistance to sulfonamide while removal of all three mutations abolished the resistance. In contrast, plasmids carrying four other mutations (A46V, E80K, Q122H, and S146G) in *folP* did not similarly confer any sulfonamide resistance to *folP*-knockout bacteria. Nevertheless, sulfonamide resistance (MIC = 50 μM) of *folP*-knockout bacteria transformed with plasmid-encoded *folP* was much less than the resistance (MIC = 4 mM) expressed by chromosomally-encoded *folP*. Therefore, *folP* point mutations only partially explain bacterial resistance to sulfonamide.

1. Introduction

Streptococcus mutans are commensal bacteria found in the oral cavity [1]. These bacteria which belong to the Viridans Streptococci Group (VSG) cause dental caries and infrequently give rise to extra oral infections like subacute bacterial endocarditis [1, 2]. Although dental caries is not usually treated by antibiotics, the VSG have attracted interest due to their potential to act as reservoirs of resistance to antibiotic determinants [3]. Additionally, in individuals taking antibiotics as prophylaxis, resistance of commensals to antibiotic determinants could be selected [4] and transferred to pathogenic organisms [5] such as *Streptococcus pneumoniae* which kills over 1,000,000 children worldwide every year [4].

Cotrimoxazole (SXT) is a combination drug (sulfamethoxazole plus trimethoprim) that is commonly used as prophylaxis in HIV/AIDS patients [6]. Sulfamethoxazole is a long-acting sulphonamide. In addition to wide usage as prophylaxis, SXT is also a highly prescribed drug especially in Sub-Saharan Africa due to its low cost and easy availability. Sub-Saharan Africa is reputed for high antibiotic abuse [7]. In Uganda, SXT is not only highly prescribed in dental practice [8], but also selected for multiple antibiotic resistance in *Streptococcus mutans* among HIV/AIDS patients [7]. Despite these findings, data on the mechanisms of SXT resistance in commensal bacteria such as *Streptococcus mutans* is still scanty. In order to better understand the mechanism of *Streptococcus mutans* resistance to SXT, we characterised the *S. mutans folP* gene that encodes dihydropteroate synthase, the target enzyme of sulfonamides [9]. Previously, we reported [7] that *folP* gene from the highly sulfonamide resistant *S. mutans* isolate 797 did not confer sulfonamide resistance to *E. coli folP* knockout bacteria and that sequencing of the *folA* gene of trimethoprim (TMP) resistant isolates did not reveal any mutations. However, the *folP* gene is very polymorphic [7], and at least one of the variants of *folP* confers high sulfonamide resistance to *E. coli folP* knockout cells. In the current study, we report site-directed mutagenesis experiments in which we altered point mutations in *S. mutans folP* gene, inserting the mutagenized *folP* DNA in pUC19 plasmids, which were then transformed into *folP* deficient *E. coli* C600 cells. By assessing the growth of the transformant *E. coli* Delta *folP* cells on media containing different levels of sulfamethoxazole, the influence of individual amino acids on sulfonamide resistance in the *folP* gene knockout bacteria was determined.

2. Methods

The mechanism of resistance to sulfamethoxazole (SMX) in *S. mutans* was characterised as summarized in Figure 1.

2.1. Bacterial Strains and Plasmids. The bacterial strains used in this study (Table 1) were previously isolated [7] from oral specimens of HIV/AIDS patients taking cotrimoxazole as prophylaxis in Kampala, Uganda. The cloning vectors pJet1.2/blunt (Fermentas, Lithuania) and pUC19 [12] were used. *E. coli* top ten cells (Invitrogen, USA) and *E. coli* recipient strain C600 Δ*folP* [13] which is a *folP* knockout strain were used in the transformation experiments. *Streptococcus pneumoniae* ATCC 49619 was used as a susceptible control when determining MICs.

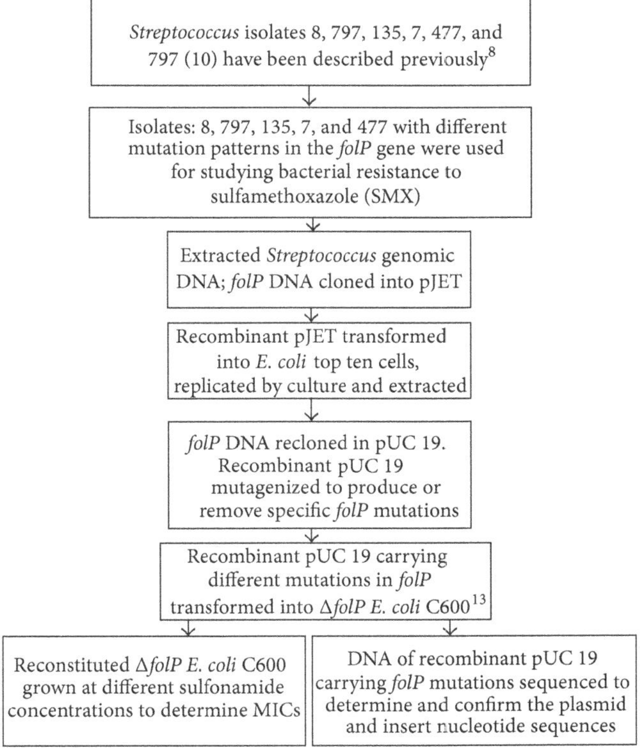

Figure 1: Flow chart showing characterization of *folP* gene. Plasmids carrying *folP* gene were transformed in *folP* gene knockout bacteria to determine the effect of different mutations in plasmid encoded *folP* on bacterial resistance to sulfonamides.

2.2. Susceptibility Testing. Minimal inhibitory concentrations (MICs) were determined by the *E*-test method (AB Biodisk, Sweden) following the manufacturer's recommendations. All tests were done on Iso-Sensitest Agar (ISA, Oxoid, UK). Plates were incubated at 37°C in 5% CO_2 for 24 h. For determination of sulfonamide resistance conferred by cloned *folP* genes, *E. coli* C600 bacteria were grown in Iso-Sensitest Broth (ISB, Oxoid, UK) to a cell density of 10^8/mL, diluted to 10^4/mL, and plated on ISA plates containing varying concentrations of sulfathiazole (Sigma Aldrich, USA).

2.3. DNA Extraction. Isolates of *Streptococcus mutans* were incubated at 37°C for 12 h on Iso-Sensitest Agar. Bacterial colonies were re-suspended in brain heart infusion broth (BioMérieux, France) and incubated at 37°C for 24 h in an atmosphere of 5% carbon dioxide. Chromosomal bacterial DNA from the cultured broth was then extracted using the Wizard Genomic DNA Purification Kit (Promega, USA).

2.4. Cloning. The PCR primer sequences used were based on the published sequence of the *folP* gene of *Streptococcus mutans* UA159 [10] (Table 1). *FolP* gene amplification was performed in 50-μL volumes containing 0.5 μM of each primer, 100 μM of the four deoxyribonucleoside triphosphates, 5 units of DNA polymerase (Pfu, Fermentas), 2 μL of template DNA (50–500 ng) preparation, and 1X reaction buffer (Pfu, Fermentas) containing 2 mM $MgSO_4$. Amplification reactions were performed with the *Eppendorf* mastercycler gradient thermocycler (Eppendorf, Germany) using the following program: heating at 94°C for 2 min, followed by 25 cycles consisting of a denaturation at 94°C for 30 s, annealing at 50°C for 30 s, and an extension at 72°C for 1.5 min. This was followed by a final extension of 72°C for 5 min and a holding step at 16°C.

The PCR products were cleaned by the PCR Cleanup kit (Omega, USA) and then used for cloning into the pJet vector using the blunt end pJET cloning kit protocol (Fermentas, Lithuania). The ligated products were introduced into *E. coli* top ten cells by heat-shock $CaCl_2$ transformation method.

Plasmids were prepared from the transformants using the plasmid preparation kit (Omega, USA) and prepared for sequencing.

2.5. Sequence Analysis. For sequencing the plasmids, the BigDye Terminator labelled cycle-sequencing kit (Applied Biosystems) and an ABI prism 310 Genetic Analyzer (Applied Biosystems) were used. The results of the *folP* gene sequence analysis were compared with database sequences of *Streptococcus mutans* UA159 [10] and NN2025 [11] using the BLAST programme at NCBI [14].

Table 1: Characteristics of bacterial isolates used in the current study, the respective genes and susceptibility to Cotrimoxazole (STX), sulfamethoxazole (SMX), and trimethoprim (TMP) as determined by *E*-test.

Isolate	Accession number of *folP* gene nucleotide sequence used	Mutations in the *folP* gene	STX	Sulfonamide susceptibility	Trimethoprim susceptibility	Mutations in the *folA* gene
S. mutans 8	Not yet submitted to gene banks but previously published [7]	A37V, N172D, and R193Q*	>32 μg/mL	>1024 μg/mL (>4 mM)	>32 μg/mL	None
S. mutans 797	HE599533.1	A46V, E80K, Q122H, and S146G**	0.5 μg/mL	>1024 μg/mL (>4 mM)	2 μg/mL	None
S. mutans 135	Not yet submitted but previously published [7]	A63S, W174LK, L175F, and M189I**	8 μg/mL	>1024 μg/mL (>4 mM)	0.38 μg/mL	None
S. sobrinus 7	HE 599535.1	None	>32 μg/mL	>1024 μg/mL (>4 mM)	>32 μg/mL	None
S. downei 477	Similar to ZP 07725257.1	None	0.125 μg/mL	Not done	Not done	Not done

*Mutations as compared to UA159 [10]. **Mutations as compared to NN2025 [11].

Table 2: Primers used for cloning and site-directed mutagenesis.

Primer name	Nucleotide sequence
Mutans DHPSph	5′-GAT CGA TCG CAT GCA CAT CAT AAC TAG GGA GCA AGC-3′
mutansDHPSBam	5′-GAT GGA TCG GAT CCA AAA TAATCT TAT CCA TAA CAC CCT CA-3′
dhpssfwph	5′-AAC CTA CTG CAT GCA TAA GAA TCA G-3′
dhpssreveco	5′-ATT GTA GGA ATT CTT CTA GAA AGA TCC-3′
downeifolpfw	5′-GCA TGC CAA AGA CAG GAA TTG CTG AC-3′
Downeifolprevps	5′-CTG CAG CCA CAA AAA TTT GCC CCA GAC-3′
	Primers for changing specific amino acids in isolate 797
DHPS46AVfw	5′-TGA AGC CAT GTT AGT AGC AGG AGC GGC TA-3′
DHPS46AVrev	5′-TAG CCG CTC CTG CTA CTA ACA TGG CTT CA-3′
DHPS80aEKfw	5′-TCG TTC CAA TTG TTA AAG CTA TTA GCG AA-3′
DHPS80aEKrev	5′-TTC GCT AAT AGC TTT AAC AAT TGG AAC GA-3′
DHPS122QHfw	5′-CTT TAT GAT GGG CAC ATG TTT CAA TTA GC-3′
DHPS122QHrev	5′-GCT AAT TGA AAC ATG TGC CCA TCA TAA AG-3′
DHPS146SGfw	5′-GTG AAG AAG TTT ATG GCA ATG TAA CAG AA-3′
DHPS146SGrev	5′-TTC TGT TAC ATT GCC ATA AAC TTC TTC AC-3′
	Primers for changing specific amino acids in isolate 8
V37Afw	5′-AAC CAA TCG ATC AGG CTC TAA AAC AGG TTG A-3′
V37Arev	5′-TCA ACC TGT TTT AGA GCC TGA TCG ATT GTT T-3′
D172Nfw	5′-GGA GTT AAA AAA GAA AAT ATT TGG CTT GAT C-3
D172Nrev	5′-GAT CAA GCC AAA TAT TTT CTT TTT TAA CTC C-3′
Q193Rfw	5′-ACA TGG AAC TTC TAC GAG GCT TAG CGG AGG T-3
Q193Rrev	5′-ACC TCC GCT AAG CCT CGT AGA AGT TCC ATG T-3′

2.6. Site-Directed Mutagenesis. Mutagenesis was carried out using a 50 μL reaction mix (Fermentas, Lithuania) containing 1X Pfu buffer with $MgSO_4$, 2.5 units Pfu DNA Polymerase (Fermentas, Lithuania), 0.1 mM dNTPs, 10–100 ng of template DNA inserted in pUC19, and 1 μM of each primer (Table 2). The PCR program started with a heating step at 95°C for 30 s, followed by 18 cycles consisting of a denaturation step of 95°C for 30 s, annealing of 50°C for 30 s, and an extension of 68°C for 7 min.

Site-directed mutagenesis products (17 μL of the PCR product) were digested with 5 units of the restriction enzyme Dpnl (Fermentas, Lithuania) in Buffer Tango to remove unchanged DNA. The mixture was incubated at 37°C for 4 h before incubating at 80°C for 20 min to inactivate Dpnl.

2.7. Transformation. Ten μL of the digested mutagenesis product were transformed into $CaCl_2$-treated *folP* knockout *E. coli* competent cells. Recombinant plasmids were prepared from part of a single transformant bacterial colony using the plasmid miniprep kit (Omega, USA). The plasmids were then sequenced as described above to confirm that site-directed mutagenesis had occurred. Bacteria from the same transformant colony were then tested for growth at different SMX concentrations.

3. Results

The results of transformation of *S. mutans folP* gene into *folP* gene knockout *E. coli* are shown in Figure 2. As previously reported [7], *S. mutans* isolate 797 carries four point mutations (A46V, E80K, Q122H, and S146G) in *folP* gene, as compared to the control strain NN2025 [11] but harbours one such mutation (S146G) if compared to control strain UA159 [12] (Table 1). Two other isolates with DHPS sequence differing from both reference strains were cloned using the same conditions as for 797. These were isolate 8 (A37V, N172D, and R193Q) and isolate 135 (A63S, W174L, L175F, and M189I). Of these, only the cloned *folP* gene from isolate 8 conferred sulfonamide resistance to the *E. coli* C600Δ*folP* recipient. In the present paper, the above-mentioned four point mutations (A46V, E80K, Q122H, and S146G), and those in *S. mutans* isolate 8 (A37V, N172D, and R193Q) (Table 1), were successfully altered or removed from *folP* gene by site-directed mutagenesis. Nonmutagenized chromosomal DNA from isolates *S. sobrinus* 7 and *S. downei* 477 was individually inserted into pUC19 as well. Effects of the different *folP* gene constructs on resistance to sulphonamide were then investigated by transforming the recombinant pUC 19 plasmids in *folP* knockout *E. coli* C600 and assessing the growth of transformant bacteria on media containing different concentrations of sulfamethoxazole (see Figure 1). Interestingly, plasmids harbouring triple mutant *folP* of *S. mutans* isolate 8 (bearing the mutations A37V, N172D, and R193Q) conferred (to *folP* knockout *E. coli*) intermediate level resistance (MIC: 50 μM) against sulfonamide (bar A in Figure 2) even though this is not equal to 4 mM, the MIC arising from the chromosomal *folP* gene in the natural isolate *S. mutans* 8 (Table 1). In addition, altering any of the three polymorphic amino acids (A37V, N172D, and R193Q) back to the UA159 sequence and transforming the double mutant *folP* gene into the *E. coli* Delta-*folP* cells produced reconstituted knockout *E. coli* cells of lower level (MIC: 30 μM) resistance to sulphonamide (Figure 2: bars B, C, D), while reversing two amino acid mutations (Figure 2 bar E) or all three amino acid mutations (Figure 2 bar F) to wild-type and transforming *folP* knockout *E. coli* likewise yielded transformant clones with low resistance to sulphonamide (MIC: 20 μM). On the other hand, as previously reported [7], cloned *folP* gene from 797 did not confer resistance of *folP* deficient *E. coli* C600 to sulphonamide. Moreover, changing the DNA encoding four amino acid mutations (A46V, E80K, Q122H, and S146G) of *folP* in *S. mutans* isolate 797 and transforming the mutant DNA in *folP* knockout *E. coli* to comply with either NN2025 or UA159 sequences (Figure 2 bar G) did not change susceptibility of *folP* knockout bacteria to sulphonamide. Controls consisting of *folP* knockout *E. coli* transformed with pUC 19 plasmids encoding mutant *folP* from *S. mutans* isolate 135 (Figure 2, bar H), wild-type *folP* from *S. sobrinus* 7 (Figure 2 bar I), or wild-type *folP* from *S. downei* 477 produced minimal or reduced resistance.

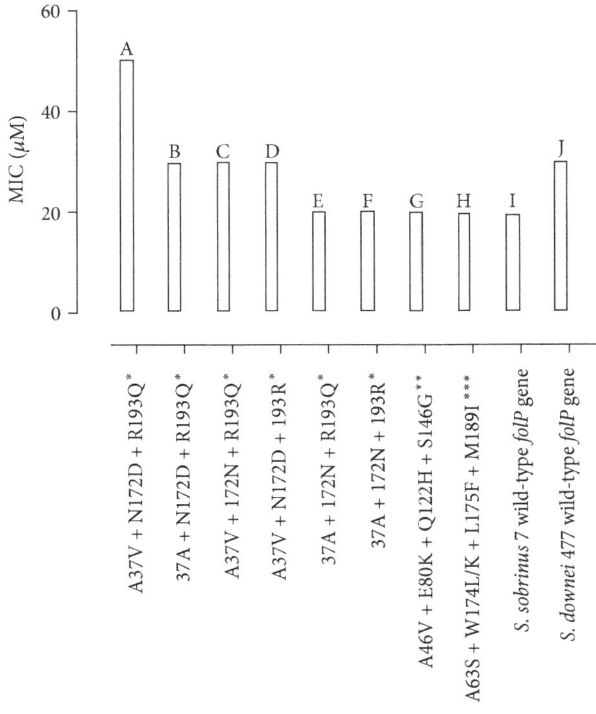

Figure 2: Comparing sulfamethoxazole minimal inhibitory concentrations (MICs) for *folP* knockout *E. coli* cells transformed with pUC19 plasmid carrying differing chromosomal *folP* genes of streptococci. To determine the effect plasmid encoded mutant *folP* has on the sulfamethoxazole resistance of transformed *C600ΔfolP E. coli* bacteria, growths (in SMX containing media) were compared for *folP*-deficient *E. coli* transformed with pUC19 carrying triple-mutant *folP* (residues 37, 172, and 193: bar A), double-mutant *folP* (bars B, C, and D), single-mutant *folP* (residue 193: bar E), and wild-type *folP* (bar F) from *S. mutans* isolate 8. The sulfamethoxazole resistance of transformed *folP* deficient cells was notably increased by transformation with plasmid encoding triple-mutated *folP* (wild-type *folP* MIC = 20 μM, triple-mutant *folP* MIC = 50 μM). Note: the MIC (sulfamethoxazole) for chromosome-encoded *folP* in *S. mutans* isolate 8 was 4 mM (see Table 1). Controls comprising *C600ΔfolP E. coli* transformed with pUC19 encoding either mutant *folP* from *S. mutans* isolates 797 (bar G) and 135 (bar H) or wild-type *folP* from *S. sobrinus* isolate 7 (bar I) or *S. downei* isolate 477 (bar J) showed basal or less sulfamethoxazole resistance (MICs = 20–30 μM). *: *S. mutans* isolate 8 mutant *folP*; **: *S. mutans* isolate 797 mutant *folP*; ***: *S. mutans* isolate 135 mutant *folP*.

4. Discussion

In the present study, we examined Ugandan *Streptococcus mutans* isolates from HIV/AIDS patients who were taking cotrimoxazole as prophylaxis [7]. The isolates were found to have different point mutations in the *folP* gene in relation to wild-type sequences found in databases. It should be noted that these same isolates lacked any mutations in the *folA* gene.

Resistance to sulfonamides in gram positive bacteria has been shown to be due to mutations in the *folP* gene that render the encoded dihydropteroate synthase insensitive to the drug [15]. In the corresponding gene of *Plasmodium*, more point mutation combinations were previously found to be associated with higher resistance rates against sulfadoxine in *P. falciparum* [16]. In the present study, we assessed the influence of different combinations of mutations in *S. mutans folP* by performing mutagenesis experiments to remove mutations in the *folP* gene and transforming the changed DNA into *folP* knockout *E. coli* cells which were subsequently tested for growth in the presence of varying levels of sulfamethoxazole. We found that the cloned *folP* gene from isolate 8 conferred substantial sulphonamide resistance to *folP* knockout *E. coli* (MIC: 50 μM) (Figure 2) but not to the level observed for the natural isolate *S. mutans* 8 (MIC: 4 mM). Changes in any of the three divergent amino acids (residues 37, 172, and 193) of *folP* reduced the level of resistance to sulfonamide and the removal of all three polymorphisms totally abolished the resistance. In contrast, no combination of the mutations in *folP* from isolate 797 (A46V, E80K, Q122H, and S146G) led to differences in susceptibility of *folP* knockout bacteria to sulfonamide. In addition, we found that isolates 135, 7, and 477 with different mutation patterns in *folP* grew to the same resistance level (MIC: 20 μM) as isolate 797. This finding corroborates the previous report [7] that the cloned *folP* gene from 797 does not confer sulfonamide resistance to *folP*-gene knockout *E. coli* cells. However, that the cloned *folP* gene from isolate *S mutans* 8 did not confer equally high sulphonamide resistance to *folP* knockout *E. coli* as shown by the natural isolate *S. mutans* 8 (MIC: 4 mM) suggests that there is another mechanism of resistance to sulfonamide other than the point mutations. One possibility may be that DHPS synthesis and expression are increased as was found in *Streptococcus agalactiae* [17] or that there may be point mutations in other folate pathway genes. We could not rule out other causes of resistance to sulfonamide in *Streptococcus mutans* since sequencing the *folA* gene of *S.mutans* 8 and flanking regions including the promoter did not reveal any mutations (results not shown). Further experiments including whole genome sequencing of *S. mutans* 8 and other sulfonamide resistant strains may enhance the understanding of sulfa resistance in Streptococci.

5. Conclusions

Point mutations are one of the explanations for the mechanism of resistance to sulfonamide in *Streptococcus mutans*. However, cloned *folP* gene did not confer full resistance to *folP* knockout cells compared to the original isolate *797*, a result which does not rule out other possible mechanisms for the resistance to sulfonamides.

Conflict of Interests

The authors report no conflict of interests.

Authors' Contribution

W. Buwembo carried out the preparation of bacterial isolates and polymerase chain reaction tests, participated in the microbiological antibiotic resistance tests, cloning, and sequencing analyses, and drafted the paper. S. Aery participated in the mutagenesis experiments. G. Swedberg designed the primers and participated in cloning and transformation experiments. F. Kironde, C. M. Rwenyonyi, and G. Swedberg conceived the study, participated in its design and coordination, and helped in writing the paper. All authors read and approved the final paper.

Acknowledgments

The authors are grateful to GS project students, namely, Mary MacRitchie, Soheila Rajabi, Marcus Andersson, and Fredrik Jonsson who assisted in this work. We sincerely thank the patients who donated clinical specimens, research associates, and administrators of TASO clinic. This work was financially supported by the Swedish Agency for Research Cooperation with Developing Countries. All the authors are responsible for the content and writing of the paper.

References

[1] W. J. Loesche, "Role of *streptococcus mutans* in human dental decay," *Microbiological Reviews*, vol. 50, no. 4, pp. 353–380, 1986.

[2] R. Facklam, "What happened to the streptococci: overview of taxonomic and nomenclature changes," *Clinical Microbiology Reviews*, vol. 15, no. 4, pp. 613–630, 2002.

[3] A. Bryskier, "Viridans group streptococci: a reservoir of resistant bacteria in oral cavities," *Clinical Microbiology and Infection*, vol. 8, no. 2, pp. 65–69, 2002.

[4] M. Wilén, W. Buwembo, H. Sendagire, F. Kironde, and G. Swedberg, "Cotrimoxazole resistance of Streptococcus pneumoniae and commensal streptococci from Kampala, Uganda," *Scandinavian Journal of Infectious Diseases*, vol. 41, no. 2, pp. 113–121, 2009.

[5] P. Echave, J. Bille, C. Audet, I. Talla, B. Vaudaux, and M. Gehri, "Percentage, bacterial etiology and antibiotic susceptibility of acute respiratory infection and pneumonia among children in rural Senegal," *Journal of Tropical Pediatrics*, vol. 49, no. 1, pp. 28–32, 2003.

[6] A. Sosa, "Who issues guidelines on use of cotrimoxazole prophylaxis," 2006, http://www.who.int/hiv/pub/guidelines/ctx/en/index.html.

[7] B. William, C. M. Rwenyonyi, G. Swedberg, and F. Kironde, "Cotrimoxazole prophylaxis specifically selects for cotrimoxazole resistance in *streptococcus mutans* and *Streptococcus sobrinus* with varied polymorphisms in the target genes *folA* and *folP*," *International Journal of Microbiology*, vol. 2012, Article ID 916129, 10 pages, 2012.

[8] A. Kamulegeya, B. William, and C. M. Rwenyonyi, "Knowledge and antibiotics prescription pattern among ugandan oral health care providers: a cross-sectional survey," *Journal of Dental Research, Dental Clinics, Dental Prospects*, vol. 5, no. 2, pp. 61–66, 2011, http://dentistry.tbzmed.ac.ir/joddd.

[9] O. Skold, "Resistance to trimethoprim and sulfonamides," *Veterinary Research*, vol. 32, no. 3-4, pp. 261–273, 2001.

[10] D. Ajdić, W. M. McShan, R. E. McLaughlin et al., "Genome sequence of *streptococcus mutans* UA159, a cariogenic dental pathogen," *Proceedings of the National Academy of Sciences of the United States of America*, vol. 99, no. 22, pp. 14434–14439, 2002.

[11] F. Maruyama, M. Kobata, K. Kurokawa et al., "Comparative genomic analyses of *streptococcus mutans* provide insights into chromosomal shuffling and species-specific content," *BMC Genomics*, vol. 10, article 358, 2009.

[12] C. Yanisch-Perron, J. Vieira, and J. Messing, "Improved M13 phage cloning vectors and host strains: nucleotide sequences of the M13mp18 and pUC19 vectors," *Gene*, vol. 33, no. 1, pp. 103–119, 1985.

[13] C. Fermér and G. Swedberg, "Adaptation to sulfonamide resistance in Neisseria meningitidis may have required compensatory changes to retain enzyme function: kinetic analysis of dihydropteroate synthases from N. meningitidis expressed in a knockout mutant of Escherichia coli," *Journal of Bacteriology*, vol. 179, no. 3, pp. 831–837, 1997.

[14] S. F. Altschul, W. Gish, W. Miller, E. W. Myers, and D. J. Lipman, "Basic local alignment search tool," *Journal of Molecular Biology*, vol. 215, no. 3, pp. 403–410, 1990.

[15] O. Sköld, "Sulfonamides and trimethoprim," *Expert review of anti-infective therapy*, vol. 8, no. 1, pp. 1–6, 2010.

[16] S. Sridaran, S. K. McClintock, L. M. Syphard, K. M. Herman, J. W. Barnwell, and V. Udhayakumar, "Anti-*folA*te drug resistance in Africa: meta-analysis of reported dihydro*folA*te reductase (dhfr) and dihydropteroate synthase (dhps) mutant genotype frequencies in African Plasmodium falciparum parasite populations," *Malaria Journal*, vol. 9, no. 1, article 247, 2010.

[17] M. Brochet, E. Couvé, M. Zouine, C. Poyart, and P. Glaser, "A naturally occurring gene amplification leading to sulfonamide and trimethoprim resistance in Streptococcus agalactiae," *Journal of Bacteriology*, vol. 190, no. 2, pp. 672–680, 2008.

21

Antimicrobial Activity of Xoconostle Pears (*Opuntia matudae*) against *Escherichia coli* O157:H7 in Laboratory Medium

Saeed A. Hayek and Salam A. Ibrahim

Food Microbiology and Biotechnology Laboratory, North Carolina Agricultural and Technical State University, Greensboro, NC 27411, USA

Correspondence should be addressed to Salam A. Ibrahim, ibrah001@ncat.edu

Academic Editor: Todd R. Callaway

The objective of this study was to investigate the antimicrobial activity of xoconostle pears (*Opuntia matudae*) against *Escherichia coli* O157:H7. Xoconostle pears were sliced, blended, and centrifuged. The supernatant was then filtered using a 0.45 μm filter to obtain direct extract. Direct extract of xoconostle pears was tested against four strains of *E. coli* O157:H7 in brain heart infusion (BHI) laboratory medium using growth over time and agar well diffusion assays. Our results showed that direct extract of xoconostle pears had a significant ($P < 0.05$) inhibitory effect at 4, 6, and 8% (v/v) concentrations and complete inhibitory effect at 10% (v/v) during 8 h of incubation at 37°C. Minimum inhibitory volume (MIV) was 400 μL mL^{-1} (v/v) and minimum lethal volume (MLV) was 650 μL mL^{-1} (v/v). The inhibitory effect of xoconostle pears found to be concentration dependent and not strain dependent. Thus, xoconostle pears extract has the potential to inhibit the growth of *E. coli* O157:H7 and could provide a natural means of controlling pathogenic contamination, thereby mitigating food safety risks.

1. Introduction

Foodborne pathogens are major concern to consumers, food industry, and food regulatory agencies. The yearly cost of foodborne illnesses in the United States as reported in 2010 was about $152 billion including $993 million caused by *Escherichia coli* O157:H7 in healthcare, workplace, and other economic losses [1]. *E. coli* O157:H7 is one of the most important foodborne pathogens in the United States, having been first identified in 1982. From 1982 to 2002, *E. coli* O157:H7 caused 73,000 illnesses annually in the United States including 8,598 infection cases and 40 deaths [2]. *E. coli* O157:H7 causes severe gastrointestinal diseases such as bloody diarrhea haemorrhagic colitis, haemolytic uremic syndrome, and traveler's diarrhea [3, 4]. Therefore, the control and prevention of *E. coli* O157:H7 in food products is an area that is receiving worldwide attention.

Many food preservation techniques have been developed to control and prevent foodborne pathogens including antimicrobials. Most antimicrobials used in the food industry are chemical preservatives [3]. Even though chemical preservatives have been approved by many countries and used for years, they are considered by most consumers to be unhealthy [3, 5]. Therefore, natural antimicrobials have become increasingly important to the food industry in order to meet the consumers demands [5]. Natural antimicrobials can be found in a variety of plants including herbs, spices, fruits, vegetables, and tropical plants [5, 6]. Plants contain an array of natural compounds with many medicinal benefits and provide about 50% of current pharmaceuticals [6]. Even though, several reports have demonstrated the efficacy of using natural ingredients [7–10] and plant extractions [11–16] to control the growth of foodborne pathogens, no plants are currently used as antimicrobials [6]. Thus, food industries are very motivated to replace chemical preservatives with natural antimicrobials.

Xoconostle pears (*Opuntia matudae*) have attracted the attention of researchers around the world due to this particular pear's strong anticancer, antidiabetic, and antioxidant characteristics [17–19]. Xoconostle pears are rich source of soluble phenolics, ascorbic acid, and betalains compared to most common fruits and vegetables [19]. Soluble phenolics, betalains, and ascorbic acid have already shown effective antimicrobial activity in many studies [6, 8, 10–12, 14–16]. Thus xoconostle pears have great potential as natural antimicrobial. To the best of our knowledge, there is no information in the literature on the antimicrobial activity of xoconostle pears against any pathogenic bacteria including *E. coli* O157:H7. Therefore, the objective of this study was to examine the antimicrobial activity of xoconostle pears direct extract against *E. coli* O157:H7 in brain heart infusion (BHI) laboratory medium.

2. Materials and Methods

2.1. Bacterial Culture Activation and Preparation. Four strains of *E. coli* O157:H7, F4546 (alfalfa sprout isolate), H1730 (lettuce isolate), 43895 (beef isolate), and 944 (salami isolate) were used in this study. These *E. coli* O157:H7 strains were selected from different isolating sources and have been associated with several outbreaks. The *E. coli* O157:H7 strains were supplied by Dr. S. S. Summer, Department of Food Science and Technology at Virginia Tech, and stored at -80°C freezer stock storage of our laboratory. Immunoblot using Protran nitrocellulose membranes (BA85, Whatman, Schleicher and Schuell, Sanford, ME) was performed to identify the *E. coli* O157:H7 strains [20]. A confirmation step using polymerase chain reaction (PCR) assay was also conducted to identify the serogroup of *E. coli* O157:H7 strains [21]. The strains were activated in BHI (Becton Dickinson, Sparks, MD, USA) broth by transferring 100 μL from the stored culture to 10 mL BHI broth and incubating at 37°C for 24 h. Activated strains were stored in a refrigerator at 4°C. Prior to each experimental replication, each individual bacterial strain was streaked on BHI agar and incubated for 24 h at 37°C. One isolated colony was transferred to 10 mL BHI broth, and incubated at 37°C for next day use.

2.2. Xoconostle Extract Preparation. Xoconostle pears were obtained from a local grocery market in Greensboro, NC. For each experiment replication, 450 g of fresh xoconostle pears were rinsed under running tap water, blotted with paper towel, sliced into small pieces, and blended in a kitchen blender for 4 min. This preparation was placed in 50 mL tubes and centrifuged at 7800 × g for 10 min using Thermo Scientific* Sorvall RC 6 Plus Centrifuge (Thermo Scientific Co., Asheville, NC, USA). The supernatant was filtered using a 0.45 μm Nalgene filter (Nalge Nunc International Corp, Rochester, NY, USA) to collect the xoconostle direct extract which was stored at 4°C until used within 12–16 h.

2.3. Bacterial Enumeration. Bacterial populations were determined by plating onto BHI agar. In this procedure, samples were individually diluted into serial of 9 mL 0.1% peptone water solution (Bacto peptone, Becton Dickinson, Sparks, MD, USA); (pH 7.25 ± 0.08); then 100 μL of appropriate dilutions were surface plated onto triplicates BHI agar plates and incubated at 37°C for 24 h. Plates with colonies ranging between 30–300 were considered for colony counting to determine the bacterial populations [3].

2.4. Growth Over Time Assay. Growth over time assay was employed following the instructions by Marwan and Nagel [22] and Parish and Carroll [23] with slight modifications. Overnight activated bacterial strains were individually diluted into serial of 9 mL 0.1% peptone water solution to obtain a bacterial population of approximately 4 log CFU mL^{-1}. For each individual strain, batches of sterilized tubes containing 9 mL BHI broth were mixed with xoconostle extract to obtain different concentrations (4, 6, 8, and 10% v/v) and inoculated with 1 mL of the previously diluted individual bacterial strains. Control samples without xoconostle extract for each individual bacterial strain and blank samples without bacterial inoculation for each treatment level were included. The initial bacterial populations for each strain were approximately 3 log CFU mL^{-1} and that was determined using the bacterial enumeration method previously described. Samples were incubated with shaking at 37°C for 8 h, and bacterial growth was monitored by measuring the optical density (O.D. 610 nm) at 2 h interval using Thermo Scientific Genesys 10S UV-Vis spectrophotometer (Thermo Fisher Scientific Co., Madison, WI, USA). The final bacterial populations were determined at the end of the incubation period.

2.5. Agar Well Diffusion Assay. Agar well diffusion assay described by Hugo and Russel [24] and Ibrahim and others [16] with slight modifications was employed. Individual strain was grown overnight then serially diluted into 9 mL 0.1% peptone water solution to obtain bacterial populations of 6 log CFU mL^{-1}approximately. Diluted strains at 10 mL each were mixed in a sterilized container. BHI agar at 500 mL with 0.2% of Tween 80 was prepared and sterilized at 121°C for 15 min. Prepared BHI agar was placed in a water bath at 48°C and allowed to cool down then inoculated with 20 mL of previously mixed culture to achieve a bacterial population of 4-5 log CFU mL^{-1}. Inoculated BHI agar was poured into Petri dishes ($15 \times 100\,mm^2$) at approximately 50 mL each and allowed to solidify in 30 min under biohazard cabinet. A sterile cork borer (8.0 mm diameter) was used to punch wells in the inoculated agar. The agar plugs were removed using a sterilized wire loop. Xoconostle extract at different volumes (200–1000 μL with 25 μL unit increase) were adjusted with sterilized distilled water to 1 mL to obtain different concentrations (v/v) and poured into the wells to the top. Plates were kept under a biohazard cabinet for 30 min for prediffusion to occur, incubated at 37°C for 12 h, and then examined for the development of clear inhibitory zone. Observed inhibitory zones were measured to the nearest 0.1 mm and reported after subtracting the well diameter from the observed zone diameter. Minimum inhibitory volume (MIV) was determined at this point. Incubation of the plates was continued for three days to determine the minimum lethal volume (MLV). MIV was defined as the lowest volume concentration that caused significant inhibitory effect during 12 h of incubation at 37°C and MLV was defined as the lowest volume concentration that showed significant inhibitory effect after three days of incubation [16, 25]. Inhibitory zone at 3 mm or larger was considered significant [25, 26].

2.6. Statistical Analysis. Each experimental test was conducted three times to determine the effect of xoconostle pears on the growth of *E. coli* O157:H7. Mean values and standard deviations were calculated from the triplicate samples. Statistical analysis system (SAS) [27] version 9.2 was used to determine significant antimicrobial activity at different concentrations of xoconostle extract and significant differences among strains at the same concentration of xoconostle extract using the data means by a factorial analysis of variance of triplicate samples at a significant level of $P < 0.05$.

3. Results

Figure 1 shows the growth of *E. coli* O157:H7 in BHI broth treated with different volumes of xoconostle extract during 8 h of incubation at 37°C. In control samples, optical density readings reached absorbance in the range of 0.654–0.812 (O.D. 610 nm). When *E. coli* O157:H7 strains were grown in BHI broth treated with 4, 6, 8, and 10% (v/v) xoconostle extract, optical density readings reached ranges of 0.512–0.668, 0.339–0.440, 0.220–0.259, and 0.036–0.103 (O.D. 610 nm) respectively. An optical density reading of 0.1 (O. D. 610 nm) or less was previously defined as the division between visual growth and no growth [23, 25]. Thus, xoconostle extract at 4, 6, and 8% (v/v) concentrations was able to slow down the growth of *E. coli* O157:H7 strains whereas 10% was enough to cause no growth. Table 1 shows the final population of *E. coli* O157:H7 strains grown in BHI broth treated with different concentrations (v/v) of xoconostle extract after 8 h of incubation at 37°C. In control samples, *E. coli* O157:H7 continued to grow and reached the stationary phase. The additions of xoconostle extract at 4, 6, 8, and 10% (v/v) caused significant ($P < 0.05$) reductions in *E. coli* O157:H7 populations at averages of 0.99 ± 0.17, 2.23 ± 0.35, 3.66 ± 0.22, and 5.78 ± 0.41 log CFU mL^{-1}, respectively. Samples treated with 10% (v/v) xoconostle extract caused final bacterial populations to remain within the initial count range (about 3 logs CFU mL^{-1}). These results indicate that xoconostle pears have a significant inhibitory effect on *E. coli* O157:H7, and 10% (v/v) concentration of xoconostle extract is required to achieve complete growth inhibition.

Table 2 shows the inhibitory zones (with 100 μL unit increase) that were formed around the wells after 12 h of incubation at 37°C. Bacterial growth developed a greenish cloud all over the agar whereas distinguishable clear zone remained around the well. The lowest concentration that shows a clear inhibitory zone was 275 $\mu L\,mL^{-1}$ (v/v) with an average of 1.0 ± 0.2 mm. MIV was recorded for a significant inhibitory effect at 400 $\mu L\,mL^{-1}$ (v/v) with an average of 2.9 ± 0.2 mm. When xoconostle extract without dilution was transferred to the well, the average inhibitory zone reached 9.8 ± 1.01 mm. After three days of incubation, MLV was recorded for a significant inhibitory effect at 650 $\mu L\,mL^{-1}$ (v/v) with an average of 2.8 ± 0.25 mm. These data support the growth over time assay results indicating that xoconostle pears had significant inhibitory effect on *E. coli* O157:H7.

4. Discussion

An increasing consumer demand for food products that are minimally processed and contain natural ingredients has been noticed recently. This demand has resulted in an effort by the food industry to search for natural antimicrobials. In the present work, the antimicrobial activity of xoconostle pears was studied using growth over time and well diffusion assays. Both assays are common for studying the antimicrobial activity in food microbiology [14, 23, 25, 26] and our laboratory has used both assays with consistent results [7, 9, 10, 13, 16]. The growth over time and well diffusion assays are practical, simple, and could be used for direct screening of direct extracts from fruits and vegetables. Direct extract of xoconostle pears was obtained mechanically without any chemical, heating, or concentration processing. Direct extraction is a simple and safe procedure that can avoid any possible alteration to or destruction of the native structure of the active ingredients. Common extraction procedures include chemical or heating treatments could alter the active ingredients total content, functionality, natural characteristics, or could produce unsafe compounds [28, 29]. In addition, direct extracts from fruits or vegetables can be applied to food products in a safe manner.

Our results showed that xoconostle pears have significant inhibitory effect on the growth of *E. coli* O157:H7. Figure 1 and Table 1 show that the increase in xoconostle extract concentration associated with a slower growth rate and more bacterial populations reductions, respectively. Table 2 shows that the gradual increase in the inhibitory zone with respect to the increase in xoconostle extract concentration. These results indicate that the inhibitory effect of xoconostle extract is concentration dependent. On the other hand, Table 1 shows no significant ($P > 0.05$) differences in the final bacterial populations among *E. coli* O157:H7 strains grown at the same concentration of xoconostle extract except for *E. coli* O157:H7 43895. However, different *E. coli* O157:H7 strains may grow at different growth rates [30] which may explain the difference in the final bacterial population of *E. coli* O157:H7 43895. Thus, xoconostle pears have the same level of antimicrobial activity on different *E. coli* O157:H7 strains. Therefore, the inhibitory effect of xoconostle pears is concentration dependent and not strains dependent.

The antimicrobial activity of xoconostle pears can thus be accounted for several active compounds including phenolics, ascorbic acid, and betalains. Xoconostle pears have a combination of phenolic compounds including gallic, vanillic, 4-hydroxybenzoic acids, catechin, epicatechin, and vanillin [19]. Even though the exact antimicrobial mechanism of phenolic compounds is not clear, phenolic compounds are commonly known for their antimicrobial effects [6]. The ability of phenolic compounds to alter microbial cell permeability, thereby permitting the loss of macromolecules from the cell interior, could help explain some of the antimicrobial activities [11]. Another explanation might be that phenolic compounds interfere with membrane function and interact with membrane proteins, causing deformation in structure and functionality [11]. A combination of phenolic compounds can provide synergistic antimicrobial

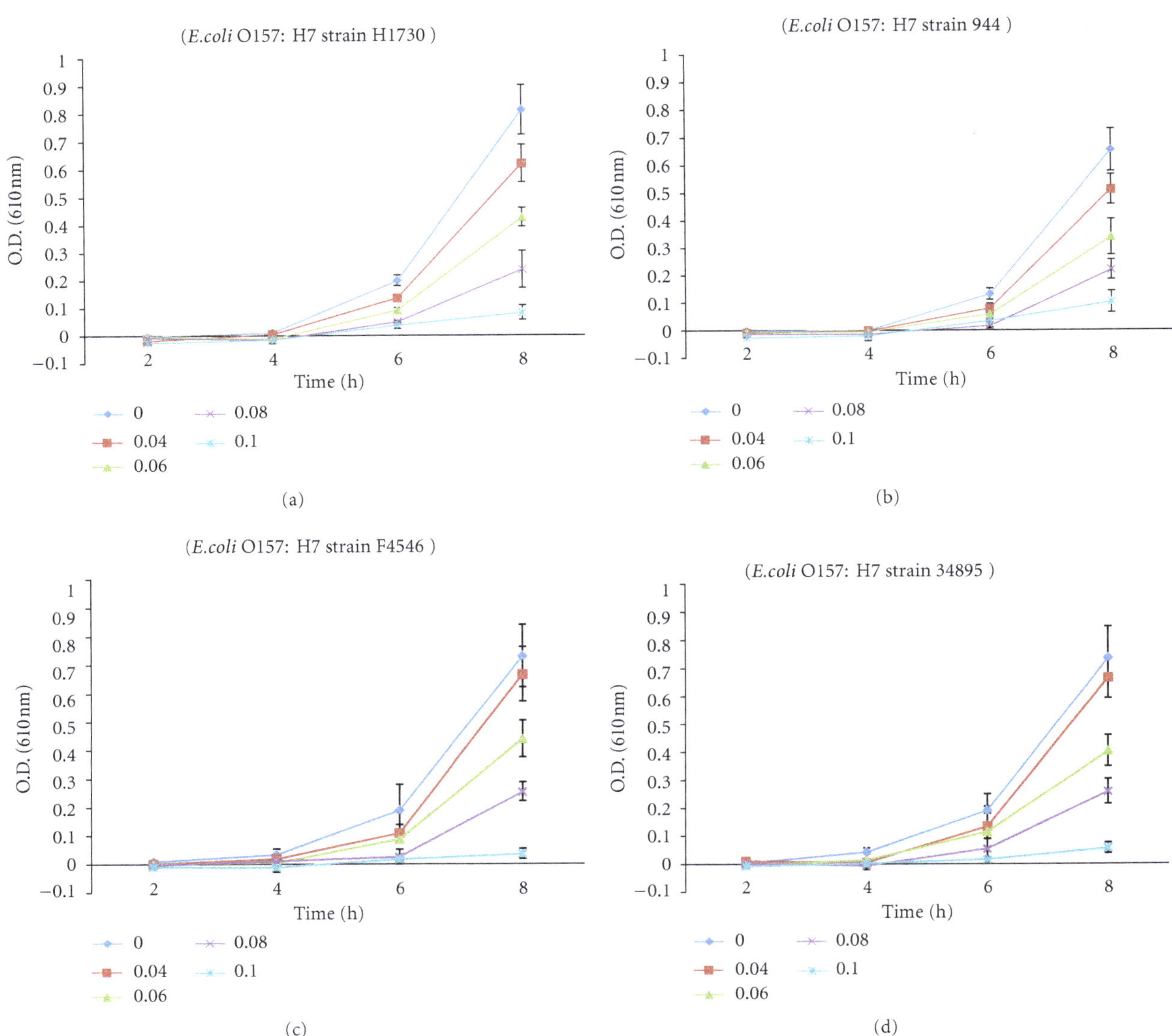

FIGURE 1: Bacterial growth curve of *E. coli* O157:H7 strains in the present of xoconostle extract at different concentrations % (v/v) in BHI broth during 8 h of incubation at 37°C measured by turbidity (optical density 610 nm). Data points are the average of 3 replicates with standard error.

TABLE 1: Final population of *E. coli* O157:H7 strains in BHI broth in the presence of xoconostle extract at different concentrations % (v/v) after 8 h of incubations at 37°C. Data represent the average of three replicates with standard error.

xoconostle extract Concentration % (v/v)	Final population (Log CFU/mL) of *E. coli* O157:H7 strains			
	H1730	944	F4546	43895
0 (control)	$9.19^{aA} \pm 0.16$	$8.66^{aA} \pm 0.12$	$9.12^{aA} \pm 0.20$	$9.31^{aA} \pm 0.13$
4	$8.09^{bA} \pm 0.09$	$7.92^{bA} \pm 0.14$	$8.08^{bA} \pm 0.17$	$8.25^{bA} \pm 0.15$
6	$6.79^{cA} \pm 0.19$	$6.75^{cA} \pm 0.14$	$6.78^{cA} \pm 0.17$	$7.04^{cA} \pm 0.15$
8	$5.32^{dA} \pm 0.15$	$5.16^{dA} \pm 0.18$	$5.32^{dA} \pm 0.11$	$5.90^{dA} \pm 0.14$
10	$3.08^{eA} \pm 0.16$	$2.98^{eA} \pm 0.16$	$3.07^{eA} \pm 0.12$	$4.02^{eB} \pm 0.09$

*Averages with different lower case letters in the same column are significantly different ($P < 0.05$).
*Averages with different upper case letters in the same row are significantly different ($P < 0.05$).

Table 2: Inhibitory zones in BHI agar inoculated with mixture of *E. coli* O157:H7 strains that formed around the wells due to the present of xoconostle extract at different concentrations (v/v) after 12 h of incubation at 37°C. Concentrations are in μL adjusted to 1 mL by distilled water and inhibitory zone = diameter of the zone −8 mm (diameter of the well). Data represent the average of three replicates with standard error.

Concentration (μL/mL)	Inhibitory zone (mm)
200	0
300	1.4 ± 0.3
400	2.9 ± 0.2
500	4.3 ± 0.45
600	5.7 ± 0.3
700	6.6 ± 0.36
800	7.7 ± 0.35
900	9.0 ± 0.45
1000	9.8 ± 1.01

effects and can contribute to better antimicrobial reaction as compare to the reaction of an individual compound [15]. The nature of xoconostle pears containing several phenolic compounds may contribute to strong antimicrobial activity. Xoconostle pears are also rich in ascorbic acid and have higher amount of ascorbic acid than most common fruits and vegetables such as raspberry, red plum, green grape, pear, apple, peach, banana, onion, spinach, green cabbage, pea, cauliflower, lettuce, and tomato [19]. Ascorbic acid is well characterized as a reducing agent with free chemical radicals in chemical and biological systems [8, 14]. In addition, ascorbic acid has the ability to absorb oxygen which might provide a barrier against available oxygen required for *E. coli* O157:H7 [8]. However, ascorbic acid alone has a weaker inhibitory effect compared to the synergetic effect of ascorbic acid and phenolic compounds [8, 10].Furthermore, xoconostle pears are rich in betalains that are well documented for excellent antiradical and antioxidant activity [31, 32]. Betalains can also act as modulators of adhesive molecule expression in endothelial cells [31]. Betalains have metal chelating activities in which they can chelate the cell's indispensable inner cations Ca^{2+}, Fe^{2+}, and Mg^{2+} [12]. These characteristics of betalains have received increased attention suggesting antiviral and antimicrobial activities [12, 32]. The strong antioxidant properties of betalains might provide an additional barrier with ascorbic acid against available oxygen required for *E. coli* O157:H7. The presence of a combination of soluble phenolics with ascorbic acid and betalains might introduce a strong synergetic antimicrobial effect. Therefore, it would be possible to suggest that antimicrobial activity of xoconostle pears is due to the synergistic effect of these active compounds.

Based on the results of these experiments, we suggest that it is possible to use direct extract from xoconostle pears as natural antimicrobial agent against *E. coli* O157:H7. Since the inhibitory effect of xoconostle pears was found to be concentration dependent and not strain dependent, the effective concentration must be determined for successful industrial applications. These findings may thus lead to more attention to the antimicrobial activity of xoconostle pears against other pathogens. Further work is needed to determine the impact of xoconostle pears on *E. coli* O157:H7 in various food applications. Future work in our laboratory will be conducted to determine the antimicrobial activity of xoconostle pears against other pathogenic bacteria including *Salmonella* and *Listeria monocytogenes* and to be tested in food model.

Acknowledgments

This work was supported by the USDA National Institute of Food and Agriculture, Hatch project no. NC.X-2345-09-170-1 in the Agricultural Research Program at North Carolina Agricultural and Technical State University. This work was poster presented at the International Association for Food Protection Annual Meeting, July 31–August 3, 2011, Milwaukee, WI, USA.

References

[1] R. Scharff, "From foodborne illness in the United States," Produce Safety Project, 2010, http://www.producesafetyproject.org/admin/assets/files/Health-Related-Foodborne-Illness-Costs-Report.pdf-1.pdf. Accessed February 13, 2011.

[2] J. M. Rangel, P. H. Sparling, C. Crowe, P. M. Griffin, and D. L. Swerdlow, "Epidemiology of *Escherichia coli* O157:H7 outbreaks, United States, 1982–2002," *Emerging Infectious Diseases*, vol. 11, no. 4, pp. 603–609, 2005.

[3] T. Montville and K. Matthews, *Food Microbiology an Introduction*, ASM Press, Washington, DC, USA, 2nd edition, 2008.

[4] A. L. Robinson and J. L. McKillip, "Biology of *Escherichia coli* O157:H7 in human health and food safety with emphasis on sublethal injury and detection," in *Current Research, Technology and Education Topics in Applied Microbiology and Microbial Biotechnology*, vol. 2, pp. 1096–1105, 2010.

[5] A. M. Tajkarimi, S. A. Ibrahim, and D. Cliver, "Antimicrobial herb and spice compounds in food," *Food Control*, vol. 21, no. 9, pp. 1199–1218, 2010.

[6] I. D. Ciocan and I. I. Bara, "Plant products as antimicrobial agents," *Analele Ştiinţifice ale Universităţii, Alexandru Ioan Cuza, Secţiunea Genetică şi Biologie Moleculară*, vol. 8, pp. 151–156, 2007.

[7] S. A. Ibrahim, H. Yang, and C. W. Seo, "Antimicrobial activity of lactic acid and copper on the growth of *Escherichia coli* O157:H7 and *Salmonella* in carrot juice," *Food Chemistry*, vol. 109, no. 1, pp. 137–143, 2008.

[8] B. Zambuchini, D. Fiorini, M. C. Verdenelli, C. Orpianesi, and R. Ballini, "Inhibition of microbiological activity during sole (*Solea solea L.*) chilled storage by applying ellagic and ascorbic acids," *LWT—Food Science and Technology*, vol. 41, no. 9, pp. 1733–1738, 2008.

[9] S. A. Ibrahim, M. M. Salameh, S. Phetsomphou, H. Yang, and C. W. Seo, "Application of caffeine, 1,3,7-trimethylxanthine, to control *Escherichia coli* O157:H7," *Food Chemistry*, vol. 99, no. 4, pp. 645–650, 2006.

[10] A. M. Tajkarimi and S. A. Ibrahim, "Antimicrobial activity of ascorbic acid alone or in combination with lactic acid on *Escherichia coli* O157:H7 in laboratory medium and carrot juice," *Food Control*, vol. 22, no. 6, pp. 801–804, 2011.

[11] V. K. Bajpai, A. Rahman, N. T. Dung, M. K. Huh, and S. C. Kang, "In vitro inhibition of food spoilage and foodborne pathogenic bacteria by essential oil and leaf extracts of *Magnolia liliflora* Desr," *Journal of Food Science*, vol. 73, no. 6, pp. M314–M320, 2008.

[12] A. Hilou, O. G. Nacoulma, and T. R. Guiguemde, "In vivo antimalarial activities of extracts from *Amaranthus spinosus* L. and *Boerhaavia erecta* L. in mice," *Journal of Ethnopharmacology*, vol. 103, no. 2, pp. 236–240, 2006.

[13] S. A. Ibrahim, G. Yang, D. Song, and T. S. F. Tse, "Antimicrobial effect of guava on *Escherichia coli* O157:H7 and *Salmonella* typhimurium in liquid medium," *International Journal of Food Properties*, vol. 14, no. 1, pp. 102–109, 2011.

[14] M. Z. Končić, D. Kremer, J. Gruz et al., "Antioxidant and antimicrobial properties of Moltkia petraea (Tratt.) Griseb. flower, leaf and stem infusions," *Food and Chemical Toxicology*, vol. 48, no. 6, pp. 1537–1542, 2010.

[15] A. Tafesh, N. Najami, J. Jadoun, F. Halahlih, H. Riepl, and H. Azaizeh, "Synergistic antibacterial effects of polyphenolic compounds from olive mill wastewater," *Evidence-Based Complementary and Alternative Medicine*, vol. 2011, Article ID 431021, 9 pages, 2011.

[16] S. A. Ibrahim, T. S. Tse, H. Yang, and A. Fraser, "Antibacterial activity of a crude chive extract against *Salmonella* in culture medium," *Food Protection Trends*, vol. 29, pp. 155–160, 2009.

[17] J. M. Feugang, P. Konarski, D. Zou, F. C. Stintzing, and C. Zou, "Nutritional and medicinal use of Cactus pear (*Opuntia* spp.) cladodes and fruits," *Frontiers in Bioscience*, vol. 11, no. 2, pp. 2574–2589, 2006.

[18] L. G. García-Pedraza, J. A. Reyes-Agüero, J. R. Aguirre-Rivera, and J. M. Pinos-Rodríguez, "Preliminary nutritional and organoleptic assessment of xoconostle fruit (*Opuntia* spp.) as a condiment or appetizer," *Italian Journal of Food Science*, vol. 17, no. 3, pp. 333–340, 2005.

[19] S. H. Guzmán-Maldonado, A. L. Morales-Montelongo, C. Mondragón-Jacobo, G. Herrera-Hernández, F. Guevara-Lara, and R. Reynoso-Camacho, "Physicochemical, nutritional, and functional characterization of fruits xoconostle (*opuntia matudae*) pears from central-México Region," *Journal of Food Science*, vol. 75, no. 6, pp. C485–C492, 2010.

[20] C. Kilonzo, E. R. Atwill, R. Mandrell, M. Garrick, V. Villanueva, and B. R. Hoar, "Prevalence and molecular characterization of *Escherichia coli* O157:H7 by multiple-locus variable-number tandem repeat analysis and pulsed-field gel electrophoresis in three sheep farming operations in California," *Journal of Food Protection*, vol. 74, no. 9, pp. 1413–1421, 2011.

[21] A. M. Valadez, C. Debroy, E. Dudley, and C. N. Cutter, "Multiplex PCR detection of shiga toxin-producing *Escherichia coli* strains belonging to serogroups o157, o103, o91, o113, o145, o111, and o26 experimentally inoculated in beef carcass swabs, beef trim, and ground beef," *Journal of Food Protection*, vol. 74, no. 2, pp. 228–239, 2011.

[22] A. G. Marwan and C. W. Nagel, "Quantitative determination of infinite inhibition concentrations of antimicrobial agents," *Applied and Environmental Microbiology*, vol. 51, no. 3, pp. 559–561, 1986.

[23] M. E. Parish and D. E. Carroll, "Minimum inhibitory concentration studies of antimicrobial combination against *Saccharomyces cerevisae*in a model broth system," *Journal of Food Science*, vol. 53, pp. 237–239, 1988.

[24] W. B. Hugo and A. Russell, *Pharmaceutical Microbiology*, Blackwell Scientific Publications, London, UK, 5th edition, 1992.

[25] A. L. Vigil, E. Palou, M. E. Parish, and P. M. Davidson, "Methods for activity assay and evaluation of results," in *Antimicrobial in Food*, P. M. Davidson, Ed., pp. 659–680, CRC Taylor & Francis, Boca Raton, Fla, USA, 3rd edition, 2005.

[26] K. Emeruwa, "Antibacterial substance from Carica papaya fruit extract," *Journal of Natural Products*, vol. 45, no. 2, pp. 123–127, 1982.

[27] SAS Institute, *SAS/STAT—User's Guide, Version 9.2*, SAS Institute, Cary, NC, USA, 2010.

[28] T. Beta, L. W. Rooney, L. T. Marovatsanga, and J. R. Taylor, "Effect of chemical treatments on polyphenols and malt quality in sorghum," *Journal of Cereal Science*, vol. 31, no. 3, pp. 295–302, 2000.

[29] Y. C. Lin and C. C. Chou, "Effect of heat treatment on total phenolic and anthocyanin contents as well as antioxidant activity of the extract from Aspergillus awamori-fermented black soybeans, a healthy food ingredient," *International Journal of Food Sciences and Nutrition*, vol. 60, no. 7, pp. 627–636, 2009.

[30] R. C. Whiting and M. H. Golden, "Variation among *Escherichia coli* O157:H7 strains relative to their growth, survival, thermal inactivation, and toxin production in broth," *International Journal of Food Microbiology*, vol. 75, no. 1-2, pp. 127–133, 2002.

[31] C. Gentile, L. Tesoriere, M. Allegra, M. A. Livrea, and P. D'Alessio, "Antioxidant betalains from cactus pear (*Opuntia ficus-indica*) inhibit endothelial ICAM-1 expression," *Annals of the New York Academy of Sciences*, vol. 1028, pp. 481–486, 2004.

[32] H. M. C. Azeredo, "Betalains: properties, sources, applications, and stability—a review," *International Journal of Food Science and Technology*, vol. 44, no. 12, pp. 2365–2376, 2009.

22

Development of Class IIa Bacteriocins as Therapeutic Agents

Christopher T. Lohans and John C. Vederas

Department of Chemistry, University of Alberta, Edmonton, AB, Canada T6G 2G2

Correspondence should be addressed to John C. Vederas, john.vederas@ualberta.ca

Academic Editor: John Tagg

Class IIa bacteriocins have been primarily explored as natural food preservatives, but there is much interest in exploring the application of these peptides as therapeutic antimicrobial agents. Bacteriocins of this class possess antimicrobial activity against several important human pathogens. Therefore, the therapeutic development of these bacteriocins will be reviewed. Biological and chemical modifications to both stabilize and increase the potency of bacteriocins are discussed, as well as the optimization of their production and purification. The suitability of bacteriocins as pharmaceuticals is explored through determinations of cytotoxicity, effects on the natural microbiota, and *in vivo* efficacy in mouse models. Recent results suggest that class IIa bacteriocins show promise as a class of therapeutic agents.

1. Introduction

Bacteriocins are natural peptides secreted by many varieties of bacteria for the purpose of killing other bacteria. This provides them with a competitive advantage in their environment, eliminating competitors to gain resources. These peptides are ribosomally synthesized, although some are extensively posttranslationally modified.

The classification system for bacteriocins has been subject to ongoing revision [1–3]. However, bacteriocins from Gram-positive bacteria are generally classified according to size, structure, and modifications. Class I bacteriocins are the lantibiotics, which are highly posttranslationally modified peptides containing lanthionine and methyllanthionine residues. Class II consists of small peptides that do not contain modified residues. Cotter et al. suggested to divide class II bacteriocins into several subclasses: class IIa (pediocin-like bacteriocins), class IIb (two-peptide bacteriocins), and class IIc (circular bacteriocins) [3]. However, others have suggested to consider circular bacteriocins as a separate class [4]. Nonbacteriocin lytic proteins, termed bacteriolysins (also referred to as class III bacteriocins), are large and heat-labile proteins with a distinct mechanism of action from other Gram-positive bacteriocins [3].

Class IIa bacteriocins are generally from 37 to 48 amino acids long, and are characterized by several features. Although they do not have broad spectrum antimicrobial activity compared to other antibiotics, they are particularly potent inhibitors of *Listeria* species, showing activity at low nanomolar concentrations [5]. They are heat-stable, and not posttranslationally modified beyond the proteolytic removal of a leader peptide and the formation of a conserved N-terminal disulfide bridge (although some members contain an additional C-terminal disulfide bridge). The N-terminal region contains a characteristic YGNGV amino acid sequence, although variants with the alternate YGNGL sequence have been classified in class IIa [6]. A representative class IIa bacteriocin is shown in Figure 1. There have been a number of thorough reviews describing aspects of the genetics, biosynthesis, immunity, structure, mode of action, and the application of class IIa bacteriocins to foods [7–13].

Briefly, class IIa bacteriocins kill susceptible bacteria by forming pores in their membranes, resulting in the loss of the proton-motive force and depletion of ATP [14]. It is thought that these cationic bacteriocins are drawn to bacterial cells through an initial electrostatic interaction [15]. Then, the amphiphilic C-terminal α-helix inserts into the membrane, wherein the bacteriocin induces the formation of hydrophilic pores. This mechanism of action is reliant on a mannose phosphotransferase (MPT) protein complex found in the membranes of susceptible organisms, but the exact nature

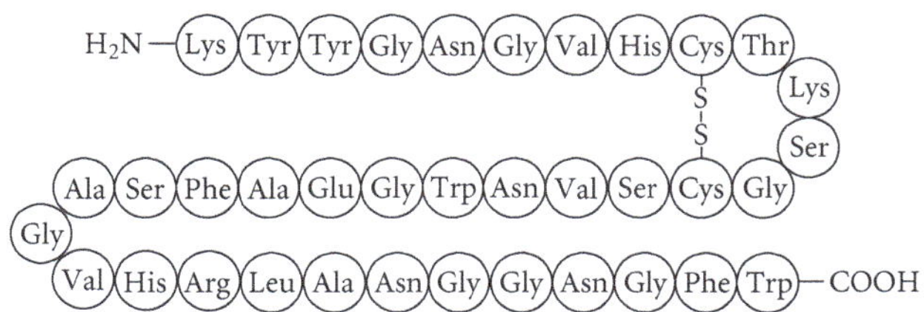

Figure 1: A representation of class IIa bacteriocin leucocin A, with the YGNGV consensus sequence and an N-terminal disulfide bridge.

Figure 2: The NMR solution structure of leucocin A [20].

of this interaction is not yet clear [16–18]. This is covered in more detail by Drider et al. [12] and Nissen-Meyer et al. [19].

Structurally, the N-termini of class IIa bacteriocins tend to exhibit a three-strand antiparallel beta-sheet structure rigidified by a disulfide bridge. The C-terminal region shows an amphiphilic helix terminating in a hairpin structure. In aqueous conditions, class IIa bacteriocins are randomly structured. However, membrane-mimicking conditions such as dodecylphosphocholine micelles or trifluoroethanol induce structure formation [20]. This is not unexpected as their mode of action involves membrane permeabilization [14]. The NMR solution structures of class IIa bacteriocins leucocin A (shown in Figure 2) [20], carnobacteriocin B2 [21] and its precursor precarnobacteriocin B2 [22], sakacin P [23], and curvacin P [24] have been solved to date.

Much of the research on class IIa bacteriocins has focused on their application for food preservation. While they may be well-suited for this purpose, there is a growing body of research exploring the prospect of using these bacteriocins as *in vivo* therapeutic agents. Bacteriocins are a promising substitute for conventional antibiotics for several reasons. The restricted target specificity of some bacteriocins minimizes their impact on commensal microbiota and may decrease the threat of opportunistic pathogens. Furthermore, most bacteriocins are active at low concentrations, and their degradation products are easily metabolized by the body. With the development of resistance to many important antibiotics, another tool for fighting bacteria is invaluable.

Class IIa bacteriocins are active against several important human pathogens. Perhaps most promising is their activity against the foodborne pathogen *Listeria monocytogenes*, the deadliest bacterial source of food poisoning [25]. Up to 30% of foodborne infections by *L. monocytogenes* in high-risk individuals are fatal. Other bacterial foodborne pathogens inhibited by some class IIa bacteriocins include *Bacillus cereus, Clostridium botulinum*, and *C. perfringens* [5].

Beyond foodborne pathogens, class IIa bacteriocins are also active against other human pathogens, such as vancomycin-resistant enterococci [26] and the opportunistic pathogen *Staphylococcus aureus* [5]. Although bacteria sensitive to class IIa bacteriocins are almost exclusively Gram-positive, the Gram-negative opportunistic pathogen *Aeromonas hydrophila* is also inhibited [27]. Bacteriocins from this class also show other potentially therapeutic properties as antineoplastic [28, 29] and antiviral [30] agents.

The potential of other groups of bacteriocins such as lantibiotics, colicins, and microcins as oral and gastrointestinal antibiotics has been reviewed by Kirkup [31]. Focusing on bacteriocins from Gram-positive bacteria, there have been successes with the administration of either lantibiotic-producing bacteria [26] or purified lantibiotics [32–38] for the treatment of infections by several different pathogens. However, less is known about the *in vivo* use of class IIa bacteriocins.

One method for the therapeutic use of bacteriocins is to introduce bacteriocinogenic bacteria to the gastrointestinal tract as probiotics, which has yielded positive results. Frequently, the mechanisms by which probiotic bacteria benefit the host are not well characterized, but convincing evidence has been put forth by Corr et al. for the production of class IIb bacteriocin Abp118 *in vivo* [39]. Generally, the introduction of bacteriocinogenic bacteria prior to infection with a pathogen has been more effective [26, 39] than the concomitant introduction of both species [40]. This suggests that probiotic strains may be valuable for prophylactic purposes, but less suited for treating preexisting infections.

Indeed, introduction of a bacteriocin either concomitantly with the infectious agent or postchallenge has proven effective [32–38]. A variety of administration methods have been used successfully: subcutaneous, intravenous, intranasal, intragastric, intraperitoneal, and topical [41]. The efficacy of the different methods has not been directly compared and likely depends on the pathogen targeted. Furthermore, some of these methods may be unnecessarily invasive for use in humans, with oral administration being preferred. Although the possibility exists of using crude bacteriocin extract instead of purified bacteriocin, the introduction of complex mixtures into a human may be hazardous and less reproducible. Instead, this paper will focus on the administration of purified class IIa bacteriocin.

Compared to other classes of Gram-positive bacteriocins, the engineering of improved class IIa bacteriocins is somewhat simplified due to their unmodified nature. Creating analogues, by biological or chemical means, does not require implementation of the thioether bridges found in lantibiotics or the cyclization required for circular bacteriocins. Nor does the recombinant expression of class IIa bacteriocins require the biosynthetic machinery, such as dehydratases and cyclases, required for some other bacteriocins. This also allows for the production of class IIa bacteriocins as fusion proteins, a means of increasing production levels and simplifying purifications.

This paper will explore different aspects of the development of class IIa bacteriocins as therapeutic agents for *in vivo* utilization. The first section discusses attempts to design bacteriocins and bacteriocin analogues with increased stability and potency. Next, methods for improving production and purification of large amounts of bacteriocin from fermentation and recombinant expression will be explored. Finally, the suitability of class IIa bacteriocins for therapeutic use, based on studies testing cytotoxicity, stability, the development of resistance, and the *in vivo* potential of class IIa bacteriocins will be examined.

2. Engineering Class IIa Bacteriocins for Increased Stability and Potency

The structure-function relationship of class IIa bacteriocins has been well studied, and its implications for their mode of action has been well reviewed [8, 12, 19]. This paper focuses on structure-function as it contributes to the development of improved therapeutics. Specifically, engineering bacteriocins to increase their stability, potency, and spectrum of activity, such that they are more suitable for *in vivo* utilization and other applications will be discussed.

The introduction of an additional disulfide bridge likely has the effect of rigidifying a specific conformation and could result in improved bacteriocin activity. There is a subgroup of class IIa bacteriocins, including pediocin PA-1, that contain an additional disulfide bridge near the C-terminus. The effect of introducing a C-terminal disulfide bridge into sakacin P, a bacteriocin containing only the conserved N-terminal disulfide bridge, was examined [42]. This modification broadened its spectrum of antimicrobial activity in addition to decreasing the detrimental effect of increased temperature on potency. The C-terminus has been otherwise associated with the target specificity of class IIa bacteriocins [43]. Notably, this sakacin P mutant was found to retain much of its activity at 37°C compared to the natural peptide, and thus is more effective at human physiological temperature [42].

The necessity of the N-terminal disulfide bridge for activity in class IIa bacteriocins has also been explored. Removal of this disulfide bridge could render bacteriocins more stable in reductive environments. Substitution of cysteines 9 and 14 of leucocin A [44] and mesentericin Y105 [45] with serines resulted in a complete loss of activity. However, replacement with hydrophobic residues such as allylglycine, norvaline, and phenylalanine resulted in retention of activity in leucocin A [46]. Furthermore, the replacement of the disulfide bridge with a carbocycle also yielded a biologically active peptide, although the activity was decreased by an order of magnitude [44]. However, the substitution of cysteines 9 and 14 of pediocin PA-1 with allylglycine and phenylalanine residues resulted in no observable activity [46]. This work has been discussed further in a mini-review by Sit and Vederas [47].

Class IIa bacteriocins may also be stabilized by simple amino acid substitutions. Methionine-31 of pediocin PA-1 was found to oxidize over time with an accompanying loss of activity [48]. Mutation of this residue to leucine, isoleucine or alanine resulted in only minor decreases in potency while stabilizing the mutant [48]. Similarly, a 4- to 8-fold decrease in activity was reported for carnobacteriocin BM1 due to an oxidized methionine residue [49], but replacement with a valine residue yielded a mutant with comparable activity [50]. However, in some cases substitution of only a single amino acid residue in class IIa bacteriocins results in dramatically decreased activity relative to their wild-type counterparts [51].

Consideration of the mode of action of class IIa bacteriocins may permit the rational design of mutants with increased potency. Enhancing the net positive charge of a bacteriocin may be expected to promote the initial electrostatic interaction with the membrane of the target and thus result in an increase in activity. Support for this was found in the 44 K (with an additional lysine introduced to the C-terminus) and T20K mutants of sakacin P, which show increased cell binding and potency relative to the wild-type peptide [52].

Approaches to stabilizing other classes of bacteriocins may have potential for use with class IIa bacteriocins. Due to their composition, proteolytic cleavage of bacteriocins in the gastrointestinal tract represents a major hurdle for any attempts to control gastrointestinal infections. Careful alteration of trypsin recognition sites in class IIb bacteriocin salivaricin P had only minor effects on activity [53]. Chemically synthesized peptides with incorporated D-amino acids may be similarly expected to render the peptide less susceptible to proteolytic cleavage. Analogues of class IIb bacteriocin lactococcin G were synthesized with the N- and C-terminal residues replaced with D-amino acids, which decreased their susceptibility to exopeptidases without much effect on activity [54]. However, the extent of incorporation of D-amino acids has limitations. The enantiomer of leucocin A was synthesized containing exclusively D-amino acids, but it was found to be largely inactive [55]. This may be rationalized based on a chiral interaction between class IIa bacteriocins with the MPT complex [16]. Nonetheless, these methods may be valuable for stabilizing class IIa bacteriocins.

Much work is focused on using biological means to create bacteriocin analogues. Mutagenesis of bacteriocins can be readily achieved, and large quantities of a desired mutant are readily available through recombinant expression. However, the biological production of analogues suffers the restriction of the proteogenic amino acid library. As a contrast, chemical peptide synthesis offers a vast array of possibilities for the introduction of nonproteogenic amino acids. Furthermore, unnatural structural features not found in class IIa bacteriocins such as carbocyclic rings and D-amino acids are feasible. However, chemical peptide synthesis is not trivial, and it is relatively time consuming and costly. For a chemically synthesized bacteriocin to be considered a viable therapeutic agent, it would have to be greatly superior to any biologically producible bacteriocins.

The rational substitution of amino acids in class IIa bacteriocins is one method of creating mutants, and this has provided much information about the structure-function relationship of these bacteriocins. However, for the most part, the mutants have had decreased activity relative to the wild-type bacteriocin. Another common approach uses error-prone PCR to randomly generate mutants in the hope of finding interesting or improved activity. However, approaches such as DNA shuffling [51] of related bacteriocins and NNK scanning [56] have been used to randomly generate vast numbers of mutants, greatly increasing the number of variants produced without requiring a proportionate amount of labour.

NNK scanning allows for the systematic examination of the role of each residue in a peptide. The native codons are replaced one by one with the NNK triplet oligonucleotide, replacing the amino acid coded for by that codon with any of the 20 proteogenic amino acids. This allows for testing a much larger number of variants without requiring the time consuming preparation of each mutant separately. Consequently, the possibility of discovering a mutant with increased potency is greater. NNK scanning has been applied to pediocin PA-1 to examine the importance of each residue for bactericidal activity and was indeed successful in creating some mutants with increased activity [56].

Often, changing one amino acid at a time is not sufficient to create improved variants. It has been suggested that bacteriocins have evolved to be as effective as possible, and so the creation of improved bacteriocins requires greater modification [51]. This is possible using an alternate approach that allows for the swapping of multiamino acid sequences between different class IIa bacteriocins to create a hybrid bacteriocin. This approach has been used to create a DNA-shuffling library in which regions of pediocin PA-1 have been shuffled with 10 other class IIa bacteriocins [51]. Some of the hybrids did indeed show increased activity relative to the wild-type bacteriocins from which they were derived [51].

Another approach explored for creating new analogues is to mix the N-terminus of one bacteriocin with the C-terminus of another, thereby creating a chimera. Some chimeras of pediocin PA-1 with other class IIa bacteriocins showed either comparable or greater bactericidal activities to the corresponding natural bacteriocins against certain indicator strains [43, 51].

These approaches to randomly generate vast numbers of mutants and hybrids may allow for simplified drug development, facilitating the discovery of novel potent bacteriocins. Furthermore, these approaches enable the development of new bacteriocins tailored towards different strains of pathogenic bacteria.

3. Methods for Improving Production of Class IIa Bacteriocins

For any potential therapeutic use of class IIa bacteriocins, an inexpensive method for the production of large quantities must be available. One possibility is to purify class IIa bacteriocins from their natural producer strain, taking advantage of the cationic and hydrophobic characters of these peptides. However, these purifications typically yield only small amounts of purified peptide, often consisting of less than a milligram per liter of culture [49, 57]. However, the outlook is not bleak, as optimizing culture conditions and improving the design of purifications maximizes bacteriocin recovery and permits increased scale.

Of the class IIa bacteriocins, pediocin PA-1 is most well characterized in terms of optimization of fermentation. Even then, reported yields must be interpreted carefully, as the sensitivity of indicator strains varies and activity tests are performed differently. A variety of different cultivation methods have been used, such as shake-flasks, batch cultures, and fed-batch cultures. Batch cultures in reactors generally allow for greater control over conditions than shake-flask cultur es, with precise control of stirring, aeration, and pH. Fed-batch cultures are similar to batch cultures, except a growth-limiting nutrient is added over time, allowing for higher cell densities.

For the large-scale production of class IIa bacteriocins to be feasible, several conditions must be met. The yield of the fermentation must be satisfactory, otherwise production costs will be high. The growth media must also be inexpensive, although this must be balanced with the bacteriocin yield as the use of more expensive media has been related to improved bacterial growth and bacteriocin production [58].

The highest reported volumetric productivity was accomplished by a repeated-cycle batch culture of *Pediococcus acidilactici* UL5 immobilized in κ-carrageenan/locust bean gum gel beads, reaching levels of 133 mg of pediocin PA-1 produced per liter per hour in complex de Man Rogosa and Sharpe (MRS) media [58]. Using less expensive supplemented whey permeate (SWP) media under otherwise identical conditions, 50 mg of pediocin PA-1 was produced per liter per hour. The production of bacteriocins has tended to be much superior in immobilized cell cultures compared to free cell cultures [59], as exemplified by the greater than tenfold increase in production of pediocin PA-1 under immobilized conditions [58]. Naghmouchi et al. have published an informative literature summary of recent work on fermentation yields of pediocin PA-1 [58].

Bacteriocin-producing fermentations have been tested in a large variety of media as an attempt to minimize production costs. Waste from the food industry especially has been investigated as an inexpensive alternative to complex growth media. Examples of this include mussel-processing waste [60, 61], whey permeate [58, 62–64], trout and squid viscera, and swordfish muscle [65]. Complex growth media tend to be composed of a mixture of nutrients tailored to certain types of bacteria to meet their specific nutritional requirements, while industrial effluents are not so optimized [63, 66].

Although the production of large amounts of bacteriocin is feasible, the purification of these peptides is another matter. A review by Carolissen-Mackay et al. discusses previous purification approaches for bacteriocins [57]. Many purification protocols provide poor yields of bacteriocin with recoveries of under 20% [57]. These poor yields are likely due to unoptimized protocols requiring a large number of steps. More recently, several general protocols have been published specifically for the purpose of purifying class IIa bacteriocins [67–69]. For the industrial-scale production of bacteriocins required for therapeutic use, an efficient, inexpensive, and scalable purification scheme with high recovery is needed.

Commonly, the purification of class IIa bacteriocins requires precipitation and centrifugation steps. The latter represents a major bottleneck when attempts are made to increase the scale of production. Furthermore, ammonium sulfate precipitations are frequently a source of loss of material, yielding only 40%± 20% for a reported pediocin PA-1 purification [67]. Using an initial ion-exchange chromatographic step to concentrate the bacteriocin directly from the culture media is a possible solution [67].

More recent general purification schemes generally follow a similar sequence, taking advantage of the cationic and hydrophobic character of class IIa bacteriocins. First, the culture supernatant is passed through a cation-exchange column [67–69] although loading the whole bacterial culture to avoid centrifugation has been reported [67]. Following this step, the eluate is further purified using hydrophobic interaction chromatography, yielding greater than 90% pure bacteriocin in only two steps [67, 69]. HPLC may also be used to further clean up the sample at this stage [68]. These purifications allow for the acquisition of purified bacteriocin in only a few hours [67], with bacteriocin recovery rates reported ranging from 60% [68] to greater than 80% [67].

The development of antibodies capable of recognizing bacteriocins has allowed for an alternate approach to purification, namely, immunoaffinity chromatography [70–73]. Indeed, this approach has been used to purify divercin V41, piscicocin V1b, enterocin P, and pediocin PA-1 from culture supernatant in a single step. Although reported yields are sparse, the recovery of enterocin P was 44% [71], while 53% of pediocin PA-1 was retained [73]. Although pure bacteriocin is obtained after a single step, superior yields have been reported for lengthier procedures [67], and the immunoaffinity purification requires costly noncommercial antibody-conjugated resins.

Another notable purification approach uses triton X-114 phase partitioning, which has been applied to the purification of divercin V41 [74]. This approach does not require removal of bacterial cells from the culture, thereby enabling collection of the bacteriocin normally lost adhered to the cell pellet. After the two phases partition, the detergent rich phase is removed, diluted, and loaded on to an ion-exchange column. Purified bacteriocin is simply eluted from the column, with a recovery of greater than 55% [74].

All of these reported purifications have unique advantages and drawbacks. However, the focus has shifted to the large-scale production of bacteriocins instead of purifying only enough for characterization. These approaches have focused on attaining improved yields in fewer steps with mostly scalable steps.

3.1. Heterologous Expression. As an alternative to purification from the natural producer, the recombinant expression of bacteriocins offers a promising means for producing the large amounts of material required for any potential therapeutic use. There have been many reports of the heterologous expression of class IIa bacteriocins in many different hosts, although the focus has been on Gram-positive lactic acid bacteria phylogenetically similar to the producer strain. Gram-negative bacteria such as *Escherichia coli* and yeast expression platforms such as *Saccharomyces cerevisiae* have also been used as expression hosts.

The subject of the heterologous expression of mature bacteriocins in lactic acid bacteria has been summarized in an excellent review by Rodríguez et al. [89], which also discusses heterologous bacteriocin production in *E. coli* and other bacterial strains. Although some of these expression systems allow for the secretion of active bacteriocin into the culture supernatant, the quantity of bacteriocin obtained from these cultures tends to be lower than from the natural producer strain. As such, this is not yet suitable for the large-scale production of bacteriocins required for any potential therapeutic use. However, these heterologous producers may be suitable for food preservation as many lactic acid bacteria are generally recognized as safe. Furthermore, these organisms are capable of simultaneously producing multiple different bacteriocins allowing for a greater spectrum of activity in addition to the possibility of overcoming the development of resistance [90–92]. However, the use of genetically modified organisms in food products is still a contentious issue.

The heterologous expression of bacteriocins as fusion proteins in *E. coli* has been successfully used for the production of larger amounts of bacteriocin than obtained using other approaches. In particular, the commercial strain *E. coli* Origami (DE3) has been used extensively in this area. Mutations in the genes encoding the glutathione and thioredoxin reductases of this strain allow for the facile formation of the conserved disulfide bridge in the host cytosol.

Additionally, the heterologous expression of class IIa bacteriocins in *E. coli* as fusion proteins offers many advantages. A summary of the reported use of fusion proteins partnered with class IIa bacteriocins is presented in Table 1. Fusions with affinity labels, such as hexahistidine tags, allow for simplified purification protocols. Additionally, some fusion partners help solubilize the bacteriocin and prevent the desired peptide from forming inclusion bodies, allowing for increased bacteriocin production. The presence of a fusion partner also decreases the antimicrobial activity, avoiding possible toxic effects on the host cell [93–95], although there are exceptions [96]. Thioredoxin in particular is useful as a fusion partner. Beyond circumventing the formation of inclusion bodies, a thermostable thioredoxin fusion allows for a thermal coagulation purification step [97]. This has been used to remove high molecular weight contaminants during the purification of carnobacteriocins BM1 and B2 [50]. Furthermore, thioredoxin may even assist in the formation of the conserved N-terminal disulfide bridge [97].

Expression of a bacteriocin solely with a hexahistidine tag has been reported for pediocin PA-1 [104]. However, this recombinant pediocin PA-1 was found to be toxic to the *E. coli* producer. The purification was complicated by the requirement for denaturing conditions to allow for immobilized-metal affinity chromatography, although the His-tagged peptide was antimicrobially active. Furthermore, expression of small-sized recombinant peptides in *E. coli* is complicated due to the presence of proteases [105]. The expression of bacteriocins with a larger fusion partner is likely to be advantageous.

The conditions used for fermentations have a significant impact on the amount of fusion protein produced. The final yield of purified bacteriocin is influenced by the purification protocol as well as the method used for fusion protein cleavage. Simple shake-flask cultures have been reported

Table 1: Fusion partners used for class IIa bacteriocins.

Fusion partner	Bacteriocin
Thioredoxin	Pediocin PA-1 [95, 98], carnobacteriocins BM1 and B2 [50], divercin V41 [96], enterocin P [72], piscicolin 126 [99]
Maltose-binding protein	Carnobacteriocin B2 [93] and its precursor [22], Pediocin AcH [100]
Intein-chitin-binding domain	Piscicolin 126, divercin V41, enterocin P, pediocin PA-1 [101]
Dihydrofolate reductase	Pediocin PA-1 [94]
Xpress tag	Pediocin PA-1 [102]
Cellulose-binding domain	Enterocin A [103]
Hexahistidine tag	Pediocin PA-1 [104]

most, although many of the reported yields are admittedly not optimized. Piscicolin 126 was cleaved from a thioredoxin fusion yielding 26 mg per liter [99], while a divercin RV41-thioredoxin fusion yielded between 18 and 23 mg of purified peptide per liter of culture [96, 106].

High-cell density *E. coli* cultures have also been explored as a means to further increase the production of bacteriocin fusion proteins [50, 106]. The level of production of a recombinant divercin V41-thioredoxin fusion in batch and fed-batch cultivation has been compared to shake-flask cultures [106]. Compared to the yield of 18 ± 1 mg obtained per liter in shake flask cultures, batch and fed-batch yielded 30 ± 2 and 74 ± 5 mg per liter, respectively. However, the highest yields reported are for carnobacteriocins BM1 and B2. These bacteriocins were expressed as thioredoxin fusions in a fed-batch fermentation induced with lactose. The final yields reported are around 320 mg of carnobacteriocin BM1 and carnobacteriocin B2 per liter of the culture, fourfold greater than previous reports [50].

A disadvantage of using bacteriocin fusion proteins is the necessary cleavage and further purification required to get pure bacteriocin. Enzymatic cleavage methods are the most common approach, while chemical methods have also been used. Enzymatic approaches offer the advantage of more specific recognition sites and are thus more compatible with most bacteriocin sequences—although the enzyme recognition is not always infallible [22].

Cyanogen bromide (CNBr) is a common chemical means of cleaving fusion proteins, selectively cleaving on the C-terminal side of methionine residues. However, methionine is found in many class IIa bacteriocins. This has been circumvented with carnobacteriocin BM1, wherein methionine-41 was substituted with a valine residue with some impact on activity [50]. However, CNBr has significant advantages over proteases: cost and cleavage efficiency. Besides being much less expensive, the cleavage efficiency of CNBr has been reported to be up to twofold higher than enterokinase [50, 95].

An alternative approach for fusion protein cleavage requires the presence of the amino acid sequence Asp-Pro just N-terminal to the desired sequence. This cleavage method requires heating under strongly acidic conditions, as has been applied for the cleavage of a divercin V41 thioredoxin fusion [106]. This offers an inexpensive method to remove the fusion tag, although these may seem like unsuitable conditions for a peptide. However, class IIa bacteriocins tend to be stable at elevated temperatures and in acidic conditions [49].

The use of the intein-chitin-binding domain as a fusion partner allows for circumvention of several of the issues related to fusion proteins. Following the binding of the fusion protein on a chitin resin, cleavage is induced with DTT, resulting in elution of purified bacteriocin without requiring purification from the fusion partner. This has been successfully applied for a variety of class IIa bacteriocins, although the yields have not been very substantial [101].

Yeast expression platforms are another option for the production of class IIa bacteriocins. *Saccharomyces cerevisiae* has been used as an expression host for pediocin PA-1 [107] and plantaricin 423 [108]. Antimicrobial activity was indeed observed, and colonies of yeast growing on agar inoculated with *Listeria* showed zones of inhibition. However, very little antimicrobial activity was observed in the supernatant [107, 108]. This low level of activity may be attributed to the bacteriocin remaining associated with the fungal cell wall [107].

The use of *Pichia pastoris* as an expression host is more promising, showing much higher levels of activity. The levels of enterocin P produced by *P. pastoris* reached levels up to 28 mg/L, almost four-fold higher than that produced by the natural producer strain, *Enterococcus faecium* P13 [109]. However, the final purified yield of enterocin P from *E. faecium* P13 was still superior, demonstrating that improved purification methods are required to take advantage of any increased production. Class IIa-like bacteriocin hiracin JM79 has also been expressed in *P. pastoris*, with similar issues [91]. Although the quantified amount of bacteriocin exceeds that of the natural producer, the observed antimicrobial activity was found to be relatively smaller. Neutral proteases have been suggested as a possible reason for this discrepancy, and bacteriocin amounts may be overestimated due to the nature of the quantitative techniques used [91, 109]. Furthermore, the activity of pediocin PA-1 produced by *P. pastoris* was found to be inhibited by the presence of a collagen-like material, which appeared to be covalently bound to the pediocin [110].

4. *In Vivo* Utilization of Class IIa Bacteriocins

As previously discussed, most published work regarding the *in vivo* use of bacteriocins has focused on the introduction of probiotic bacteria to the gastrointestinal tract, where they will potentially secrete bacteriocins. Considerably less research has been done on the administration of purified bacteriocin. The use of probiotic strains may prove beneficial as a prophylactic, but the use of purified bacteriocins appears to be superior for countering an established infection. This has been demonstrated by the administration of either pediocin

PA-1 or *Pediococcus acidilactici* UL5, a producer of pediocin PA-1, to mice infected with *L. monocytogenes* [40].

An important concern regarding the use of antibiotics is the effects they have on the microbiota of the body. The presence of commensal bacteria offers an invaluable barrier to infection by opportunistic pathogens. Ideally, an antimicrobial agent should specifically target the pathogenic bacteria with only minimal impact on the natural flora. In fact, the spectrum of activity for class IIa bacteriocins may be extremely well suited for targeting specific pathogens such as *L. monocytogenes in vivo*. Pediocin PA-1 has been tested *in vitro* against screens of common human intestinal bacteria such as bifidobacteria [75, 76], and at the concentrations tested, no antagonistic activity was observed against any of the assayed organisms. This differs from class I lantibiotics nisin A and nisin Z, both of which inhibited the majority of Gram-positive strains tested [75, 76]. Similarly, culture supernatant containing pediocin PA-1 was found to only inhibit one strain of a screen of common gut bacterial species [77]. Furthermore, an *in vivo* study of pediocin PA-1 in a mouse model showed no effect on the composition of the mouse intestinal flora. Likewise, purified pediocin PA-1 fed to rats did not affect the majority of their microbiota [77]. As a contrast, antibiotics such as penicillin and tetracycline strongly inhibited most of the common intestinal microbiota tested [76].

Two different routes of bacteriocin administration to fight *L. monocytogenes* have been tested in mouse models: intravenous [78, 79] and intragastric [40]. The effects of pediocin PA-1 have also been studied in uninfected mice [80], rabbits [80], and rats [77]. The suitability of the route depends on the nature of the pathogen being targeted, as well as the stage of the infection. However, as peptides, bacteriocins face challenges related to their structure not shared by many antibiotics.

Piscicolin 126, recombinant divercin RV41 (DvnRV41), and structural variants of DvnRV41 were all administered intravenously to mice previously or soon to be infected with *L. monocytogenes* [78, 79]. In the control, the intravenous and intraperitoneal injection of these bacteriocins into healthy mice resulted in no visible ill effects [78, 79]. The efficacy of intravenous administration of bacteriocin was tested both prior to and after the intravenous introduction of *Listeria*. Injection of bacteriocins was effective both 15 minutes prechallenge and 30 minutes postchallenge. However, administration of piscicolin 126 24 hours postchallenge showed no significant reduction in listerial counts. Both of these experiments used only 2 μg of purified bacteriocin. The intracellular nature of *Listeria* as a pathogen may explain the lack of sensitivity observed following bacteriocin administration 24 hours postchallenge [25].

A possible concern with the intravenous administration of peptides is the possibility of an immune response. Foreign peptides are often antigenic, and the introduction of these peptides could trigger an immune response. To test this, pediocin AcH was intraperitoneally introduced into mice and rabbits to determine its antigenic properties. However, it did not elicit an antibody response and appears to be nonimmunogenic [80]. In fact, approaches to develop antibodies to class IIa bacteriocins have required conjugation to polyacrylamide gel [81] or carrier proteins such as keyhole limpet hemocyanin [70, 71, 82, 83].

The intragastric administration of bacteriocins suffers from its own set of problems. Bacteriocins are subjected to harsh environments designed precisely for the proteolytic cleavage of peptides and proteins. Class IIa bacteriocins are susceptible to common digestive proteases. Furthermore, the stomach is a highly acidic environment. However, class IIa bacteriocins tend to be relatively stable to acidic conditions, and pediocin PA-1 was stable at pH 2.5 for at least two hours [84].

The stability of bacteriocins in the gastrointestinal tract has been examined by passing purified pediocin PA-1 through an artificial system mimicking the human stomach and small intestine [85]. Pediocin PA-1 retained some activity after 90 minutes in the artificial gastric conditions, while all activity was lost once the sample was in the duodenal compartment. It was suggested that pancreatin in the duodenum was responsible for the ultimate cleavage of the pediocin PA-1, while a combination of pepsin and low pH may be responsible for the decrease in activity observed in the gastric chamber. This is in agreement with *in vivo* results, as pediocin PA-1 fed to rats was not detected in their fecal samples [77]. Despite this, the intragastric administration of pediocin PA-1 has been proven effective for decreasing the load of *L. monocytogenes* in a mouse model [40]. Furthermore, enca-psulation may preserve bacteriocin potency in the gastrointestinal tract, although this has not been reported for class IIa bacteriocins as of yet. However, encapsulating the lantibiotic nisin in liposomes has shown some success [86–88].

The intragastric administration of pediocin PA-1 to mice infected with *L. monocytogenes* has been examined [40]. Treatment with 250 μg of pediocin PA-1 a day for three consecutive days resulted in a 2-log reduction in fecal listerial counts. *L. monocytogenes* generally crosses the epithelial barrier once it enters the small intestine and then spreads to the liver, spleen, and central nervous system [25]. This bacteriocin treatment was found to decrease the amount of *L. monocytogenes* reaching the liver and spleen [40].

4.1. Toxicity. An advantage that bacteriocins hold over some other antimicrobial therapies is their composition. These peptides can be easily broken down to simple nontoxic amino acids that are metabolized, although this also means that they may not be as long-lasting compared to antibiotics. However, information regarding the *in vitro* cytotoxicity of class IIa bacteriocins is relatively limited. The cytotoxicity of pediocin PA-1 was tested against simian virus 40-transfected human colon cells and Vero monkey kidney cells [111]. At the levels tested, pediocin PA-1 did show cytotoxic effects on both cell lines, with a bacteriocin dose of 700 AU/mL (likely around 10–20 mg/mL) causing a decrease of greater than 50% on the viable cell counts. Lower dosages also affected the viable cell count, although this was not as dramatic. However, combinations of carnobacteriocins BM1 and B2 at concentrations 100-fold higher than required for antimicrobial activity displayed no significant cytotoxic effects to the human gastrointestinal Caco-2 cell line [112]. The means of bacteriocin production and purification must also be considered with respect to potential toxic effects. Although this paper focuses on the administration of purified bacteriocin only, there still may be the possibility of toxic contaminants retained in the bacteriocin sample, which could confuse any toxicity results obtained.

Based on the differing results obtained from these two *in vitro* studies, further work must be done to carefully examine what amounts of class IIa bacteriocins can be used safely without cytotoxic effects. However, it is promising that mouse and rabbit models did not show detrimental effects from bacteriocin introduction [40, 79, 80].

4.2. Resistance Mechanisms. As with all therapeutic antibiotics, the development of resistance to class IIa bacteriocins in pathogenic bacteria is a critical issue to consider. This topic has been the subject of a recent review by Kaur et al. [113]. Much evidence has shown that the sensitivity of a bacterial strain to class IIa bacteriocins is dependent on the presence of a mannose phosphotransferase (MPT) transporter system [16, 114–116]. Additionally, there is evidence that nonclass IIa bacteriocin lactococcin A also requires MPT as a receptor [17]. Decreased expression levels of MPT have been implicated in resistance to class IIa bacteriocins in many strains of *L. monocytogenes* insensitive to bacteriocins [114].

Beyond decreased receptor expression, *L. monocytogenes* and other susceptible strains have developed other resistance mechanisms. Multiple mechanisms may be operative at once contributing to an overall resistant phenotype. Modifications of the bacterial membrane have been implicated as another source of bacterial resistance. Alterations of the bacterial membrane, such that the acyl chains of phosphotidylglycerols are shorter and more unsaturated, affect membrane fluidity and the efficiency of bacteriocin insertion [117, 118]. Several other observed cell surface adaptations have been implicated in resistance, such as increasing the net positive charge on the membrane and lysinylation of membrane phospholipids [119].

Of special concern is the cross-resistance that has been observed for bacteriocins from different classes. For example, a strain of *L. monocytogenes* has shown resistance to nisin, pediocin PA-1, and leuconocin S, bacteriocins from three separate classes [120]. Based on this, the prospect of using multiple bacteriocins to overcome resistant strains may not be entirely feasible. Like other antibiotics, bacteriocins need to be used judiciously to minimize the spread of resistant phenotypes.

5. Conclusions

Class IIa bacteriocins are antagonistic to many important human pathogens. These bacteriocins have the ability to target a relatively narrow range of bacteria without affecting much of the natural microbiota of the body, which is an important advantage, especially when compared to other antibiotics. Although these bacteriocins do not target as many pathogens as other antibiotics, they have the potential to perform a very specific role. Having another tool to combat infections is especially important with consideration of the ever-growing problem of antibiotic resistance.

Although relatively little has been published about the actual *in vivo* use of class IIa bacteriocins to control bacterial infections, what is known is promising. Preliminary experiments have shown these bacteriocins to be effective at fighting *L. monocytogenes* infections in mouse models.

Now, more is known about the mode of action of bacteriocins, and attempts at engineering bacteriocins with greater potency and stability have been successful. Compared to some other classes of bacteriocins, class IIa bacteriocins are especially suitable for facile recombinant production and the preparation of analogues. Improved fermentation conditions in combination with scalable efficient purifications are now known, allowing for the industrial-scale production of pure bacteriocin. The recombinant production of class IIa bacteriocins as a variety of fusion proteins in *E. coli* has also been successful, allowing for the production of even greater amounts of bacteriocin.

The application of class IIa bacteriocins as therapeutic agents is a rapidly developing area, and there is still much to investigate. In particular, determination of their efficacy against pathogens other than *L. monocytogenes* is open for exploration and would further reveal their potential for therapeutic use. In addition, it would be informative to test these bacteriocins against a wider range of targets beyond Gram-positive bacteria, as they have displayed unexpected activity.

The methodology is now in place to produce and purify large amounts of class IIa bacteriocins. The preliminary characterization that has been done reveals that this class of bacteriocins possesses several desirable and useful properties as *in vivo* antimicrobial agents. What remains now is to use that knowledge to fully explore the suitability of these peptides as *in vivo* antibiotics.

Acknowledgments

This work was supported by the Natural Sciences and Engineering Research Council of Canada (NSERC) and the Canada Research Chair in Bioorganic and Medicinal Chemistry. The authors thank Marco van Belkum and Avena Ross for helpful suggestions.

References

[1] T. Klaenhammer, "Genetics of bacteriocins produced by lactic acid bacteria," *FEMS Microbiology Reviews*, vol. 12, no. 1–3, pp. 39–86, 1993.

[2] I. F. Nes, D. B. Diep, L. S. Håvarstein, M. B. Brurberg, V. Eijsink, and H. Holo, "Biosynthesis of bacteriocins in lactic acid bacteria," *Antonie van Leeuwenhoek*, vol. 70, no. 2–4, pp. 113–128, 1996.

[3] P. D. Cotter, C. Hill, and R. P. Ross, "Bacteriocins: developing innate immunity for food," *Nature Reviews Microbiology*, vol. 3, no. 10, pp. 777–788, 2005.

[4] M. J. van Belkum, L. A. Martin-Visscher, and J. C. Vederas, "Structure and genetics of circular bacteriocins," *Trends in Microbiology*, vol. 19, no. 8, pp. 411–418, 2011.

[5] L. M. Cintas, P. Casaus, M. F. Fernandez, and P. E. Hernandez, "Comparative antimicrobial activity of enterocin L50, pediocin PA-1, nisin A and lactocin S against spoilage and foodborne pathogenic bacteria," *Food Microbiology*, vol. 15, no. 3, pp. 289–298, 1998.

[6] K. Yamazaki, M. Suzuki, Y. Kawai, N. Inoue, and T. J. Montville, "Purification and characterization of a novel class IIa bacteriocin, piscicocin CS526, from surimi-associated *Carnobacterium piscicola* CS526," *Applied and Environmental Microbiology*, vol. 71, no. 1, pp. 554–557, 2005.

[7] M. J. van Belkum and M. E. Stiles, "Nonlantibiotic antibacterial peptides from lactic acid bacteria," *Natural Product Reports*, vol. 17, no. 4, pp. 323–335, 2000.

[8] S. Ennahar, T. Sashihara, K. Sonomoto, and A. Ishizaki, "Class IIa bacteriocins: biosynthesis, structure and activity," *FEMS Microbiology Reviews*, vol. 24, no. 1, pp. 85–106, 2000.

[9] D. B. Diep and I. F. Nes, "Ribosomally synthesized antibacterial peptides in gram positive bacteria," *Current Drug Targets*, vol. 3, no. 2, pp. 107–122, 2002.

[10] V. Eijsink, L. Axelsson, D. Diep, L. S. Håvarstein, H. Holo, and I. F. Nes, "Production of class II bacteriocins by lactic acid bacteria; an example of biological warfare and communication," *Antonie van Leeuwenhoek*, vol. 81, no. 1–4, pp. 639–654, 2002.

[11] J. M. Rodríguez, M. I. Martínez, and J. Kok, "Pediocin PA-1, a wide-spectrum bacteriocin from lactic acid bacteria," *Critical Reviews in Food Science and Nutrition*, vol. 42, no. 2, pp. 91–121, 2002.

[12] D. Drider, G. Fimland, Y. Hechard, L. M. McMullen, and H. Prevost, "The continuing story of class IIa bacteriocins," *Microbiology and Molecular Biology Reviews*, vol. 70, no. 2, pp. 564–582, 2006.

[13] M. Papagianni and S. Anastasiadou, "Pediocins: the bacteriocins of pediococci. Sources, production, properties and applications," *Microbial Cell Factories*, vol. 8, no. 3, 2009.

[14] M. L. Chikindas, M. J. Garcia-Garcera, A. Driessen et al., "Pediocin PA-1, a bacteriocin from *Pediococcus acidilactici* PAC1.0, forms hydrophilic pores in the cytoplasmic membrane of target cells," *Applied and Environmental Microbiology*, vol. 59, no. 11, pp. 3577–3584, 1993.

[15] Y. Chen, R. Ludescher, and T. Montville, "Electrostatic interactions, but not the YGNGV consensus motif, govern the binding of pediocin PA-1 and its fragments to phospholipid vesicles," *Applied and Environmental Microbiology*, vol. 63, no. 12, pp. 4770–4777, 1997.

[16] K. Dalet, Y. Cenatiempo, P. Cossart et al., "A $\sigma(54)$-dependent PTS permease of the mannose family is responsible for sensitivity of Listeria *monocytogenes* to mesentericin Y105," *Microbiology*, vol. 147, no. 12, pp. 3263–3269, 2001.

[17] D. B. Diep, M. Skaugen, Z. Salehian, H. Holo, and I. F. Nes, "Common mechanisms of target cell recognition and immunity for class II bacteriocins," *Proceedings of the National Academy of Sciences of the United States of America*, vol. 104, no. 7, pp. 2384–2389, 2007.

[18] M. Kjos, Z. Salehian, I. F. Nes, and D. B. Diep, "An extracellular loop of the mannose phosphotransferase system component IIC is responsible for specific targeting by class IIa bacteriocins," *Journal of Bacteriology*, vol. 192, no. 22, pp. 5906–5913, 2010.

[19] J. Nissen-Meyer, P. Rogne, C. Oppergard, H. S. Haugen, and P. E. Kristiansen, "Structure-function relationships of the non-lanthionine-containing peptide (class II) bacteriocins produced by gram-positive bacteria," *Current Pharmaceutical Biotechnology*, vol. 10, no. 1, pp. 19–37, 2009.

[20] N. L. Gallagher, M. Sailer, W. P. Niemczura, T. T. Nakashima, M. E. Stiles, and J. C. Vederas, "Three-dimensional structure of leucocin a in trifluoroethanol and dodecylphosphocholine micelles: spatial location of residues critical for biological activity in type IIa bacteriocins from lactic acid bacteria," *Biochemistry*, vol. 36, no. 49, pp. 15062–15072, 1997.

[21] Y. Wang, M. E. Henz, N. L. Gallagher et al., "Solution structure of carnobacteriocin B2 and implications for structure-activity relationships among type IIa bacteriocins from lactic acid bacteria," *Biochemistry*, vol. 38, no. 47, pp. 15438–15447, 1999.

[22] T. Sprules, K. E. Kawulka, A. C. Gibbs, D. S. Wishart, and J. C. Vederas, "NMR solution structure of the precursor for carnobacteriocin B2, an antimicrobial peptide from *Carnobacterium piscicola*: implications of the α-helical leader section for export and inhibition of type IIa bacteriocin activity," *European Journal of Biochemistry*, vol. 271, no. 9, pp. 1748–1756, 2004.

[23] M. Uteng, H. H. Hauge, P. Markwick et al., "Three-dimensional structure in lipid micelles of the pediocin-like antimicrobial peptide sakacin P and a sakacin P variant that is structurally stabilized by an inserted C-terminal disulfide bridge," *Biochemistry*, vol. 42, no. 39, pp. 11417–11426, 2003.

[24] H. S. Haugen, G. Fimland, J. Nissen-Meyer, and P. E. Kristiansen, "Three-dimensional structure in lipid micelles of the pediocin-like antimicrobial peptide curvacin A," *Biochemistry*, vol. 44, no. 49, pp. 16149–16157, 2005.

[25] V. Ramaswamy, V. M. Cresence, J. S. Rejitha et al., "*Listeria*—Review of epidemiology and pathogenesis," *Journal of Microbiology, Immunology and Infection*, vol. 40, no. 1, pp. 4–13, 2007.

[26] M. Millette, G. Cornut, C. Dupont, F. Shareck, D. Archambault, and M. Lacroix, "Capacity of human nisin- and pediocin-producing lactic acid bacteria to reduce intestinal colonization by vancomycin-resistant enterococci," *Applied and Environmental Microbiology*, vol. 74, no. 7, pp. 1997–2003, 2008.

[27] F. B. Elegado, W. J. Kim, and D. Y. Kwon, "Rapid purification, partial characterization, and antimicrobial spectrum of the bacteriocin, Pediocin AcM, from *Pediococcus acidilactici* M," *International Journal of Food Microbiology*, vol. 37, no. 1, pp. 1–11, 1997.

[28] L. Beaulieu, *Production, Purification et Caracterisation de la Pediocine PA-1 Naturelle et de ses Formes Recombiantes: Contribution a la Mise en Evidence d'une Nouvelle Activite Biologique*, Universite Laval, Quebec, Canada, 2004.

[29] G. Cornut, C. Fortin, and D. Soulières, "Antineoplastic properties of bacteriocins revisiting potential active agents," *American Journal of Clinical Oncology*, vol. 31, no. 4, pp. 399–404, 2008.

[30] S. D. Todorov, M. Wachsman, E. Tomé et al., "Characterisation of an antiviral pediocin-like bacteriocin produced by *Enterococcus faecium*," *Food Microbiology*, vol. 27, no. 7, pp. 869–879, 2010.

[31] B. C. Kirkup Jr, "Bacteriocins as oral and gastrointestinal antibiotics: theoretical considerations, applied research, and practical applications," *Current Medicinal Chemistry*, vol. 13, no. 27, pp. 3335–3350, 2006.

[32] S. Chatterjee, D. K. Chatterjee, R. H. Jani et al., "Mersacidin, a new antibiotic from *Bacillus in vitro* and *in vivo* antibacterial activity," *Journal of Antibiotics*, vol. 45, no. 6, pp. 839–845, 1992.

[33] B. P. Goldstein, J. Wei, K. Greenberg, and R. Novick, "Activity of nisin against *Streptococcus pneumoniae, in vitro*, and

in a mouse infection model," *Journal of Antimicrobial Chemotherapy*, vol. 42, no. 2, pp. 277–278, 1998.

[34] D. Kruszewska, H. G. Sahl, G. Bierbaum, U. Pag, S. O. Hynes, and A. Ljungh, "Mersacidin eradicates methicillin-resistant *Staphylococcus aureus* (MRSA) in a mouse rhinitis model," *Journal of Antimicrobial Chemotherapy*, vol. 54, no. 3, pp. 648–653, 2004.

[35] M. Mota-Meira, H. Morency, and M. C. Lavoie, "*In vivo* activity of mutacin B-Ny266," *Journal of Antimicrobial Chemotherapy*, vol. 56, no. 5, pp. 869–871, 2005.

[36] A. M. Brand, M. de Kwaadsteniet, and L. M. T. Dicks, "The ability of nisin F to control *Staphylococcus aureus* infection in the peritoneal cavity, as studied in mice," *Letters in Applied Microbiology*, vol. 51, no. 6, pp. 645–649, 2010.

[37] S. Y. Kim, S. Shin, H. C. Koo, J. H. Youn, H. D. Paik, and Y. H. Park, "*In vitro* antimicrobial effect and *in vivo* preventive and therapeutic effects of partially purified lantibiotic lacticin NK34 against infection by *Staphylococcus* species isolated from bovine mastitis," *Journal of Dairy Science*, vol. 93, no. 8, pp. 3610–3615, 2010.

[38] D. Jabes, C. Brunati, G. Candiani, S. Riva, G. Romano, and S. Donadio, "Efficacy of the new lantibiotic NAI-107 in experimental infections induced by MDR Gram positive pathogensEfficacy of the new lantibiotic NAI-107 in experimental infections induced by MDR Gram positive pathogens," *Antimicrobial Agents and Chemotherapy*, vol. 55, no. 4, pp. 1671–1676, 2011.

[39] S. C. Corr, Y. Li, C. U. Riedel, P. W. O'Toole, C. Hill, and C. G. M. Gahan, "Bacteriocin production as a mechanism for the antiinfective activity of *Lactobacillus salivarius* UCC118," *Proceedings of the National Academy of Sciences of the United States of America*, vol. 104, no. 18, pp. 7617–7621, 2007.

[40] N. Dabour, A. Zihler, E. Kheadr, C. Lacroix, and I. Fliss, "*In vivo* study on the effectiveness of pediocin PA-1 and *Pediococcus acidilactici* UL5 at inhibiting *Listeria monocytogenes*," *International Journal of Food Microbiology*, vol. 133, no. 3, pp. 225–233, 2009.

[41] C. Valenta, A. Bernkop-Schnürch, and H. P. Rigler, "The antistaphylococcal effect of nisin in a suitable vehicle: a potential therapy for atopic dermatitis in man," *Journal of Pharmacy and Pharmacology*, vol. 48, no. 9, pp. 988–991, 1996.

[42] G. Fimland, L. Johnsen, L. Axelsson et al., "A C-terminal disulfide bridge in pediocin-like bacteriocins renders bacteriocin activity less temperature dependent and is a major determinant of the antimicrobial spectrum," *Journal of Bacteriology*, vol. 182, no. 9, pp. 2643–2648, 2000.

[43] L. Johnsen, G. Fimland, and J. N. Meyer, "The C-terminal domain of pediocin-like antimicrobial peptides (class IIa bacteriocins) is involved in specific recognition of the C-terminal part of cognate immunity proteins and in determining the antimicrobial spectrum," *Journal of Biological Chemistry*, vol. 280, no. 10, pp. 9243–9250, 2005.

[44] D. J. Derksen, J. L. Stymiest, and J. C. Vederas, "Antimicrobial leucocin analogues with a disulfide bridge replaced by a carbocycle or by noncovalent interactions of allyl glycine residues," *Journal of the American Chemical Society*, vol. 128, no. 44, pp. 14252–14253, 2006.

[45] Y. Fleury, M. A. Dayem, J. J. Montagne et al., "Covalent structure, synthesis, and structure-function studies of mesentericin Y 105(37), a defensive peptide from gram-positive bacteria *Leuconostoc mesenteroides*," *Journal of Biological Chemistry*, vol. 271, no. 24, pp. 14421–14429, 1996.

[46] D. J. Derksen, M. A. Boudreau, and J. C. Vederas, "Hydrophobic interactions as substitutes for a conserved disulfide linkage in the type IIa bacteriocins, leucocin A and pediocin PA-1," *ChemBioChem*, vol. 9, no. 12, pp. 1898–1901, 2008.

[47] C. S. Sit and J. C. Vederas, "Approaches to the discovery of new antibacterial agents based on bacteriocins," *Biochemistry and Cell Biology*, vol. 86, no. 2, pp. 116–123, 2008.

[48] L. Johnsen, G. Fimland, V. Eijsink, and J. Nissen-Meyer, "Engineering increased stability in the antimicrobial peptide pediocin PA-1," *Applied and Environmental Microbiology*, vol. 66, no. 11, pp. 4798–4802, 2000.

[49] L. Quadri, M. Sailer, K. L. Roy, J. C. Vederas, and M. E. Stiles, "Chemical and genetic characterization of bacteriocins produced by *Carnobacterium piscicola* LV17B," *Journal of Biological Chemistry*, vol. 269, no. 16, pp. 12204–12211, 1994.

[50] J. Jasniewski, C. Cailliez-Grimal, E. Gelhaye, and A. Revol-Junelles, "Optimization of the production and purification processes of carnobacteriocins Cbn BM1 and Cbn B2 from *Carnobacterium maltaromaticum* CP5 by heterologous expression in *Escherichia coli*," *Journal of Microbiological Methods*, vol. 73, no. 1, pp. 41–48, 2008.

[51] T. Tominaga and Y. Hatakeyama, "Development of innovative pediocin PA-1 by DNA shuffling among class IIa bacteriocins," *Applied and Environmental Microbiology*, vol. 73, no. 16, pp. 5292–5299, 2007.

[52] M. Kazazic, J. Nissen-Meyer, and G. Fimland, "Mutational analysis of the role of charged residues in target-cell binding, potency and specificity of the pediocin-like bacteriocin sakacin P," *Microbiology*, vol. 148, no. 7, pp. 2019–2027, 2002.

[53] E. F. O'Shea, P. M. O'Connor, P. D. Cotter, R. Ross, and C. Hill, "Synthesis of trypsin-resistant variants of the *Listeria* bacteriocin salivaricin P," *Applied and Environmental Microbiology*, vol. 76, no. 16, pp. 5356–5362, 2010.

[54] C. Oppegård, P. Rogne, P. E. Kristiansen, and J. Nissen-Meyer, "Structure analysis of the two-peptide bacteriocin lactococcin G by introducing D-amino acid residues," *Microbiology*, vol. 156, no. 6, pp. 1883–1889, 2010.

[55] L. Z. Yan, A. C. Gibbs, M. E. Stiles, D. S. Wishart, and J. C. Vederas, "Analogues of bacteriocins: antimicrobial specificity and interactions of leucocin a with its enantiomer, carnobacteriocin B2, and truncated derivatives," *Journal of Medicinal Chemistry*, vol. 43, no. 24, pp. 4579–4581, 2000.

[56] T. Tominaga and Y. Hatakeyama, "Determination of essential and variable residues in pediocin PA-1 by NNK scanning," *Applied and Environmental Microbiology*, vol. 72, no. 2, pp. 1141–1147, 2006.

[57] V. Carolissen-Mackay, G. Arendse, and J. W. Hastings, "Purification of bacteriocins of lactic acid bacteria: problems and pointers," *International Journal of Food Microbiology*, vol. 34, no. 1, pp. 1–16, 1997.

[58] K. Naghmouchi, I. Fliss, D. Drider, and C. Lacroix, "Pediocin PA-1 production during repeated-cycle batch culture of immobilized *Pediococcus acidilactici* UL5 cells," *Journal of Bioscience and Bioengineering*, vol. 105, no. 5, pp. 513–517, 2008.

[59] J. Huang, C. Lacroix, H. Daba, and R. E. Simard, "Pediocin 5 production and plasmid stability during continuous free and immobilized cell cultures of *Pediococcus acidilactici* UL5," *Journal of Applied Bacteriology*, vol. 80, no. 6, pp. 635–644, 1996.

[60] J. A. V. Alvarez, M. P. Gonzalez, and M. A. Murado, "Pediocin production by *Pediococcus acidilactici* in solid state culture on a waste medium: process simulation and experimental results," *Biotechnology and Bioengineering*, vol. 85, no. 6, pp. 676–682, 2004.

[61] N. P. Guerra and L. P. Castro, "Production of bacteriocins from *Lactococcus lactis* subsp. lactis CECT 539 and *Pediococcus acidilactici* NRRL B-5627 using mussel-processing wastes," *Biotechnology and Applied Biochemistry*, vol. 36, no. 2, pp. 119–125, 2002.

[62] F. Goulhen, J. Meghrous, and C. Lacroix, "Production of a nisin Z/pediocin mixture by pH-controlled mixed-strain batch cultures in supplemented whey permeate," *Journal of Applied Microbiology*, vol. 86, no. 3, pp. 399–406, 1999.

[63] N. P. Guerra, M. L. Rua, and L. Pastrana, "Nutritional factors affecting the production of two bacteriocins from lactic acid bacteria on whey," *International Journal of Food Microbiology*, vol. 70, no. 3, pp. 267–281, 2001.

[64] N. P. Guerra, P. F. Bernardez, and L. P. Castro, "Fed-batch pediocin production on whey using different feeding media," *Enzyme and Microbial Technology*, vol. 41, no. 3, pp. 397–406, 2007.

[65] J. A. Vazquez, M. P. Gonzalez, and M. A. Murado, "Preliminary tests on nisin and pediocin production using waste protein sources: factorial and kinetic studies," *Bioresource Technology*, vol. 97, no. 4, pp. 605–613, 2006.

[66] R. Yang and B. Ray, "Factors influencing production of bacteriocins by lactic acid bacteria," *Food Microbiology*, vol. 11, no. 4, pp. 281–291, 1994.

[67] M. Uteng, H. H. Hauge, I. Brondz, J. Nissen-Meyer, and G. Fimland, "Rapid two-step procedure for large-scale purification of pediocin-like bacteriocins and other cationic antimicrobial peptides from complex culture medium," *Applied and Environmental Microbiology*, vol. 68, no. 2, pp. 952–956, 2002.

[68] D. Guyonnet, C. Fremaux, Y. Cenatiempo, and J. M. Berjeaud, "Method for rapid purification of class IIa bacteriocins and comparison of their activities," *Applied and Environmental Microbiology*, vol. 66, no. 4, pp. 1744–1748, 2000.

[69] L. Beaulieu, H. Aomari, D. Groleau, and M. Subirade, "An improved and simplified method for the large-scale purification of pediocin PA-1 produced by *Pediococcus acidilactici*," *Biotechnology and Applied Biochemistry*, vol. 43, no. 2, pp. 77–84, 2006.

[70] J. Gutierrez, R. Criado, R. Citti et al., "Performance and Applications of Polyclonal Antipeptide Antibodies Specific for the Enterococcal Bacteriocin Enterocin P," *Journal of Agricultural and Food Chemistry*, vol. 52, no. 8, pp. 2247–2255, 2004.

[71] C. Richard, D. Drider, I. Fliss, S. Denery, and H. Prevost, "Generation and utilization of polyclonal antibodies to a synthetic C-terminal amino acid fragment of divercin V41, a class IIa bacteriocin," *Applied and Environmental Microbiology*, vol. 70, no. 1, pp. 248–254, 2004.

[72] S. Cuozzo, S. Calvez, H. Prévost, and D. Drider, "Improvement of enterocin P purification process," *Folia Microbiologica*, vol. 51, no. 5, pp. 401–405, 2006.

[73] K. Naghmouchi, D. Drider, and I. Fliss, "Purification of pediocin PA-1 by immunoaffinity chromatography," *Journal of AOAC International*, vol. 91, no. 4, pp. 828–832, 2008.

[74] A. Métivier, P. Boyaval, F. Duffes, X. Dousset, J. P. Compoint, and D. Marion, "Triton X-114 phase partitioning for the isolation of a pediocin-like bacteriocin from Carnobacterium divergens," *Letters in Applied Microbiology*, vol. 30, no. 1, pp. 42–46, 2000.

[75] E. Kheadr, N. Bernoussi, C. Lacroix, and I. Fliss, "Comparison of the sensitivity of commercial strains and infant isolates of bifidobacteria to antibiotics and bacteriocins," *International Dairy Journal*, vol. 14, no. 12, pp. 1041–1053, 2004.

[76] G. le Blay, C. Lacroix, A. Zihler, and I. Fliss, "*In vitro* inhibition activity of nisin A, nisin Z, pediocin PA-1 and antibiotics against common intestinal bacteria," *Letters in Applied Microbiology*, vol. 45, no. 3, pp. 252–257, 2007.

[77] N. Bernbom, B. Jelle, C. H. Brogren, F. K. Vogensen, B. Nørrung, and T. R. Licht, "Pediocin PA-1 and a pediocin producing *Lactobacillus plantarum* strain do not change the HMA rat microbiota," *International Journal of Food Microbiology*, vol. 130, no. 3, pp. 251–257, 2009.

[78] A. Ingham, M. Ford, R. J. Moore, and M. Tizard, "The bacteriocin piscicolin 126 retains antilisterial activity *in vivo*," *Journal of Antimicrobial Chemotherapy*, vol. 51, no. 6, pp. 1365–1371, 2003.

[79] J. Rihakova, J. M. Cappelier, I. Hue et al., "*In vivo* activities of recombinant divercin V41 and its structural variants against *Listeria monocytogenes*," *Antimicrobial Agents and Chemotherapy*, vol. 54, no. 1, pp. 563–564, 2010.

[80] A. K. Bhunia, M. C. Johnson, B. Ray, and E. L. Belden, "Antigenic property of Pediocin AcH produced by *Pediococcus acidilactici* H," *Journal of Applied Bacteriology*, vol. 69, no. 2, pp. 211–215, 1990.

[81] A. K. Bhunia, "Monoclonal antibody-based enzyme immunoassay for pediocins of *Pediococcus acidilactici*," *Applied and Environmental Microbiology*, vol. 60, no. 8, pp. 2692–2696, 1994.

[82] M. I. Martinez, J. M. Rodriguez, A. Suarez, J. M. Martinez, J. I. Azcona, and P. E. Hernandez, "Generation of polyclonal antibodies against a chemically synthesized N-terminal fragment of the bacteriocin pediocin PA-1," *Letters in Applied Microbiology*, vol. 24, no. 6, pp. 488–492, 1997.

[83] J. M. Martinez, J. Kok, J. W. Sanders, and P. E. Hernandez, "Heterologous coproduction of enterocin A and pediocin PA-1 by *Lactococcus lactis*: detection by specific peptide-directed antibodies," *Applied and Environmental Microbiology*, vol. 66, no. 8, pp. 3543–3549, 2000.

[84] A. K. Bhunia, M. C. Johnson, and B. Ray, "Purification, characterization and antimicrobial spectrum of a bacteriocin produced by *Pediococcus acidilactici*," *Journal of Applied Bacteriology*, vol. 65, no. 4, pp. 261–268, 1988.

[85] E. Kheadr, A. Zihler, N. Dabour, C. Lacroix, G. le Blay, and I. Fliss, "Study of the physicochemical and biological stability of pediocin PA-1 in the upper gastrointestinal tract conditions using a dynamic *in vitro* model," *Journal of Applied Microbiology*, vol. 109, no. 1, pp. 54–64, 2010.

[86] R. O. Benech, E. E. Kheadr, R. Laridi, C. Lacroix, and I. Fliss, "Inhibition of *Listeria innocua* in cheddar cheese by addition of nisin Z in liposomes or by *in situ* production in mixed culture," *Applied and Environmental Microbiology*, vol. 68, no. 8, pp. 3683–3690, 2002.

[87] L. M. Were, B. D. Bruce, P. M. Davidson, and J. Weiss, "Size, stability, and entrapment efficiency of phospholipid nanocapsules containing polypeptide antimicrobials," *Journal of Agricultural and Food Chemistry*, vol. 51, no. 27, pp. 8073–8079, 2003.

[88] L. Were, B. Bruce, P. M. Davidson, and J. Weiss, "Encapsulation of nisin and lysozyme in liposomes enhances efficacy against *Listeria monocytogenes*," *Journal of Food Protection*, vol. 67, no. 5, pp. 922–927, 2004.

[89] J. M. Rodríguez, M. I. Martínez, N. Horn, and H. M. Dodd, "Heterologous production of bacteriocins by lactic acid

bacteria," *International Journal of Food Microbiology*, vol. 80, no. 2, pp. 101–116, 2003.

[90] J. Gutierrez, R. Larsen, L. M. Cintas, J. Kok, and P. E. Hernandez, "High-level heterologous production and functional expression of the sec-dependent enterocin P from *Enterococcus faecium* P13 in *Lactococcus lactis*," *Applied Microbiology and Biotechnology*, vol. 72, no. 1, pp. 41–51, 2006.

[91] J. Sanchez, J. Borrero, B. Gomez-Sala et al., "Cloning and heterologous production of hiracin JM79, a Sec-dependent bacteriocin produced by *Enterococcus hirae* DCH5, in lactic acid bacteria and *Pichia pastoris*," *Applied and Environmental Microbiology*, vol. 74, no. 8, pp. 2471–2479, 2008.

[92] C. Reviriego, L. Fernández, and J. M. Rodríguez, "A food-grade system for production of pediocin PA-1 in nisin-producing and non-nisin-producing *Lactococcus lactis* strains: application to inhibit *Listeria* growth in a cheese model system," *Journal of Food Protection*, vol. 70, no. 11, pp. 2512–2517, 2007.

[93] L. E. N. Quadri, L. Z. Yan, M. E. Stiles, and J. C. Vederas, "Effect of amino acid substitutions on the activity of carnobacteriocin B2. Overproduction of the antimicrobial peptide, its engineered variants, and its precursor in *Escherichia coli*," *Journal of Biological Chemistry*, vol. 272, no. 6, pp. 3384–3388, 1997.

[94] G. S. Moon, Y. R. Pyun, and W. J. Kim, "Expression and purification of a fusion-typed pediocin PA-1 in *Escherichia coli* and recovery of biologically active pediocin PA-1," *International Journal of Food Microbiology*, vol. 108, no. 1, pp. 136–140, 2006.

[95] L. Beaulieu, D. Tolkatchev, J. F. Jetté, D. Groleau, and M. Subirade, "Production of active pediocin PA-1 in *Escherichia coli* using a thioredoxin gene fusion expression approach: cloning, expression, purification, and characterization," *Canadian Journal of Microbiology*, vol. 53, no. 11, pp. 1246–1258, 2007.

[96] C. Richard, D. Drider, K. Elmorjani, D. Marion, and H. Prévost, "Heterologous expression and purification of active divercin V41, a class IIa bacteriocin encoded by a synthetic gene in *Escherichia coli*," *Journal of Bacteriology*, vol. 186, no. 13, pp. 4276–4284, 2004.

[97] E. R. LaVallie, E. A. DiBlasio, S. Kovacic, K. L. Grant, P. F. Schendel, and J. M. McCoy, "A thioredoxin gene fusion expression system that circumvents inclusion body formation in the *E. coli* cytoplasm," *Bio/Technology*, vol. 11, no. 2, pp. 187–193, 1993.

[98] S. N. Liu, Y. Han, and Z. J. Zhou, "Fusion expression of pedA gene to obtain biologically active pediocin PA-1 in *Escherichia coli*," *Journal of Zhejiang University: Science B*, vol. 12, no. 1, pp. 65–71, 2011.

[99] G. M. Gibbs, B. E. Davidson, and A. J. Hillier, "Novel expression system for large-scale production and purification of recombinant class IIa bacteriocins and its application to piscicolin 126," *Applied and Environmental Microbiology*, vol. 70, no. 6, pp. 3292–3297, 2004.

[100] K. W. Miller, R. Schamber, Y. Chen, and B. Ray, "Production of active chimeric pediocin AcH in *Escherichia coil* in the absence of processing and secretion genes from the *Pediococcus* pap operon," *Applied and Environmental Microbiology*, vol. 64, no. 1, pp. 14–20, 1998.

[101] A. B. Ingham, K. W. Sproat, M. L. V. Tizard, and R. J. Moore, "A versatile system for the expression of nonmodified bacteriocins in *Escherichia coli*," *Journal of Applied Microbiology*, vol. 98, no. 3, pp. 676–683, 2005.

[102] P. M. Halami and A. Chandrashekar, "Heterologous expression, purification and refolding of an anti-listerial peptide produced by *Pediococcus acidilactici* K7," *Electronic Journal of Biotechnology*, vol. 10, no. 4, pp. 563–569, 2007.

[103] M. Klocke, K. Mundt, F. Idler, S. Jung, and J. E. Backhausen, "Heterologous expression of enterocin A, a bacteriocin from *Enterococcus faecium*, fused to a cellulose-binding domain in *Escherichia coli* results in a functional protein with inhibitory activity against *Listeria*," *Applied Microbiology and Biotechnology*, vol. 67, no. 4, pp. 532–538, 2005.

[104] G. S. Moon, Y. R. Pyun, and W. J. Kim, "Characterization of the pediocin operon of *Pediococcus acidilactici* K10 and expression of his-tagged recombinant pediocin PA-1 in *Escherichia coli*," *Journal of Microbiology and Biotechnology*, vol. 15, no. 2, pp. 403–411, 2005.

[105] S. C. Makrides, "Strategies for achieving high-level expression of genes in *Escherichia coli*," *Microbiological Reviews*, vol. 60, no. 3, pp. 512–538, 1996.

[106] S. Yildirim, D. Konrad, S. Calvez, D. Drider, H. Prevost, and C. Lacroix, "Production of recombinant bacteriocin divercin V41 by high cell density *Escherichia coli* batch and fed-batch cultures," *Applied Microbiology and Biotechnology*, vol. 77, no. 3, pp. 525–531, 2007.

[107] H. Schoeman, M. A. Vivier, M. du Toit, L. M. T. Dicks, and I. S. Pretorius, "The development of bactericidal yeast strains by expressing the *Pediococcus acidilactici* pediocin gene (pedA) in *Saccharomyces cerevisiae*," *Yeast*, vol. 15, no. 8, pp. 647–656, 1999.

[108] C. A. van Reenen, M. L. Chikindas, W. H. van Zyl, and L. M. T. Dicks, "Characterization and heterologous expression of a class IIa bacteriocin, plantaricin 423 from *Lactobacillus plantarum* 423, in *Saccharomyces cerevisiae*," *International Journal of Food Microbiology*, vol. 81, no. 1, pp. 29–40, 2003.

[109] J. Gutierrez, R. Criado, M. Martin, C. Herranz, L. M. Cintas, and P. E. Hernandez, "Production of enterocin P, an antilisterial pediocin-like bacteriocin from *Enterococcus faecium* P13, in *Pichia pastons*," *Antimicrobial Agents and Chemotherapy*, vol. 49, no. 7, pp. 3004–3008, 2005.

[110] L. Beaulieu, D. Groleau, C. B. Miguez, J. F. Jetté, H. Aomari, and M. Subirade, "Production of pediocin PA-1 in the methylotrophic yeast *Pichia pastoris* reveals unexpected inhibition of its biological activity due to the presence of collagen-like material," *Protein Expression and Purification*, vol. 43, no. 2, pp. 111–125, 2005.

[111] S. E. Murinda, K. A. Rashid, and R. F. Roberts, "*In vitro* assessment of the cytotoxicity of nisin, pediocin, and selected colicins on simian virus 40-transfected human colon and Vero monkey kidney cells with trypan blue staining viability assays," *Journal of Food Protection*, vol. 66, no. 5, pp. 847–853, 2003.

[112] J. Jasniewski, C. Cailliez-Grimal, I. Chevalot, J. B. Milliere, and A. M. Revol-Junelles, "Interactions between two carnobacteriocins Cbn BM1 and Cbn B2 from *Carnobacterium maltaromaticum* CP5 on target bacteria and Caco-2 cells," *Food and Chemical Toxicology*, vol. 47, no. 4, pp. 893–897, 2009.

[113] G. Kaur, R. K. Malik, S. K. Mishra et al., "Nisin and class IIa bacteriocin resistance among *Listeria* and other foodborne pathogens and spoilage bacteria," *Microbial Drug Resistance*, vol. 17, no. 2, pp. 197–205, 2011.

[114] A. Gravesen, M. Ramnath, K. B. Rechinger et al., "High-level resistance to class IIa bacteriocins is associated with one general mechanism in *Listeria monocytogenes*," *Microbiology*, vol. 148, no. 8, pp. 2361–2369, 2002.

[115] M. Ramnath, S. Arous, A. Gravesen, J. W. Hastings, and Y. Héchard, "Expression of mptC of *Listeria monocytogenes* induces sensitivity to class IIa bacteriocins in *Lactococcus lactis*," *Microbiology*, vol. 150, no. 8, pp. 2663–2668, 2004.

[116] M. Kjos, I. F. Nes, and D. B. Diep, "Mechanisms of resistance to bacteriocins targeting the mannose phosphotransferase system," *Applied and Environmental Microbiology*, vol. 77, no. 10, pp. 3335–3342, 2011.

[117] V. Vadyvaloo, J. W. Hastings, M. J. van der Merwe, and M. Rautenbach, "Membranes of class IIa bacteriocin-resistant *Listeria monocytogenes* cells contain increased levels of desaturated and short-acyl-chain phosphatidylglycerols," *Applied and Environmental Microbiology*, vol. 68, no. 11, pp. 5223–5230, 2002.

[118] Y. Chen, R. Ludescher, and T. J. Montville, "Influence of lipid composition on pediocin PA-1 binding to phospholipid vesicles," *Applied and Environmental Microbiology*, vol. 64, no. 9, pp. 3530–3532, 1998.

[119] V. Vadyvaloo, S. Arous, A. Gravesen et al., "Cell-surface alterations in class IIa bacteriocin-resistant *Listeria monocytogenes* strains," *Microbiology*, vol. 150, no. 9, pp. 3025–3033, 2004.

[120] A. D. Crandall and T. J. Montville, "Nisin resistance in *Listeria monocytogenes* ATCC 700302 is a complex phenotype," *Applied and Environmental Microbiology*, vol. 64, no. 1, pp. 231–237, 1998.

Effect of Citrus Byproducts on Survival of O157:H7 and Non-O157 *Escherichia coli* Serogroups within *In Vitro* Bovine Ruminal Microbial Fermentations

Heather A. Duoss-Jennings,[1] Ty B. Schmidt,[2] Todd R. Callaway,[3] Jeffery A. Carroll,[4] James M. Martin,[5] Sara A. Shields-Menard,[6] Paul R. Broadway,[7] and Janet R. Donaldson[6]

[1] *Department of Animal Science, Iowa State University, Ames, IA, USA*
[2] *Animal Science Department, University of Nebraska, Lincoln, NE, USA*
[3] *Food and Feed Safety Research Unit, ARS, USDA, College Station, TX 79403, USA*
[4] *Livestock Issues Research Unit, ARS, USDA, Lubbock, TX 77845, USA*
[5] *Department of Animal and Dairy Sciences, Mississippi State University, Mississippi State, MS, USA*
[6] *Department of Biological Sciences, Mississippi State University, Mississippi State, MS 39762, USA*
[7] *Department of Animal and Food Sciences, Texas Tech University, Lubbock, TX, USA*

Correspondence should be addressed to Janet R. Donaldson; donaldson@biology.msstate.edu

Academic Editor: Giuseppe Comi

Citrus byproducts (CBPs) are utilized as a low cost nutritional supplement to the diets of cattle and have been suggested to inhibit the growth of both *Escherichia coli* O157:H7 and *Salmonella*. The objective of this study was to examine the effects *in vitro* that varying concentrations of CBP in the powdered or pelleted variety have on the survival of Shiga-toxin *Escherichia coli* (STEC) serotypes O26:H11, O103:H8, O111:H8, O145:H28, and O157:H7 in bovine ruminal microorganism media. The O26:H11, O111:H8, O145:H28, and O157:H7 serotypes did not exhibit a change in populations in media supplemented with CBP with either variety. The O103:H8 serotype displayed a general trend for an approximate 1 $\log_{10}$ reduction in 5% powdered CBP and 20% pelleted CBP over 6 h. There was a trend for reductions in populations of a variant form of O157:H7 mutated in the *stx*1 and *stx*2 genes in higher concentrations of CBP. These results suggest that variations exist in the survival of these serotypes of STEC within mixed ruminal microorganism fluid media when supplemented with CBP. Further research is needed to determine why CBPs affect STEC serotypes differently.

1. Introduction

Shiga-toxin producing *Escherichia coli* (STEC) is capable of naturally colonizing within the gastrointestinal tract of cattle without causing illness [1]. Human consumption of products contaminated with STEC can cause the severe illnesses hemorrhagic colitis and hemorrhagic uremic syndrome [2, 3]. The most notorious STEC within the meat industry has been *E. coli* O157:H7. Due to increased surveillance and pre- and post-harvest intervention, the occurrence of O157:H7 infections in the United States has been reduced to ≤1 case per 100,000 people. However, there now appears to be an increase in the occurrence of foodborne outbreaks due to non-O157 STEC serotypes. According to the Center for Disease Control (CDC) an estimated 265,000 cases of STEC infections were reported a year; of these, approximately 67% are attributed to non-O157 STEC [4]. With increased concerns related to the prevalence of non-O157 outbreaks, the United States Department of Agriculture Food Safety and Inspection Service (USDA-FSIS) has recently labeled the non-O157 STEC serogroups O26, O45, O103, O111, O121, and O145 as adulterants in fresh nonintact beef products [5].

The production of citrus for various food and nonfood products generates byproducts, such as the pulp and peel from citrus fruit. These citrus byproducts (CBPs) have been utilized by dairy and beef cattle producers in some regions of the United States as an inexpensive nutritionally dense feed source [6]. The incorporation of CBP into diets for cattle may also aid in the reduction of foodborne pathogens due to antimicrobial aspects of the byproducts. Citrus products and by-products contain essential oils that possess antimicrobial activities that can damage the cell wall of gram-negative bacteria [7]. The change in the fluidity of the membranes due to the permeabilization allows essential oils to coagulate in the cytoplasm [8], depleting ATP [9] and resulting in lysis of the cell [9, 10]. The rumen and intestinal gram-negative microbial populations of cattle can be altered due to this antimicrobial activity within cattle [11]. Since CBPs contain antimicrobial properties and are readily available at low costs within citrus-producing areas and has nutritional value, it is being investigated as a potential preharvest pathogen intervention strategy to reduce STEC concentrations within the gastrointestinal tract of cattle. Therefore, the objective of this study was to examine the effects that powdered and pelleted citrus by-products have on growth of the STEC serotypes O26:H11, O103:H8, O111:H8, O145:H28, O157: H7, and O157:H7 Δ*stx*1*stx*2 in bovine ruminal microorganism media.

2. Materials and Methods

2.1. Ruminal Fluid Collection and Medium Preparation. Ruminal contents (1 L) were collected from the rumen ventral sac of a 544 kg cannulated steer at the Henry Leveck Animal Research Center at Mississippi State University. Rumen particles were separated from the ruminal fluid by passing contents through nylon paint strainers as previously described by others [12]. After separation, rumen fluid was incubated for 30 min at 37°C, to allow the fluid to separate into three distinct layers. The middle layer of the rumen fluid was extracted and utilized for the mixed ruminal microorganism medium. The base medium utilized for the mixed ruminal microorganism fluid contained (per liter): 6.0 g KH_2PO_4, 6.0 g KH_2PO_4, 12.0 g $(NH_4)_2SO_4$, 12.0 g NaCl, 2.5 g $MgSO_4{\cdot}7H_2O$, 1.6 g $CaCl_2{\cdot}2H_2O$, 0.04 g cysteine HCl; base medium was sterilized by autoclaving [13]. To the base medium, 33.0 mL of an 8% solution of Na_2CO_3 and 333 mL of ruminal fluid were added and homogenized by mixing. The pH was adjusted to 6.5 with a 1 M NaOH solution and bubbled with CO_2. The fully prepared medium was incubated in an orbital shaker incubator at 140 rpm at 37°C for 12 h.

2.2. Bacterial Serotypes and Cultivation Conditions. Six strains of *E. coli* included in this study were purchased from the American Type Culture Collection (ATCC): O157:H7 (ATCC 43895), O103:H8 (ATCC 23982), O145:H28 (BAA-2129), O26:H11 (BAA-1653), O111:H8 (BAA-179), and O157:H7 Δ*stx*1*stx*2 (ATCC 43888). All *E. coli* strains were routinely cultured in Luria-Bertani medium (LB; Difco Co.; Corpus Christi, TX, USA) at 37°C. All strains were transformed with the bioluminescent pXEN-13 plasmid (Caliper Life Sciences; Hopkinton, MA, USA) as a means for identification after incubation. For transformation with the pXEN-13 plasmid, all strains were made competent by washing midlog cultures four times with ice cold 10% glycerol. Competent cells were then transformed with pXEN-13 by electroporation and cultured in LB supplemented with 100 μg/mL of ampicillin (AMP) using standard techniques [14].

2.3. Citrus Byproduct Trial. Isolates from fresh streaks of each strain transformed with pXEN-13 were used to inoculate a 5 mL starter culture of LB broth supplemented with AMP for 16 h at 37°C. Cultures were then diluted 1 : 100 in 5 mL LB broth and allowed to grow to mid-log phase (OD_{600} = 0.05), after which cultures were pelleted, residual medium was removed, and cells were resuspended in 5 mL of mixed ruminal microorganism fluid medium supplemented with 0, 5, 10, or 20% CBP (w/v). Cultures were incubated at 37°C in a shaker incubator for 6 h. Aliquots (0.1 mL) were removed at 0, 2, 4, and 6 h after incubation in ruminal medium, serially diluted in 1X phosphate buffered saline (PBS), plated onto LB agar supplemented with AMP, and incubated overnight at 37°C. The pH values were measured from each strain at each time interval at the various CBP concentrations recorded.

2.4. Statistical Analysis. Data were analyzed as a completely randomized design with repeated measures using PROC MIXED in SAS (SAS Inst. Inc., Version 9.2; Cary, NC). Experimental unit was defined as tube, and significance was declared at $P < 0.05$. Pair-wise differences among least squares means at various sample times were evaluated with the PDIFF statement.

3. Results and Discussion

A study recently conducted by our group using the serotypes analyzed in this study (O26:H11, O103:H8, O111:H8, O145:H28, O157:H7, and O157:H7 Δ*stx*1*stx*2) suggested that all serotypes were capable of growing within mixed ruminal microorganism fluid media; however decreased populations of serotypes O103:H8 and O145:H28 were observed after 24 h in comparison to O157:H7 [15]. These data suggest the possibility that not all non-O157 serotypes function similarly within cattle. To expand upon this previous study, the effect of CBP on non-O157 STEC was analyzed *in vitro*.

The growth of the various non-O157 STEC ($\log_{10}$ CFU/mL) was analyzed within mixed ruminal microorganism fluid medium supplemented with powdered CBP (Table 1). The O26:H11 and O145:H28 serotypes grew similar ($P < 0.11$) within the powdered CBP, with an exception of a decrease in O26:H11 populations ($P < 0.006$) at 4 h in the presence of 20% powdered CBP. The O103:H8 serotype exhibited approximately a 1 $\log_{10}$ reduction in populations over the 6 h study in the presence of 5% powdered CBP. The O157:H7 Δ*stx*1*stx*2 and O157:H7 serotypes had decreased populations ($P < 0.04$ and $P < 0.05$, resp.) in comparison to the other serotypes at 0 h. Although both O157 serotypes tended to grow similarly ($P < 0.06$) for 4 h, there was a difference observed at 6 h when O157:H7 Δ*stx*1*stx*2 had significantly lower populations ($P < 0.03$) in comparison

TABLE 1: Least squares means for growth of STEC O26:H11, O103:H8, O111:H8, O145:H28, O157:H7, and O157:H7 Δ*stx*1*stx*2 within bovine mixed rumen microorganisms fluid medium, supplemented with 0%, 5%, 10%, and 20% powdered citrus by-products (CBPs).

CBP	Time (h)	*E. coli* serotype (Log_{10} CFU/mL)					
		O26:H11	O103:H8	O111:H8	O145:H28	O157:H7	O157:H7 Δ*stx*1*stx*2
0%	0 h	7.55	7.11	6.93	7.55	6.03^{v}	6.46
	2 h	7.68	7.05	6.70	7.58	6.91^{w}	7.00
	4 h	7.60	6.81	6.92	8.06	6.91^{w}	7.19
	6 h	7.73	6.65	7.07	7.83	7.24^{w}	6.98
	Δ[a]	0.18	−0.046	0.14	0.28	1.21	0.52
5%	0 h	7.40	7.34^{y}	7.31^{v}	7.57	6.73	6.63
	2 h	7.47	6.86^{yz}	7.06^{v}	7.64	6.65	6.41
	4 h	7.43	6.95^{yz}	5.85^{w}	7.80	6.77	7.09
	6 h	7.59	6.55^{z}	7.22^{v}	7.80	7.02	6.74
	Δ	0.23	−0.59	0.61	−0.1	0.15	−2.63
10%	0 h	7.43	7.28	6.64	7.73	6.83	6.82^{vw}
	2 h	7.43	6.90	7.33	7.56	7.22	6.18^{w}
	4 h	7.36	6.99	7.18	7.64	6.96	7.13^{v}
	6 h	7.66	6.69	7.25	7.63	6.98	4.19^{x}
	Δ	0.07	−0.47	−0.32	0.06	−0.03	−5.04
20%	0 h	7.12^{v}	7.13	7.02	7.60	7.03	7.03^{v}
	2 h	7.54^{v}	7.33	6.78	7.52	6.94	6.53^{v}
	4 h	5.96^{w}	7.07	6.65	7.44	7.30	7.34^{v}
	6 h	7.19^{v}	6.66	6.70	7.66	7.00	1.99^{w}

[a]Change in concentration between 0 h and 6 h.
[v,w,x]Lsmeans within a column, within a treatment, without a common subscript are different if $P \leq 0.05$.
[y,z]Lsmeans within a column, within a treatment, without a common subscript tend to be different if $P < 0.09$.

TABLE 2: Least squares means for growth of STEC O26:H11, O103:H8, O111:H8, O145:H28, O157:H7, and O157:H7 Δ*stx*1*stx*2 within bovine mixed rumen microorganisms fluid medium, supplemented with 0%, 5%, 10%, and 20% pelleted citrus by-products (CBPs).

CBP	Time (h)	*E. coli* serotype (Log_{10} CFU/mL)					
		O26:H11	O103:H8	O111:H8	O145:H28	O157:H7	O157:H7 Δ*stx*1*stx*2
0%	0 h	9.27^{x}	7.01^{x}	8.47^{x}	9.61	8.48	6.46
	2 h	8.08^{y}	7.01^{x}	9.01^{xy}	9.48	8.24	6.27
	4 h	9.45^{x}	8.64^{y}	8.43^{x}	9.67	8.79	6.26
	6 h	9.64^{x}	8.77^{y}	9.48^{y}	9.50	9.01	6.51
	Δ[a]	0.37	1.76	1.01	−0.11	0.53	0.05
5%	0 h	9.27^{xy}	8.21	8.40^{xy}	9.51	9.23^{x}	6.07
	2 h	8.71^{x}	8.26	8.93^{x}	9.60	7.94^{y}	5.92
	4 h	7.29^{z}	7.93	7.98^{y}	9.59	7.96^{y}	6.23
	6 h	8.71^{y}	8.28	8.77^{x}	9.69	8.93^{x}	5.97
	Δ	−0.56	0.07	0.37	0.18	−0.3	−0.1
10%	0 h	9.07^{xz}	8.76	8.76^{xy}	8.98	8.73^{xy}	6.83
	2 h	8.43^{xy}	8.50	9.03^{xy}	9.01	8.73^{xy}	6.94
	4 h	7.90^{y}	8.37	8.66^{x}	9.55	8.40^{xy}	6.64
	6 h	8.59^{z}	8.07	9.73^{y}	9.71	9.20^{y}	6.68
	Δ	−0.48	−0.69	0.97	0.73	0.47	−0.15
20%	0 h	9.47^{x}	8.52^{x}	9.03	9.29^{x}	8.00^{x}	6.78
	2 h	8.98^{x}	8.51^{x}	8.92	9.79^{xy}	8.00^{y}	6.76
	4 h	7.69^{y}	8.57^{x}	9.09	8.95^{y}	8.71^{xy}	6.76
	6 h	9.54^{x}	7.43^{y}	9.46	9.48^{xy}	9.05^{y}	7.13
	Δ	−0.07	−1.09	−0.43	0.19	1.05	0.35

[a]Change in concentration between 0 h and 6 h.
[x,y,z]Lsmeans within a column, within a treatment, without a common subscript are different if $P \leq 0.05$.

to O157:H7. When supplemented with 10% powdered CBP, the O157:H7 Δ*stx1stx2* strains displayed approximately a 1.5 $\log_{10}$ reduction in populations, and at 20% powdered CBP there was approximately a 5 $\log_{10}$ reduction in populations over 6 h.

Variations were also observed in certain STEC serotypes when grown in the presence of mixed ruminal microorganism fluid medium supplemented with pelleted CBP (Table 2). The pelleted CBP tended to have no change ($P < 0.07$) in populations of O145:H28 from 0 h to 6 h. While O157:H7 had decreased populations ($P < 0.02$) at 0 h, there were no differences in populations observed between the O103:H8 and O157:H7 ($P < 0.11$) the remainder of the study. Populations of O103:H8 were decreased ($P < 0.02$) at 0 h, while populations tended to be similar ($P < 0.06$) to O111:H8 during the 6 h analyzed in this study. When O103:H8 was cultured in the mixed ruminal microorganism medium supplemented with 20% pelleted CBP, there was a general trend for approximately a 1 $\log_{10}$ reduction observed over the 6 h study. The O26:H11 serotype populations decreased ($P < 0.05$) throughout the study. *Escherichia coli* O157:H7 Δ*stx1stx2* exhibited the lowest populations ($P < 0.05$) throughout the study.

The O157:H7 Δ*stx1stx2* had reduced populations in mixed rumen microorganism fluid medium supplemented with powdered CBP, while O103:H8 had decreased populations within both varieties of CBP. These results are in accordance with previous studies that have suggested a decrease in O157:H7 populations using other varieties of CBP. A study conducted by Callaway et al. supplemented mixed ruminal microorganism fluid media with 0%, 0.5%, 1%, and 2% dried orange pulp and *E. coli* O157:H7 populations decreased according to increasing concentrations [6]. Reductions in *E. coli* O157:H7 populations have also been observed when sheep rations were supplemented with 5% or 10% orange peel [16].

The CBP was added to the mixed ruminal microorganism fluid media 2 h before the serotypes were added to the mixture (0 h). Although a decrease in O103:H8 and O157:H7 Δ*stx1stx2* populations was observed, other STEC populations were not affected. In the presence of CBP in either pelleted or powdered form, the pH for all strains was reduced from ~5.0 to ~4.0, while in control groups the pH increased from ~6.6 to 7.3. Therefore, the reductions in populations must be attributed to the CBP, as the pH variations were consistent between all strains. Given that this study was only conducted for 6 h, the effects of CBP within the mixed ruminal microorganism fluid media and STEC serotypes may not had been fully observed within the short time frame. This is potentially due to the diffusion properties of CBP across the cell envelope of *E. coli*. Others studies indicate that CBPs decrease *E. coli* O157:H7 populations from 24 h to 72 h [6, 16]. This study was only conducted for 6 h; an increased duration of the study could have been more beneficial to observe the effects of CBP on the various serotypes.

The essential oils within the CBP can permeabilize the bacterial cells walls and cytoplasm, leading to bacterial lysis, thus shifting the rumen environment leading to an increase in short-chain fatty acids while decreasing the pH. The acidic environment creates less favorable conditions for microbial populations to survive and replicate within, thus decreasing the possibility of *E. coli* O157:H7 populations. Although our research has reported a decrease in pH values with increasing CBP concentrations and an observed decrease in O103:H8 and O157:H7 Δ*stx1stx2* populations, this same trend was not observed within other STEC serotypes. Further research is needed to determine how the various STEC serotypes affect *E. coli* populations within mixed ruminal microorganism fluid media when supplemented with CBP.

Conflict of Interests

The authors declare that they have no conflict of interests.

Acknowledgment

This work was supported by the MAFES Special Research Initiative at Mississippi State University.

References

[1] J. G. Wells, L. D. Shipman, K. D. Greene et al., "Isolation of *Escherichia coli* serotype O157:H7 and other Shiga-like-toxin-producing *E. coli* from dairy cattle," *Journal of Clinical Microbiology*, vol. 29, no. 5, pp. 985–989, 1991.

[2] P. M. Griffin and R. V. Tauxe, "The epidemiology of infections caused by *Escherichia coli* O157:H7, other enterohemorrhagic *E. coli*, and the associated hemolytic uremic syndrome," *Epidemiologic Reviews*, vol. 13, pp. 60–98, 1991.

[3] K. Stanford, S. J. Bach, T. H. Marx et al., "Monitoring *Escherichia coli* O157:H7 in inoculated and naturally colonized feedlot cattle and their environment," *Journal of Food Protection*, vol. 68, no. 1, pp. 26–33, 2005.

[4] CDC, *Escherichia Coli O157:H7 and other Shiga-Toxin Producing Escherichia Coli (STEC)*, Division of Foodborne, Bacterial, and Mycotic Diseases. National Center for Zoonotic, Vector-Borne, and Enteric Diseases, Atlanta, Ga, USA, 2011.

[5] USDA-FSIS, *FSIS Verification Testing for Non-O157 Shiga Toxin-Producing Escherichia Coli (Non-O157 STEC) Under MT60, MT52, and MT53 Sampling Programs*, 2012.

[6] T. R. Callaway, J. A. Carroll, J. D. Arthington et al., "Citrus products decrease growth of *E. coli* O157:H7 and *Salmonella typhimurium* in pure culture and in fermentation with mixed ruminal microorganisms *in vitro*," *Foodborne Pathogens and Disease*, vol. 5, no. 5, pp. 621–627, 2008.

[7] F. Bakkali, S. Averbeck, D. Averbeck, and M. Idaomar, "Biological effects of essential oils—a review," *Food and Chemical Toxicology*, vol. 46, no. 2, pp. 446–475, 2008.

[8] J. E. Gustafson, Y. C. Liew, S. Chew et al., "Effects of tea tree oil on *Escherichia coli*," *Letters in Applied Microbiology*, vol. 26, no. 3, pp. 194–198, 1998.

[9] R. Di Pasqua, G. Betts, N. Hoskins, M. Edwards, D. Ercolini, and G. Mauriello, "Membrane toxicity of antimicrobial compounds from essential oils," *Journal of Agricultural and Food Chemistry*, vol. 55, no. 12, pp. 4863–4870, 2007.

[10] K. Fisher and C. A. Phillips, "The effect of lemon, orange and bergamot essential oils and their components on the survival of *Campylobacter jejuni*, *Escherichia coli* O157, *Listeria monocytogenes*, *Bacillus cereus* and *Staphylococcus aureusin*

vitro and in food systems," *Journal of Applied Microbiology*, vol. 101, no. 6, pp. 1232–1240, 2006.

[11] I. S. Nam, P. C. Garnsworthy, and J. H. Ahn, "Supplementation of essential oil extracted from citrus peel to animal feeds decreases microbial activity and aflatoxin contamination without disrupting *in vitro* ruminal fermentation," *Asian-Australasian Journal of Animal Sciences*, vol. 19, no. 11, pp. 1617–1622, 2006.

[12] S. A. Leyendecker, T. R. Callaway, R. C. Anderson, and D. J. Nisbet, "Technical note on a much simplified method for collecting ruminal fluid using a nylon paint strainer," *Journal of the Science of Food and Agriculture*, vol. 84, no. 4, pp. 387–389, 2004.

[13] M. A. Cotta and J. B. Russell, "Effect of peptides and amino acids on efficiency of rumen bacterial protein synthesis in a continuous culture," *Journal of Dairy Science*, vol. 65, no. 2, pp. 226–234, 1982.

[14] J. Sambrook and D. W. Russell, *Molecular Cloning: A Laboratory Manual*, Cold Spring Harbor Laboratory Press, New York, NY, USA, 3rd edition, 2001.

[15] A. L. Free, H. A. Duoss, L. V. Bergeron, S. A. Shields-Menard, E. Ward, T. R. Callaway et al., "Survival of O157:H7 and non-O157 serogroups of *Escherichia coli* in bovine rumen fluid and bile salts," *Foodborne Pathogens and Disease*, vol. 9, no. 11, pp. 1010–1014, 2012.

[16] T. R. Callaway, J. A. Carroll, J. D. Arthington, T. S. Edrington, M. L. Rossman, M. A. Carr et al., "*Escherichia coli* O157:H7 populations in ruminants can be reduced by orange peel product feeding," *Journal of Food Protection*, vol. 74, no. 11, pp. 1917–1921, 2011.

Social Behaviours under Anaerobic Conditions in *Pseudomonas aeruginosa*

Masanori Toyofuku, Hiroo Uchiyama, and Nobuhiko Nomura

Graduate School of Life and Environmental Sciences, University of Tsukuba, Tsukuba, Ibaraki 305-8572, Japan

Correspondence should be addressed to Nobuhiko Nomura, nomunobu@sakura.cc.tsukuba.ac.jp

Academic Editor: Robert P. Gunsalus

Pseudomonas aeruginosa is well adapted to grow in anaerobic environments in the presence of nitrogen oxides by generating energy through denitrification. Environmental cues, such as oxygen and nitrogen oxide concentrations, are important in regulating the gene expression involved in this process. Recent data indicate that *P. aeruginosa* also employs cell-to-cell communication signals to control the denitrifying activity. The regulation of denitrification by these signalling molecules may control nitric oxide production. Nitric oxide, in turn, functions as a signalling molecule by activating certain regulatory proteins. Moreover, under denitrifying conditions, drastic changes in cell physiology and cell morphology are induced that significantly impact group behaviours, such as biofilm formation.

1. Introduction

It is well acknowledged that bacteria exhibit social behaviours by communicating with each other through signalling molecules or by developing a community known as biofilm. The social behaviour of bacteria is of great interest to researchers, and *Pseudomonas aeruginosa* is one of the most studied bacterial model organisms.

P. aeruginosa has a flexible metabolism that can utilise nitric oxides as alternative electron acceptors to produce energy when oxygen is depleted [1]. This process is called denitrification and is also performed by many other bacteria. The stepwise process of denitrification in *P. aeruginosa* is as follows: $NO_3^- \rightarrow NO_2^- \rightarrow NO \rightarrow N_2O \rightarrow N_2$. The sequential steps are catalysed by the enzymes NO_3^- reductase (NAR), NO_2^- reductase (NIR), NO reductase (NOR), and N_2O reductase (N_2OR), respectively [2]. This process is important in the nitrogen cycle to produce nitrogen gases from NO_3^- and NO_2^-. Moreover, recent studies indicate that the denitrification process is related to the virulence of this bacterial species. *P. aeruginosa* is notorious as an opportunistic pathogen that infects immunocompromised patients, such as cystic fibrosis (CF) patients. How the bacteria adapt to the host environment is important in terms of its pathogenesis. The CF airway has been described as a microaerobic to anaerobic environment [3, 4]. Independent studies indicate the expression of denitrifying genes in the CF lung, suggesting that denitrification is important for the pathogenicity of *P. aeruginosa* [5, 6]. Thus, an understanding of the physiology under anaerobic conditions is important for the understanding of bacterial virulence under such conditions.

While there are many excellent reviews available about the social behaviours of *P. aeruginosa* under aerobic conditions, few have focused on anaerobic conditions. In this paper, we examine the social behaviour of *P. aeruginosa* under anaerobic conditions.

2. Denitrification Regulation by Physicochemical Conditions

The expression of denitrifying enzymes is controlled by a sophisticated regulatory network that responds to low oxygen conditions and the availability of nitrate or nitrite. The master regulator that monitors oxygen concentration is the ANR (anaerobic regulation of arginine deiminase and nitrate reduction) regulatory protein [7]. The active form of ANR contains an $[4Fe\text{-}4S]^{2+}$ cluster that is destroyed in the presence of oxygen [8]. ANR induces genes that are involved in producing energy under low-oxygen conditions or anaerobic conditions. One of these genes, the *cbb*$_3$-2 terminal oxidase, has a high affinity for oxygen, indicating that it plays a role under low-oxygen conditions [9, 10]. Other genes induced by ANR include the genes involved in fermentation [11]. In addition to these genes, ANR induces other transcriptional regulators involved in denitrification, NarXL, and DNR (dissimilative nitrate respiration regulator) [12, 13]. NarX and NarL comprise a two-component regulatory system that responds to nitrate. The sensor kinase NarX detects nitrate and activates the response regulator NarL, which regulates the transcription of *narK1*, *nirQ*, and *dnr* [12]. In addition, NarL partially represses the expression of arginine fermentation genes, enabling the bacteria to benefit from the more energetically efficient denitrification instead of low energy-yielding fermentation in the presence of nitrate under anaerobic conditions [14]. DNR is activated by binding to NO, and the active DNR regulator activates transcription of all four denitrifying reductases [12, 13, 15].

3. Cell-Cell Communication Signals in *P. aeruginosa*

In *P. aeruginosa*, two chemically distinct types of signalling molecules have been characterised in detail. One type consists of the *N*-acyl-L-homoserine lactone (AHL) signals. AHLs are produced widely in gram-negative bacteria, and *P. aeruginosa* is known to possess two AHL signaling systems, the LasR-LasI (*las*) and the RhlR-RhlI (*rhl*) systems [16]. LasI directs the synthesis of the AHL signal *N*-(3-oxododecanoyl)-L-homoserine lactone (3-oxo-C_{12}-HSL) [17, 18], and RhlI directs the synthesis of another AHL signal, *N*-butyryl-L-homoserine lactone (C_4-HSL) [19, 20]. These AHL signals have cognate receptors (LuxR proteins), LasR [21] and RhlR [22], that are activated by 3-oxo-C_{12}-HSL and C_4-HSL, respectively. In addition to the LasR and RhlR receptors, two additional LuxR homologues, QscR and VqsR, which are orphan LuxR proteins and not associated with a cognate AHL synthase, have been found. QscR binds to a variety of AHL molecules with different side chains, while the AHL molecule that binds to VqsR is unknown [23, 24]. The other type of signalling molecules in *P. aeruginosa* include 2-alkyl-4-quinolone (AQ). The AQ signalling molecules are produced by the product of the *pqsABCDE* operon together with the product of the *pqsH* gene and converts 2-heptyl-4-quinolone (HHQ) to the *Pseudomonas* quinolone molecule (PQS; 2-heptyl-3-hydroxy-4-quinolone) [25]. HHQ is also produced in bacteria other than *P. aeruginosa* [26]. In *P. aeruginosa*, both PQS and HHQ are able to regulate the *pqsABSDE* operon via a transcriptional regulator, PqsR (MvfR) [27]. Interestingly, PQS is carried by outer membrane vesicles (OMVs), which are thought to target neighbouring cells [28, 29].

4. Denitrification Regulation by Signalling Molecules

As mentioned above, denitrification is well regulated by the physiochemical environment. In addition, denitrification is regulated by cell-cell communication molecules. Interestingly, all three cell-cell signalling molecules, C_4-HSL, 3-oxo-C_{12}-HSL, and PQS, repress denitrification. AHLs and PQS affect denitrification in different manners. In the study conducted by Yoon et al. [30], it was demonstrated that the activity of the denitrifying enzymes is higher in *rhlR* mutants compared to its parent strain. A microarray study suggested that the denitrifying genes are regulated by AHLs [31]. Following these observations, it was demonstrated in detail that, indeed, AHLs regulate denitrifying activity. Both C_4-HSL and 3-oxo-C_{12}-HSL repressed denitrifying activity via their cognate regulator, RhlR or LasR, by regulating the expression of the denitrifying genes [32]. Regulation by the *las* quorum-sensing system was dependent on the *rhl* quorum-sensing system, suggesting hierarchical regulation by the *las* system over the *rhl* system in denitrification regulation, although the precise mechanism of denitrification regulation by AHLs has yet to be identified. In *P. aeruginosa* isolates from CF patients, mutations in the *lasR* gene are often found [33, 34]. Considering the fact that there are microenvironments with low oxygen tension inside the CF lung, it is possible that the *lasR* mutation confers a growth advantage by the activation of denitrification Figure 1 [35].

While the AHLs regulate the transcription of denitrifying genes, PQS affects the activity of denitrifying enzymes posttranscriptionally. It has been shown that NAR and NOR activity is repressed and NIR activity is increased in the presence of PQS [36]. Furthermore, when the PQS molecule was added to a crude extract containing denitrifying enzymes, NO_3^--respiration activity and NOR activity were repressed [36]. This was the first study to demonstrate that a signalling molecule affects enzyme activity in a direct manner. The transcription of denitrifying genes was unaffected by PQS under anaerobic conditions [36], while microarray studies suggested that PQS represses *nar* gene transcription under aerobic conditions [37], indicating that there may be several pathways in denitrification regulation by PQS. In addition to working as a signalling molecule that activates its cognate receptor, the PQS molecule is shown to have multiple functions, such as chelating iron, balancing the production of reacting oxygen species, and inducing outer membrane production [28, 38–40]. Iron concentration is a key environmental condition for the PQS effect on denitrification because excess amounts of iron inhibit this effect [36]. Interestingly, PQS affects the aerobic and anaerobic growth of bacterial species other than *P. aeruginosa*, indicating its broad impact on the bacterial community [41].

5. Cell-Cell Signalling under Denitrifying Conditions

While cell-cell communication has a wide impact on cell physiology, most of the studies in *P. aeruginosa* have been performed under aerobic conditions, and there have been only a limited number of studies concerning the impact of the cell-cell communication systems under anaerobic conditions. Expression of *lasR*, *lasI*, *rhlR*, and *rhlI* was

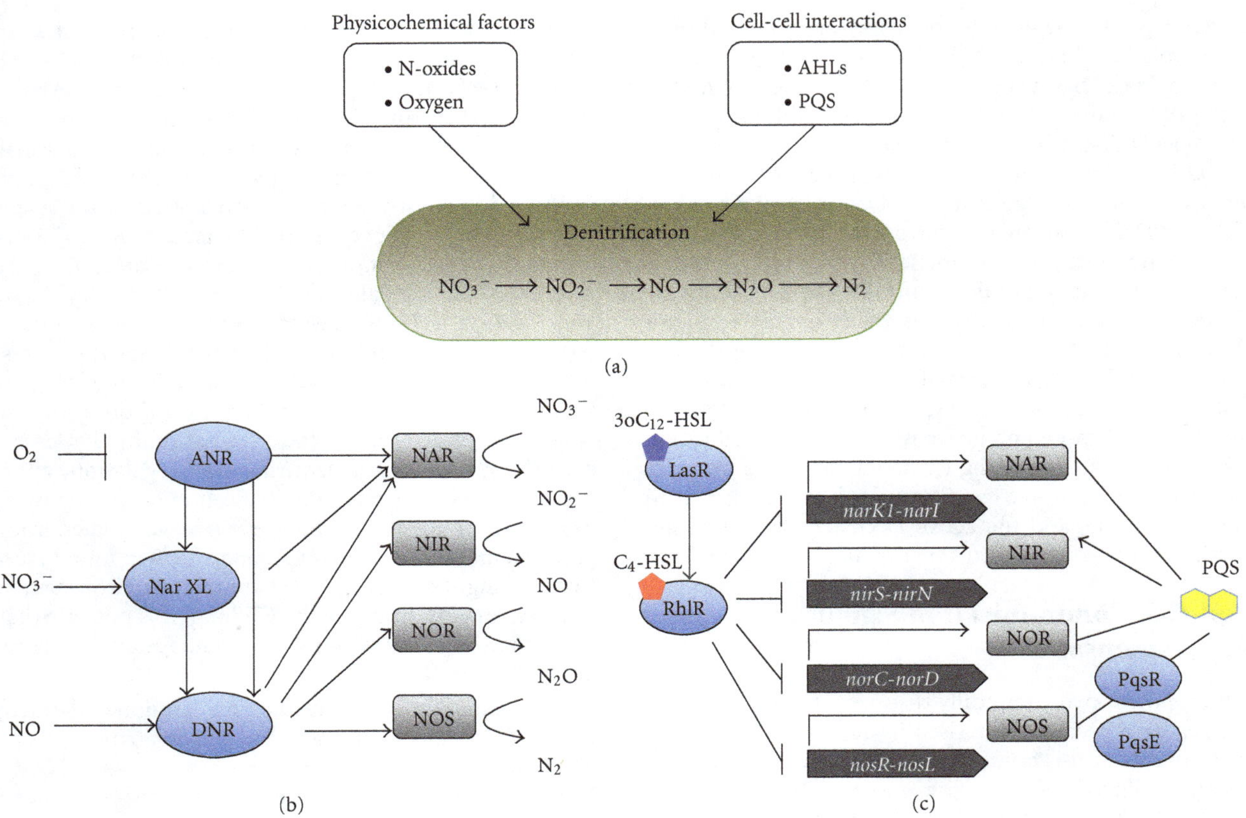

Figure 1: (a) Denitrification regulation in *P. aeruginosa*. Denitrification is regulated by physiochemical conditions, such as oxygen concentration and the availability of nitric oxides as well as cell-cell communication molecules. (b) Denitrification regulation by physiochemical factors in *P. aeruginosa*. ANR is activated under low-oxygen tension conditions, which increases the transcription of the NarXL two-component system and DNR. NarXL responds to nitrate and activates the expression of DNR together with ANR. DNR is activated in the presence of NO and induces the transcription of the four reductases. The *narK1K2GHJI* operon, which encodes nitrate/nitrite transporters and the structural gene for NAR, is also induced by ANR and NarL. (c) Denitrification regulation by cell-cell communication. C_4-HSL and 3-o-C_{12}-HSL repress the transcription of denitrifying reductases via their cognate receptors RhlR and LasR. Regulation by LasR is dependent on RhlR. PQS represses the activity of the NAR, NOR, and NOS enzymes while activating the NIR enzyme. PqsR and PqsE are involved in the PQS effect on NOR, while the effect on the other three enzymes does not require PqsR or PqsE.

shown to be altered under denitrifying conditions [31, 42]. Furthermore, a recent study measuring the AHLs under denitrifying conditions revealed that the production of AHLs is significantly lower compared to aerobic cultures [43]. The exact mechanism of this attenuation in signal production has not been revealed, but it is proposed that the limitation of the acyl carrier proteins leads to the low level of signalling molecules [43]. Interestingly, the AHL signalling systems under denitrifying conditions still actively regulate genes to some extent, as has been demonstrated by the regulation of denitrifying activity using AHL production-defective mutant strains [30, 32]

PQS production is also suppressed under anaerobic conditions [36, 44]. In this case, the enzymes that convert HHQ to PQS require oxygen. Therefore, under anaerobic conditions, hydroxylation of HHQ does not occur, and thus, PQS synthesis is prevented [44]. However, under these conditions, a sufficient amount of HHQ that can induce *pqsABCDE* transcription is present [44]. It was shown that both PQS and HHQ bind to the PqsR transcriptional regulator, although PQS is approximately 100-fold more potent in stimulating the activation of PqsR [27]. These results imply that HHQ may play an important role in cell-cell communication under denitrifying conditions. In fact, HHQ is used as a signalling molecule in several other bacteria [26], but the impact of HHQ as a signalling molecule has yet to be fully understood in *P. aeruginosa*.

Although production of all three signalling molecules is attenuated under anaerobic conditions, the exogenous addition of these signalling molecules restore the transcription of target genes [36, 43]. These results indicate that the cells under denitrifying conditions are altered in producing signalling molecules, but they are still able to respond to them. It is important to consider that the natural habitat of bacteria is not a stable environment and conditions, such as oxygen concentration, are likely to fluctuate. Moreover, even under the same environmental conditions, oxygen-limited patches are produced by the cells itself as observed in biofilms [45]. In these cases, it is possible that the signalling molecules produced in one growth condition affect the cells in other growth conditions. One example is the PQS effect on denitrification. As mentioned above, PQS is not produced at a sufficient amount under anaerobic conditions [36, 44]; however, when bacteria were grown under oxygen-limiting conditions in which oxygen was depleted depending on cell growth, denitrifying activity was higher in the non-PQS-producing mutants [36]. This result demonstrates that a signalling molecule produced in one environment can affect the cells in another environment. Because it is known that cell-cell signalling is modulated by environmental factors [46], it would also be interesting to further investigate whether there are any differences in the regulon regulated by the same molecules under aerobic and anaerobic conditions.

6. NO Signalling in *P. aeruginosa*

NO has been studied extensively as a signalling molecule in eukaryotic cells that plays important roles in many biological processes. However, the role of NO in bacteria has not been fully understood. Some recent studies have demonstrated that NO is produced through the oxidation of L-arginine in certain gram-positive bacteria [47, 48], as it is in mammals, and it can protect the bacteria from reactive oxygen species. During denitrification, NO is produced as an intermediate by the reduction of nitrite, a process catalysed by the NIR enzyme. In *P. aeruginosa*, the denitrifying pathway is the only biological pathway known to produce NO. This implies that under denitrifying conditions, NO may become a signal that, when produced by one cell, can affect the other cell. A number of regulators that respond to NO have been revealed. As explained above, NO is an important signal for inducing the denitrifying genes. In this case, NO is recognised by the regulatory protein DNR, which regulates the denitrifying genes [15, 49]. The activated DNR recognises the conserved DNA binding site (ANR box) in promoters to regulate transcription. A recent comprehensive study to determine ANR and DNR regulons suggested that, in addition to the denitrifying genes, the transcription of three genes, C4-dicarboxylate transport (PA1183), a hypothetical protein (PA3519), and the RND-type efflux pump *mexG* (PA4205), is influenced by DNR, although no ANR box has been found in the promoter regions of these genes [50]. Still, the roles of these genes under denitrifying conditions or in NO response are not known. Another type of RND efflux pump (*mexEF-oprN*) was suggested to be induced by NO [51]. In this study, the MexT regulator was required for *mexEF-oprN* induction by NO, while the mechanism is still obscure. Nevertheless, it is becoming evident that the expression of efflux pumps is involved in diverse cellular activities that affect the expression of several genes. A recent study has demonstrated that the expression of the MexEF-oprN efflux pump downregulates several genes [52]; thus, NO may impact the expression of these genes through the expression of this efflux pump.

FhpR is another NO-responsive regulator that, under aerobic conditions, regulates the flavohaemoglobin (*fhp*) gene, the product of which oxidises NO. FhpR is an orthologue of NorR found in *E.coli* and belongs to the σ^{54}-dependent family of transcriptional activators [53].

While high amounts of NO induce genes for NO detoxification [54] and the DNR and FhpR regulatory proteins are likely to regulate these genes, a series of studies have revealed that NO in nontoxic levels regulates the social behaviour of bacteria, such as the dispersal of *P. aeruginosa* in biofilms [55]. The underlying mechanism is not yet fully understood, but a secondary messenger (c-di-GMP) is involved in this process. Low levels of c-di-GMP induce bacterial motility in *P. aeruginosa*, which in turn induce dispersal in biofilms. NO has been shown to increase the activity of enzymes that degrade c-di-GMP. This process requires the chemotaxis transducer BdlA [56]. These data indicate that there is a biological pathway independent of the toxic response that responds to NO.

Although it is not fully characterised, another NO-responsive gene that regulates biofilm formation has been suggested. This gene product (PA2663) increases biofilm formation by inducing the production of psl exopolysaccharides while reducing swarming and swimming motility [57]. It also increases the production of virulence-related factors, such as pyoverdine, PQS, and elastase. The gene is located within the same operon as the *fhp* gene, indicating that NO will induce its expression. Interestingly, PA2663 has two transmembrane regions and may sense NO, although further study is needed.

NO has also been suggested to induce virulence factors. A *nirS* mutant deficient in NO production was shown to have a reduced amount of type III secretion system-dependent exotoxin production compared to wild type. This NO-dependent exotoxin induction was confirmed by the addition of exogenous NO donors [58].

One of the important factors that determines the cellular response to NO is the concentration of NO. High levels of NO are toxic to the cell, but low levels of NO could function as a signal. How NO concentration is modulated in the cell during denitrification is not fully understood, but one possibility is denitrification control by the AHL and PQS molecules because they affect NIR and NOR activity to a different extent. For example, rhl quorum-sensing mutants increase NO production due to the imbalance of NIR and NOR activity, and this induces cell death under denitrifying conditions [30]. PQS upregulates NO-producing NIR enzyme activity, while it downregulates the NO-reducing NOR enzyme activity, suggesting that PQS would induce NO accumulation [36].

7. Biofilm Formation under Denitrifying Conditions

A study by Yawata et al. [59] showed that filamentous cells emerge extensively in biofilms formed under denitrifying conditions. This was followed by the study by Yoon et al., which indicated that the filamentation is due to a defect in cell division, and filamentation of the cells is correlated with biofilm formation [60]. It was also suggested in this study that the filamentation is induced by NO [60]. Filamentation of cells has long been characterised as a trait of cell morphology under stressful conditions, such as UV light exposure, antibiotic treatment, and nutrient deprivation. In contrast, it is known that filamentation of the cell is a part of cell differentiation in bacteria, such as *Proteus mirabilis* [61]. Currently, it is not known whether the filamentation in denitrifying biofilms is a response to NO toxicity or an adaptive response to denitrifying conditions or whether there is a specific signalling pathway that initialises cell filamentation in *P. aeruginosa*. Nevertheless, morphological changes may impact cell physiology and behaviours and provide advantages for survival under certain circumstances [62].

Most of the studies investigating biofilm formation in *P. aeruginosa* have been performed under aerobic conditions. The morphological change in cells under denitrifying conditions causes us to question how the biofilm development process differs under aerobic conditions. Studies using a selected number of mutants have indicated that the same gene may have different effects on biofilm formation under aerobic and anaerobic conditions [30]. These studies suggest the possibility that the biofilm development procedure differs under denitrifying conditions.

8. Perspectives

As our knowledge of energy metabolism under anaerobic conditions expands, new questions arise with respect to the physiology of the cells under those conditions. Under denitrifying conditions, the cells undergo morphological changes [59, 60], and cell physiology is likely to change dynamically with the shift in respiration systems [42]. Understanding the physiology under denitrifying conditions in *P. aeruignosa* is also important from clinical perspectives because anaerobiosis leads to the alteration in antibiotic tolerance, the mechanism of which is not fully understood [45]. Attenuation of the AHL and PQS signalling systems and different biofilm development processes raise further questions about the social behaviours of the bacteria under anaerobic conditions. It will be interesting to investigate whether additional signals or systems that coordinate group behaviours under anaerobic conditions may exist.

Acknowledgments

The authors thank Core Research for Evolutional Science and Technology (CREST) and the Ministry of Education, Culture, Sports, and Technology of Japan for financial support.

References

[1] H. Arai, "Regulation and function of versatile aerobic and anaerobic respiratory metabolism in *Pseudomonas aeruginosa*," *Frontiers in Microbiology*, vol. 2, no. 103, 2011.

[2] W. G. Zumft, "Cell biology and molecular basis of denitrification?" *Microbiology and Molecular Biology Reviews*, vol. 61, no. 4, pp. 533–616, 1997.

[3] D. Worlitzsch, R. Tarran, M. Ulrich et al., "Effects of reduced mucus oxygen concentration in airway *Pseudomonas* infections of cystic fibrosis patients," *Journal of Clinical Investigation*, vol. 109, no. 3, pp. 317–325, 2002.

[4] K. Sanderson, L. Wescombe, S. M. Kirov, A. Champion, and D. W. Reid, "Bacterial cyanogenesis occurs in the cystic fibrosis lung," *European Respiratory Journal*, vol. 32, no. 2, pp. 329–333, 2008.

[5] C. Beckmann, M. Brittnacher, R. Ernst, N. Mayer-Hamblett, S. I. Miller, and J. L. Burns, "Use of phage display to identify potential *Pseudomonas aeruginosa* gene products relevant to early cystic fibrosis airway infections," *Infection and Immunity*, vol. 73, no. 1, pp. 444–452, 2005.

[6] M. S. Son, W. J. Matthews, Y. Kang, D. T. Nguyen, and T. T. Hoang, "In vivo evidence of *Pseudomonas aeruginosa* nutrient acquisition and pathogenesis in the lungs of cystic fibrosis patients," *Infection and Immunity*, vol. 75, no. 11, pp. 5313–5324, 2007.

[7] A. Zimmermann, C. Reimmann, M. Galimand, and D. Haas, "Anaerobic growth and cyanide synthesis of *Pseudomonas aeruginosa* depend on *anr*, a regulatory gene homologous with *fnr* of Escherichia coli," *Molecular Microbiology*, vol. 5, no. 6, pp. 1483–1490, 1991.

[8] S. S. Yoon, A. C. Karabulut, J. D. Lipscomb et al., "Two-pronged survival strategy for the major cystic fibrosis pathogen, *Pseudomonas aeruginosa*, lacking the capacity to degrade nitric oxide during anaerobic respiration," *EMBO Journal*, vol. 26, no. 15, pp. 3662–3672, 2007.

[9] J. C. Comolli and T. J. Donohue, "Differences in two *Pseudomonas aeruginosa* cbb3 cytochrome oxidases," *Molecular Microbiology*, vol. 51, no. 4, pp. 1193–1203, 2004.

[10] T. Kawakami, M. Kuroki, M. Ishii, Y. Igarashi, and H. Arai, "Differential expression of multiple terminal oxidases for aerobic respiration in *Pseudomonas aeruginosa*," *Environmental Microbiology*, vol. 12, no. 6, pp. 1399–1412, 2010.

[11] M. Gamper, A. Zimmermann, and D. Haas, "Anaerobic regulation of transcription initiation in the arcDABC operon of *Pseudomonas aeruginosa*," *Journal of Bacteriology*, vol. 173, no. 15, pp. 4742–4750, 1991.

[12] K. Schreiber, R. Krieger, B. Benkert et al., "The anaerobic regulatory network required for *Pseudomonas aeruginosa* nitrate respiration," *Journal of Bacteriology*, vol. 189, no. 11, pp. 4310–4314, 2007.

[13] H. Arai, T. Kodama, and Y. Igarashi, "Cascade regulation of the two CRP/FNR-related transcriptional regulators (ANR and DNR) and the denitrification enzymes in *Pseudomonas aeruginosa*," *Molecular Microbiology*, vol. 25, no. 6, pp. 1141–1148, 1997.

[14] B. Benkert, N. Quäck, K. Schreiber, L. Jaensch, D. Jahn, and M. Schobert, "Nitrate-responsive NarX-NarL represses arginine-mediated induction of the *Pseudomonas aeruginosa* arginine fermentation *arcDABC* operon," *Microbiology*, vol. 154, no. 10, pp. 3053–3060, 2008.

[15] G. Giardina, S. Rinaldo, K. A. Johnson, A. Di Matteo, M. Brunori, and F. Cutruzzolà, "NO sensing in *Pseudomonas aeruginosa*: structure of the Transcriptional Regulator DNR," *Journal of Molecular Biology*, vol. 378, no. 5, pp. 1002–1015, 2008.

[16] E. C. Pesci, J. P. Pearson, P. C. Seed, and B. H. Iglewski, "Regulation of las and rhl quorum sensing in *Pseudomonas aeruginosa*," *Journal of Bacteriology*, vol. 179, no. 10, pp. 3127–3132, 1997.

[17] J. P. Pearson, K. M. Gray, L. Passador et al., "Structure of the autoinducer required for expression of *Pseudomonas aeruginosa* virulence genes," *Proceedings of the National Academy of Sciences of the United States of America*, vol. 91, no. 1, pp. 197–201, 1994.

[18] L. Passador, J. M. Cook, M. J. Gambello, L. Rust, and B. H. Iglewski, "Expression of *Pseudomonas aeruginosa* virulence genes requires cell-to- cell communication," *Science*, vol. 260, no. 5111, pp. 1127–1130, 1993.

[19] J. P. Pearson, L. Passador, B. H. Iglewski, and E. P. Greenberg, "A second N-acylhomoserine lactone signal produced by *Pseudomonas aeruginosa*," *Proceedings of the National Academy of Sciences of the United States of America*, vol. 92, no. 5, pp. 1490–1494, 1995.

[20] U. A. Ochsner and J. Reiser, "Autoinducer-mediated regulation of rhamnolipid biosurfactant synthesis in *Pseudomonas aeruginosa*," *Proceedings of the National Academy of Sciences of the United States of America*, vol. 92, no. 14, pp. 6424–6428, 1995.

[21] M. J. Gambello and B. H. Iglewski, "Cloning and characterization of the *Pseudomonas aeruginosa lasR* gene, a transcriptional activator of elastase expression," *Journal of Bacteriology*, vol. 173, no. 9, pp. 3000–3009, 1991.

[22] U. A. Ochsner, A. K. Koch, A. Fiechter, and J. Reiser, "Isolation and characterization of a regulatory gene affecting rhamnolipid biosurfactant synthesis in *Pseudomonas aeruginosa*," *Journal of Bacteriology*, vol. 176, no. 7, pp. 2044–2054, 1994.

[23] S. A. Chugani, M. Whiteley, K. M. Lee, D. D'Argenio, C. Manoil, and E. P. Greenberg, "QscR, a modulator of quorum-sensing signal synthesis and virulence in *Pseudomonas aeruginosa*," *Proceedings of the National Academy of Sciences of the United States of America*, vol. 98, no. 5, pp. 2752–2757, 2001.

[24] M. Juhas, L. Wiehlmann, B. Huber et al., "Global regulation of quorum sensing and virulence by VqsR in *Pseudomonas aeruginosa*," *Microbiology*, vol. 150, no. 4, pp. 831–841, 2004.

[25] L. A. Gallagher, S. L. McKnight, M. S. Kuznetsova, E. C. Pesci, and C. Manoil, "Functions required for extracellular quinolone signaling by *Pseudomonas aeruginosa*," *Journal of Bacteriology*, vol. 184, no. 23, pp. 6472–6480, 2002.

[26] S. P. Diggle, P. Lumjiaktase, F. Dipilato et al., "Functional genetic analysis reveals a 2-Alkyl-4-quinolone signaling system in the human pathogen *Burkholderia pseudomallei* and related bacteria," *Chemistry and Biology*, vol. 13, no. 7, pp. 701–710, 2006.

[27] G. Xiao, E. Déziel, J. He et al., "MvfR, a key *Pseudomonas aeruginosa* pathogenicity LTTR-class regulatory protein, has dual ligands," *Molecular Microbiology*, vol. 62, no. 6, pp. 1689–1699, 2006.

[28] L. M. Mashburn and M. Whiteley, "Membrane vesicles traffic signals and facilitate group activities in a prokaryote," *Nature*, vol. 437, no. 7057, pp. 422–425, 2005.

[29] Y. Tashiro, S. Ichikawa, M. Shimizu et al., "Variation of physiochemical properties and cell association activity of membrane vesicles with growth phase in *Pseudomonas aeruginosa*," *Applied and Environmental Microbiology*, vol. 76, no. 11, pp. 3732–3739, 2010.

[30] S. S. Yoon, R. F. Hennigan, G. M. Hilliard et al., "*Pseudomonas aeruginosa* anaerobic respiration in biofilms: relationships to cystic fibrosis pathogenesis," *Developmental Cell*, vol. 3, no. 4, pp. 593–603, 2002.

[31] V. E. Wagner, D. Bushnell, L. Passador, A. I. Brooks, and B. H. Iglewski, "Microarray analysis of *Pseudomonas aeruginosa* quorum-sensing regulons: effects of growth phase and environment," *Journal of Bacteriology*, vol. 185, no. 7, pp. 2080–2095, 2003.

[32] M. Toyofuku, N. Nomura, T. Fujii et al., "Quorum sensing regulates denitrification in *Pseudomonas aeruginosa* PAO1," *Journal of Bacteriology*, vol. 189, no. 13, pp. 4969–4972, 2007.

[33] D. A. D'Argenio, M. Wu, L. R. Hoffman et al., "Growth phenotypes of *Pseudomonas aeruginosa lasR* mutants adapted to the airways of cystic fibrosis patients," *Molecular Microbiology*, vol. 64, no. 2, pp. 512–533, 2007.

[34] L. R. Hoffman, H. D. Kulasekara, J. Emerson et al., "*Pseudomonas aeruginosa lasR* mutants are associated with cystic fibrosis lung disease progression," *Journal of Cystic Fibrosis*, vol. 8, no. 1, pp. 66–70, 2009.

[35] L. R. Hoffman, A. R. Richardson, L. S. Houston et al., "Nutrient availability as a mechanism for selection of antibiotic tolerant *Pseudomonas aeruginosa* within the CF airway," *PLoS Pathogens*, vol. 6, no. 1, Article ID e1000712, 2010.

[36] M. Toyofuku, N. Nomura, E. Kuno, Y. Tashiro, T. Nakajima, and H. Uchiyama, "Influence of the *Pseudomonas* quinolone signal on denitrification in *Pseudomonas aeruginosa*," *Journal of Bacteriology*, vol. 190, no. 24, pp. 7947–7956, 2008.

[37] G. Rampioni, C. Pustelny, M. P. Fletcher et al., "Transcriptomic analysis reveals a global alkyl-quinolone-independent regulatory role for PqsE in facilitating the environmental adaptation of *Pseudomonas aeruginosa* to plant and animal hosts," *Environmental Microbiology*, vol. 12, no. 6, pp. 1659–1673, 2010.

[38] S. P. Diggle, S. Matthijs, V. J. Wright et al., "The *Pseudomonas aeruginosa* 4-quinolone signal molecules HHQ and PQS play multifunctional roles in quorum sensing and iron entrapment," *Chemistry and Biology*, vol. 14, no. 1, pp. 87–96, 2007.

[39] F. Bredenbruch, R. Geffers, M. Nimtz, J. Buer, and S. Häussler, "The *Pseudomonas aeruginosa* quinolone signal (PQS) has an iron-chelating activity," *Environmental Microbiology*, vol. 8, no. 8, pp. 1318–1329, 2006.

[40] S. Häussler and T. Becker, "The pseudomonas quinolone signal (PQS) balances life and death in *Pseudomonas aeruginosa* populations," *PLoS Pathogens*, vol. 4, no. 9, Article ID e1000166, 2008.

[41] M. Toyofuku, T. Nakajima-Kambe, H. Uchiyama, and N. Nomura, "The effect of a cell-to-cell communication molecule, Pseudomonas quinolone signal (PQS), produced by p. aeruginosa on other bacterial species," *Microbes and Environments*, vol. 25, no. 1, pp. 1–7, 2010.

[42] M. J. Filiatrault, V. E. Wagner, D. Bushnell, C. G. Haidaris, B. H. Iglewski, and L. Passador, "Effect of anaerobiosis and nitrate on gene expression in *Pseudomonas aeruginosa*," *Infection and Immunity*, vol. 73, no. 6, pp. 3764–3772, 2005.

[43] K. M. Lee, M. Y. Yoon, Y. Park, J. H. Lee, and S. S. Yoon, "Anaerobiosis-induced loss of cytotoxicity is due to inactivation of quorum sensing in *Pseudomonas aeruginosa*," *Infection and Immunity*, vol. 79, no. 7, pp. 2792–2800, 2011.

[44] J. W. Schertzer, S. A. Brown, and M. Whiteley, "Oxygen levels rapidly modulate *Pseudomonas aeruginosa* social behaviours via substrate limitation of PqsH," *Molecular Microbiology*, vol. 77, no. 6, pp. 1527–1538, 2010.

[45] G. Borriello, E. Werner, F. Roe, A. M. Kim, G. D. Ehrlich, and P. S. Stewart, "Oxygen limitation contributes to antibiotic tolerance of *Pseudomonas aeruginosa* in biofilms," *Antimicrobial Agents and Chemotherapy*, vol. 48, no. 7, pp. 2659–2664, 2004.

[46] P. Williams and M. Cámara, "Quorum sensing and environmental adaptation in *Pseudomonas aeruginosa*: a tale of regulatory networks and multifunctional signal molecules," *Current Opinion in Microbiology*, vol. 12, no. 2, pp. 182–191, 2009.

[47] I. Gusarov, K. Shatalin, M. Starodubtseva, and E. Nudler, "Endogenous nitric oxide protects bacteria against a wide spectrum of antibiotics," *Science*, vol. 325, no. 5946, pp. 1380–1384, 2009.

[48] K. Shatalin, I. Gusarov, E. Avetissova et al., "*Bacillus anthracis*-derived nitric oxide is essential for pathogen virulence and survival in macrophages," *Proceedings of the National Academy of Sciences of the United States of America*, vol. 105, no. 3, pp. 1009–1013, 2008.

[49] H. Arai, M. Mizutani, and Y. Igarashi, "Transcriptional regulation of the *nos* genes for nitrous oxide reductase in *Pseudomonas aeruginosa*," *Microbiology*, vol. 149, no. 1, pp. 29–36, 2003.

[50] K. Trunk, B. Benkert, N. Quäck et al., "Anaerobic adaptation in *Pseudomonas aeruginosa*: definition of the Anr and Dnr regulons," *Environmental Microbiology*, vol. 12, no. 6, pp. 1719–1733, 2010.

[51] H. Fetar, C. Gilmour, R. Klinoski, D. M. Daigle, C. R. Dean, and K. Poole, "*mexEF-oprN* multidrug efflux operon of *Pseudomonas aeruginosa*: regulation by the MexT activator in response to nitrosative stress and chloramphenicol," *Antimicrobial Agents and Chemotherapy*, vol. 55, no. 2, pp. 508–514, 2011.

[52] Z. X. Tian, E. Fargier, M. Mac Aogáin, C. Adams, Y. P. Wang, and F. O'Gara, "Transcriptome profiling defines a novel regulon modulated by the LysR-type transcriptional regulator MexT in *Pseudomonas aeruginosa*," *Nucleic Acids Research*, vol. 37, no. 22, Article ID gkp828, pp. 7546–7559, 2009.

[53] H. Arai, M. Hayashi, A. Kuroi, M. Ishii, and Y. Igarashi, "Transcriptional regulation of the flavohemoglobin gene for aerobic nitric oxide detoxification by the second nitric oxide-responsive regulator of *Pseudomonas aeruginosa*," *Journal of Bacteriology*, vol. 187, no. 12, pp. 3960–3968, 2005.

[54] A. M. Firoved, S. R. Wood, W. Ornatowski, V. Deretic, and G. S. Timmins, "Microarray analysis and functional characterization of the nitrosative stress response in nonmucoid and mucoid *Pseudomonas aeruginosa*," *Journal of Bacteriology*, vol. 186, no. 12, pp. 4046–4050, 2004.

[55] N. Barraud, D. J. Hassett, S. H. Hwang, S. A. Rice, S. Kjelleberg, and J. S. Webb, "Involvement of nitric oxide in biofilm dispersal of *Pseudomonas aeruginosa*," *Journal of Bacteriology*, vol. 188, no. 21, pp. 7344–7353, 2006.

[56] N. Barraud, D. Schleheck, J. Klebensberger et al., "Nitric oxide signaling in *Pseudomonas aeruginosa* biofilms mediates phosphodiesterase activity, decreased cyclic di-GMP levels, and enhanced dispersal," *Journal of Bacteriology*, vol. 191, no. 23, pp. 7333–7342, 2009.

[57] C. Attila, A. Ueda, and T. K. Wood, "PA2663 (PpyR) increases biofilm formation in *Pseudomonas aeruginosa* PAO1 through the psl operon and stimulates virulence and quorum-sensing phenotypes," *Applied Microbiology and Biotechnology*, vol. 78, no. 2, pp. 293–307, 2008.

[58] N. E. Van Alst, M. Wellington, V. L. Clark, C. G. Haidaris, and B. H. Iglewski, "Nitrite reductase NirS is required for type III secretion system expression and virulence in the human monocyte cell line THP-1 by *Pseudomonas aeruginosa*," *Infection and Immunity*, vol. 77, no. 10, pp. 4446–4454, 2009.

[59] Y. Yawata, N. Nomura, and H. Uchiyama, "Development of a novel biofilm continuous culture method for simultaneous assessment of architecture and gaseous metabolite production," *Applied and Environmental Microbiology*, vol. 74, no. 17, pp. 5429–5435, 2008.

[60] M. Y. Yoon, K. M. Lee, Y. Park, and S. S. Yoon, "Contribution of cell elongation to the biofilm formation of *Pseudomonas aeruginosa* during anaerobic respiration," *PLoS ONE*, vol. 6, no. 1, Article ID e16105, 2011.

[61] R. M. Morgenstein, B. Szostek, and P. N. Rather, "Regulation of gene expression during swarmer cell differentiation in *Proteus mirabilis*," *FEMS Microbiology Reviews*, vol. 34, no. 5, pp. 753–763, 2010.

[62] S. S. Justice, D. A. Hunstad, L. Cegelski, and S. J. Hultgren, "Morphological plasticity as a bacterial survival strategy," *Nature Reviews Microbiology*, vol. 6, no. 2, pp. 162–168, 2008.

Permissions

The contributors of this book come from diverse backgrounds, making this book a truly international effort. This book will bring forth new frontiers with its revolutionizing research information and detailed analysis of the nascent developments around the world.

We would like to thank all the contributing authors for lending their expertise to make the book truly unique. They have played a crucial role in the development of this book. Without their invaluable contributions this book wouldn't have been possible. They have made vital efforts to compile up to date information on the varied aspects of this subject to make this book a valuable addition to the collection of many professionals and students.

This book was conceptualized with the vision of imparting up-to-date information and advanced data in this field. To ensure the same, a matchless editorial board was set up. Every individual on the board went through rigorous rounds of assessment to prove their worth. After which they invested a large part of their time researching and compiling the most relevant data for our readers. Conferences and sessions were held from time to time between the editorial board and the contributing authors to present the data in the most comprehensible form. The editorial team has worked tirelessly to provide valuable and valid information to help people across the globe.

Every chapter published in this book has been scrutinized by our experts. Their significance has been extensively debated. The topics covered herein carry significant findings which will fuel the growth of the discipline. They may even be implemented as practical applications or may be referred to as a beginning point for another development.

The editorial board has been involved in producing this book since its inception. They have spent rigorous hours researching and exploring the diverse topics which have resulted in the successful publishing of this book. They have passed on their knowledge of decades through this book. To expedite this challenging task, the publisher supported the team at every step. A small team of assistant editors was also appointed to further simplify the editing procedure and attain best results for the readers.

Our editorial team has been hand-picked from every corner of the world. Their multi-ethnicity adds dynamic inputs to the discussions which result in innovative outcomes. These outcomes are then further discussed with the researchers and contributors who give their valuable feedback and opinion regarding the same. The feedback is then collaborated with the researches and they are edited in a comprehensive manner to aid the understanding of the subject.

Apart from the editorial board, the designing team has also invested a significant amount of their time in understanding the subject and creating the most relevant covers. They scrutinized every image to scout for the most suitable representation of the subject and create an appropriate cover for the book.

The publishing team has been involved in this book since its early stages. They were actively engaged in every process, be it collecting the data, connecting with the contributors or procuring relevant information. The team has been an ardent support to the editorial, designing and production team. Their endless efforts to recruit the best for this project, has resulted in the accomplishment of this book. They are a veteran in the field of academics and their pool of knowledge is as vast as their experience in printing. Their expertise and guidance has proved useful at every step. Their uncompromising quality standards have made this book an exceptional effort. Their encouragement from time to time has been an inspiration for everyone.

The publisher and the editorial board hope that this book will prove to be a valuable piece of knowledge for researchers, students, practitioners and scholars across the globe.

List of Contributors

Monde Ntwasa
School of Molecular and Cell Biology, University of the Witwatersrand, Wits 2050, South Africa

Akira Goto and Shoichiro Kurata
Graduate School of Pharmaceutical Sciences, Tohoku University, Aoba 6-3, Aramaki, Aoba-ku, Sendai 980-8578, Japan

Indira T. Kudva
Food Safety and Enteric Pathogens Research Unit, National Animal Disease Center, Agricultural Research Service, U.S. Department of Agriculture, Ames, IA 50010, USA

Margaret A. Davis
Department of Microbiology, Molecular Biology and Biochemistry, University of Idaho, Moscow, ID 83843, USA
Department of Veterinary Microbiology and Pathology, College of Veterinary Medicine, Washington State University, Pullman, WA 99164, USA

Robert W. Griffin and Jeonifer Garren
Division of Infectious Diseases, Massachusetts General Hospital, Boston, MA 02114, USA

Megan Murray
Division of Infectious Diseases, Massachusetts General Hospital, Boston, MA 02114, USA
Department of Epidemiology, Harvard School of Public Health, Boston, MA 02115, USA

Manohar John
Division of Infectious Diseases, Massachusetts General Hospital, Boston, MA 02114, USA
Department of Medicine, Harvard Medical School, Boston, MA 02115, USA
Pathovacs Inc., Ames, IA 50010, USA

Carolyn J. Hovde
School of Food Sciences, University of Idaho, Moscow, ID 83843, USA

Stephen B. Calderwood
Division of Infectious Diseases, Massachusetts General Hospital, Boston, MA 02114, USA
Department of Medicine, Harvard Medical School, Boston, MA 02115, USA
Department of Microbiology and Immunobiology, Harvard Medical School, Boston, MA 02115, USA

Goualie Gblossi Bernadette
Laboratoire de Biotechnologies, Filiere Biochimie-Microbiologie, Unite de Formation et de Recherche en Biosciences, Universite de Cocody-Abidjan, 01 BP 582, Abidjan, Cote divoire
Institut Pasteur de Cote divoire, 01 BP 490, Abidjan, Cote divoire

Dosso Mireille, Kakou-N'Gazoa Elise Solange, Guessennd Natalie and Bakayoko Souleymane
Institut Pasteur de Cote divoire, 01 BP 490, Abidjan, Cote d'Ivoire

Akpa Eric Essoh and Niamke Lamine Sebastien
Laboratoire de Biotechnologies, Filiere Biochimie-Microbiologie, Unite de Formation et de Recherche en Biosciences, Universite de Cocody-Abidjan, 01 BP 582, Abidjan, Cote d'Ivoire

Paola Papoff, Carla Cerasaro, Elena Caresta, Fabio Midulla and Corrado Moretti
Pediatric Emergency and Intensive Care Division, Department of Pediatrics, Sapienza University of Rome, Viale Regina Elena 324, 00161 Rome, Italy

Giancarlo Ceccarelli and Gabriella d'Ettorre
Department of Public Health and Infectious Diseases, Sapienza University of Rome, 100161 Rome, Italy

Elisa Korenblum, Diogo Bastos Souza and Lucy Seldin
Laboratorio de Genetica Microbiana, Instituto de Microbiologia Prof. Paulo de Goes, Universidade Federal do Rio de Janeiro, Centro de Ciencias da Saude, Bloco I, Ilha do Fundao, 21941-590 Rio de Janeiro, RJ, Brazil

Monica Penna
Gerencia de Biotecnologia e Tratamentos Ambientais, CENPES-PETROBRAS, Ilha do Fundao, 21949-900 Rio de Janeiro, RJ, Brazil

Rob de Jonge
Laboratory for Zoonoses and Environmental Microbiology, National Institute for Public Health and the Environment (RIVM), 3720 BA Bilthoven, The Netherlands

Aarieke E.I. de Jong
Laboratory for Zoonoses and Environmental Microbiology, National Institute for Public Health and the Environment (RIVM), 3720 BA Bilthoven, The Netherlands
Division Consumer and Safety, New Food and Consumer Product Safety Authority (nVWA), 1018 BK Amsterdam, The Netherlands

Esther D. van Asselt
Laboratory for Zoonoses and Environmental Microbiology, National Institute for Public Health and the Environment (RIVM), 3720 BA Bilthoven, The Netherlands
Rikilt, Institute of Food Safety, 6700 AE Wageningen, The Netherlands

Marcel H. Zwietering
Laboratory of Food Microbiology, Wageningen University, 6700 EV Wageningen, The Netherlands

Maarten J. Nauta
Laboratory for Zoonoses and Environmental Microbiology, National Institute for Public Health and the Environment (RIVM), 3720 BA Bilthoven, The Netherlands
National Food Institute, Technical University of Denmark, 1790 Copenhagen V, Denmark

Valeria Souza and Luis E. Eguiarte
Departamento de Ecologıa Evolutiva, Instituto de Ecologıa, Universidad Nacional Autonoma de Mexico, Apartado Postal 70-275, 04510 Mexico, DF, Mexico

Alejandra Rodrıguez-Verdugo
Departamento de Ecologıa Evolutiva, Instituto de Ecologıa, Universidad Nacional Autonoma de Mexico, Apartado Postal 70-275, 04510 Mexico, DF, Mexico
Department of Ecology and Evolutionary Biology, University of California, Irvine, CA 92091, USA

Ana E. Escalante
Departamento de Ecologıa Evolutiva, Instituto de Ecologıa, Universidad Nacional Autonoma de Mexico, Apartado Postal 70-275, 04510 Mexico, DF, Mexico
Departamento de Ecologıa de la Biodiversidad, Instituto de Ecologıa, Universidad Nacional Autonoma de Mexico, Apartado Postal 70-275, 04510 Mexico, DF, Mexico

Gary P. Moran, David C. Coleman and Derek J. Sullivan
Division of Oral Biosciences, Dublin Dental University Hospital, Trinity College Dublin, Dublin 2, Ireland

Piotr Nowak and Samir Abdurahman
Department of Infectious Diseases, Institution of Medicine, Karolinska University Hospital and Karolinska Institutet, 14186 Stockholm, Sweden

Anders Sonnerborg
Department of Infectious Diseases, Institution of Medicine, Karolinska University Hospital and Karolinska Institutet, 14186 Stockholm, Sweden
Department of Clinical Microbiology, Institution of Laboratory Medicine, Karolinska University Hospital and Karolinska Institutet, 14186 Stockholm, Sweden

Annica Lindkvist
Department of Clinical Microbiology, Institution of Laboratory Medicine, Karolinska University Hospital and Karolinska Institutet, 14186 Stockholm, Sweden

Marius Troseid
Department of Infectious Diseases, Oslo University Hospital, Ulleval, 0424 Oslo, Norway

Mircea Radu Mihu
Department of Medicine, Sound Shore Medical Center of Westchester, New Rochelle, NY 10802, USA

Joshua Daniel Nosanchuk
Division of Infectious Diseases, Departments of Medicine and Microbiology and Immunology, Albert Einstein College of Medicine, Bronx, NY 10461, USA

Vanessa da Silveira Duarte and Italo Delalibera Júnior
Department of Entomology and Acarology, ESALQ/University of S˜ao Paulo, 13418-900 Piracicaba, SP, Brazil

Karin Westrum and Ingeborg Klingen
Plant Health and Plant Protection Division, Norwegian Institute for Agricultural and Environmental Research (Bioforsk), 1432 As, Norway

Ana Elizabete Lopes Ribeiro and Manoel Guedes Corrêa Gondim Junior
Department of Agronomy, Federal Rural University of Pernambuco, 52171-900 Recife, PE, Brazil

Manisha Vaish and Vineet K. Singh
Microbiology and Immunology, Kirksville College of Osteopathic Medicine, A.T. Still University of Health Sciences, 800West Jefferson Street, Kirksville, MO 63501, USA

Sujatha Kabilan, Mahalakshmi Ayyasamy, Sridhar Jayavel and Gunasekaran Paramasamy
UGC-Networking Resource Centre in Biological Sciences, School of Biological Sciences, Madurai Kamaraj University, Madurai 625021, India

S. P. Gautam
Central Pollution Control Board, New Delhi, India

P. S. Bundela and M. K. Awasthi
Regional office, M. P. Pollution Control Board, Vijay Nagar, Jabalpur, India

A. K. Pandey
Mycological Research Laboratory, Department of Biological Sciences, Rani Durgavati University, Jabalpur, India

Jamaluddin
Yeast and Mycorrhiza Biotechnology Laboratory, Department of Biological Sciences, Rani Durgavati University, Jabalpur, India

S. Sarsaiya
Regional office, M. P. Pollution Control Board, Vijay Nagar, Jabalpur, India
International Institute of Waste Management (IIWM), Bhopal, India

Rebekah A. Frampton and Peter C. Fineran
Department of Microbiology & Immunology, University of Otago, P.O. Box 56, Dunedin 9054, New Zealand

Andrew R. Pitman
New Zealand Institute for Plant & Food Research, Private Bag 4704, Christchurch 8140, New Zealand

Giulia Morace and Elisa Borghi
Department of Public Health, Microbiology, and Virology, Universita degli Studi di Milano, Via C. Pascal 36, 20133 Milan, Italy

Fabien Cottier
School of Biosciences, University of Kent, Canterbury, Kent CT2 7NJ, UK
Singapore Immunology Network, Agency for Science, Technology, and Research, Singapore 138648

Fritz A. Muhlschlegel
School of Biosciences, University of Kent, Canterbury, Kent CT2 7NJ, UK
Clinical Microbiology Service, William Harvey Hospital, East Kent Hospitals University NHS Foundation Trust, Ashford, Kent TN24 0LZ, UK

Theodore C. White
Seattle Biomedical Research Institute, 307 Westlake Avenue North, Suite 500, Seattle, WA 98109-5219, USA
School of Biological Sciences, University of Missouri-Kansas City, 5007 Rockhill Road, BSB 404, Kansas City, MO 64110, USA

Rebecca Rashid Achterman
Seattle Biomedical Research Institute, 307 Westlake Avenue North, Suite 500, Seattle, WA 98109-5219, USA

Donna M. MacCallum
Aberdeen Fungal Group, School of Medical Sciences, Institute of Medical Sciences, University of Aberdeen, Foresterhill, Aberdeen, AB25 2ZD, UK

W. Buwembo
Department of Anatomy, Makerere University, P.O. Box 7072, Kampala, Uganda

S. Aery and G. Swedberg
Department of Medical Biochemistry and Microbiology, Uppsala University, Husargaten 3, Building D7 Level 3, P.O. Box 582, SE-75123 Uppsala, Sweden

C. M. Rwenyonyi
Department of Dentistry, Makerere University, P.O. Box 7072, Kampala, Uganda

F. Kironde
Department of Biochemistry, Makerere University, P.O. Box 7072, Kampala, Uganda

Saeed A. Hayek and Salam A. Ibrahim
Food Microbiology and Biotechnology Laboratory, North Carolina Agricultural and Technical State University, Greensboro, NC 27411, USA

Christopher T. Lohans and John C. Vederas
Department of Chemistry, University of Alberta, Edmonton, AB, Canada

Heather A. Duoss-Jennings
Department of Animal Science, Iowa State University, Ames, IA, USA

Ty B. Schmidt
Animal Science Department, University of Nebraska, Lincoln, NE, USA

Todd R. Callaway
Food and Feed Safety Research Unit, ARS, USDA, College Station, TX 79403, USA

Jeffery A. Carroll
Livestock Issues Research Unit, ARS, USDA, Lubbock, TX 77845, USA

James M. Martin
Department of Animal and Dairy Sciences, Mississippi State University, Mississippi State, MS, USA

Sara A. Shields-Menard and Janet R. Donaldson
Department of Biological Sciences, Mississippi State University, Mississippi State, MS 39762, USA

Paul R. Broadway
Department of Animal and Food Sciences, Texas Tech University, Lubbock, TX, USA

Masanori Toyofuku, Hiroo Uchiyama and Nobuhiko Nomura
Graduate School of Life and Environmental Sciences, University of Tsukuba, Tsukuba, Ibaraki 305-8572, Japan

www.ingramcontent.com/pod-product-compliance
Lightning Source LLC
Chambersburg PA
CBHW080651200326
41458CB00013B/4814

* 9 7 8 1 6 3 2 3 9 0 4 7 9 *